INTERMOLECULAR FORCES

THE JERUSALEM SYMPOSIA ON
QUANTUM CHEMISTRY AND BIOCHEMISTRY

Published by the Israel Academy of Sciences and Humanities,
distributed by Academic Press (N.Y.)

1st JERUSALEM SYMPOSIUM: *The Physicochemical Aspects of Carcinogenesis* (October 1968)

2nd JERUSALEM SYMPOSIUM: *Quantum Aspects of Heterocyclic Compounds in Chemistry and Biochemistry* (April 1969)

3rd JERUSALEM SYMPOSIUM: *Aromaticity, Pseudo-Aromaticity, Antiaromaticity* (April 1970)

4th JERUSALEM SYMPOSIUM: *The Purines: Theory and Experiment* (April 1971)

5th JERUSALEM SYMPOSIUM: *The Conformation of Biological Molecules and Polymers* (April 1972)

Published by the Israel Academy of Sciences and Humanities,
distributed by D. Reidel Publishing Company (Dordrecht, Boston and London)

6th JERUSALEM SYMPOSIUM: *Chemical and Biochemical Reactivity* (April 1973)

Published and distributed by D. Reidel Publishing Company
(Dordrecht, Boston and London)

7th JERUSALEM SYMPOSIUM: *Molecular and Quantum Pharmacology* (March/April 1974)

8th JERUSALEM SYMPOSIUM: *Environmental Effects on Molecular Structure and Properties* (April 1975)

9th JERUSALEM SYMPOSIUM: *Metal-Ligand Interactions in Organic Chemistry and Biochemistry* (April 1976)

10th JERUSALEM SYMPOSIUM: *Excited States in Organic Chemistry and Biochemistry* (March 1977)

11th JERUSALEM SYMPOSIUM: *Nuclear Magnetic Resonance Spectroscopy in Molecular Biology* (April 1978)

12th JERUSALEM SYMPOSIUM: *Catalysis in Chemistry and Biochemistry Theory and Experiment* (April 1979)

13th JERUSALEM SYMPOSIUM: *Carcinogenesis: Fundamental Mechanisms and Environmental Effects* (April/May 1980)

VOLUME 14

INTERMOLECULAR FORCES

PROCEEDINGS OF THE FOURTEENTH JERUSALEM SYMPOSIUM ON QUANTUM CHEMISTRY AND BIOCHEMISTRY HELD IN JERUSALEM, ISRAEL, APRIL 13–16, 1981

Edited by

BERNARD PULLMAN

Université Pierre et Marie Curie (Paris VI)
Institut de Biologie Physico-Chimique
(Fondation Edmond de Rothschild) Paris, France

D. REIDEL PUBLISHING COMPANY

DORDRECHT : HOLLAND / BOSTON : U.S.A.
LONDON : ENGLAND

6575-511X

CHEMISTRY

Library of Congress Cataloging in Publication Data

CIP

Jerusalem symposium on quantum chemistry and biochemistry
(14th: 1981)
Intermolecular forces.

(The Jerusalem symposia on quantum chemistry and biochemistry ;
v. 14)
Includes index.
1.   Intermolecular forces—Congresses.   I.   Pullman, Bernard,
1919-   .   II.   Title.   III.   Series.

QD461.J47   1981   541.2'26   81–12000
ISBN 90–277–1326–X   AACR2

Published by D. Reidel Publishing Company,
P.O. Box 17, 3300 AA Dordrecht, Holland.

Sold and distributed in the U.S.A. and Canada
by Kluwer Boston Inc.,
190 Old Derby Street, Hingham, MA 02043, U.S.A.

In all other countries, sold and distributed
by Kluwer Academic Publishers Group,
P.O. Box 322, 3300 AH Dordrecht, Holland.

D. Reidel Publishing Company is a member of the Kluwer Group.

Printed in The Netherlands

TABLE OF CONTENTS

PREFACE

The 14th Jerusalem Symposium continued the tradition of the pleasant and exciting meetings which once a year gather distinguished scientists, the world's most renowned experts in specific fields of quantum chemistry and biochemistry, in the impressive surroundings of the Israël Academy of Sciences and Humanities. The subject discussed this year – Intermolecular forces – is one of the utmost interest for all molecular sciences.

I wish to thank all those who made this meeting possible and contributed to its success: the Baron Edmond de Rothschild whose continuous generosity guarantees the perenniality of our venture, the Israël Academy of Sciences and in particular its Vice-President, Professor Yoshua Jortner for his devoted contribution to the organization and holding of this meeting, the high authorities of the Hebrew University of Jerusalem and in particular the Rector Meshulam for their constant support and Dr. Pierre Claverie for his efficient help in the preparation of the program. Mrs Abigail Hyam and Mrs Myriam Yogev must be thanked for their contribution to the efficiency and success of the local arrangements.

Bernard Pullman

B. Pullman (ed.), Intermolecular Forces, ix.
Copyright © 1981 by D. Reidel Publishing Company.

INTERMOLECULAR FORCES:
WHAT CAN BE LEARNED FROM AB INITIO CALCULATIONS?

Ad van der Avoird
Institute of Theoretical Chemistry,
University of Nijmegen,
Toernooiveld, Nijmegen, The Netherlands.

1. INTRODUCTION

Various experiments, such as elastic or rotationally inelastic molecular beam scattering[1,2] and spectroscopic studies of so-called Van der Waals molecules[3,4], have been designed especially to provide information about the Van der Waals interactions between molecules. The results of these measurements, as well as other experimental data obtained on bulk materials, e.g. the phonon frequencies in molecular crystals[5], depend very sensitively on the shape of the intermolecular potentials. Still, it is not easy to extract the potentials from these data. One has to assume parametrized model potentials of a certain analytic form and to fit the parameters such that calculations yield the best agreement with the measured quantities. Often, this does not lead to unique and accurate values for all the parameters and, moreover, the model potentials assumed may have a form which is not completely correct and not sufficiently flexible. Therefore, it is very useful that, for smaller systems, information about the intermolecular potentials can also be obtained from ab initio quantumchemical calculations.

In this survey, we shall not describe any of the technical details of these calculations, Let us just emphasize that truncations of the expansions used for the wave functions (on the one-electron level the atomic orbital bases, on the many-electron level the electronic configuration function bases) must be carried out very carefully. Otherwise, the calculated intermolecular potentials become artifacts of the calculations rather than physically meaningful results, just as it has occurred in the past that many of the intermolecular potentials obtained by semi-empirical methods are mainly determined by the approximations made in these methods as well as by the parametrization of the remaining interactions (approximations and parametrizations which were sometimes completely inappropriate for the weak intermolecular Van der Waals forces). The basis sets required for accurate calculations of Van der Waals potentials between molecules appear to be rather large, which makes the computations expensive even for small systems.

1

B. Pullman (ed.), Intermolecular Forces, 1–14.
Copyright © 1981 by D. Reidel Publishing Company.

The interaction potential between two (rigid) molecules A and B depends on the relative position of their mass centers, given by the vector $R_{AB} = (R,\Omega) = (R,\Theta,\Phi)$, and on their orientations, determined by three Euler angles for each molecule: $\omega_A = (\alpha_A,\beta_A,\gamma_A)$ and $\omega_B = (\alpha_B,\beta_B,\gamma_B)$. All these quantities are defined with respect to some arbitrary space fixed frame. The distance and orientational dependence of the potential are given explicitly by the so-called spherical expansion[6]:

$$V(R,\underline{\omega}_A,\underline{\omega}_B,\underline{\Omega}) = \sum_{L_A,L_B,L=0}^{\infty} \sum_{K_A=-L_A}^{L_A} \sum_{K_B=-L_B}^{L_B} v_{L_A K_A L_B K_B L}(R) \times$$

$$A_{L_A K_A L_B K_B L}(\underline{\omega}_A,\underline{\omega}_B,\underline{\Omega}) \tag{1}$$

The angular functions are given by:

$$A_{L_A K_A L_B K_B L}(\underline{\omega}_A,\underline{\omega}_B,\underline{\Omega}) = \sum_{M_A=-L_A}^{L_A} \sum_{M_B=-L_B}^{L_B} \sum_{M=-L}^{L} D_{M_A K_A}^{L_A}(\underline{\omega}_A)^* \; D_{M_B K_B}^{L_B}(\underline{\omega}_B)^* \times$$

$$c_M^L(\underline{\Omega}) \begin{pmatrix} L_A & L_B & L \\ M_A & M_B & M \end{pmatrix} \tag{2}$$

The "rotation" functions $D_{M_A K_A}^{L_A}$ and $D_{M_B K_B}^{L_B}$ are the irreducible matrix representations of the rotation group SO(3)[7], c_M^L is a spherical harmonic in the Racah normalization and the expression in brackets is a Wigner 3-j symbol[7]. Since the angular functions (2) are invariant under overall rotations of the system and also the intermolecular potential (1) depends only on the "internal" angles, one can omit three angles from the expressions (1) and (2) and define the remaining angles with respect to a body-fixed frame (for instance, with the z-axis lying along R_{AB} and the xz-plane coinciding with the local xz-plane of molecule A, i.e. $\Theta=\Phi=0$ and $\alpha_A=0$). The expansion coefficients $v_{L_A K_A L_B K_B L}(R)$, which depend only on the distance R, but which determine completely the orientation dependence of the intermolecular potential V, can be obtained by integration over all (internal) angles, if the potential V is known:

$$v_{L_A K_A L_B K_B L}(R) = (256\pi^5)^{-1}(2L_A+1)(2L_B+1)(2L+1) \int_{\underline{\omega}_A,\underline{\omega}_B,\underline{\Omega}} d\underline{\omega}_A d\underline{\omega}_B d\underline{\Omega} \times$$

$$A_{L_A K_A L_B K_B L}(\underline{\omega}_A,\underline{\omega}_B,\underline{\Omega}) \; V(R,\underline{\omega}_A,\underline{\omega}_B,\underline{\Omega}) \tag{3}$$

The coefficient $v_{0,0,0,0,0}(R)$ is the isotropic potential, the terms in (1) with $L_A,L_B\neq0$ describe the anisotropic interactions.

The ab initio calculations are carried out for fixed nuclear geometry (using the Born-Oppenheimer approximation). Choosing a mesh of geometries $\beta_A,\gamma_A,\alpha_B,\beta_B,\gamma_B$, for given R, that permits the integral in (3) to be performed numerically (e.g. by Gauss type quadratures[8]), yields the explicit orientational dependence of the ab initio potential. In practice this has been done for linear (N_2) molecules[9], where $K_A=K_B=0$, the rotation functions $D_{M_A K_A}^{L_A}{}^*$ and $D_{M_B K_B}^{L_B}{}^*$ are equal to spherical harmonics $C_{M_A}^{L_A}$ and $C_{M_B}^{L_B}$, and only three internal angles occur ($\beta_A, \alpha_B, \beta_B$). In the long range (i.e. for large R), analytic expressions for the coefficients $v_{L_A K_A L_B K_B L}(R)$, power series in R^{-1}, can be calculated directly by perturbation theory, if the interaction operator is written in the multipole expansion and some angular momentum recoupling techniques are used[6,10,11].

2. RESULTS

2.1. Quantitative knowledge about mechanisms contributing; shape of model potentials

Although it is known in principle which mechanisms can contribute to the Van der Waals potential between (closed shell) molecules: electrostatic, induction and dispersion forces, exchange and charge transfer effects[12,13], it is often not clear a priori what is the relative importance of these contributions. One must realize, moreover, that some of the contributions, such as charge transfer interactions or exchange-induction and exchange-dispersion interactions, are not uniquely defined. Having made an ab initio calculation which is sufficiently flexible to include all these interactions, one can always make a decomposition of the interaction energy, however. We find it practical to start in the long range, where the molecular wave functions do not overlap and the multipole expansion for the interaction operator can be used, and to calculate first the interactions by perturbation theory. This yields explicit expressions in powers of R^{-1} for the (1st order) electrostatic and (2nd order) induction and dispersion interactions with the orientational dependence given by formula (1). Here, it is relevant to note that not only the leading terms in the powers series, e.g. the R^{-6} term in the dispersion energy, but also higher powers in R^{-1} are important for distances around the Van der Waals minimum[14-17]. Next, we perform full ab initio calculations for smaller distances R, around the Van der Waals minimum and in the repulsive region, and we look how far the interaction energies deviate from the power series. Since these deviations are all caused by overlap effects, they should have an exponential dependence on R. If the available computer time permits, such as in the case of N_2-N_2[9], we also determine the explicit orientational dependence of these exponential terms by calculating the potential for a sufficient number of (suitably chosen) points to perform the numerical integration (3).

Sometimes, the results are somewhat unexpected. So, it has been found[9,18], for instance, for two (parallel) N_2 molecules that the electrostatic interactions between the charge clouds as determined by their multipole moments, is completely cancelled by penetration effects at the Van der Waals minimum (see fig. 1). At smaller distances these penetration effects (mainly incomplete screening of the nuclei due to the overlap between the electron clouds) dominate: the electrostatic interaction between two parallel N_2 molecules becomes attractive. In

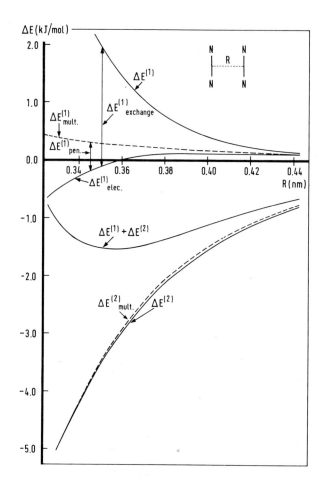

Figure 1. Different contributions to the interaction energy
between two parallel N_2 molecules; definitions and
results from ref. ([9]).

the (second order) dispersion energy these penetration effects are much
smaller. For the induction energy they become considerable at small distances, but this induction energy itself remains very small (the leading
term in the long range behaves as R^{-8}). The dominating overlap effect
is the (first order) exchange repulsion, which is 5 to 10 times larger
than the attractive electrostatic penetration effects.

Another interesting result for the N_2 dimer concerns the orientational dependence of the potential (see fig. 2). In various model po-
tentials([5]) it has been assumed that the main (and sometimes only) anisotropic interaction between N_2 molecules is the electrostatic quadrupole-quadrupole R^{-5} term. Looking at fig. 2, we observe that the aniso-

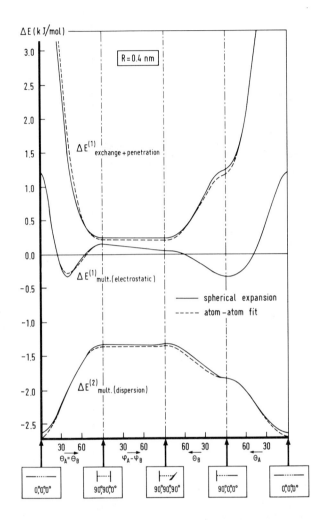

Figure 2. Orientational dependence of different long range
(multipole) and short range (exchange and penetration) contributions to the N_2-N_2 interaction energy,
at R=4Å, from ref. ([9]). The curves are generated by
the spherical expansion (1) of the ab initio results[9] and by fitting an atom-atom model potential
to these results.

tropy in the dispersion interactions (R^{-6},R^{-8} and R^{-10} terms) is equally
important, near the Van der Waals minimum (although, of course, the
total dispersion energy is always attractive). But, the dominant anisotropic contribution at this distance is clearly caused by overlap effects (mainly exchange repulsion). In a spherical expansion (1) of the

N_2-N_2 potential one has to include terms up to $L_A, L_B=4$ inclusive for
the exchange repulsion and the electrostatic contribution and terms up
to $L_A, L_B=2$ inclusive for the dispersion interactions[9].

Having made such an analysis of the interaction energy in terms of
power series and exponential terms in R and knowing also their angular
behaviour, it becomes clear what should be the shape of model potentials.
If one wishes to use model potentials of the atom-atom type[13,19]
rather than using molecular paramters, one can, for instance, represent
the electrostatic multipole-multipole interactions by a Coulombic interaction between atomic point charges. Such a point charge model cannot
represent the charge cloud penetration effects, however. Ab initio calculations[9,20] have taught us that the distance and orientational dependence of these penetration effects are similar to the behaviour of
the exchange effects. So, they can be added to the latter and represented by the same type of analytic functions (exponentially dependent
on the atom-atom distances). Complete analytic relations between atom-
atom type model potentials and molecular expansions have been given
recently[21,22]

2.2. Parametrization of model potentials

For small systems, all the parameters in an analytic representation
of the potential can be directly derived from ab initio calculations,
as indicated in the preceding section. For somewhat larger systems, still
part of the parameters in a model potential of well-chosen shape can be
obtained from ab initio calculations, for instance, the coefficients in
the R^{-n} expansion which can be calculated from monomer wave functions
by perturbation theory. Several of the more recent semi-empirical potentials[3,23,24] are actually based on the availability of ab initio data.
For large molecules, even ab initio calculations on the monomers are
not feasible anymore, and one has traditionally resorted to the use of
atom-atom potentials[13,19] (pairwise additive isotropic potentials between atoms in different molecules), mostly of Lennard-Jones (12-6)·
or Buckingham (exp-6) type, with parameters derived from experimental
data.

The most informative tests on these atom-atom potentials are made
by applying them to small molecules and invoke ab initio results, for
comparison. It is hard to draw general conclusions about the quality of
empirical atom-atom potentials since they show such a wide scatter. The
more extensively calibrated hydrocarbon potentials which include Coulombic point charge interactions in addition to exp-6 terms[25,26] give a
reasonable representation of the distance and orientational dependence
of the interaction between ethene (C_2H_4) molecules[20]; the point charges
yield too small a quadrupole moment, however. Apparently, the empirical
quantities used for calibrating the potentials are not very sensitive
to the molecular multipole moments: many empirical atom-atom potentials
do not contain any electrostatic interactions at all. The parameters in
the atom-atom model potentials can also be fitted directly to the ab

initio data. Calculations on C_2H_4 dimers[20] and N_2 dimers[9] have
shown us that this can lead to fairly accurate representations of the
ab initio results (see figs. 2 and 3 for N_2), if the force centers for
the atom-atom interactions (or rather site-site interactions, then) are
allowed to shift away from the nuclei. This makes it difficult to transfer these atom-atom potentials to larger molecules, however.

2.3. Testing and calibration of approximate methods

For larger systems one can also invoke less expensive, but more
approximate, methods to calculate the interaction potential. These
methods can again be tested by comparison with good ab initio results
for small systems. One such method, which has become rather popular is
the so-called Gordon-Kim method[27,28] which uses expressions for the
interaction energy contributions in terms of the electron density
function, expressions which are based on a homogeneous electron gas
model. As Marc van Hemert will show in his contribution to this Symposium[29], this method yields fairly accurate results for the distance
and orientational dependence of the (first order) electrostatic and
exchange interactions in several systems, provided that an appropriate
correction is made for the electron self-exchange[30,31].

2.4. Interpretation of scattering and spectroscopic experiments

As mentioned in the introduction, the interpretation of molecular
beam scattering experiments[1,2] and spectra of Van der Waals molecules
is done[3,4] by using model potentials and fitting the parameters so
that calculated cross sections and frequencies give the best agreement
with the measured data. In recent work, the choice of the model potentials has been guided by the experience obtained from ab initio calculations on small molecules ($He-H_2$,H_2-H_2)[15,33-35] and, moreover, part
of the parameters are fixed on ab initio calculated values, which is
very helpful in obtaining more unique values for the remaining parameters in the fit. Complete ab initio potentials have been used in some
cases, too[36,37]. Not only, this helps in interpreting the experimental
data, but it also gives a check on the accuracy of the ab initio results.
By mutual stimulation both the experiments and the ab initio calculations become more refined, leading to more detailed and accurate knowledge of the intermolecular potentials.

In this section, two examples are presented where somewhat unexpected structures of Van der Waals molecules are predicted or confirmed
by ab initio calculations. The first one is the N_2 dimer which is stable
in the gas phase at 77K. In a first interpretation of its infrared spectrum one has assumed[38], mainly guided by looking at the quadrupole-
quadrupole interactions (see section 2.1.), that this dimer has a T-
shaped equibrium structure. Subsequent calculations[17], which include
the higher multipole interactions as well as the (anisotropic) long
range dispersion forces, confirm this possibility, but also show that
a shifted parallel structure, which leads to attractive quadrupole-

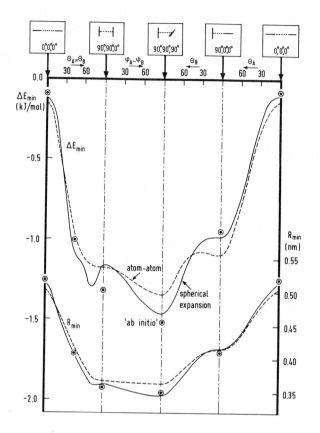

Figure 3. Orientational dependence of the Van der Waals well
depth, ΔE_{min}., and equilibrium distance, R_{min}., in
the N_2-N_2 system. Ab initio results, the spherical
expansion of these results and an atom-atom model
potential fit are shown (see ref. ([9])).

quadrupole forces, too, is about equally stable (cf. the electrostatic
and dispersion contributions in fig. 2). Approximate calculations[39,40]
which add the short range repulsions by a rather arbitrary geometrical
model do not lead to a unique answer. An ab initio calculation of the
full N_2-N_2 potential ([9]) predicts that the "crossed" structure is most
stable (see fig. 3), mainly because the exchange repulsion is weakest
for this structure while the electrostatic interactions are more favourable than for the parallel structure (see fig. 2). This N_2 dimer
shows interesting dynamical behaviour, which must be known before the
theoretical results can be directly compared with the measured infrared
spectrum[38]. Internal rotations of the N_2 monomers around the axis
which connects their mass centers, from the crossed equilibrium structure through the parallel structure to an equivalent crossed structure,
corresponds to a barrier of about 15 to 20 cm^{-1} (according to the ab
initio results, see fig. 3). A preliminary study of the nuclear dynamics

on the ab initio potential surface, in the harmonic approximation, yields a fundamental frequency of 20 cm^{-1} for this rotational vibration. Obviously, the harmonic approximation does not work in this case, but the result demonstrates that the N_2 molecules in the dimer exhibit large amplitude rotational oscillations or hindered rotations, at least in one direction. Similar motions probably occur in the orientationally disordered β-phase of solid N_2, which is stable above 35.6K.

The second example concerns the KCN molecule, which is not a Van der Waals molecule in the full sense, as its binding energy (relative to K^+ and CN^-) is rather large, 4.84 eV. Ab initio calculations of the interaction energy and the dipole moment by Wormer and Tennyson[41] indicate, however, that both these quantities are very well described by assuming electrostatic, induction and exchange interactions between the closed shell constituents K^+ and CN^-. So, if we extend the definition of Van der Waals forces to include the Coulombic forces between monopoles, we can consider $K^+ - CN^-$ to be a Van der Waals molecule. Also its floppiness in the bending direction, which emerges from the ab initio calculations[41], is characteristic of a Van der Waals molecule. For some time, this molecule was believed to be linear, just as HCN and LiCN. Recent microwave and molecular beam experiments[42] have shown that the molecule has a triangular structure, however. The ab initio calculations[41] which were carried out subsequently, confirmed this finding (see fig. 4) and gave very good agreement with the meas-

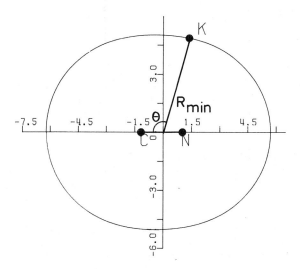

Figure 4. Equilibrium structure of KCN (indicated by dots), according to ref. (41). Distances in bohrs, origin in the CN mass center. The nearly elliptical path shows R_{min}. as a function the angle Θ.

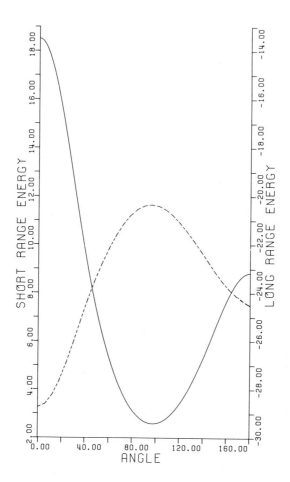

Figure 5. Short range (full line) and long range (dashed line) contributions to the K^+-CN^- interaction energy in 10^{-2} hartree, as functions of the angle Θ for $R=R_{min.}=5.055$ bohr, according to ref. (41).

ured properties of the molecule. It turned out that the long range (multipole-multipole and multipole-induced multipole) interactions do favour a linear geometry (see fig. 5). But, the exchange repulsion is weakest for K^+ approaching the CN^- ion in a nearly perpendicular direction and, just as for the N_2 dimer, this term dominates in determining the equilibrium geometry (see fig. 6).

Both for K^+-CN^- and for N_2-N_2 the full interaction potential has been obtained from ab initio calculations and represented in the form (1). Studies of the nuclear motions in these systems which should yield the rotational-vibrational spectra are in progress.

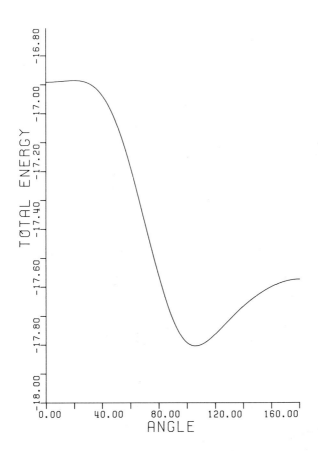

Figure 6. Total K^+-CN^- interaction energy (in 10^{-2} hartree)
along the path of minimum energy shown in fig. 4,
according to ref. [41].

2.5. Information about bulk properties

The intermolecular potentials obtained from ab initio calculations
can give much interesting and useful information about the properties
of non-ideal gases and condensed systems. One can use them to calculate
transport properties, virial coefficients and other characteristics of
the equation of state. Also, one can apply them in Monte Carlo or
Molecular Dynamics simulation studies of liquids. In our group, we have
used ab initio potentials in lattice dynamics calculations which yield
the structure and the phonon dispersion relations for C_2H_4[20,43] and
N_2[44] crystals (the latter in two ordered phases, α and γ). Thus, one
can directly compute thermodynamic and optical properties of these
molecular crystals, as functions of the temperature and pressure, without using any empirical data or assuming model potentials. The final

results have been compared with the experimental data available and the agreement is very satisfactory[43,44]. In the near future, we wish to pay special attention to the description of the librations (rotational oscillations) of the molecules in such crystals and to related properties (e.g. orientational order-disorder phase transitions), which depend in particular on the anisotropy in the intermolecular potentials. It is interesting to observe that even for a gas phase property as the second virial coefficient in N_2 (up to rather high temperatures) the contributions from the anisotropy in the potential are of essential importance[29,45].

3. CONCLUSION

This paper tries to show that good ab initio calculations of intermolecular potentials can be useful in many respects. Perhaps, their main purpose is to correct some of the current beliefs about the shape of these potentials (especially about the contributions which determine their anisotropic behaviour) which have entered into various model potentials.

ACKNOWLEDGEMENT

I like to thank (in alphabetical order) R. Berns, M. van Hemert, T. Luty, F. Mulder, J. Tennyson, T. Wasiutynski and P. Wormer, who all have contributed to the results mentioned in this paper.

REFERENCES

1. Atom-molecule collision theory, (Bernstein, R.B. ed.): 1979, Plenum, New York.
2. Andres, J., Buck, U., Huisken, F., Schleusener, J., and Torello, F.: 1980, J. Chem. Phys. 73, pp. 5620-5630.
3. Le Roy, R.J., and Carley, J.S.: 1980, Advan. Chem. Phys. 42, pp. 353-420, and references therein.
4. Waayer, M.: 1981, Thesis, Nijmegen.
5. Raich, J.C., and Gillis, N.S.: 1977, J. Chem. Phys. 66, pp. 846-861, and references therein.
6. Van der Avoird, A., Wormer, P.E.S., Mulder, F., and Berns, R.M.: 1980, Topics in Current Chemistry 93, Springer, Berlin, pp. 1-51, and references therein.
7. Brink, D.M., and Satchler, G.R.: Angular momentum, 2 ed., 1975, Clarendon, Oxford.
8. Abramowitz, M., and Stegun, I.A.: Handbook of mathematical functions, 1964, National Bureau of Standards, Washington, D.C.
9. Berns, R.M., and Van der Avoird, A.: 1980, J. Chem. Phys. 72, pp. 6107-6116.
10. Wormer, P.E.S.: 1975, Thesis, Nijmegen.

11. Wormer, P.E.S., Mulder, F., and Van der Avoird, A.: 1977, Intern. J. Quantum Chem. 11, pp. 959-970.
12. Hirschfelder, J.O., and Meath, W.J.: 1967, Advan. Chem. Phys. 12, pp. 3-106.
13. Claverie, P. in: Intermolecular interactions: From diatomics to biopolymers (Pullman, B., ed.): 1978, Wiley, New York, pp. 69-306.
14. Mulder, F., Van Hemert, M., Wormer, P.E.S., and Van der Avoird, A.: 1977, Theor. Chim. Acta 46, pp. 39-62.
15. Mulder, F., Van der Avoird, A., and Wormer, P.E.S.: 1979, Mol. Phys. 37, pp. 159-180.
16. Mulder, F., Van Dijk, G., and Huiszoon, C.: 1979, Mol. Phys. 38, pp. 577-603.
17. Mulder, F., Van Dijk, G., and Van der Avoird, A.: 1980, Mol. Phys. 39, pp. 407-425.
18. Ng, K.C., Meath, W.J., and Allnatt, A.R., 1977, Mol. Phys. 33, pp. 699-715.
19. Kitaigorodsky, A.I.: Molecular crystals and molecules, 1973, Academic, New York.
20. Wasiutynski, T., Van der Avoird, A., and Berns, R.M.: 1978, J. Chem. Phys. 69, pp. 5288-5300.
21. Downs, J., Gubbins, K.E., Murad, S., and Gray, C.G.: 1979, Mol. Phys. 37, pp. 129-140.
22. Briels, W.J.: 1980, J. Chem. Phys. 73, pp. 1850-1861.
23. Tang, K.T., and Toennies, J.P.: 1977, J. Chem. Phys. 66, pp. 1496-1506, 1978, J. Chem. Phys. 68, pp. 5501-5517, 1981, J. Chem. Phys. 74, pp. 1148-1161.
24. Ahlrichs, R., Penco, R., and Scoles, G.: 1977, Chem. Phys. 19, pp. 119-130.
25. Hall, D., and Williams, D.E.: 1975, Acta Cryst. A31, pp. 56-58.
26. Starr, T.L., and Williams, D.E.: 1977, Acta Cryst. A33, pp. 771-776.
27. Gordon, R.G., and Kim, Y.S.: 1972, J. Chem. Phys. 56, pp. 3122-3133.
28. Waldman, M., and Gordon, R.G.: 1979, J. Chem. Phys. 71, pp. 1325-1358.
29. Van Hemert, M.C.: 1981, Proc. 14th Jeruzalem Symposium (Pullman, B., ed.), Reidel, Dordrecht.
30. Rae, A.I.M.: 1973, Chem. Phys. Letters 18, pp. 574-577.
31. Gazquez, J.L. and Ortiz, E.: 1981, Chem. Phys. Letters 77, pp. 186-189.
32. Meyer, W.: 1976, Chem. Phys. 17, pp. 27-33.
33. Tsapline, B., and Kutzelnigg, W.: 1973, Chem. Phys. Letters 23, pp. 173-177.
34. Geurts, P.J.M., Wormer, P.E.S., and Van der Avoird, A.: 1975, Chem. Phys. Letters 35, pp. 444-449.
35. Meyer, W., Hariharan, P.C., and Kutzelnigg, W.: 1980, J. Chem. Phys. 73, pp. 1880-1897.
36. Schaefer, J., and Meyer, W.: 1979, J. Chem. Phys. 70, pp. 344-360.
37. Monchick, L., and Schaefer, J.: 1980, J. Chem. Phys. 73, pp. 6153-6161.
38. Long, C.A., Henderson, G., and Ewing, G.E.: 1973, Chem. Phys. 2, pp. 485-489.

39. Koide, A., and Kihara, T.: 1974, Chem. Phys. 5, pp. 34-48.
40. Sakai, K., Koide, A., and Kihara, T.: 1977, Chem. Phys. Letters 47, pp. 416-420.
41. Wormer, P.E.S., and Tennyson, J.: 1981, J. Chem. Phys., in press.
42. Törring, T., Bekooy, J.P., Meerts, W.L., Hoeft, J., Tiemann, E., and Dymanus, A.: 1980, J. Chem. Phys. 73, pp. 4875-4882.
43. Luty, T., Van der Avoird, A., Berns, R.M., and Wasiutynski, T.: 1981, J. Chem. Phys., in press.
44. Luty, T., Van der Avoird, A., and Berns, R.M.: 1980, J. Chem. Phys. 73, pp. 5305-5309.
45. Van Hemert, M., and Berns, R.M.: 1981, J. Chem. Phys., in press.

QUANTUM MECHANICAL DETERMINATION OF INTERMOLECULAR INTERACTIONS. AB
INITIO STUDIES

E. Kochanski
Equipe de recherche n° 139 du CNRS - Institut Le Bel
Université Louis Pasteur, B.P. 296/R8, 67008 Strasbourg Cédex

Abstract : using some approximate equivalencies between the supermolecule and the perturbation methods, the intermolecular energies are computed as the sum of an S.C.F. supermolecule determination and of a
dispersion term obtained from a perturbation procedure. The main energetic contributions (electrostatic, repulsion, induction, charge transfer,
dispersion) are compared, according to the nature (ionic or neutral) of
the interacting molecules (or atoms). Some systems of non polar molecules are specially studied.

I. INTRODUCTION

    One of the main advantages of ab initio calculations is that no
experimental information is required. Furthermore, very accurate results
may be obtained. However, high accuracy generally implies sophisticated
procedures and the use of large basis sets. The high cost of such
refined treatments leads to investigate the nature of the approximations
and the possibilities of rather small basis sets which can give a
reasonable accuracy at a reasonable cost. Within the framework of ab
initio calculations, our own studies have been a contribution to this
investigation.

II. THE METHOD

    Let us first consider some quantum mechanical methods proposed to
treat intermolecular interactions. Most of them belong to one of the
two following categories :
- the supermolecule treatment where the whole system of the interacting
molecules (or atoms) is treated as one molecule (this can be for
instance an S.C.F. treatment, followed -or not- by a configuration
interaction (C.I.) calculation).
- a perturbation method where the intermolecular energy is considered
as a small perturbation with respect to the energy of the isolated
molecules. Generally, the zeroth order wavefunction and the zeroth

15

B. Pullman (ed.), Intermolecular Forces, 15–31.

order hamiltonian are obtained from the wavefunctions and the hamiltonians of the isolated molecules. Several procedures have been proposed. My purpose is not to discuss them, this has been done extensively (see for instance Ref. [1] and also the recent note in Ref. [2]). I just recall that an accurate treatment is not an easy task, specially in the region of small overlap (that is in the region of the energy minimum) : besides the problems of the antisymmetrization of the zeroth order wavefunction, there are also those of the convergence of the perturbation series, the relative importance of the higher order terms when we stop to second order, the effect of the overlap between the orbitals of the interacting molecules. If we use the S.C.F. functions of the isolated molecules, we have to account for the intramolecular correlation effect. Also, the validity of the multipole expansion, which is very often used in such treatments, must be considered.

There are some approximate equivalencies between these two categories of methods. As long as a very accurate determination is not required, it can be suitable to take advantage of these equivalencies. It is relevant to have a quick look on them before discussing the results obtained for some interacting systems. In all cases, we shall consider two interacting molecules (or atoms).

1) The supermolecule treatment

Let us consider an S.C.F. calculation followed by a C.I. treatment. It can be suitable to start the S.C.F. calculation with the S.C.F. vectors of the isolated molecules, reorthogonalized over the whole system for each intermolecular distance. We call ΔE_1 the difference between the energy obtained at the first iteration of the calculation and the S.C.F. energy of the isolated molecules system.

We call ΔE_{SCF} the difference between the S.C.F. energy of the interacting system and the S.C.F. energy of the isolated molecules system.

We call E_{deloc} the difference between the S.C.F. energy of the interacting system and the energy obtained at the first iteration of the S.C.F. calculation. This is a delocalization term. ΔE_{SCF} is then the sum of ΔE_1 and E_{deloc}.

If we use molecular orbitals localized on each molecule, the C.I. correlation energy may be decomposed into three terms : i) the intramolecular part which is the largest contribution and which varies only slightly with the intermolecular separation ; ii) the intermolecular part ; iii) a term which describes the effect of the intramolecular correlation on the intermolecular interaction. The difference between the C.I. energy of the interacting systems and that of the isolated molecules is largely due to the second term and partly to the third term ; the variation in the first term is generally smaller than the other two contributions.

2) The perturbation theory

Assuming that the zeroth order wavefunction is an antisymmetrized product of the wavefunctions of the isolated molecules, the first order energy includes an electrostatic term (corresponding to the single product of the wavefunctions) and a repulsive term (corresponding to the antisymmetrization).

The main contributions to second order are the induction and dispersion energy (long range terms) and the charge transfer energy which is an exchange term. The dispersion-exchange term is much smaller than the others [3]. If the molecular wavefunctions are the S.C.F. wavefunctions of the isolated molecules, we have also to take account of the effect of the intramolecular correlation energy. The overlap between the molecular orbitals of the interacting molecules, which can involve qualitative changes in the description of the first order term, is less important for second order terms [4],[3].

It is generally assumed that the higher order terms of the perturbation series have a much smaller contribution to the intermolecular energies.

3) Equivalencies between these methods

As written above, there are some approximate equivalencies between these two categories of methods and we have checked some of them. We give some numerical comparisons.

ΔE_1 : the first order perturbation term is about equal to ΔE_1. Table 1 and 2 produce the results obtained for $Li^+ + H_2$ and $H_2 + H_2$, using a perturbation theory based on the use of biorthogonal orbitals [4],[5],[3]. E_1^{BO} is given by the expression [3] :

$$E_1^{BO} = 2\sum_{a_o}(\underline{a_o}|V_{NB}|a_o) + 2\sum_{b_o}(\underline{b_o}|V_{NA}|b_o) + \sum_{a_o b_o}4(\underline{a_o}a_o|\underline{b_o}b_o) - \sum_{a_o b_o}2(\underline{a_o}b_o|\underline{b_o}a_o)$$

a_o and b_o refer to the S.C.F. molecular occupied orbitals of molecules A or B ; V_{NA} and V_{NB} are the nuclear attraction hamiltonians. $\underline{a_o}$ and $\underline{b_o}$ are biorthogonal orbitals defined by

$$|\underline{i}> = \sum_{j}|j>(S^{-1})_{ji}$$

S being the total overlap matrix built from the set of molecular orbitals of A and B. The three first terms of E_1^{BO} corresponds to the electrostatic energy, the last one to the repulsive interaction.

If we neglect the overlap between the molecular orbitals of the interacting molecules, the biorthogonal orbitals $\underline{a_o}$ are identical to a_o and $\underline{b_o}$ to b_o. The corresponding expression of E_1^{BO} is then called E_1^{NO}. Comparison of E_1^{BO} and E_1^{NO} (Tables 1 and 2) shows that this

Table 1 : Li^+ + H_2, first order energy in 10^{-4} hartrees [a]

geometry	d(bohrs)	ΔE_1	E_1^{BO}	E_1^{NO}
C_{2v}	3	163.8	163.55	-196.32
	4	- 6.6	- 6.77	- 49.70
	4.5	-15.7	-15.73	- 29.87
	6	-10.8	-10.79	- 10.52
	10	- 2.4	- 2.46	- 1.87
linear	2.5		1527.65	-126.67
	3.5	249.4	261.71	61.82
	4.7	60.0	60.12	45.55
	6	24.0	24.00	24.04
	10	5.0	4.96	5.59

[a] The basis sets used in these calculations include [4] :
- 6s and 4p gaussian functions for H, kept uncontracted,
- 9s and 4p gaussian functions for Li^+, with the contraction 5s 4p.

Table 2 : H_2 + H_2, first order energy in 10^{-4} hartrees

geometry	d(bohrs)	ΔE_1	E_1^{BO}	E_1^{NO}
rectangular	5	18.62	19.05	-35.46
	6.5	1.52	1.61	- 2.49
linear	5	38.58	39.32	-59.64
	6.5	3.44	3.64	- 3.79
T shape	5	21.09	21.71	-51.55
	6.5	.85	.97	- 4.54
non planar	5	17.48	17.91	-36.38
	6.5	1.13	1.28	- 2.82

For the two closed-shell molecules, the dispersion energy, E_{disp} is obtained from the expression [9] :

$$E_{disp} = 4 \sum_{a_o b_o} \sum_{a_v b_v} |(a_o b_o | a_v b_v)|^2$$

$$x \; (\varepsilon_{a_o}^A + \varepsilon_{b_o}^B - \varepsilon_{a_v}^A - \varepsilon_{b_v}^B + J_{a_o a_v} - 2K_{a_o a_v} + J_{b_o b_v} - 2K_{b_o b_v})^{-1}$$

this expression corresponds to an Epstein-Nesbet partition of the molecular hamiltonians (state energy differences). The neglect of the J and K terms corresponds to a Moller-Plesset partition of the hamiltonians (orbital energy differences). As previously, we have used the S.C.F. occupied and virtual molecular orbitals. In the expression of E_{disp} given above, we neglect the overlap of the molecular orbitals of the interacting molecules. In the case of H_2+H_2, we have studied [3] the effect of this overlap through the use of biorthogonal orbitals and found that it generally decreases the values of the dispersion energy at short distances (repulsive part of the total energy curves).

If one of the molecule has an half-closed shell, the expression of E_{disp} is somewhat more complicated but may be obtained within a similar treatment [10]. Table 4 allows a comparison between the C.I. and the perturbation results [8,11] in the case of H_2+H_2. We can see that E_{disp} is

Table 4 : $H_2 + H_2$, dispersion and C.I. contribution in 10^{-4} hartrees

d(bohrs)	ΔE_{correl}	E_{disp}	ΔE_{correl}	E_{disp}
	rectangular geometry		T shape	
5.5	-4.00	-4.55	-4.91	-6.24
6.5	-1.52	-1.73	-1.82	-2.32
7	- .97	-1.11	-1.15	-1.47
7.5	- .64	- .73	- .73	- .98
10	- .11	- .13	- .09	- .15
	linear geometry		non planar geometry	
5.5	-7.26	-9.57	-3.84	-4.23
6.5	-2.83	-3.43	-1.44	-1.64
7	-1.81	-2.13	- .91	-1.06
7.5	-1.18	-1.36	- .60	- .70
10	- .18	- .21	- .10	- .12

overlap effect may be crucial for the description of the first order
energy.

Similar comparisons for other systems and perturbation procedures
have also been done by other authors.

E_{deloc} : this term is about equivalent to the induction and charge
transfer energy of a perturbation treatment. We have partly checked
this equivalency [6] in the case of $Li^+ + H_2$ where the charge transfer
energy is weak (Table 3). The term E_{ind} is given by [6] :

$$E_{ind} = 2 \sum_{a_o} \sum_{a_v} \left[(a_o|V_{NB}|a_v) + 2 \sum_{b_o} (a_o b_o|a_v b_o) \right]^2 \times (\varepsilon_{a_o}^A - \varepsilon_{a_v}^A + J_{a_o,a_v} - 2K_{a_o,a_v})^{-1}$$

a_o, b_o and V_{NB} have been defined above ; a_v are the S.C.F. virtual
orbitals of molecule A ; $\varepsilon_{a_o}^A$ and $\varepsilon_{a_v}^A$ are the energies of the S.C.F.
occupied and virtual orbitals of A ; J and K are the usual Coulomb
and exchange molecular integrals. We can see from the denominator that
we have used an Epstein-Nesbet partition of the molecular hamiltonians
(state energy differences). The overlap between the molecular orbilals
of the interacting molecules is neglected.

Table 3 : $Li^+ + H_2$, E_{deloc} in 10^{-4} hartrees

geometries	C_{2v}		linear	
d(bohrs)	E_{deloc}	E_{ind}	E_{deloc}	E_{ind}
3.5	-.01141	-.01501		
4	-.00812	-.00914	-.01292	-.01614
4.5	-.00568	-.00560	-.00883	-.00985
5	-.00395	-.00382	-.00611	-.00631
6	-.00196	-.00184	-.00297	-.00292
10	-.00022	-.00024	-.00034	-.00035

C.I. contribution and dispersion energy : the C.I. contribution
to the intermolecular interaction may be compared to the dispersion
energy. Table 4 allows such a comparison in the case of H_2+H_2. The
C.I. calculations have been performed using the perturbation method
developped by B. Roos [7], which is equivalent to a C.I. calculation
including mono and diexcited configurations. Four geometric configurations of the system H_2+H_2 have been studied [8], using the 6s4p basis set
defined above.

systematically larger than ΔE_{correl}, but both terms give the same
relative behaviour for the four configurations. A similar comparison
has been done in the case of two Neon atoms [12] where our values of
E_{disp} are compared to the intermolecular correlation energies obtained
by Das and Wahl from M.C.S.C.F. calculations [13]. If we compare E_{disp}
with the values obtained from expressions using the multipole expansion,
a good agreement can generally be found for large intermolecular (atomic) distances while it fails at short distances, leading to think that
the use of the multipole expansion may be questionable in the region of
the minimum of the total energy curves.

Total energies : we can then expect that the addition of either
ΔE_{correl} or E_{disp} to the S.C.F. determinations [8] will give similar
results, the perturbation results giving somewhat deeper minima, but
with no significant qualitative change in the relative behaviour of the
four configurations. Since only averaged energies are available for
$H_2 + H_2$, the total energies have been averaged with a weight of 0.25,
0.085, 0.415 and 0.25 for the rectangular, linear, T shape and non planar geometric configurations, respectively [14]. Table 5 gives the values
obtained [8], compared with some experimental determinations. In the present case the use of the perturbation dispersion energy gives better
results than the C.I. treatment. Since the equivalencies are not rigourous, we may suppose that some compensation of errors occurs and that
the agreement will not be so beautiful with any system. However, the

Table 5 : $H_2 + H_2$, averaged energies in 10^{-4} hartrees

d(bohrs)	SCF + ΔE_{correl}	SCF + E_{disp}	Experimental data [a]		
			Ref. [15]	Ref. [16]	Ref. [17]
5.5	2.63	1.65	.18	1.78	–
6.	.11	– .44	−1.11	– .65	– .41
6.5	– .59	– .95	−1.10	−1.06	−1.10
(6.73)	–	$(-1.10)^b$	–	–	–
7.	– .67	– .90	– .86	– .93	– .98
7.5	– .57	– .74	– .63	– .70	– .75
10.	– .11	–	– .13	– .12	– .12

a) Other determinations are available, for instance Ref. [18] gives a
minimum $\varepsilon = -1.14 \times 10^{-4}$ hartrees for $r = 6.46$ bohrs.
b) Obtained from a parabolic interpolation.

addition of the dispersion energy to the S.C.F. supermolecule determinations seems a quite reasonable way to characterize such complexes.
From a computational point of view, the evaluation of the dispersion

energy may be easier and faster (and then less expensive) than a C.I. treatment.

The basis sets problem (superposition error and need for diffuse polarization functions) : as mentioned in the introduction, very accurate results may be obtained with sophisticated methods and large basis sets. Unfortunately, the use of large basis sets leads to expensive calculations. We have then investigated the possibility to use smaller basis sets and found that the characteristics of such basis sets are different for the S.C.F. and for the dispersion determinations. In our calculations, the molecular orbitals are linear combinations of gaussian functions.

For the S.C.F. supermolecule treatment, we generally use a double zeta contraction for the two highest shells of the atoms (quantum numbers n and n-1) while the inner shells functions may be contracted according to a single zeta type. A polarization function, optimized to obtain the lowest molecular energies, is added on each atom (however, it may happen, in the case of rather large systems, that our computer memory size does not allow the use of such polarization functions). An important difficulty in the S.C.F. supermolecule treatment is the problem of the so-called "superposition error". This is due to the possibility, at intermediate distances, that molecule A may use the basis set of molecule B (and vice versa) to improve the description of its own energy. This leads to a lowering of the total S.C.F. energy of the interacting system with respect to the energy of the isolated molecules which must not be considered as intermolecular interaction. Even with large basis sets, this error may exist, but it is generally not too large [19]. With smaller basis sets, such as those described above, this error may become important, maybe of the same order of magnitude as the intermolecular energies. It is difficult to exactly evaluate this error. The most known prodedure to correct it, proposed by Boys et al [20], includes ghost functions located on molecule B to compute the molecular energy of molecule A, and vice-versa. In fact, this procedure gives an upper limit of the error and may lead to an overcorrection. However, it may be wise to use this information ; in particular, we have seen that, in some cases [21], this so-called "counterpoise correction" may invert the relative stabilities of two different geometric configurations of a system of interacting molecules.

For the dispersion energy, the set of functions is generally contracted into a double zeta set for the valence shells of the atom, and into a single zeta set for the inner shells. We have checked [22] that we loose only some percents of the dispersion energy when the valence shells basis set is contracted according to a single zeta set. For neutral atoms, linear and planar molecules, polarization functions are absolutely necessary. Except for low-polarizability ions, these are diffuse functions. This need for polarization functions seems less crucial for non planar molecules. Several calculations [6,9,11,10,12,23] have shown the special importance of one polarization function among functions of the same type : the use of one adequate polarization function may give more than 80 % of the dispersion energy obtained with a basis set including several polarization functions of the same type. This is an interesting characteristic for small basis sets and many of our studies have been done with basis sets including one optimized polarization function on each atom. Table 6 gives the values of the corresponding exponents.

Table 6 : exponents of the polarization function used to compute the dispersion energy.

atom	H	He	Li^+	Li	C	O	Ne	Mg^{++}
type p	0.2	0.3	0.775	0.04			0.2	
type d		0.14		0.035	0.17	0.287	0.45	1.02

atom	Mg^+	Mg	Cl	Ar	Ca^{++}	Ca^+	Ca	Br	I
type p	0.138	0.12				0.09	0.08		
type d	0.17	0.15	0.22	0.28	0.56	0.17	0.11	0.15	0.14

One particular case occurs with the OH + H_2 system [24]. Due to very high electronic asymmetry along the OH bond, the optimized polarization functions have a somewhat different exponent according to the geometric configurations considered. For collinear configurations, we obtain $\alpha_p(H) = 0.335160$ and $\alpha_d(O) = 0.25$ if H_2 is in the region of H and $\alpha_p(H) = 0.065787$, $\alpha_d(O) = 0.367487$ if H_2 is in the region of O. If the molecular axes are parallel, we have $\alpha_p(H) = 0.25$, $\alpha_d(O) = 0.299486$. The largest error involved in the dispersion energy is about 20 % in the worst case at short distances, and much smaller in the region of the total energy minimum of the curves if we do not use the best of these three basis sets.

The basis sets used for the dispersion energy generally give acceptable values of the polarizability for each component [25]. The procedure used to compute the polarizability component is a perturbative treatment, analogous to that used for the dispersion energy. It shows that the "state energy differences" expression generally gives better results that the "orbital energy differences" expression, with a tendancy of an overestimation in the first case and an underestimation in the last one. However, in some cases, Moller-Plesset partition gives a better agreement with experimental data. Such conclusions are also found in Ref. [26]. We expect that they are also valid for the dispersion energy.

III. RELATIVE IMPORTANCE OF THE DIFFERENT ENERGETIC CONTRIBUTIONS

It is interesting to compare the different energetic contributions, defined in the previous section, in some significant cases. We shall

do this comparison in the case of a neutral atom which looses one and
two electrons, and interacts with one H_2O molecule. We shall consider
the atoms Mg and Ca. Before commenting Table 7, it is good to keep in
mind the physical meaning of the different terms. The use of the multipole expansion shows that :
- the electrostatic energy involves the molecular moments (dipoles,
quadrupoles, octupoles, etc...) or the ionic charges of both molecules
- the induction energy involves "the molecular moments or ionic charge
of one molecule and the polarizability of the other molecule
- the dispersion energy involves the polarizabilities of both molecules.

Table 7 : interaction of M with H_2O (M = Mg^{++}, Mg^+, Mg, Ca^{++}, Ca^+, Ca),
energy contribution in kcal/mole.

M	d(Å)	ΔE_1	E_{deloc}	ΔE_{SCF}	E_{disp}	E_{TOT}
Mg^{++}	1.8	-63.99	-19.33	-83.32	- 1.45	-84.76
	2.	-67.48	-16.57	-84.04	- .84	-84.88
	2.2	-62.87	-13.67	-76.52	-· .49	-77.02
Mg^+	1.8	- 8.97	-19.68	-28.65	- 8.67	-37.32
	2.	-19.62	-15.05	-34.67	- 6.39	-41.06
	2.2	-22.33	-10.84	-33.18	- 4.64	-37.82
Mg	1.8	+30.86	-19.62	+11.23	-15.92	- 4.69
	2.	+17.20	-14.76	+ 2.44	-12.16	- 9.72
	2.2	+10.85	-10.35	+ .50	- 9.11	- 8.61
	2.4	+ 7.38	- 6.91	+ .46		
Ca^{++}	2.1	-39.96	-11.12	-51.08	- 3.60	-54.68
	2.3	-47.73	- 9.73	-57.46	- 2.22	·-59.69
	2.5	-47.29	- 8.35	-55.64	- 1.38	-57.02
Ca^+	2.1	- 6.45	-13.35	-19.79	- 6.69	-26.48
	2.3	-16.50	-11.04	-27.54	- 5.04	-32.58
	2.5	-19.00	- 8.77	-27.77	- 3.78	-31.54
	2.7	-18.64	- 6.63	-25.28	- 2.81	-28.09
Ca	2.3	+ 6.07	-11.96	- 5.88	- 8.89	-14.78
	2.5	+ 2.46	- 9.42	- 6.96	- 6.94	-13.90
	2.7	+ 1.39	- 7.06	- 5.67	- 5.38	-11.05

H$_2$O being a polar molecule, it is then easy to understand, that ΔE_1 (Table 7) is very attractive with dications, less attractive with monocations and repulsive with neutral atoms. E_{deloc} is nearly independent of the charge. E_{disp} increases from dications to neutral atoms. Compared to the total energy, E_{disp} is small for dications, justifying that such systems may be studied through an S.C.F. treatment only. With monocations, E_{disp} may represent 25 % of E_{TOT} ; its neglect will then depend on the accuracy needed. With neutral atoms, the dispersion energy is largely or mainly responsible for the stability of the system. It then must not be neglected in these cases.

IV. SYSTEMS INVOLVING LOW-POLARITY MOLECULES

We have been specially interested in the study of systems involving at least one, or two, low-polarity molecules. As seen, E_{disp} has to be included in the treatment.

1) Charge-transfer complexes

Charge-transfer complexes between two non-polar molecules represent an interesting category. It is known that charge-transfer molecules are characterized by a charge-transfer band in U.V. spectra. However, little is still known on the origin of their stability in the ground state, though much work has been devoted to understand this point.

(Cl$_2$)$_2$: the study of this system [19] was motivated by some experimental work on the dimer, but also in connection with the structure of the crystal. It is experimentally known that the crystal has an ortho-rhombic structure. A simple Lennard-Jones potential is not able to predict the right structure. Some authors have shown the importance of the quadrupole-quadrupole electrostatic interaction and of the charge transfer contribution but fail to give a correct description of the structure of the crystal. Previous ab initio calculations were also in desagreement with experimental data (a review of many of the studies on this system may be found in Ref. [19]). We have then studied 4 planar configurations with the following geometries : T, L, open L with an angle of 60° and linear. The stabilization energy is very weak (-0.13, -0.18, -0.11 kcal/mole for the T, L, open L configurations) or repulsive (linear geometry) at the S.C.F. level and is due mainly to the dispersion energy, as for van der Waals molecules. The L structure is slightly favored with respect to the other configurations, with a stabilization energy of 2.09 kcal/mole for an intermolecular distance of 3.27 Å (d is the distance between the two nearest non bonded atoms -linear, L, open L geometry- or between the nearest atom and the middle of a Cl-Cl bond in the T configuration [19]). This would be compatible with the experimental work of Klemperer et al [27] who found that the dimer is polar. In the crystal, the experimental geometry between two nearest Cl$_2$ molecules is close to an L or an open L structure, with an intermolecular distance of 3.34 Å. Using some analogy between the pheno-enon of indirect exchange in ionic solids with paramagnetic cations and weak binding energies of molecular complexes, Jansen and Black [28] suggest that the most stable geometry of the dimer could be non planar. We have not studied such configurations and plane to check this possibility.

C_2H_4 + Cl_2, Br_2, I_2 : these charge-transfer complexes are assumed to be the first step in the trans-addition reaction of molecular halogens with ethylene, their formation being then considered as a prereaction. Some experimetnal and theoretical studies concerning these systems are mentioned in Ref. [22].

It is often assumed, without real prooves, that the halogen molecule is perpendicular to the plane of C_2H_4, its axis being collinear with the C_{2v} axis of C_2H_4. This is called the "axial model". Using matrix isolation technics, Fredin and Nelander [29] showed that the ethylene-chlorine complex has a C_{2v} symmetry. This would be compatible with the axial model or with a "resting model" where the Cl_2 molecule is in a plane parallel to the C_2H_4 plane, the molecular axis of Cl_2 being parallel (R) or orthogonal (X) to the C-C bond. We have then studied these three configurations, along with a non symmetrical one by analogy with the L configuration of $(Cl_2)_2$. In this last case, the Cl_2 molecular axis is perpendicular to the C_2H_4 plane and passes through a C atom. We found that, again, the main contribution to the stability of the ground state is due to the dispersion energy. The more stable configuration is the axial model with a stabilization energy of 3.7 kcal/mole for an intermolecular distance of 3.0 Å (d is the distance between the nearer halogen atom and the C_2H_4 plane). The resting models are much less stable (1.56 and 1.25 kcal/mole).

We have then extended our studies to the case of the Bromine and Iodine complexes, assuming an axial model geometry. The general behaviour is the same for the three molecules Cl_2, Br_2, I_2, though the charge-transfer contribution increases from chlorine to bromine and to iodine. In all cases, the main element of the stability of the ground state is the dispersion energy.

Benzene-molecular halogens complexes : another example of very well known charge-transfer complexes are Benzene-molecular halogens complexes. Lot of work has been done on these systems since the first publications of Mulliken [30]. In crystals and in liquids, it is generally taken for granted that the halogen axis is along the C_6 benzene axis (axial model). The situation is less clear for 1:1 complexes. The recent work of Nelander et al [31] using matrix isolation technics leads to think that, in 1:1 complexes, the I_2 molecular axis would lie along the C_6 benzene axis while Cl_2 would rather interact preferently with one bond as in the C_2H_4 + Cl_2 complex. We have then started calculations [21] on the benzene-chlorine and benzene-iodine complexes, considering two geometric configurations : the axial one (A) and a configuration which we called "perpendicular-to-bond" model (PB). In this last case, the halogen molecular axis is perpendicular to the benzene plane and passes through the middle of a C-C bond. For such large

systems, we could not include polarizations functions in the basis
sets used for the S.C.F. supermolecule treatment. We are then in a case
where the superposition error may be qualitatively important ; this
point will be discussed in due time.

In all cases, the main element of the stability of the complexes
is the dispersion energy [21]. For the chlorine complex, the most stable
configuration is the perpendicular-to-bond one. If the S.C.F. supermolecule values are not corrected for the superposition error, the
total energy stabilization is -2.30 and -3.09 kcal/mole for configurations (A) and (PB), with intermolecular distances d of 3.440 and
3.175 Å, respectively (d is the distance between the benzene plane and
the nearer halogen atom). For these equilibrium distances, a counterpoise correction of the S.C.F. supermolecules values leads to -1.94
and -2.57 kcal/mole, respectively : this correction decreases the
depths of the minima but does not modify the relative order of the
stability of the two geometric configurations.

The results obtained for the benzene-iodine complex are quite similar to those concerning the benzene-chlorine complexes as long as the
uncorrected S.C.F. contributions are considered : the total energy stabilization is -3.73 and -4.58 kcal/mole for configurations (A) and (PB)
with intermolecular distances d of 3.440 and 3.175 Å, respectively.
However, the counterpoise correction leads to a completely different
situation : we have then -1.94 and -1.87 kcal/mole respectively. We can
see that the correction is very large in this case (1.79 and 2.71 kcal/
mole, respectively), and much larger for configuration (PB) than for
(A). Both configurations are now nearly degenerate, (A) being slightly
more stable than (PB). This would be in agreement with the work of
Fredin and Nelander [31] which notes a different behaviour between
chlorine and iodine complexes.

2) Systems of astrophysical interest : $CO + H_2$, $OH + H_2$

In collaboration with astrophysicits and collisionists, we have
engaged some work on the interaction of CO and OH with H_2 at intermediate and large distances. These molecules are relatively abondant in
the interstellar medium. Our work has been done in connection with
theoretical and experimental studies on the rotational excitation of CO
and OH by collision with H_2 molecules. The theoretical determinations
require, as a first step, the evaluation of the intermolecular potential at any point of a surface. Special attention was paid to the effect
of the anisotropy of the surface, in particular the anisotropy due to
the orientation of the H_2 molecular axis : no potential surface giving
information on this last point were available, even from semi-empirical
determinations. The different steps of the complete work are the following : a) the intermolecular energy is computed [32,24] for some geometric
configurations of the system and some intermolecular distances from ab
initio calculations (S.C.F. determinations and dispersion energy) ;
b) appropriate formulas using Legendre polynomials and spherical harmonics are proposed to fit the ab initio calculations values [33]. Such a

functional form is then directly usable in collision calculations ;
c) differential or total scattering cross sections and pressure broadening coefficients are computed [33-35], using the potential parameters
deduced from a) and b), and compared to experimental data [35].

CO + H_2 : given the large number of calculations necessary, we
could not afford the use of very large basis sets. We have chosen "double zeta plus one appropriate polarization function" basis sets [32], as
described in section 2. In a first step [32], we have studied 7 non equivalent configurations, which corresponds to all configurations having
the CO or H_2 molecular axes and the intermolecular axis along the x, y
or z axes. In the region of the van der Waals minimum, the dispersion
energy is generally important [32]. The anisotropy due to H_2 is large
for collinear configurations. At short distances, on the contrary, the
anisotropy due to H_2 is not large, justifying an average over the H_2
orientations as it is done in some determinations [36]. In order to fit
the intermolecular potential values with an appropriate functional
form, as described above, we have had to consider other configurations,
which correspond to other regions of the space. This finally leads to
19 non equivalent configurations.

Comparison between theoretical and experimental collision parameters [35] shows that the overall agreement is generally satisfactory,
with an overestimation of the theoretical absolute values of cross
sections which could be explained by a too deep van der Waals minimum.
Since our S.C.F. values are not corrected for the superposition error,
we do expect the computed minimum to be too low. D. Diercksen and
coll. [37] are presently refining the potential energy surface determinations, using large C.I. calculations and extended basis sets. They
paid special attention to the superposition error and found that for
some geometric configurations, a counterpoise correction could decrease
the minimum depth by a factor of about 1/2. Accurate determinations of
cross sections would require more accurate intermolecular energies.
However, the qualitative information obtained from these "relatively"
unexpensive potentials encourages us to use similar approximations and
basis sets for other systems.

OH + H_2 : this system is more complicated than the previous one,
which involves closed-shell molecules, because of the half-closed shell
in OH. The single electron may be in one of the two π orbitals, leading
to two degenerate configurations. We have now 12 non equivalent configurations [24] corresponding to the 7 configurations considered in the
first step of our work on CO + H_2. Also, as explained in section 2, the
high electronic disymmetry along the OH bond leads to use different
polarization functions according to the geometric configurations studied [24]. As in the previous case, the dispersion energy is generally
an important contribution to the stabilization of the system [24]. These
potentials have been used [34] to compute cross sections for the rotational excitation of OH by para H_2 in its ground rotational state
(J = 0). These studies will be extended to the case of the rotational
excitation of OH by ortho H_2 by Dewangan and Flower.

3) Weakly bound systems involving transition metal complexes : Mo(CO)$_5$Kr

We have also been interested in the study of the interaction between the transition metal complex Mo(CO)$_5$ and the rare gas atom Kr.
In the U.V.-visible spectra of the matrix-generated pentacarbonyls
M(CO)$_5$ the position of the visible band is very sensitive to the matrix
material used [38]. It has been assumed that the square pyramid M(CO)$_5$
forms a weak bond with a rare gas atom via the vacant coordination
site [38]. Turner et al postulated that this bond was of the donor-
acceptor type [38].

Veillard and coll. have long been involved in the theoretical study
of transition metal complexes [39] and considered worthwhile to investigate
through ab initio calculations, the nature of the interaction between
M(CO)$_5$ and a rare gas atom [40]. In the case of M = Mo and Kr, a double
zeta quality basis set gives repulsive S.C.F. supermolecule energies,
which implies that the charge-transfer energy is not a dominant contribution to the stabilization of the ground state of the system. The
dispersion energy added to the S.C.F. determinations leads to a stabilization energy of 5.3 kcal/mole for a Mo-Kr distance of 3.0 bohr. A
counterpoise correction of the S.C.F. energies would probably lead to
more repulsive values. On the other hand, the dispersion energy may be
underestimated, owing to practical limitations in the calculation
(no f functions, which might be necessary ; use of the orbital energy
differences expression). Compensation of errors may then occur and this
value of 5.3 kcal/mole seems reasonable. This energy is neither very
small nor very large and we can probably speak of a metal-rare gas
bond due to the dispersion energy.

V. CONCLUSION

I have tried to show what information could be obtained through
the use of the procedure described in section 2. As written in the
Introduction, these are ab initio calculations and, consequently, no
experimental data is required. The dispersion energy obtained from a
perturbative procedure is added to the S.C.F. supermolecule treatment,
taking advantage of some equivalencies between both categories of
methods. This is not a rigourous treatment and we cannot expect accurate results. Furthermore, in order to reduce the cost of the calculations, special attention has been paid to the possibility of the use
of relatively small basis sets. This, also, is a source of inaccuracy.
Such calculations are then highly qualitative, and even the qualitative conclusions must be taken with care. However, within all these
uncertainties, it seems that they are able to give important information
and we hope that they can be a not negligible contribution in the
general investigation on this field.

REFERENCES

1. Claverie, P. : 1978, "Intermolecular Interactions from Diatomics to Biopolymers", Ed. B. Pullman, Wiley, N.Y., p. 69.
2. Sadlej, A.J. : 1980, Mol. Phys., 39, p. 1249.
3. Kochanski, E., Gouyet, J.F. : 1975, Mol. Phys., 29, p. 693.
4. Kochanski, E., Gouyet, J.F. : 1975, Theor. Chim. Acta, 39, p. 329.
5. Gouyet, J.F. : 1973, J. Chem. Phys., 59, p. 4637 ; 1974, J. Chem. Phys., 60, p. 3690.
6. Kochanski, E. : 1974, Chem. Phys. Letters, 28, p. 471.
7. Roos, B. : 1972, Chem. Phys. Letters, 15, p. 153.
8. Jaszunski, M., Kochanski, E., Siegbahn, P. : 1977, Mol. Phys., 33, p. 139.
9. Kochanski, E. : 1973, J. Chem. Phys., 58, p. 5823.
10. Seger, G., Kochanski, E. : 1980, Chem. Phys. Letters, 76, p. 568.
11. Kochanski, E. : 1975, Theor. Chim. Acta, 39, p. 339.
12. Kochanski, E. : 1977, Chem. Phys. Letters, 47, p. 391.
13. Das, G., Wahl, A.C. : 1974, J. Chem. Phys., 60, p. 2195.
14. Evett, A.A., Margenau, H. : 1953, Phys. Rev., 90, p. 1021.
15. Hirschfelder, J.O., Curtiss, C.F., Bird, R.B. : 1964, Molecular theory of gases and liquids, Wiley, N.Y.
16. Dondi, M.G., Valbusa, U., Scoles, G. : 1972, Chem. Phys. Letters, 17, p. 137.
17. Farrar, J.M., Lee, Y.T. : 1972, J. Chem. Phys., 57, p. 5492.
18. Gengenbach, R., Hahn, C., Schrader, W., Toennies, J.P. : 1974, Theor. Chim. Acta, 34, p. 199.
19. Prissette, J., Kochanski, E. : 1978, J. Am. Chem. Soc., 100, p. 6609 ; 1977, ibid, p. 7352.
20. Boys, S.F., Bernardi, F. : 1970, Mol. Phys., 19, p. 553.
21. Kochanski, E., Prissette, J. : 1980, Nouv. J. Chimie, 4, p. 509.
22. Prissette, J., Seger, G., Kochanski, E. : 1978, J. Am. Chem. Soc., 100, p. 6941 ; Prissette, J., Kochanski, E. : 1978 : Nouv. J. Chim., 2, p. 107 ; Jaszunski, M., Kochanski, E. : 1977, J. Am. Chem. Soc., 99, p. 4624 ; Kochanski, E. : 1980, in "Quantum Theory of Chemical Reactions", Vol. 2, p. 177, R. Daudel, A. Pullman, L. Salem, A. Veillard, eds., D. Reidel publishing company.
23. Kochanski, E. : 1974, Int. J. Quantum Chem., S8, p. 219.
24. Kochanski, E., Flower, D.R. : 1981, Chem. Physics, to be published.
25. Seger, G., Kochanski, E. : 1980, Int. J. Quantum Chem., XVII, p. 955.
26. Mulder, F., van Dijk, G., Huiszoon, C. : 1979, Mol. Phys., 38, p. 577 ; Mulder, F. : 1978, Thesis, Nijmegen (The Netherlands).
27. Harris, S.J., Novick, S.E., Winn, J.S., Klemperer, W. : 1974, J. Chem. Phys., 61, p. 3866.
28. Jansen, L., Block, R. : 1981, submitted to J. Chem. Phys.
29. Fredin, L., Nelander, B. : 1973, J. Mol. Structure, 16, p. 205.
30. Mulliken, R.S. : 1950, J. Am. Chem. Soc., 72, p. 600 ; 1951, J. Chem. Phys., 19, p. 514 ; 1952, J. Am. Chem. Soc., 74, p. 811 ; 1952, J. Phys. Chem., 56, p. 801.

31. Fredin, L., Nelander, B. : 1974, Mol. Physics, 27, p. 885 ;
    1974, J. Am. Chem. Soc., 96, p. 1672.
32. Prissette, J., Kochanski, E., Flower, D.R. : 1978, Chem. Physics,
    27, p. 373.
33. Flower, D.R., Launay, J.M., Kochanski, E., Prissette, J. : 1979,
    Chem. Physics, 37, p. 355.
34. Dewangan, D.P., Flower, D.R. : 1981, J. Phys. B, to be published.
35. Brechignac , Ph., Picard-Bersellini, A., Charneau, R., Launay,
    J.M. : 1980, Chem. Physics, p. 165.
36. Green, S., Thaddeus, P. : 1976, Astrophys. J., 205, p. 766.
37. Diercksen, G.H.F., and collaborators : 1980, private communication.
38. Perutz, R.N., Turner, J.J. : 1975, J. Am. Chem. Soc., 97, p. 4791.
39. Veillard, A., Demuynck, J. : 1977, in "Modern Theoretical Chemistry",
    Vol. 4, H.F. Schaefer ed., Plenum Publishing, N.Y., p. 187.
40. Demuynck, J., Kochanski, E., Veillard, A. : 1979, J. Am. Chem.
    Soc., 101, p. 3467 ; Rossi, A., Kochanski, E., Veillard, A. : 1979,
    Chem. Phys. Letters, 66, p. 13.

COMPLEXES OF NEUTRAL MOLECULES ONTO NEGATIVE IONS

Alberte Pullman and Hélène Berthod
Institut de Biologie Physico-Chimique
Laboratoire de Biochimie Théorique
Associé au C.N.R.S.
13, rue P. et M. Curie - 75005 Paris -

I. INTRODUCTION

A large amount of information has been accumulated in the last decade on the thermodynamics of association of molecules and ions in the gas phase, owing in particular to the development of high pressure mass-spectroscopy [1]-[3] as well as ion-cyclotron resonance spectroscopy [4]. The data obtained on the formation and stability of small clusters, in conjonction with theoretical computations on the corresponding complexes can provide an understanding of the intrinsic nature of the forces involved between the interacting entities, a fundamental prerequisite to the elaboration of proper theories of nucleation and condensation phenomena [3] [5], of solvation, and ultimately of crystal growth.

Although the major attractive forces between an ion and a neutral ligand are expected to be electrostatic in nature, the detailed variations of the binding properties will depend on the molecular structure of both the ion and the ligand : thus the most interesting informations on the nature of the binding come from a comparison of clustering reactions of different neutrals on the same ion or of the same neutral on different ions. Numerous studies, both experimental [6 - 12] and theoretical [13 - 20] have dealt with the rationalization of cation-ligand interactions. Less has been done until now towards a systematic study of anion-molecule attachement. Very recently, a set of experimental data was made available on the clustering of H_2O, CO_2 and SO_2 on various anions [21]. The wide range of anions studied and the regularities (and irregularities) observed in the experimental data make it worth while to attempt a large-scale theoretical study. Aside from its general interest, the detailed understanding of anion-ligand interactions has a particular relevance for the work carried out in our laboratory on the properties of biopolymers : the phosphodiester anion, a repetitive building block of the structure of the nucleic acids, is a fondamental determining element in their overall properties, in particular in their interactions with ions, solvent and larger molecules [22]. Similarly, the carboxylate anion, a fundamental constituent of the side chains of proteins and peptides, appears to play a central role in numerous molecular interactions invol-

B. Pullman (ed.), Intermolecular Forces, 33–48.

ved in the cellular machinery. We have carried out detailed theoretical
analysis of the binding properties of these groups for water and cations
[23] [24]. In the absence of gas phase experiments on these systems, we
think that the now existing data on other polyatomic anions can provide
a test of the reliability of our theoretical computations.

In this paper, we limit ourselves to the interaction of NO_2 with H_2O,
CO_2 and SO_2. The nitrite anion is intermediate in molecular complexity
between the oxygenated anions of biological interest and the spherical
halide ions, and is likely to present structural features of its own
which will govern to a certain extent the directionality of the clustering of neutral molecules. On the other hand the three ligands H_2O, CO_2
and SO_2 are sufficiently different from one another to cluster differently on the same ion. Indeed the experimental data [21] show striking
differences, their enthalpies of attachement on NO_2 being - 15.2, - 9.3
and - 25.9 kcal/mole respectively for the first clustering molecule.

II. COMPUTATIONAL DETAILS

It is now well known that the gas-phase experimental value of the enthalpy of binding of two entities can be reproduced quite accurately by theoretical computations using the supermolecule approach, provided the
computation is carried out at the extreme limit of accuracy by using the
Hartree-Fock procedure with very extended basis set, supplemented with
a sufficient amount of configuration interaction.

Unfortunately, this extreme level of methodological refinement cannot be
utilized on a very large scale for obvious economical reasons and attempt
to reproduce at a lower cost the essential features of the binding have
been actively pursued. In the area of hydrogen bonding and cation binding to neutral ligands, we had found [25] that the utilization of the
Hartree-Fock procedure using a minimal, but good quality, basis set, led
to very satisfactory results. We have recently shown [23] that the same
basis set reproduced quite satisfactorily the results obtained with an
extended polarized basis in the case of anion binding (to $HCOO^-$). We
have tested the possibilities of this basis set to reproduce the experimental data on the clustering of H_2O, CO_2 and SO_2 on anions.

The basis set utilized has 7s, 3p and 10s, 6p Gaussian primitives on
the first-row and second-row atoms respectively and 3s functions on hydrogen with a scaling factor 1.2. These are contracted to a minimal basis : exponents and contraction coefficients are those of reference
[24] for H, C, O, N and P. For sulfur, the 10s, 6p basis [26] was contracted similarly to that of phosphorus in reference [24] (see table I).
For the third-row atoms which are in a state of hypervalency (phosphorus
in PO_4H_2, sulfur in SO_2) a set of d functions is added with an appropriately optimized exponent (0.38 for P [24], 0.48 for S) (note that
this d exponent is optimized for association with the above-defined minimal basis and need not be identical with that found in different associations). The effect of this d orbital addition on the results will be

TABLE I : Exponents [26] and contraction coefficients of the basis set
on the sulfur atom.

TYPE	EXPONENT	COEFFICIENT
s	25 506,3	0,0015775
s	3 812,82	0,0122172
s	860,556	0,0611657
s	242,940	0,211761
s	79,0448	0,452013
s	27,5705	0,400193
s	6,49476	0,382204
s	2,41078	0,657797
s	0,469815	0,463545
s	0,173396	0,579568
p	129,088	0,0290689
p	29,6305	0,179893
p	8,84715	0,478170
p	2,85576	0,496736
p	0,626108	0,441036
p	0,175233	0,664344
d	0,48	1,0

discussed in the section on SO_2.

The geometries adopted for the computations are the experimental values :
ONO = 115°.4, NO = 1.236 A for NO_2^- [27], OHO = 104°.52, OH = 0.957 A for
H_2O [28], OCO = 180°, CO = 1.165 A for CO_2 [29] and OSO = 119°.5, SO =
1.432 A for SO_2 [30]. These geometries have been kept frozen in this ex-
ploratory study of the supersystems NO_2^-...H_2O, NO_2^-...CO_2 and NO_2^-...
SO_2.

A first exploration of the interaction hypersurface was performed in all
cases using an electrostatic procedure developed in our laboratory [31],
in which an approximation of the electrostatic component of the interaction energy is computed using for each partner of the supersystem an
overlap multipole expansion (OMTP) of its ab initio density distribution.
Provided that the limits of approach of two atoms on the two partners
are set appropriately, this technique has proven very useful in the finding of the most important binding regions around a given entity [32]
[33]. After this preliminary scanning, supermolecule optimizations were
performed in each interesting region. Since in the present work, we are
interested not only in the order of magnitude of the binding energies
but in their numerical values, we have taken into account the basis set
superposition error inherent to all supermolecule computations, correc-
ting it by the Boys-Bernardi counterpoise procedure [34] recently reem-
phisized [35], where each molecule is computed in the presence of the
empty atomic orbitals of its partner in the complex.

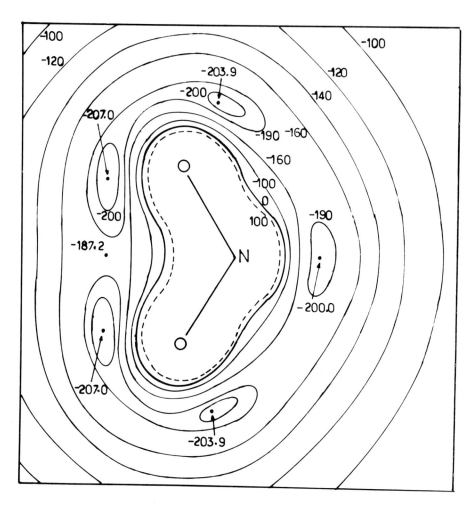

FIGURE 1 : The molecular electrostatic potential of the nitrite ion in
the plane of the molecule.

III. RESULTS AND DISCUSSION

A) The nitrite anion

Fig. 1 gives the distribution of the molecular electrostatic potential
[36] of NO_2^- in the molecular plane. It is seen that the negative potential surrounds entirely the molecule and that a very deep minimum of
- 200 kcal/mole appears in the neighbourhood of the nitrogen atom, nearly as deep as the two minima (- 207 and - 204 kcal/mole) flanking each
of the oxygen atoms. It is interesting to compare this rather isotropic
distribution of the potential to the image of the electronic structure
given by the population analysis where the nitrogen atom appears sligh-

tly positive (+ 0.017 e) with the negative charge practically localized on the two oxygens (- 0.508 e each). Note that the distribution of the potential around the oxygens in NO_2^- is very similar to that observed around the oxygens of the isoelectronic species $HCOO^-$ [23]. But in this last case the potential minimum on the opposite side of the molecule reached only - 93 kcal/mole. In the case of NO_2^-, the rather small anisotropy of the potential around the molecular periphery indicates that the extra situm lability [14] of the binding of an electrophile may be relatively large. This will be discussed in connection with the detailed results on the interactions.

Concerning the representation of NO_2^- given by the present basis set we would like to stress that, as in the case of the phosphate [24] and of the formate ion [23], the highest occupied molecular orbital is correctly found bonding, a requisite for a satisfactory solution of the Hartree-Fock problem in the case of anions [37].

B) Binding of H_2O

Since a very detailed SCF ab initio study of the hypersurface of interaction of NO_2^- and H_2O was made recently [38], in view of Monte Carlo calculations, using a double zeta basis set augmented by diffuse p functions, we have limited our own computations on this system to a search of the most stable position of the water molecule and to testing whether, the numerical values of the binding energy measured in the gas-phase experiments [21] could be reproduced with our basis. The exploration of the interaction hypersurface with the OMTP procedure indicates the electrostatic interaction to be best when the two molecules are coplanar, with a clear maximum when H_2O takes up a bridged position between the two anionic oxygens of NO_2^- (Fig. 2(a)). The whole molecular periphery

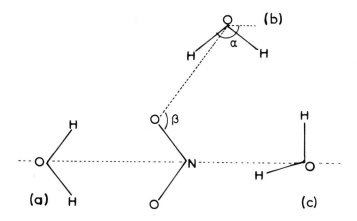

FIGURE 2 : Definition of the positions of the water molecule in the plane of the nitrite ion.

of the nitrite ion shows a favorable electrostatic interaction with water, although less favorable than in the bridge position. (Thus, for the
bridge, the electrostatic energy is − 20.6 kcal/mole ; for the arrangement (b) of figure 2, the optimal value is − 15.9 kcal/mole at $\beta = 110°$,
$\alpha = 140°$. For that of (c) the corresponding value is − 15.8 kcal/ mole ;
all values computed for 0 ... 0 = 2.8 Å). These results being in very
satisfactory agreement with those of reference [37], we have limited our
SCF computation of the complex to the bridge position ; the results before and after counterpoise correction are given in Table II : after
applying the correction, the binding energy is − 15.4 kcal/mole for an
0 ... 0 distance of 2.92 Å. Compared to the experimental value of − 15.2
$\mp$ 0.1 [21] the agreement is striking. The amount of charge transferred
as measured by the electron population is reduced from 0.030 e before CP
correction, to 0.008 e after the correction, a very small value characteristic of hydrogen bonding.

TABLE II : Energy values and counterpoise correction for NO_2^- ... H_2O
in the best bridge site : distances in Å, energies in kcal/
mole. ΔE_{SCF} is computed with respect to the energies of the
isolated species ; CP is the counterpoise correction (lowering of the energy of one species computed with the basis
set of the whole complex) ; ΔE is the final binding energy.

d	ΔE_{SCF}	$CP_{NO_2^-}$	CP_{H_2O}	ΔE
2.7	− 19.65	5.40	0.56	− 13.69
2.8	− 19.95	4.60	0.41	− 14.94
2.9	− 19.46	3.80	0.29	− 15.37

To conclude this section we predict in agreement with ref. [38] the most
stable position of water on NO_2^- to be the bridge position. In view of
the numerical agreement that we obtain with the experimental value we
feel fairly confident that this is indeed the most stable position. We
also agree with ref. [38] that the rest of the molecular periphery presents a number of binding possibilities for water with similar energies,
a few kcal/mole less favorable than the bridge position.

C) Binding of CO_2

A more complete exploration was performed in this case, where, to our
knowledge, no theoretical computations existed before the present one.

A rather complete spanning of the hypersurface of interaction using the
OMTP procedure was done for the following mutual dispositions of the two
molecules a) CO_2 lying entirely in the plane of NO_2^-, b) CO_2 perpendicular to the same plane with the carbon atom in the plane, c) CO_2 entirely

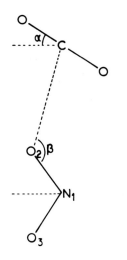

FIGURE 3 : Angles defining the position of the CO_2 molecule with respect to NO_2^- for the coplanar arrangement.

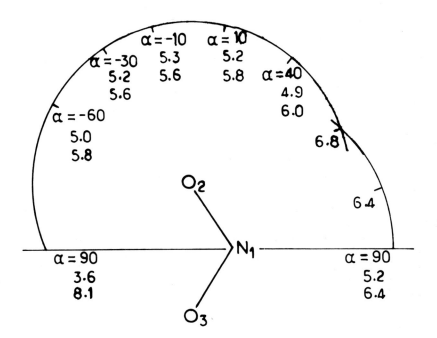

FIGURE 4 : OMTP values of the electrostatic interaction energies between CO_2 and NO_2^- at constant O ... C or N ... C distance (= 2.8 Å). First value : CO_2 in-plane, α as indicated, second value : CO_2 perpendicular to the plane.

out of the NO_2^- plane, either parallel or perpendicular to it, with the
carbon atom at a fixed distance of the plane. The essential outcome of
this exploration is that the perpendicular arrangement is always more
favorable than the in-plane one for the same CO distance. This is illustrated by the data in figure 4 (see figure 3 for the definitions of the
angles α and β). On the other hand, the completely out-of-plane positions of CO_2 are always less favorable. Finally, for the most favorable
(perpendicular) arrangement, three interesting regions appear : one is
clearly favored, on the internal bisectrix of the ONO angle. Another
one exists on the external bisectrix. But it is also clear that a rather wide region of possible binding exists in the external region of
the ONO angle and in its symmetrical counterpart.

Whereas the internal bisecting position stands out clearly, it is not
possible on the basis of the electrostatic data alone to decide precisely on the relative energies of the external bisecting position and of
the intermediate region because of the different nature of the atoms
(O and N) involved in the binding and of the proximity of the electrostatic energies computed at constant distance [31].

TABLE III : SCF binding energies (uncorrected) of CO_2 to NO_2^-.

	LOCATION	ΔE_{SCF}(a)	CONFIGURATION OF CO_2
B_1	O_2C = 2.5 Å ★	− 12.5	perpendicular
B_2	NC = 2.6 Å ★	− 7.5	perpendicular
E_1	O_2C = 2.4 Å ★ β = 105° ★	− 8.8	perpendicular
E'_1	O_2C = 2.5 Å ★ β = 105° ★	− 7.8	in-plane α = 40°
E_2	O_2C = 2.4 Å ★ β = 125°	− 8.4	perpendicular
E'_2	O_2C = 2.5 Å ★ β = 125°	− 8.0	in-plane α = 20°

(a) kcal/mole
(★) optimized value.

Thus, an SCF optimization was done in these three regions varying the
O_2 ... C or N ... C distance and the β angle. The main results at the
SCF level (without counterpoise correction) are given in Table III : it
is observed that the regions of favorable binding are indeed those indicated by the electrostatic data. The best energy of interaction is
found for position B_1 on the internal bisectrix of the ONO angle. The
position on the external bisectrix, B_2, is less favorable by a few ki-
localories/mole. Another equilibrium position E_1 is also found in the

exterior of the ONO angle, intermediate in energy and endowed with a relatively large lability (compare the optimal position E_1 (β = 105°) to position E_2 (β = 125°). In all sites the perpendicular arrangement is preferred to the in-plane one (compare for instance E_1 to E'_1 and E_2 to E'_2 in table III).

Since we are primarily interested in the most favorable binding site, we have computed the counterpoise correction for position B only. The computations, reported in table IV, yield a final corrected value of − 8.2 kcal/mole for a distance of 2.7 Å from the carbon atom to the two oxygens of NO_2^-. The experimental enthalpy of binding on this case being 9.3 ∓ 0.1 [21], the agreement, although somewhat less perfect than for water, is still quite satisfactory and lends credibility to the prevision that the most stable position is in the B_1 region. Furthermore the results seem to indicate that the nature of the binding between the nitrite ion and CO_2 is dominated by the electrostatic forces : the value 2.7 Å of the equilibrium distance O_2C shows that no chemical bond tends to be formed in this case, contrary to what occurs probably with OH^- [21]. A similar indication is given by comparing the electronic populations in the complex to those of the separate entities : the global apparent charge transfer to CO_2 is only 0.048 e and it is reduced further to 0.030 e after counterpoise correction. This value is however larger than that found for H_2O in the bridge position (0.008 e) indicating the increased weight of the charge transfer character in CO_2 binding.

TABLE IV : Energy values and counterpoise correction for NO_2^-...CO_2 in the best bisecting site B_1 (same notations as in table II).

d	ΔE_{SCF}	$CP_{NO_2^-}$	CP_{CO_2}	ΔE
2.4	− 11.93	5.96	0.81	− 5.16
2.5	− 12.47	4.88	0.51	− 7.08
2.6	− 12.19	3.93	0.32	− 7.94
2.7	− 11.48	3.10	0.20	− 8.18

D) Binding of SO_2

As was done for CO_2 binding, we have first made an exploration of the hypersurface of interaction with NO_2^- using the OMTP procedure for the following mutual dispositions of the two molecules :
i) SO_2 completely in the plane of NO_2^- (α = 0, β variable in fig. 5)
ii) the sulfur atom only in the plane of NO_2 with various relative inclinations (α and θ of figure 5) of the planes of the two molecules,
iii) SO_2 out of the plane of NO_2^- with the S atom moving in a plane parallel to that of NO_2^- or in the bisector plane of NO_2^-. These explorations indicating a very clear preference for type ii) interactions we

have limited our further search to this type of geometries. In that case, it was then found that the most favorable arrangement always occurs for a perpendicular disposition of the two molecules, with the plane of NO_2^- bisecting the OSO angle (Fig. 5 (b) (e)). Figure 6 gives, for this perpendicular arrangement, the variation, according to the angle β, of the electrostatic interaction energy and of the most favorable orientation (θ) computed with the OMTP approximation. Worth noting is the progressive variation of the inclination of SO_2 with respect to the nitrite ion which results from the desire to insure the best possible electrostatic interaction. The complex nature of this interaction is very apparent in these data which cannot be rationalized using simple global dipole-dipole or higher multipole interactions : the overlap multipole expansion utilized here which uses a multipolar expansion of each overlap distribution of the electron density function translates more accurately the interplay of the various terms.

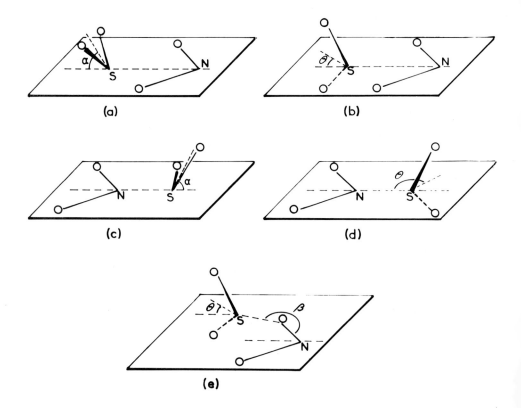

FIGURE 5 : The different orientations considered for SO_2 with the sulfur atom in the plane of the nitrite ion. α and θ are the angles of the bisectrix of SO_2 with N ... S for bisectrix positions, with the direction of the bisectrix of ONO for the more general case (e).

As in the case of CO_2, these electrostatic results indicate three do-
mains of favorable interaction : the inner bisectrix region, the external bisectrix region and an intermediate domain in the exterior of
the ONO angle. We do not expect, however, that the $SO_2...NO_2^-$ binding
will be entirely governed by the electrostatic attraction since SO_2 is
known to have a very high electron affinity [21] which should make it
very apt to charge transfer interactions. Indeed the charge-transfer
character of its strong complex with trimethylamine was stressed recently [39], [40]. In agreement with the high electron affinity of SO_2 we
find that the energy of the lowest empty molecular orbital computed with
our basis set has the extremely low value of 0.0222 a.u. Furthermore it
is a π orbital, thus oriented perpendicularly to the plane of the molecule, a feature which will make it particularly apt to accept electrons
from the nitrite ion in the configurations favored by the pure multipole-multipole interaction (Fig. 6). It can be expected that the decision
as to the best binding positions around the nitrite ion will strongly
depend on the charge transfer component of the binding energy.

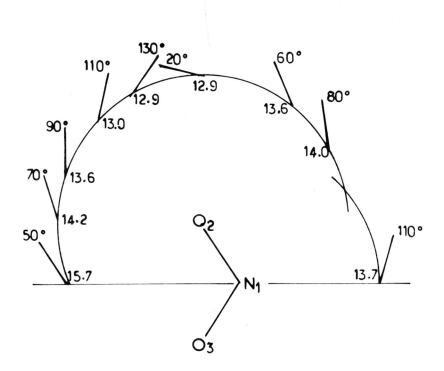

FIGURE 6 : The preferred orientation of the plane of SO_2 in the mutual
            perpendicular arrangement when the S atom moves around the
            periphery of the nitrite ion. (Electrostatic interaction on-
            ly (values as indicated)) ; O_2 ... S or N ... S = 2.6 Å).

A set of SCF computations was thus carried out to find the best binding sites. Due to the cost of the computations with the d orbitals, a first, rather detailed, optimization was performed deleting the d orbitals on the sulfur atom. (The location on the energy scale of the lowest empty molecular orbital on SO_2 is not very affected by this deletion (0.0103 instead of 0.0222) thus the charge transfer character of the binding should not be drastically altered. Moreover the electrostatic interactions computed without the d orbitals show an evolution qualitatively similar to that obtained with the d orbitals in the most attractive regions). After this optimization, the interaction energies were recomputed including the d orbitals with partial reoptimization only : the results of both computations are given in table V. It appears that among the essential three regions B_1, B_2, and E defined earlier, the external site E is by far the most favorable : the strong shortening of the equilibrium distance in this region indicates the sizeable influence of the charge transfer component which is illustrated further by the values of the global electron population transferred to SO_2 in the complex (uncorrected, vide infra). Another position of equilibrium, I, appears in table V, which was not found in the case of CO_2 : it corresponds to a site practically symmetrical to site E with respect to the NO bond. In this region, the tendancy to favor the charge transfer allows the SO_2 molecule to come closer to the oxygen atom, thereby reaching a better energy value than in site B_1. (We have tested by explicit computations that this situation does not occur with CO_2).

TABLE V : The most favorable SCF binding sites for $SO_2...NO_2^-$. (Energies uncorrected).
   (a) all parameters optimized for each position
   (b) distance reoptimized, β and θ assumed
   d is the O_2 ... S or N ... S distance ; β and θ as defined in figure 5.
   CT represents in 10^{-3} electron units the global non-corrected amount of electron population transferred.

	Without d orbitals (a)					With d orbitals (b)				
	d	β	θ	ΔE_{SCF}	CT	d	β	θ	ΔE_{SCF}	CT
B_1	2.7	–	70	– 21.9	135	2.6	–	50	– 23.8	169
B_2	2.4	–	110	– 21.7	187	2.4	–	110	– 26.7	211
E	2.3	110	50	– 25.6	177	2.0	110	60	– 35.65	331
I	2.4	250	70	– 25.8	180	2.2	250	60	– 32.7	273

Finally, with the d orbitals included, the sites of binding appear in the order $E > I > B_2 > B_1$. Furthermore, the strong binding character of SO_2 relative to CO_2 and H_2O is evidenced.

Concerning the numerical values of the corresponding energies, the counterpoise correction must be applied. The computations, (with d orbitals

included), for the best binding position are reported in table VI. The
final corrected value of the binding energy is 26.1 kcal/mole for an
O_2 ... S distance of 2.2 Å. Note that no angular variation of the counterpoise correction has been performed although it is likely that a small
correction would somewhat modify the final values of the angles θ and β
and of the energy. At any rate such refinements would not modify the essential results concerning the location of the best binding site in the
E region, nor the order of magnitude of the binding energy. Considering

TABLE VI : Energy values and counterpoise correction for $SO_2 \cdots NO_2^-$ in
the best external site.

d	ΔE_{SCF}	$CP_{NO_2^-}$	CP_{SO_2}	ΔE
2.1	- 35.6	6.68	2.84	- 26.0
2.2	- 34.3	5.81	2.34	- 26.1
2.3	- 32.3	5.07	1.88	- 25.4

that the experimental value is 25.9 ∓ 0.2 kcal/mole [21] the agreement
is gratifying and lends support to the prediction of the position in the
external region. The amount of electron transfer after counterpoise correction is 0.201 e. This, together with the very short distance O_2 ... S
in the complex indicates the charge transfer character of the adduct.

TABLE VII : Atom populations in the isolated molecules (without and with
counterpoise) and in the most favorable complex.
(a) population on the basis of S with no corresponding nucleus.

	ISOLATED	NO_2^- WITH CP	SO_2 WITH CP	COMPLEX
N	6.983	6.968	0.0	6.912
O_1	8.508	8.488	0.0	8.436
O_2	8.508	8.497	0.0	8.404
S	15.275	0.047(a)	15.278	15.302
O	8.363	0	8.361	8.473

The distribution of the electron populations in the complex given in Table VII in comparison to those of the isolated molecules (without and
with CP correction) shows that by the interplay of the internal electron
redistribution, the electrons transferred to SO_2 end up essentially on
its oxygen atoms and that, similarly, the transfer is made at the expense of the oxygens of NO_2, which at the end, appear globally less negative than those of SO_2. This is an interesting indication that the attachement of a second SO_2 molecule in a cluster might prefer to occur in a

chain-like manner rather than on the nitrite ion itself. Such a possibility has been evoked in reference [21] and deserves a theoretical investigation which we are pursuing.

A side conclusion of this set of computations is that an adequate description of the SO_2 complex is obtained only if d orbitals are added to the minimal basis set on the sulfur atom. Although the trends in the binding energies and characteristics are present in the results obtained without the d orbitals, the latter are necessary to insure a numerical agreement with the experimental values. The d orbitals are essentially necessary to account for the hypervalency of the third-row atom, and as was observed in the case of the pentavalent phosphorus atom in the phosphate anion [41] [24], their presence insures a more reasonable distribution of the electrons between the central atom and the bound oxygens : thus without d functions the sulfur atom in SO_2 carries a net positive charge of 1.175 e and this is reduced to 0.725 e in the presence of the d functions. Hence the relative weight of the electrostatic component in the interaction decreases while that of the charge transfer is obviously increased.

IV. CONCLUDING REMARKS

We have shown in this work that the numerical values of the binding enthalpies of single attachment of H_2O, CO_2 and SO_2 could be reproduced with a striking accuracy by SCF ab initio computations using a reasonable minimal basis set with counterpoise correction. The positions of binding to NO_2 are predicted to be an in-plane bridge for H_2O, a bisecting position perpendicular to the plane for CO_2 and an external position for SO_2, perpendicular to the plane and inclined. The charge-transfer character of the binding, very small for water, is slightly increased for CO_2 and very appreciable for SO_2.

The results obtained and the agreement observed with the experimental data indicate clearly that, although the precise weights of the different components of the binding energies cannot be assessed exactly in calculations with a minimal basis set, the gross relative importance of the electrostatic and charge transfer factors are correctly represented. A more precise assessment would require computations with a much more extended basis set and the introduction of correlation effects. Only this level of accuracy will clarify the precise role of the dispersion contribution in the binding energy. Computations in this direction are being performed in our group. At the present stage and for practical purposes it seems that our minimal basis set with the counterpoise correction is a reliable tool of exploration. We have good reasons to believe that this is valid for anions other than NO_2^-. It remains to be seen whether a satisfactory description of the successive clustering reactions can be obtained as well.

REFERENCES

[1] Kebarle, P. and Hogg, A.M., 1965, J. Chem. Phys., 42, pp. 668-675.
[2] Kebarle, P., 1977, Ann. Rev. of Phys. Chem. 28, pp. 445-476.
[3] Tang, I.N. and Castleman, Jr., A.W., 1972, J. Chem. Phys. 57, pp. 3638-3642.
[4] Beauchamp, J.L., 1971, Ann. Rev. Phys. Chem., 22, pp. 527-561.
[5] Castleman, Jr. A.W., 1979, in Advances in Colloid and Interface Science Nucleation, Vol. 10, A. Zettlemeyer, Edr. Elsevier, Oxford, pp. 73-128.
[6] Dzidic, I. and Kebarle, P., 1970, J. Phys. Chem., 74, pp. 1466-1474.
[7] Payzant, J.D., Cunningham, A.J. and Kebarle, P., 1973, Can. J. of Chemistry, 51, pp. 3242-3249.
[8] Hiraoka, K. and Kebarle, P., 1975, J. Am. Chem. Soc., 97, pp. 4179-4184.
[9] Tang, I.N. and Castleman, Jr. A.W., 1975, J. Chem. Phys., 62, pp. 4576-4578.
[10] Castleman, Jr. A.W., Holland, P.M., Lindsay, D.M. and Peterson, K. J., 1978, J. Am. Chem. Soc., 100, pp. 6039-6045.
[11] Woodin, R.L. and Beauchamp, J.L., 1978, J. Am. Chem. Soc. 100, pp. 501-508.
[12] Staley, R.H. and Beauchamp, J.L., 1975, J. Am. Chem. Soc. 97, pp. 5920-5925.
[13] Pullman, A., and Armbruster, A.M., 1975, Chem. Phys. Letters, 36, pp. 558-560.
[14] Pullman, A., 1976, In the New World of Quantum Chemistry, Procee-dings of the Second International Congress of Quantum Chemistry, R. Parr and B. Pullman Edrs., Reidel, Dordrecht, pp. 149-188.
[15] Kistenmacher, H., Popkie, H. and Clementi, E., 1973, J. Chem. Phys., 58, pp. 1689-1696.
[16] Kollman, P. and Rothenberg, S., 1977, J. Am. Chem. Soc., 99, pp. 1333-1340.
[17] Kollman, P., 1978, Chem. Phys. Letters, 55, pp. 555-560.
[18] Woodin, R.L., Houle, F.P. and Goddard III, W.A., 1976, Chem. Phys., 14, pp. 461-468.
[19] Berthod, H. and Pullman, A., 1980, Israel J. of Chemistry, 19, pp. 299-304.
[20] Berthod, H. and Pullman, A., 1980, Chem. Phys. Letters, 70, pp. 434-437.
[21] Keese, R.G., Lee, N. and Castleman, Jr. A.W., 1980, J. Chem. Phys., 73, pp. 2195-2202.
[22] Pullman, A. and Pullman, B., in Quarterly Rev. of Biophysics, in press.
[23] Berthod, H. and Pullman, A., 1981, J. of Computational Chemistry, 2, pp. 87-95, and references therein.
[24] Pullman, B., Gresh, N., Berthod, H. and Pullman, A., 1977, Theoret. Chim. Acta, (Berl.), 44, pp. 151-163.
[25] Pullman, A., Berthod, H. and Gresh, N., 1976, Int. J. of Quantum Chem., S 10, pp. 59-76.
[26] Roos, B. and Siegbahn, R.F., 1970, Theoret. Chim. Acta, (Berl.), 17, pp. 209-215.

[27] Carpenter, G.B., 1955, Acta. Cryst., 8, pp. 852-857.

[28] Benedict, W.S., Gailar, N., Plyler, E.K., 1956, J. Chem. Phys., 24, pp. 1139-1165.

[29] Jönnson, B., Karlström, G., Wennerström, H., 1975, Chem. Phys. Letters, 30, pp. 58-59.

[30] Kivelson, D., 1954, J. Chem. Phys., 22, pp. 904-908.

[31] for details see Pullman, A. and Perahia, D., 1978, Theoret. Chim. Acta, (Berl.), 48, pp. 29-33 and references therein.

[32] Goldblum, A., Perahia, D. and Pullman, A., 1979, Int. J. of Quant. Chem., 15, pp. 121-129.

[33] Pullman, A. and Demoulin, D., 1979, Int. J. of Quant. Chem., 16, pp. 641-653.

[34] Boys, S.F. and Bernardi, F., 1970, Mol. Phys., 19, pp. 553-560.

[35] Kolos, W., 1979, Theoret. Chim. Acta, (Berl.), 51, pp. 219-240.

[36] Bonaccorsi, R. Scrocco, F. and Tomasi, J., 1970, J. Chem. Phys., 52, pp. 5270-5277.

[37] Ahlrichs, R., 1975, Chem. Phys. Letters, 34, pp. 570-574.

[38] Banerjee, A., Shepard, R. and Simons, J., 1980, J. Chem. Phys., 73, pp. 1814-1826.

[39] Lucchese, R.R., Haber, K. and Schaeffer III, H.F., 1976, J. Am. Chem. Soc.,98, pp. 7617-7620.

[40] Douglas, J.E. and Kollman, P.A., 1978, J. Am. Chem. Soc., 100, pp. 5226-5227.

[41] Perahia, D., Pullman, A. and Berthod, H., 1975, Theoret. Chim. Acta, (Berl.), 40, pp. 47-60.

A COMPARISON OF THE AB INITIO SUPERMOLECULE AND INTERACTION APPROACHES: MULTIPOLE MOMENTS, HYDROGEN BONDING AND ION PAIRS

Robert Rein and Masayuki Shibata
Department of Experimental Pathology and
Department of Biophysics
Roswell Park Memorial Institute
Buffalo, New York 14263

ABSTRACT

The molecular multipole moments calculated by the IEHT wavefunctions are compared with the ab initio results and the experimental values available in the literature. The results indicate that the molecular multipole moments obtained by the IEHT wavefunctions are in good agreement with the experimental values. The basis set dependence of the multipole moments calculated from ab initio wavefunctions is discussed. Hydrogen bond energies for several systems have been calculated using intermolecular interaction theory based on IEHT wavefunctions. These hydrogen bonding energies are compared with available results obtained from ab initio supermolecule calculations. The results indicate that our approach , which is computationally more practical for the applications to large biological systems, can be used to reproduce the results of the ab initio supermolecule approach.

I. INTRODUCTION

It is well recognized that intermolecular forces play important roles in biological systems such as drug-nucleic acid interactions, protein-nucleic acid recognitions, enzyme-substrate interactions, as well as the structual organization of biopolymers. These topics have been extensively studied by our group using empirical energy functions or the quantum mechanical perturbation approach(1-7).

Hydrogen bonding is one of the most commomly observed interactions in chemical and biological systems and has been studied extensively by various experimental and theoretical methods in the past(8,9). The well-established intermolecular interaction theory based on the Rayleigh-Schrödinger perturbation theory has proven to be a powerful tool and has been successfully applied to various systems. On the other hand, recent progress in computational chemistry provides us with accurate descriptions of hydrogen bonding for small systems using the supermolecule approach within the framework of the ab initio method.

49

B. Pullman (ed.), Intermolecular Forces, 49–63.

Energy decomposition studies have provided some insight into the relative importance of various terms contributing to binding and into the relation between supermolecule and perturbation approaches. Because of the size of the molecule, however, the _ab initio_ supermolecule approach is not practical in many areas of interest in molecular biology. Thus recently several papers appeared which aimed to overcome this difficulty by developing schemes which simulate the _ab initio_ results with considerable reduction of required computational time(10,11). This type of approach, which reduces computational time but retains the quality of _ab initio_ calculations, is inevitable in the study of large complex systems. We have recently presented a preliminary report which compares the interaction study of the active site of α-chymotrypsin with the result obtained by the _ab initio_ method(21) and demonstrates that for a system of this size the values predicted by these two methods are in good agreement(12).

The objective of this paper is three fold. First, we reexamine the accuracy of the charge distribution obtained from the semi-empirical MO, IEHT, wavefunctions for use in electrostatic energy calculations. For this purpose multipole moments calculated using the IEHT densities are compared with those obtained from various levels of _ab initio_ wavefunctions on the one hand and with experimental values on the other. This analysis also permits a discussion of the basis set effect on the accuracy of predicting charge distributions, multipole moments and electrostatic energies. Second, we compare our interaction approach with _ab initio_ supermolecule results in the studies of hydrogen bonding systems. Third, we extend our discussion of the comparison of the interaction and _ab initio_ supermolecule approaches to the treatment of ion pairs.

II. METHODS

A detailed description of the methodology was given in a series of papers from this laboratory and was summarized in several review articles(13-15). Therefore, only a brief description is given here.

Molecular multipole moments, following the definition of Buckingham, were calculated from the IEHT wavefunctions with the Mulliken type division of overlap charges. The geometries which were taken from relevant literature sources are given in each table.

The interaction scheme for hydrogen bonding studies includes the following components: electrostatic, polarization, dispersion and short range overlap repulsion energies. The segmental multipole moments up to octopoles were included in the calculation of electrostatic and polarization components. The bond polarizability approximation was used for polarization and dispersion components where the bond polarizabilities were taken from LeFavre(16) and the ionization potentials from Gordon and Ford(17). The short range repulsion term was taken from Claverie(18) which is based on the empirical Kitaigorodsky function. This parameter was adjusted for the study of stacking interactions(18) and no attempt was made to reparametrize it for this hydrogen bonding study. The geometries were taken from relevant literature except for the case of cytosine-glutamine where the monomer geometries were taken from Voet and Rich(19) for cytosine and from Browman et al.(20) for glutamine. Acetamide was used as a model for glutamine by replacing α carbon with hydrogen using an appropriate bond length. The relative orientations of monomers in the complex are given in the results and discussion section and Table IX.

For ion pair studies, the interaction energies were calculated for different models of the active site of α-chymotrypsin(12) by using the geometries of the idealized triads given by Hayes and Kollman (21). The comparison was made using the scheme described in our previous paper(12) and summarized in Table X.

III. RESULTS and DISCUSSION

A. Multipole Moments.

In the interaction approach, an accurate description of charge distribution is required for a reliable evaluation of the electrostatic component. The expectation values of multipole moments provide sensitive probes of the accuracy of the electronic distribution throughout space and the IEHT wavefunction was shown to satisfy this requirement about a decade ago(22,23). Since then several papers appeared on the calculation of one electron properties within the framework of the ab initio method, thus it is very interesting to

TABLE I. Comparison of NH_3 Multipole Moments

			μ^a	θ^b_{zz}	Refs.
Experiment			−1.468	−1.0	25
IEHT[c]			−1.840	−1.90	
STO−3G	(6s3p/3s)	[2s1p/1s]	−1.787	−1.60	24
STO−6G	(12s6p/6s)	[2s1p/1s]	−1.812	−1.63	24
4−31G	(8s4p/4s)	[3s2p/2s]	−2.300	−2.59	24
6−31G	(10s4p/4s)	[3s2p/2s]	−2.324	−2.67	24
Dunning IA	(9s5p/4s)	[4s2p/2s]	−2.346	−2.92	24
Huzinaga III	(11s6p/4s)	[4s2p/2s]	−2.233	−2.21	24
KC	(13s7p1d/8s1p)	[6s2p1d/3s1p]	−1.946	−2.58	24
LSL	(5s4p1d/2s1p)	Slater	−1.687	−−−	26

a) in Debye. b) in 10^{-26} esu cm^2. c) Geometry from Ref(24).

TABLE II.   Comparison of Formic Acid
Multipole Moments at Different Geometries

	Geom.[d]	IEHT	MBS[e]	SV[e]	DZ[e]	DZ+P[e]	IEHT[c]	Exp.[c]
μ^a	M	1.21	1.08	1.29	1.48	1.63	1.21	1.35
	D	1.48	1.49	1.64	1.82	1.98		
	C	1.72	1.64	1.92	2.17	2.34		
θ^b_{xx}	M	−5.80	−8.34	−7.80	−8.47	−7.13	−7.06	−5.24
	D	−5.90	−8.34	−7.80	−8.61	−7.13		
	C	−5.63	−8.07	−7.66	−8.34	−5.63		
θ^b_{yy}	M	5.34	6.86	6.32	7.13	6.46	6.60	5.24
	D	5.52	7.13	6.59	7.40	6.72		
	C	5.27	6.86	6.32	7.13	6.46		
θ^b_{zz}	M	0.46	1.48	1.48	1.34	0.67	0.46	0.13
	D	0.38	1.21	1.21	1.21	0.54		
	C	0.35	1.21	1.21	1.21	0.54		
θ^b_{xy}	M	−3.95	−5.24	−5.11	−5.38	−4.71		
	D	−3.79	−5.38	−5.24	−5.51	−4.84		
	C	−3.68	−5.24	−5.11	−5.38	−4.70		
Refs.	27		27	27	27	27		28

a) in Debye.   b) in 10^{-26} esu cm^2.   c) in a principle axis.
d) Description of geometries.
      M: The gas-phase monomer geometry.
      D: The gas-phase geometry of half a dimer
      C: The geometry in the formic acid crystal
e) Description of basis sets.
      MBS   (6s3p/3s)   [2s1p/1s]
      SV   (6s3p/3s)   [3s2p/2s]
      DZ   (9s5p/4s)   [4s2p/2s]
      DZ+P (9s5p1d/4s1p) [4s2p1d/2s1p]

reexamine the quality of the IEHT densities by comparing them with the
ab initio results. The results obtained by the ab initio method are
ultimately governed by the choice of basis set, and the significant
basis set dependence of one electron propertiies is well known. One of
the most systematic studies on basis set dependence was recently
presented by Goddard and Csizmadia(24) for ammonia and some of their
values are listed in Table I to compare with the IEHT result.
Summarizing their results, it was clearly shown that increasing in the
number of primitive functions to improve the calculated energy does not
necessarily improve the overall representation of the charge
distribution. Multipole moments are very sensitive to the choice of
basis set. The split valence and double zeta basis sets tend to
overemphasize charge separation, resulting in large magnitudes of
moments. The inclusion of polarization functions is necessary to
compensate for this trend. It is interesting to note that the minimal

basis set gave better results than the extended basis set. It can be recognized that the IEHT value is better than those of Goddard and Csizmadia's polarization function calculation but slightly larger than the experimental and the near Hartree-Fock limit values.

Although the above calculations were performed on a fixed geometry, the effect of geometry on the multipole moments was assumed to be small at near experimental geometries. This can be examined by using different geometries as shown by Smit et al. for the case of formic acid(27). They have applied three different experimental geometries to calculate mutipole moments within the framework of the _ab initio_ method and the values are shown in Table II for comparison with our IEHT results. The general trend of basis set dependence described in the case of ammonia is observed for formic acid. Although the dipole moments exhibit a slight dependence on geometry, quadrupole moments show even less geometry dependence. This table also shows that the IEHT wavefunction correctly follows the trends of _ab initio_ calculations at each different geometry. Furthermore, the IEHT values indicate better agreement with experimental values than those of the _ab initio_ method.

In Table I, the near Hartree-Fock limit calculation for ammonia showed better agreement with experiment than the IEHT result, therefore, we compared the IEHT multipole moments with those of very extended basis set caculations for water(29,30) in Table III. These calculations include two d orbitals on the oxygen atom and showed good agreement with experiment. It should be noted that the Goddard and Csizmadia calulation with one d orbital for ammonia did not give a better agreement with experiment than the IEHT result. This may indicate that it might be neccessary to include at least two polarization functions on a heavy atom to produce better multipole moments than those of IEHT. It should also be emphasized that the IEHT results are still very close to those of the very extended basis set calculations.

TABLE III. Comparison of H_2O Multipole Moments

	μ^a	θ_{xx}^b	θ_{yy}^b	θ_{zz}^b	Ω_{xxz}^c	Ω_{yyz}^c	Ω_{zzz}^c	Refs.
Experiment	1.85	-2.50	2.63	-0.13	---	---	---	25
IEHT^d	2.23	-1.73	1.71	0.02	-0.76	1.68	-0.92	
(9s5p2d/3s2p) Gaussian	1.98	-2.36	2.45	-0.10	-0.97	2.30	-1.33	29
(10s6p2d/4s2p) Gaussian	1.99	-2.44	2.53	-0.11	-0.96	2.30	-1.34	29
(10s5p2d/4s1p)[5s3p2d/2s1p]	2.09	-2.39	2.49	-0.11	-0.89	2.19	-1.30	29
(11s7p2d/5s1p)[6s5p2d/3s1p]	2.03	-2.47	2.66	-0.20	-0.84	2.19	-1.35	30
(5s4p1d/3s1p) Slater	2.05	-2.54	2.61	-0.07	-0.98	2.39	-1.44	30

a)in Debye. b)in 10^{-26}esu cm^2. c)in 10^{-34}esu cm^3. d)Geom. from Ref(29).

TABLE IV.   Comparison of CH_4 Octopole Moments

			Ω_{xyz}^a	Refs.
Experiment	Virial Coefficient		3.7	31
	Phase Transition		1.6	34
	Far Infrared	Kihara Pot.	2.13~2.25	32,33
		L.J. Pot.	1.80~1.89	32,33
IEHT			2.7	23
Slater MBS			2.8	36
[5s3p/3s]	SCF		1.82	35
	CI		1.73	35
[5s3p1d/3s]	SCF		2.13	35
	CI		2.21	35
[5s3p1d/3s1p]	SCF		2.03	35
	CI		1.96	35
[5s3p2d/3s1p]	SCF		1.98	35
	CI		1.93	35
[6s4p3d/3s1p]	SCF		1.86	35
	CI		1.82	35

a) in 10^{-34} esu cm^3.

Table IV shows the results of recent calculations of methane octopole moments by the HF-SCF and CI methods(35) along with the IEHT and experimental values. Because of difficulty in direct experimental observation, there are slight variations in the experimental values, but the IEHT value is within the acceptable range.

Finally Table V shows the multipole moments of pyridine as an example for the case of a large molecule. Although there are some differences between the IEHT result and the others, it should be

TABLE V.   Comparison of Pyridine Multipole Moments

	μ^a	θ_{xx}^b	θ_{yy}^b	θ_{zz}^b	Ω_{xxz}^c	Ω_{yyz}^c	Ω_{zzz}^c	Refs.
Experiment	-2.2	-6.2	9.7	-3.5	---	---	---	25,28
IEHT[d]	-1.59	-3.09	5.94	-2.85	8.11	6.58	-14.68	
(9s5p/4s) [4s2p/2s]	-2.98	-4.8	8.8	-4.0	11.45	13.57	-25.02	37
Gaussian Lobe	-2.61	-8.5	10.7	-2.2	---	---	---	38
Xα (IV)	-2.78	-4.2	7.1	-2.9	---	---	---	39

a)in Debye. b)in 10^{-26} esu cm^2. c)in 10^{-34} esu cm^3. d)Geom. from Ref(39).

remembered that the double zeta basis set usually overestimates the magnitude of moments. However it is also possible that the aromatic molecule may have a different basis set dependence, and further investigation is necessary to elucidate the general characterization of basis set dependence of the multipole moments for larger molecules.

In summary, the IEHT wavefunction gives a good description of charge distribution in space which is at least as reliable if not better than the double zeta quality basis sets calculations if judged by comparison with experimental values of multipoles.

B. Hydrogen Bonding

The charge distributions obtained by the IEHT wavefunctions were applied to the studies of nucleic acid base interactions and successfully reproduced experimental results(1). It is interesting to compare the numerical values with those obtained by the _ab initio_ supermolecule approach available in the literature for small hydrogen bonding systems. In Table VI the results obtained for linear hydrogen bonding of a water dimer and of an ammonia dimer are summarized. There are several different values available ranging from 3.9 to 9.6 kcal/mol for the water dimer and 2.4 to 4.5 kcal/mol for the ammonia dimer showing dependence on the choices of geometries and basis sets. It is interesting to note that the interaction approach gives similar results to those of near Hartree-Fock calculations. It is also shown that the STO-3G results are better than split valence ones. Dill et al.(40) emphasized, however, that the STO-3G basis set overestimates the dimerization energy in some systems but underestimates it in others. Therefore a great deal of care must be taken in the interpretation of STO-3G results.

TABLE VI. Comparison of Dimerization Energies[a] in kcal/mol.

		Geometry	NH_3	H_2O	Refs.
Experiment			4.5	5.1	40
IEHT		Exp.	1.6	4.1	
STO-3G		STO-3G	3.8	6.0	40
4-31G		STO-3G	4.5	9.6	40
4-31G		4-31G	4.1	8.1	40
HFAO		EXP.	2.7	5.3	40
6-31G*		6-31G*	2.9	5.6	40
(11s6p2d/5s1p)	[5s4p1d/3s1p]	Exp.	2.4	---	41
(11s7p2d/6s1p)	[4s3p1d/3s1p]	Exp.	---	4.6	42
(11s7p2d/6s2p)	[4s3p2d/2s2p]	Exp.	---	3.9	42
(13s8p2d1f/6s2p1d)	[8s5p2d1f/4s2p1d]	Exp.	---	3.88	42

a) at linear configulation.

TABLE VII.   Comparison of Energy Components in kcal/mol
for a Linear Water Dimer at Experimental Geometry.

	ES	REP	EX	POL	DISP	CT	MIX	Total	Total–REP	Ref.
IEHT	–4.22	1.05	----	–0.29	–0.60	----	----	–4.06	–5.11	
STO–3G	–4.2	----	4.0	–0.1	----	–4.8	–0.1	–5.1	----	43
4–31G	–8.9	----	4.2	–0.5	----	–2.1	–0.3	–7.7	----	43
6–31G**	–7.5	----	4.3	–0.5	----	–1.8	–0.1	–5.6	----	43

A furthere insight into the basis set dependence can be obtained
by examining the energy decomposition studies. Umeyama and
Morokuma(43) showed that generally the STO–3G basis set tends to
overemphasize the charge transfer term and the split valence set
overestimates the electrostatic component. Some of their results are
reproduced here for comparison with the interaction approach for the
case of the linear water dimer at experimental geometry in Table VII.
Although there is no direct correspondence between the terms obtained
by the supermolecule and interaction approaches(due to their different
foundations), it is still interesting to note that there is a good
agreement in electrostatic terms obtained with the STO–3G basis set and
with the interaction approach.

The interaction approach is not valid when there is an appreciable
amount of overlap in the interacting region which may cause a
significant charge redistribution upon dimer formation. The formic
acid dimer results which are shown in Table VIII may be an example of
this situation. Significantly different experimental geometries are
reported for the monomer and the half a dimer in gas phase which could
be an indication of a large overlap in the interacting region. Despite

TABLE VIII.   Comparison of Energy Components in kcal/mol
for a Linear Formic Acid Dimer at Experimental Geometry.

	R^a	ES	REP	EX	POL	DISP	CT	Total	Total–REP	Ref.
IEHT	2.7	–7.38	16.99	----	–1.63	–2.68	----	5.30	–11.69	
MBS[b]	2.7	–30.15	----	21.45	–3.99	----	–11.04	–23.73	----	44
SV[b]	2.7	–32.07	----	28.63	–4.71	----	–13.39	–21.54	----	44
DZ[b]	2.7	–36.21	----	34.16	–6.58	----	–10.30	–18.93	----	44
IEHT	3.0	–5.47	3.27	----	–0.83	–1.51	----	–4.54	–7.81	
MBS[b]	3.0	–17.22	----	5.63	–2.00	----	–3.66	–17.25	----	44
SV[b]	3.0	–17.74	----	8.66	–2.33	----	–6.54	–17.95	----	44

a) Separation distance of O–O in Angström.
b) Description of basis sets is given in Table II.

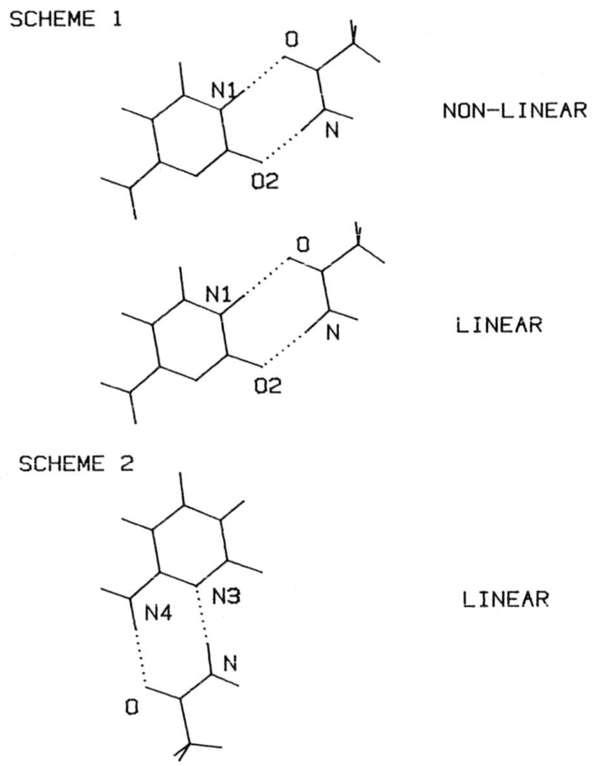

SCHEME 1

NON-LINEAR

LINEAR

SCHEME 2

LINEAR

FIGURE 1.   CYTOSINE-ACETAMIDE HYDROGEN BONDING SCHEMES.

the fact that the IEHT gives better multipole moments for this molecule than ab initio calculations as shown in Table II, its dimerization energy calculated by the interaction scheme is underestimated.   It should be noted that our short range repulsion term is not optimized at this stage, and thus the interaction energy without the repulsion term is also included in the tables.   It can be concluded that the interaction approach is less than optimal for treatment of strong hydrogen bonds, i.e.   hydrogen bonds which are significantly shorter than usual hydrogen bonding distance.

To study the interaction between cytosine and acetamide(as a model of glutamine), we have considered two different possible hydrogen bonding schemes 1 and 2, which are shown in Figure 1.   Since there is no description of the geometry of this complex available in the literature, we examined several configurations.   First, we considered geometries in which the two hydrogen bonds are linear.   The interaction energies were examined at several different separation distances which are shown in Table IX.   Our calculation indicates that the hydrogen bonding scheme 1 is slightly more favorable than scheme 2.   Due to the

TABLE IX.   Interaction Energy Components for Cytosine-Acetamide System (kcal/mol).

	R^a	ES	REP	POL	DISP	Total	Total-REP	Ref.
S (7s3p/3s) [2s1p/1s]	–	---	---	---	---	-12.5	---	45
c								
h	2.8	-11.63	8.82	-3.17	-2.50	-8.48	-17.30	
e Linear	3.0	-10.05	2.97	-2.05	-1.78	-10.91	-13.88	
m	3.2	-8.57	1.00	-1.38	-1.29	-10.24	-11.24	
e								
1 Nonlinear	b	-12.88	5.34	-2.17	-1.98	-11.69	-17.03	
	c	-10.52	1.81	-1.48	-1.43	-11.62	-13.43	
S								
c (7s3p/3s) [2s1p/1s]	–	---	---	---	---	-12.3	---	45
h								
e	2.8	-9.99	11.83	-2.06	-2.53	-2.75	-14.58	
m Linear	3.0	-9.10	4.07	-1.44	-1.81	-8.28	-12.35	
e	3.2	-7.99	1.41	-1.04	-1.31	-8.95	-10.36	
2								

a) O-O and O-N separation distances in angstrom.
b,c) Geometrical parameters for Nonlinear model
     (distances in angstrom, angles in degree).

	O_1-O	O_2-N	$\angle N_1$HO	$\angle O_2$HN
b	2.85	2.77	154.1	151.1
c	3.04	2.97	153.7	150.9

inadequacy of our repulsion term, the total energies may not represent the correct trends.   A preliminary attempt was made to examine the orientational effect for hydrogen bonding scheme 1 by rotating acetamide to give an N-H-O2 angle of about 150° with average hydrogen bond distances of about 2.8 and 3.0 Å.   The interaction energies for these cases in Table IX indicate that all components of the energy are similar to those calculated for the linear cases except for the repulsion term.   This may suggest the necessity of improving the representation of short range repulsion terms.   The importance of a careful choice of the form of the repulsion term was also pointed out by Gresh et al.(11).   The corresponding energies by Gresh and Pullman(45) are -12.5 kcal/mol for scheme 1 and -12.3 kcal/mol for scheme 2.   Our overall results are in this range and that scheme 1 is slightly favored over scheme 2 is in agreement with the results of the ab initio based calculations of Gresh and Pullman(45).

In the past, it was shown in both small systems(46) and large biological systems(1) that the electrostatic component obtained by a rigorous multipole expansion method dominates the energetics of the complex at experimental geometry.   This is due to the systematic cancellation of the other smaller repulsive and attractive components. This conclusion was also confirmed by the supermolecule approach(47,48). However, this conclusion is based on minimal basis set calculations. Recent ab initio calculations have shown a significant basis set dependence for the relative magnitude of each component in the energy decomposition scheme(43). Therefore further experience with the energy decomposition analysis with near Hartree-Fock quality basis sets is desired before ultimate conclusions can be drawn. In addition, whether these trends of basis set dependence can be directly applied to larger aromatic molecules should be further examined.

However, at this stage with some caution, the minimal basis set results can be accepted. This is based on the fact that the STO-3G

TABLE X. Scheme for a Comparison of
Supermolecule and Interaction Approaches.

The interaction energy difference, Δ INT, between neutral(AB) and ionic(A^+B^-) complexes can be calculated as follows:

I. Ab Initio Supermolecule Approach(MO).

Interaction energies(INTs) are obtained by,

$$INT_{AB}^{MO} = E_{AB}^{MO} - E_{A}^{MO} - E_{B}^{MO} ,$$

$$INT_{A^-B^+}^{MO} = E_{A^-B^+}^{MO} - E_{A^-}^{MO} - E_{B^+}^{MO} ,$$

where the Es are ab initio total energies.

The difference in interaction energies can be obtained by,

$$\Delta INT^{MO} = INT_{A^-B^+}^{MO} - INT_{AB}^{MO}$$

$$= (E_{A^-B^+}^{MO} - E_{AB}^{MO}) - (E_{A^-}^{MO} - E_{A}^{MO}) + (E_{B}^{MO} - E_{B^+}^{MO})$$

$$= (E_{A^-B^+}^{MO} - E_{AB}^{MO}) + (-PA_{A^-}^{MO} + PA_{B}^{MO}) ,$$

where Proton Affinities(PAs) are defined as follows:

$$A^- + H^+ ---> AH \qquad PA_{A^-}^{MO} = E_{A^-}^{MO} - E_{A}^{MO} ,$$

$$B + H^+ ---> BH^+ \qquad PA_{B}^{MO} = E_{B}^{MO} - E_{B^+}^{MO} .$$

II. Interaction Approach(IA).

$$\Delta INT^{IA} = INT_{A^-B^+}^{IA} - INT_{AB}^{IA} .$$

Relation between Two Approaches.

$$(E_{A^-B^+}^{MO} - E_{AB}^{MO}) + (-PA_{A^-}^{MO} + PA_{B}^{MO}) + Dispersion$$

$$= (INT_{A^-B^+}^{IA} - INT_{AB}^{IA}) + Charge\ Transfer.$$

---

description of charge distribution is not far from experimental multipole moments and therefore by fortunate cancellation effects the STO-3G may give a good agreement with experimental dimerization energies. At this stage, we still believe that the multipole interaction energy based on the good quality wavefunction(IEHT) gives a qualitatively correct picture of hydrogen bonding and further work toward quantitative agreement is now in progress. In summary, the calculated IEHT multipole moments have been shown to be in somewhat better agreement with experimental values than the STO-3G values. Thus based on similar arguments as above, the interaction energies can be accepted as giving a correct description of hydrogen bonding. An exception to this conclusion is a system with short and strong hydrogen bonds, it appeared that in such systems the above perturbation approach may be breaking down.

C. Ion Pairs.

Another aspect of general interest in this study is the comparison of ab initio supermolecule and interaction approaches on the evaluation of interaction energies of ion pairs. In order to compare interaction energies with the ab initio results, one has to consider what process is involved in the ab initio supermolecule calculations of neutral and ionic complexes. The detailed description has been reported previously(12) and the exact relations between the terms in the ab initio supermolecule and interaction calculations are listed in Table X. The final equation shows that by using calculated total energies of complexes and proton affinities one can obtain the interaction energy difference of the neutral and ionic complexes from ab initio calculations. As it is well known, the supermolecule approach cannot take into account the dispersion interaction. On the other hand, the interaction approach lacks the charge transfer term. For the chymotrypsin charge relay triad, the ab initio energy

TABLE XI. Comparison of Supermolecule and Interaction
Approaches for the Charge Relay Triad (in kcal/mol).

Systems	Differences in Toatl Energies[a] $E_{A^-B^+}^{MO} - E_{AB}^{MO}$	Calculated Proton Affinities[a] $-PA_{A^-}^{MO} + PA_{B}^{MO}$	Interaction Energy Differences ΔINT^{MO}	ΔINT^{IA}	Differences between Two Approaches
HIS SER	0.0	0.0	0.0	0.0	0.0
HIS$^+$SER$^-$	114.3	−257.0	−142.7	−103.28	39.4
ASP$^-$HIS SER	0.0	0.0	0.0	0.0	0.0
ASP$^-$HIS$^+$SER$^-$	92.9	−257.0	−164.1	−131.24	32.9
ASP HIS SER$^-$	38.7	−69.0	−30.3	−11.39	18.9

a) Ref.21

components as defined in Table X were obtained from the calculation by Hayes and Kollman(21). These values are listed in Table XI along with our results obtained by the interaction approach.

The agreement between the two approaches is very rewarding. The same trend of interactions is appearent in both sets of calculations. The remaining difference between the two sets of calculations is (in the range of 18.9 - 39.4 kcal/mole) due to one of the following reasons. The charge transfer term which is missing in our calculation, differences in the electron densities (ab initio versus IEHT) implicit in the electrostatic terms obtained by the two approaches, and possiblely the basis set error involved in the ab initio results obtained with small basis set calculations. In fact it is known that limited basis set calculations exaggerate energy state differences of neutral and ionic species.

IV. CONCLUSION

From this study it appears that an interaction approach to the study of ion pairs and large molecular complexes can be based on much less computationaly expensive interaction schemes without significant loss of accuracy. Furthermore, the use of electron densities obtained from the semi-empirical MO method (IEHT) in the interaction calculations are of equal quality to more sophisticated ab initio calculations. The results of this study together with those of Gresh and Pullman(45) are very encouraging for the study of large biological systems such as proteins and nucleic acids in using the computationally more feasible intermolecular interaction approach.

ACKNOWLEDGEMENTS

This work was supported by NASA Grant NSG-7305. We thank the computer centers at RPMI and SUNY/AB for their generous allotment of computer time.

REFERENCES

1. Rein, R.(1978) in 'Intermolecular Interactions:From Diatomics to Biopolymers', Pullman, B., Ed., Jhon Wiley and Sons, pp.308 .
2. Rein, R., Renugopalakrishnan, V. and Bernard, E.A.(1971) First Europ. Biophys. Congress Proc. vol.6, pp.35 .
3. Renugopalakrishnan, V. and Rein, R.(1976) Biochi. Biophys. Acta 434, pp.164 .
4. Ornstein, R. and Rein, R.(1979) Biopolymers 18, pp.2821 .
5. Ornstein, R. and Rein, R.(1980) Chemico-Biolog. Interactions 30, pp.87 .

6. Rein, R., Garduno, R., Colombano, S., Nir, S., Haydock, K. and
    MacElroy, R.D.(1981) in 'Biomolecular Structure, Conformation,
    Function and Evolution', Srinivasav, R., Ed., Pergamon Press
    pp.387 .
7. Garduno, R., Haydock, K. MacElroy, R.D. and Rein, R.
    Ann. N.Y. Acad. Sci. (in press).
8. Kollman, P.A.(1977) in 'Application of Electronic Structure
    Theory, Modern Theoretical Chemistry Vol.4',
    Schaefer, III, H.F, Ed., Plenum Press, pp.109 .
9. Schuster, P.(1978) in 'Intermolecular Interactions: From
    Diatomics to Biopolymers', Pullman, B., Ed., John Wiley and Sons,
    pp.363 .
10. Clementi, E. in 'Computational Aspects for Large Chemical
    Systems, Lecture Notes in Chemistry Vol.19', Springer-Verlag,
    Berlin, pp.1 .
11. Gresh, N, Claverie, P. and Pullman, A.(1979) Int. J. Quantum
    Chem. QCS 13, pp.243 .
12. Shibata, M., Kieber-Emmons, T., Dutta, S. and Rein, R.
    Int. J. Quantum Chem. QBS (accepted).
13. Rein, R., Clarke, G.A. and Harris, F.E.(1970) in 'Quantum Aspects
    of Heterocyclic Compounds in Chemistry and Biochemistry',
    Bergman, E.D. and Pullman, B., Eds., The Israeli
    Acad. Sci. Humanities, Jerusalem, pp.86 .
14. Rein, R.(1973) Adv. Quantum Chem. 7, pp.335 .
15. Rein, R.(1975) in 'Electronic Structure of Polymers and Molecular
    Crystals', Andre, J.-P., and Ladik, J., Eds., Plenum Press,
    pp.505 .
16. LeFavre, R.J.W.(1965) Adv. Phys. Org. Chem. 3, pp.1 .
17. Gordon, A.J. and Ford, R.A.(1972) in 'The Chemists Companion, A
    Handbook of Practical Data, Technique and References', John Wiley
    and Sons, pp.236 .
18. Claverie, P.(1968) in 'Molecular Associations in Biology',
    Pullman, B., Ed., Academic Press, pp.115 .
19. Voet, D. and Rich, A.(1970) Prog. Nucleic Acids Res.
    and Mol. Biol. 10, pp.183 .
20. Browman, M.J., Carruthers, L.M., Kashuba, K.L., Momany, F.A.,
    Pottle, M.S., Rosen, S.M. and Rumsey, S,M.(1975) Quantum
    Chemistry Program Exchange 286 .
21. Hayes, D.M. and Kollman, P.A.(1979) in 'Catalysis in Chemistry
    and Biochemistry, Theory and Experiment', Pullman, B., Ed.,
    D. Reidel Pub. Comp., pp.77 .
22. Rabinowitz, J.R., Swissler, T.J. and Rein, R.(1972)
    Int. J. Quantum Chem. S6, pp.353 .
23. Swissler, T.J. and Rein, R.(1972) Chem. Phys. Lett. 15, pp.617 .
24. Goddard, J.D. and Csizmadia, I.M.(1978) J. Chem. Phys. 68,
    pp.2172 .
25. Stogryn, D.E. and Stogryn, A.P.(1966) Mol. Phys. 11, pp.371 .
26. Laws, E.A., Stevens, R.M. and Lipscomb, W.N.(1972)
    J. Chem. Phys. 56, pp.2029 .
27. Smit, P.H., Derissen, J.L. and van Duijneveldt, F.B.(1977)
    J. Chem. Phys. 67, pp.274 .

28. Flygare, W.H. and Benson, R.C.(1971) Mol. Phys. 20, pp.225 .
29. Neumann, D. and Moskowitz, J.W.(1968)
     J. Chem. Phys. 49, pp.2056 .
30. Dunning,Jr., T.H., Pitzer, R.M. and Aung, S.(1972)
     J. Chem. Phys. 57, pp.5044 .
31. Spurling, T.H., De Rocco, A.G. and Storvick, T.S.(1968)
     J. Chem. Phys. 48, pp.1006 .
32. Akhmedzhanov, R., Gransky, P.V. and Bulanin, M.O.(1976)
     Can. J. Chem. 54, pp.519 .
33. Cohen, E.R. and Birnbaum, G.(1977) J. Chem. Phys. 66, pp.2443 .
34. James, H.M. and Keenan, T.A.(1959) J. Chem. Phys. 31, pp.12 .
35. Amos, R.D.(1979) Mol. Phys. 38, pp.33 .
36. Sinai, J.J.(1964) J. Chem. Phys. 40, pp.3596 .
37. von Niessen, W., Dierksen, G.H.F. and Cederbaum, L.S.(1975)
     Chem. Phys. 10, pp.345 .
38. Ha, T.-K. and O'Konski, C.T.(1973) Int. J. Quantum Chem. 7,
     pp.609 .
39. Case, D.A., Cook, M. and Karplus, M.(1980) J. Chem. Phys. 73,
     pp.3294 .
40. Dill, J.D., Allen, L.C., Topp, W.C. and Pople, J.A.(1975)
     J. Amer. Chem. Soc. 97, pp.7220 .
41. Hinchliffe, A., Bounds, D.G., Klein, M.L., McDonald, I.R. and
     Righini, R.(1981) J. Chem. Phys. 74, pp.1211 .
42. Popkie, H., Kistenmacher, H. and Clementi, E.(1973)
     J. Chem. Phys. 59, pp.1325 .
43. Umeyama, H. and Morokuma, K.(1977) J. Amer. Chem. Soc. 99,
     pp.1316 .
44. Smit, P.H., Derissen, J.L. and van Duijneveldt, F.B.(1979)
     Mol. Phys. 37, pp.501 .
45. Gresh, N. and Pullman, B.(1980)
     Biochim. Biophys. Acta 608, pp.47 .
46. van Duijneveldt-van de Rijdt, J.A.C.M. and
     van Duijneveldt, F.B.(1971) J. Amer. Chem. Soc. 93, pp.5644 .
47. Morokuma, K.(1971) J. Chem. Phys. 55, pp.1236 .
48. Dreyfus, M. and Pullman, B.(1970) Theor. Chim. Acta 19, pp.20 .

THE EXTRACTION OF INTERMOLECULAR POTENTIALS FROM MOLECULAR SCATTERING DATA: DIRECT INVERSION METHODS

R.B. Gerber
Department of Physical Chemistry, The Hebrew University
of Jerusalem, Jerusalem, Israel

ABSTRACT

It is shown that intermolecular potentials can be extracted very accurately and efficiently from molecular beam scattering data by direct inversion methods, without any trial-and-error fitting. Inversion methods are described that yield the dependence of the intermolecular potentials on the internal coordinates from measured state-selected elastic and inelastic differential cross sections. The following problems are discussed: (1) Determination of anisotropic potentials from the rotationally inelastic and elastic cross sections; (2) Obtaining nonadiabatic interactions for atom-atom systems in which a crossing of electronic curves occurs; (3) Extraction of atom-(solid) surface interactions from measured diffraction peak intensities. Physical approximations are used in all cases to simplify the scattering dynamics and provide non-complicated inversion schemes. It is concluded that the combination of molecular beam experiments with direct inversion methods is a powerful tool for obtaining intermolecular forces.

1.   INTRODUCTION

Despite the important progress in quantum-chemical calculations of intermolecular potentials, only in very few cases can the theory, at the present day, provide an interaction compatible with the accuracy of the best experiments available. Indeed, for most systems the only realistic way to consider for obtaining accurate interaction potentials is by extracting these quantities from experimental data. Both physical considerations and practical experience indicate that molecular beam scattering data is one of the most sensitive experimental probes of intermolecular forces[1]. The differential cross sections measured in such experiments cary detailed information on the interaction potential over a large range of intermolecular distances. However, the relation between the interaction and the observable quantities is very complicated, and the task of extracting the potential from the data poses a very difficult problem the general solution of which is still not

B. Pullman (ed.), Intermolecular Forces, 65–78.

available. Until recently, all efforts in this direction were based on
trial-and-error approaches in which the scattering equations are solved
repeatedly for different trial potentials, until good agreement with
the experimental cross sections are found. However, such methods are
inherently inefficient and complicated, and depend on arbitrary guesses
of trial potentials. The difficulties of such approaches are extremely
severe when the interaction sought is a potential energy surface, depending on internal coordinates of the molecules involved. Clearly,
there is great current interest in determining such interactions.

In this lecture, a description will be given of progress made in
recent years in developing methods that extract the interaction potential by direct, explicit inversion of the scattering data. Important
early steps in this field were made by Buck, Pauly, and others[1,2],
leading to a partial solution of the inversion problem for elastic
atom-atom scattering. Later, a first complete inversion of elastic
scattering data was carried out by Gerber et al.[3] The developments
which are the subject-matter of the present lecture deal, however,
with inversion methods for interactions that depend on the internal coordinates of the molecules involved, i.e. with the determination of
intermolecular potential energy surfaces. Physical approximations are
invoked in these methods in order to simplify the complicated collision
dynamics. In Sec. 2, as a brief introduction to the techniques involved,
the inversion problem for a purely elastic atom-atom collision is discussed. In Sec. 3 results are presented on an inversion method for inelastic atom-atom collisions leading to determination of the nonadiabatic interaction in this case[4]. Sec. 4 describes an inversion scheme
for obtaining the anisotropic atom-molecule interaction potentials from
rotationally elastic and inelastic differential cross sections[5]. In
Sec. 5 we briefly comment on an inversion method for a very different,
yet somewhat related problem: Extraction of the interaction between an
atom and a solid surface from diffraction peak intensities in beam
scattering[6]. Some of these results were already recently reviewed[7].

2.   INVERSION PROBLEM FOR ELASTIC ATOM-ATOM COLLISIONS

While this problem may be considered solved, we discuss it briefly here
in order to introduce some of the ideas and methods that play a role
also in the inversion scheme for inelastic collision problems.

2.1. The two steps of an inversion problem

Consider two atoms interacting through a potential $V(r)$ corresponding
to the electronic ground state curve. If $V(r)$ is known then the calculation of the differential cross section $\sigma(\theta)$ for each angle θ is carried out as follows: For each relative angular momentum ℓ of the collision partners, a conserved quantity here, the Schrödinger equation
for the corresponding partial wave is solved, yielding the elastic
partial-wave scattering amplitude $S(\ell) = \exp[2i\eta_\ell]$ where the η_ℓ is the
ℓ-phase shift. From these quantities one obtains the scattering amplitude $f(\theta)$:

$$f(\theta) = \frac{1}{2ik} \sum_{\ell=0}^{\infty} (2\ell+1)[e^{2i\eta_\ell}-1]P_\ell(\cos\theta) \tag{1}$$

where k is the collision wave number. The cross section is given by:

$$\sigma(\theta) = |f(\theta)|^2 \tag{2}$$

An inversion process must reverse the above-mentioned two stages, proceeding therefore in the sequence:

$$\sigma(\theta) \overset{I}{\to} S(\ell) = e^{2i\eta_\ell} \overset{II}{\to} V(r) \tag{3}$$

Thus an inversion scheme should solve two different problems: (1) Determination of all the partial-wave components $S(\ell)$ of the scattering from the cross section $\sigma(\theta)$. (2) Obtain the potential $V(r)$ from all the $S(\ell)$.

2.2. The Unitarity Method for obtaining the $S(\ell)$

Determination of $f(\theta)$ readily yields also all the η_ℓ, by Eq. (1) and the orthogonality property of the Legendre polynomials. The question is therefore whether knowledge of $\sigma(\theta)$ for all   determines also the phase of $f(\theta)$ as well as its magnitude. The partial-wave analysis of the scattering from a given $\sigma(\theta)$ is thus equivalent to solving the phase problem mentioned above. An important result in scattering theory, the unitarity equation[8], derived as a consequence of flux conservation in scattering, relates the phase $\alpha(\theta)$ of the scattering amplitude to the cross section $\sigma(\theta) = |f(\theta)|^2$. This relation becomes in fact a nonlinear integral equation for the unknown phase $\alpha(\theta)$ in terms of the input $\sigma(\theta)$. Mathematical conditions were established on $\sigma(\theta)$ under which the unitarity equation determines the phase $\alpha(\theta)$ uniquely[9],[10]. Gerber and Shapiro[11] showed for simulated examples that phase determination by the unitarity equation can be accomplished for typical cases, having provided a suitable numerical method for solving the unitarity integral equations in such conditions. The unitarity method was applied to real scattering data for the He-Ne system[3], the first system for which a full inversion was carried out. Although the method is in principle exact, it suffices in practice to compute $\alpha(\theta)$ to an accuracy of the order of a few percent, as this is consistent with the accuracy limits of real data. The partial wave amplitudes $S(\ell)$ are then accurate to within $\sim$1%. The method is restricted to cross sections where all quantum oscillations are experimentally well-resolved. In practice this is realized mainly for systems exhibiting diffraction oscillations[2].

2.3. The peeling method for obtaining the potential

The peeling method[12] extracts the potential from the $S(\ell)$ by a sequence of steps such that the high ℓ data are used to determine the potential at long distance ranges, and the potential at successively shorter ranges is determined from $S(\ell)$ for successively decreasing ℓ values. In a sense, the peeling scheme is a solution of the inversion problem by a certain perturbation expansion, one that is especially suitable for con-

ditions realized in molecular scattering. The physical considerations
on which the peeling method is based are the following: (a) The number
of partial waves that contribute significantly to the elastic scattering is very large ($\sim 10^2$-10^3). (b) for high partial waves the short-
range part of the interaction is screened off by the centrifugal
barrier. Since the long range part of $V(r)$ is weak, the scattering dynamics for very high ℓ values can be treated perturbatively (Born expansion). (c) As ℓ decreases, the scattering process probes the potential at increasingly shorter ranges. Excluding cases such as orbiting
resonances, the additional, shorter, range probed as one shifts from ℓ
to ℓ-1 is very small and the modification in the scattering due to the
new range can be treated perturbatively, by a distorted wave approach.

The details of the peeling method will not be given here, only the
qualitative essence of the approach will be described. The potential is
expanded in a series:

$$V(r) = \sum_{i=1}^{\infty} V_i(r) \tag{4}$$

chosen to have the property: range of $[V_i(r)] >$ range of $[V_j(r)]$ if $i<j$.
One example of such a series is a Laplace expansion:

$$V_i(r) = C_i e^{-\alpha_i r} \tag{5-a}$$

and another is an inverse power expansion:

$$V_i(r) = C_i r^{-\alpha_i} \tag{5-b}$$

In such cases the inversion procedure is to determine all the C_i, α_i of
the terms that contribute non-negligibly to (4). The procedure
begins by considering $S(\ell)$ for the highest ℓ's available in the data
set and which still differ (within a preset tolerance) from zero. For
these very high partial waves the Born approximation is valid, and can
be used to extract, say, N terms $V_1(r),\ldots,V_N(r)$ from the corresponding
$S(\ell)$. As lower ℓ-values are then considered, shorter ranges of the
interaction begin to make a contribution, and the Born expansion fails
due to the stronger potential felt. At this, N+1 step, one employs a
distorted-wave approximation, with zero-order wave functions determined
by the already known part of the interaction $V_1(r)+\ldots+V_N(r)$, to account for the contribution of $V_{N+1}(r)$ to $S(\ell_{N+1})$. This makes it possible
to determine $V_{N+1}(r)$ from the data $S(\ell_{N+1})$. The distorted-wave treatment
is valid in this case since the distance range of the potential that
contributes to $S(\ell_{N+1})$ is only slightly larger (in the direction of
decreasing r) than that which contributes to $S(\ell_N)$. One continues in
this way for $i>N+1$ until all the significant terms $V_i(r)$ are determined.

The peeling method can be formulated in terms of an expansion sequence that is, in principle, exact. However the main importance of
this method lies in the fact that it is based on general physical considerations of the scattering dynamics (the role of the angular momen-

tum ℓ in the process), and therefore it can be extended also to inelastic collision systems. As will become evident later, the peeling
method is essentially a general approach to inversion problems in
molecular scattering theory.

2.4. The Firsov inversion transform

In the framework of the WKB approximation, $V(r)$ can be written as an
explicit simple integral transform of the phase shofts η_ℓ.[1,13] The
Firsov inversion was successfully applied by Buck to several systems[1,2]
(e.g. Li-Hg), the phase shifts η_ℓ input having been extracted from the
experimental cross section by a procedure that combines a semiclassical analysis with a fitting of parametrized functions used to represent η_ℓ. Tests have shown that the accuracy of the Firsov transform is
excellent, and it may always be used for elastic scattering. The dis-
advantage of the Firsov method is that it is limited to elastic scattering only - it is not   possible to extend this transform to systems involving internal degrees of freedom.

2.5. Application to the He-Ne system

The He-Ne system is the first case for which a complete inversion, in-
cluding both step $\sigma(\theta) \to S(\ell)$ and $S(\ell) \to V(r)$ was carried out[3] without
any recourse to some aspect of trial and error fitting. The unitarity
method was used to extract the $S(\ell)$ (hence the η_ℓ), and then $V(r)$ was
obtained by Firsov transform[3]. A major motivation for choosing He-Ne
is that the differential cross section for this system exhibits well-
resolved quantum-mechanical diffraction oscillations and non of the
latter is "washed out" by the experimental averaging. Also, for He-Ne
a very elaborate trial-and-error potential was independently available.
The agreement between the fitted potential and the inverted one was
excellent, being within 5% for the well-depth; 0.7% for the zero of
the potential, and 1.3% for the position of the minimum. The deviations are all in the range consistent with the experimental uncertainty of the data.

3.   INVERSION OF INELASTIC ATOM-ATOM SCATTERING DATA

We consider in this Section the inversion problem for one of the simplest inelastic scattering systems: A collision between two atoms governed by two electronic energy curves (in the diabatic representation
[14]) that cross at some point r_0. An interaction $V_{12}(r)$ couples the two
electronic states leading to inelastic transitions. The objective of
the inversion scheme described here[15] is to determine the interaction
function $V_{12}(r)$ from the inelastic differential cross section. It is
assumed for simplicity that the two (diabatic) curves $V_1(r), V_2(r)$ are
known. Such information might be available either from an electronic
energy calculation or from experiment. For instance, if the inelastic
coupling is so weak that its effect on the elastic scattering may be
neglected to first approximation, then the electronic (diabatic)

curves may be obtained from experiment, by inversion of the elastic cross sections measured in the ground and in the excited electronic states.

The inversion method given by Child and Gerber[15] takes advantage of a basic feature in the structure of the inelastic differential cross section for such systems, the familiar Stuckelberg oscillations[14], the physical origin of which can be seen from Fig. 1. In a situation where the two curves $V_i(r)$ as the distance from the crossing point increases, the mutual positioning of the two effective potential energy curves:

$$U_{i\ell}(r) = V_i(r) + \frac{\hbar^2 \ell(\ell+1)}{2\mu r^2} \quad ; \quad i = 1,2 \tag{6}$$

(μ– the relative mass) varies strongly with ℓ. The local overlap between the corresponding zero-order wave functions $\psi_i^{(0)}(r)$ defined for these energy curves (with coupling neglected) is thus very sensitive to the value of ℓ.

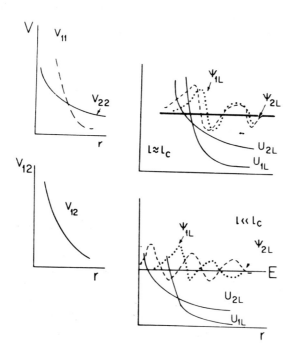

Figure 1. Zero-Order Wave Functions for Curve Crossing System

Depending on the existence of good or poor overlap, the partial-wave amplitudes $S_{12}(\ell)$ oscillate with ℓ. The cross section is related to $S_{12}(\ell)$ by:

$$\sigma_{1 \to 2}(\theta) = (k_2/k_1)|f_{12}(\theta)|^2 \tag{7}$$

where k_i is the collision wave number for state i (i=1,2), and:

$$f_{12}(\theta) = \frac{1}{2i(k_1k_2)^{1/2}} \sum_{\ell=0}^{\infty} (2\ell+1)S_{12}(\ell)P_\ell(\cos\theta) \tag{8}$$

and it can be shown that the oscillations of $S_{12}(\ell)$ in ℓ are reflected in oscillations of $\sigma_{1\to2}(\theta)$ in θ- the Stuckelberg oscillations. The inversion procedure that follows based on the Stuckelberg oscillations essentially employs the variations with ℓ of the overlap between the wave functions $\psi_{1\ell}^{(0)}(r),\psi_{2\ell}^{(0)}(r)$ as an "interferometry" to measure the curve $V_{12}(r)$. We now proceed to give an outline of the essential features of the inversion procedure[15].

3.1.Recovery of $S_{12}(\ell)$ from $\sigma_{1\to2}(\theta)$

To determine the partial wave amplitudes $S_{12}(\ell)$ from the inelastic cross section Child and Gerber invoked a semiclassical analysis of the oscillations in these two functions. For $S_{12}(\ell)$ use was made of the semiclassical approximation of Bandrauk and Child[16] that provides an Airy function representation of the partial-wave amplitude:

$$S_{12}(\ell) = A^{1/2}(\ell)A_i[\xi(\ell)]\exp[2i\bar{\eta}(\ell)] \tag{9}$$

where $\bar{\eta}(\ell)$ is the average of the elastic phase shifts $\eta_1(\ell)$ and $\eta_2(\ell)$ pertaining to the two curves $V_i(r)$, i=1,2. $A(\ell)$ is a slowly varying function, so the oscillations of $S_{12}(\ell)$ are determined entirely by $A_i[\xi(\ell)]$. Child and Gerber[15] gave a semiclassical uniform (Airy) representation of the cross section $\sigma_{1\to2}(\theta)$. To first approximation one has from that expression

$$\sin\theta\sigma_{1\to2}(\theta) \propto A_i^2[\xi(\theta)] \tag{10}$$

From the positions of the extrema in the experimental cross section, compared with those of the Airy function $A_i^2(\xi)$, $\xi(\theta)$ can be recovered as a function of θ. The oscillations of $|S_{12}(\ell)|$ in ℓ and of $\sigma_{1\to2}(\theta)$ in θ stem from the same physical origin. On that basis a transformation was given[15] that constructs $\xi(\ell)$ from the corresponding function $\xi(\theta)$. The comparison of the absolute magnitude of $\sigma_{1\to2}(\theta)$ with expressions (8),(9) leads to the determination of $A(\ell)$. With that $S_{12}(\ell)$ has been constructed.

3.2. Determination of $V_{12}(r)$ from $S_{12}(\ell)$

This inversion step can be done by a peeling method, very similar in concept to that discussed in 2.3. The starting point is again an expansion of the unknown interaction in a series of terms that are of successively decreasing range. In the present case we use an expansion in step functions:

$$V_{12}(r) = \sum_{j=0}^{\infty} V_{12}^j \theta_j(r) \tag{11}$$

where the constants V_{12}^j are to be determined, and:

$$\theta_j(r) \; \{ \begin{array}{l} 1 \text{ for } r_{1j} \leqslant r \leqslant r_{j+1} \\ 0 \text{ for } r > r_{1j+1} \text{ or } r < r_{1j} \end{array} \tag{12}$$

with r_{1j} taken as the turning point for the j-th partial wave on the steeper of the two diabatic curves. The justification is (i) that at typical molecular beam scattering energies the separation between adjacent turning points is typically of the order of $10^{-2} A^2$ or less; and (ii) the value of $S_{12}(\ell)$ is expected to be relatively insensitive to the variation of $V_{12}(r)$ over this small region.

The peeling method proceeds as follows: Let L be the largest value for which $S_{12}(\ell)$ still differs significantly from zero. $S_{12}(L)$ is then used to determine the longest range contribution to the interaction We denote by V_{12}^L the corresponding coefficient of that term in (12), and then $V_{12}^j=0$ for $j>L$. One then proceeds to determine the next term in the hierarchy of ranges using the partial-wave element $S_{12}(L-1)$ and the previously determined V_{12}^L and so on. The problem is greatly simplified in the weak-coupling cases, when the distorted-wave approximation can be used. In this framework $S_{12}(\ell)$ is given by:[15]

$$S_{12}(\ell) = -2i(\mu/h^2)(k_1 k_2)^{-1/2} \exp[2i\bar{n}(\ell)] \int_o^{\infty} \psi_{2\ell}^{(0)}(r) V_{12}(r) \psi_{1\ell}^{(0)} dr \tag{13}$$

where $\bar{n}(\ell), \psi_{i\ell}^{(0)}(r)$ are defined as in Sec. 3.1. From (11)-(13) with some algebra and with further approximations, the following equations are obtained for the peeling heirarchy:

$$V_{12}^L = S_{12}(L)/A_L^L$$

$$V_{12}^{L-i} = [S_{12}(L-i) - \sum_{i=1}^{i} V_{12}^{L-i+q} A_{L-i}^{L-i+q}]/A_{L-i}^{L-i} \qquad i=1,\ldots L-1 \ . \tag{14}$$

The A_ℓ^j are overlap integrals over peeling intervals:

$$A_\ell^j = -(4i\mu/\hbar^2)(k_1 k_2)^{-1/2} \exp[2i\bar{n}(\ell)] \int_{r_{1j}}^{r_{1j+1}} \psi_{2\ell}^{(0)}(r) \psi_{1\ell}^{(0)}(r) dr \tag{15}$$

The above scheme, with certain improvements, was applied to a test case of simulated data computed for the Olson-Smith[16] model of the Ne+He+ system. The results are shown in Fig. 2. Not only does the inversion scheme reproduce the true interaction to very good accuracy (inversion results are given by solid line, dashed line is exact potential), but also the method provides insight into the relation between specific data points and corresponding parts of the inverted potential.

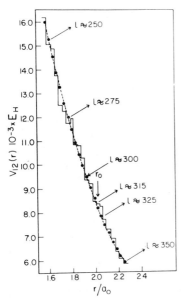

Figure 2. The Inverted and True $V_{12}(r)$ for He$^+$+Ne.

4.   INVERSION METHOD FOR ANISOTROPIC INTERMOLECULAR POTENTIALS

In the present Section we shall outline a direct inversion scheme for obtaining interaction potential surfaces between atoms and diatomic molecules. The orientation dependence of intermolecular potentials is of major chemical interest, and an inversion scheme for the accurate extraction of this property is therefore of considerable potential importance.

The interaction potential between an atom and a diatomic molecule (treated as a rigid rotor) depends on r, the distance between the atom and the molecular c.m., and γ, the angle between the distance vector and the molecular axis. The inversion method described below is developed for weakly anisotropic systems. Also, we take for simplicity the molecule to be a homonuclear one. There is substantial evidence that this low-anisotropy interaction can be represented to sufficient accuracy in the form:

$$V(r,\gamma) = V_0(r) + V_2(r)P_2(\cos\gamma) \qquad (16)$$

The aim of an inversion scheme in this case is to determine the functions $V_0(r)$ and $V_2(r)$. Clearly this cannot be done from the elastic cross section alone. We assume the availability of input data consisting of two rotationally state-selected differential cross sections $\sigma_{j \to j'}(\theta)$. Such cross sections that were recently measured for several systems[17,18]. A brief outline will now be given of the main features of the inversion scheme given by Gerber, Buch and Buck[19].

4.2. Extraction of $S_{jj'}(\ell)$ from $\sigma_{j \to j'}(\theta)$

Four essential simplifications were made in handling this problem, based on physical properties of the collision processes considered: (i) Transitions involving considerable rotational energy transfer require hard impact, "head on" collisions that take place for low ℓ values. These collisions lead to high angle scattering. Therefore, relatively large inelasticity, and thus most of the sensitivity to the anisotropic part of the potential, is found in large-angle scattering. The inversion procedure employs only that data range to obtain information on the anisotropic interaction term. The large angle scattering is such that both the elastic and the inelastic cross sections have a smooth, interference-free structure, indicative of an essentially classical behavior. The conclusion that can be drawn is that a classical-limit approximation can be applied to the scattering cross sections $\sigma_{0 \to j}(\theta)$. Such an approximation is conveniently introduced by applying the stationary phase approximation[14] the partial-wave expansion of the scattering amplitude. The main simplification gained is that a one-to-one classical type correspondence is obtained between θ and the angular momentum ℓ[19]. (ii) A restriction will now be made to collision energies that are not very high (e.g. <1eV for X+H$_2$ systems). This, together with the assumption of weak anisotropy imply that approximations aimed at the range of moderate coupling should be adequate. A suitable choice appears to be the Exponential Distorted Wave Approximation[14] that improves on the weak coupling Distorted Wave approximation by unitarizing the latter. Adequacy of EDW for the type of systems considered here has been demonstrated in the literature[20]. (iii) We assume the J_z conserving approximation[21], in which framework ℓ, the relative angular momentum is conserved during the collision. Tests have shown that the J_z-conserving approximation leads to results of good accuracy. (iv) A two-state approximation is adopted, in which only the initial and the final states are included in the treatment of the collision dynamics. In most applications (e.g. for Ne+D$_2$) the relevant states are j=0 and j=2. Based on these approximations, it is found that the partial-wave amplitudes are related to the cross sections by[19]:

$$|S_{02}(\ell)|^2 = \sin^2[a(\ell)], \quad |S_{00}(\ell)|^2 = 1 - |S_{02}(\ell)|^2 \tag{17}$$

where

$$\tan^2[a(\ell_{av})] = \frac{\sigma_{0 \to 2}(\theta)}{\sigma_{0 \to 0}(\theta)} \frac{\ell_{av}}{\ell_0} \frac{|\theta'_0(\ell_0 + 1/2)|}{|\theta'_{av}(\ell_{av} + 1/2)|} \tag{18}$$

where $\theta_{av}(\ell)=1/2[\theta_o(\ell)+\theta_2(\ell)]$, and $\theta_o(\ell)$, $\theta_2(\ell)$ are the classical deflection functions calculated from the isotropic potential term at collision energies E_0 and E_2 respectively (corresponding to the channel energies) ℓ_{av} and ℓ_o are defined as solutions of the following equations:

$$\theta_o(\ell_o+1/2) = \theta \; ; \qquad \theta_{av}(\ell_{av}+1/2) = \theta \qquad (19)$$

where θ is the scattering angle. The determination of $|S_{j,j'}(\ell)|^2$ is thus easy if $V_o(r)$ is known. It is therefore very important that one can show that if $\sigma_{tot}(\theta) \equiv \sigma_{0\to0}(\theta)+\sigma_{0\to2}(\theta)$ is inverted as though it were an elastic d.c.s., the corresponding potential obtained will be, to excellent approximation, $V_o(r)$.

4.3. Extraction of $V_2(r)$.

The determination of $V_2(r)$ from the partial-wave quantities is done, as in the previously-discussed inversion problem, by a peeling method. The quantities $a(\ell)$, simply related by (17) to $S_{02}(\ell)$, are in fact the distorted wave integrals for the $0\to2$ transition, i.e.:

$$a(\ell) = \text{const} \; x \; _0\!\int^{\infty}\psi_{o\ell}^{(0)}(r)V_2(r)\psi_{2\ell}^{(0)}(r)\,dr \qquad (20)$$

A peeling procedure, very similar to that of Sec. 3 can now be used to extract $V_2(r)$ from $a(\ell)$.

4.4. Tests and applications

To test the approximations involved in the two steps of this inversion procedure, it was first applied to a simulated, but realistic example, Ne+D$_2$ [19]. The cross sections $\sigma_{0\to0}(\theta)$, $\sigma_{0\to2}(\theta)$ were calculated from an assumed potential surface, and then inverted by the method of Secs. 4.2, 4.3. To test the accuracy of the first part of the inversion process $\sigma_{j\to j'}(\ell)\to S_{jj'}(\ell)$ (or the related quantities $a(\ell)$), the inversion results for $a(\ell)$ were compared with those calculated directly from the potential. The agreement found was within $\sim$3% of accuracy. The second part of the procedure $a(\ell)\to V_2(r)$, gave an anisotropic potential of the accuracy of $\sim$5%. Peeling was attempted with two different expansion bases, and in both cases gave a similar inverted potential $V_2(r)$. The inversion result is therefore only weakly sensitive to the expansion basis employed for peeling. Finally, the inversion scheme was applied to real data, the $\sigma_{0\to0}(\theta)$, $\sigma_{0\to2}(\theta)$ measured for Ne+D2 by Buck and collaborators at the Max-Planck-Institut, Göttingen. The results of the inversion are shown on Figs. 3 and 4[22]. To test the consistency of the results, after the potential was obtained we recalculated the $a(\ell)$ and compared with the results of the first part of the inversion (Fig.3). As for $V_2(r)$, it is compared in Fig.4 with the results of a fitting procedure[22]. We also recalculated the cross sections from the $V_2(r)$ obtained and found that the inverted $V_2(r)$ reproduces the experimental data better than the one from trial-and-error one. Note that at the

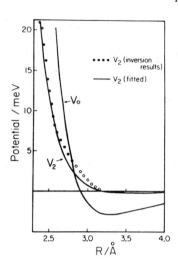

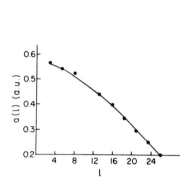

Figure 3. Inverted $a(\ell)$
for Ne+D$_2$. Solid line
shows values recalcu-
lated from inverted $V_2(r)$.

Figure 4. The Inverted Anisotropic Potential for Ne+D$_2$.

same time the inversion process is computationally far simpler than
the fitting calculation!

5. INVERSION METHOD FOR ATOM-SURFACE SCATTERING

Molecular interactions with solid surfaces are of basic significance
to a wide range of physical and chemical processes (heterogeneous catalysis, absorption). Quantitative knowledge on these interaction potentials is, however, lacking at present. Great importance is therefore
attached to the efforts of pursuing this information by molecular beam
studies. The dynamics of molecule-surface scattering differs from that
of gas-phase intermolecular collisions in several fundamental aspects.
In the context of the inversion problem one basic point is the absence
of rotational invariance for molecule-solid collisions; This destroys
the physical relevance of the variable ℓ, the collisional angular momentum, and eliminates the possibility of using a Peeling Method - the
very procedures by which most of the progress on the inversion problem
for gas-phase molecular scattering was made. Nevertheless, very recently preliminary progress was made in providing an inversion method for
atom-surface scattering[23]. Let (x,y) be coordinates parallel to the
surface, z the coordinate perpendicular to the plane. Suppose that the
atom-surface potential is of the form:

$$V(x,y,z) = V_o(z) + V_1(z)Q(x,y)$$

where $Q(x,y)$ is a multi-periodic function in the surface-plane coordinates. Such a form is definitely sufficiently flexible enough for

the needs and the accuracy range relevant to the field now. To determine the potential, the functions $V_0(z)$, $V_1(z)Q(x,y)$ must all be quantitatively obtained. For a periodic stationary (phononless) solid, atomic beam scattering necessarily leads to sharp discrete diffraction peaks: The component of the incident momentum in parallel to the surface (k_x,k_y) can change only by discrete, reciprocal space vectors

$$\underline{G}_{mn} = (\frac{2\pi}{a_x}m, \frac{2\pi}{a_y}n)\cdot$$

where a_x, a_y are the lattice constants. The relative intensity of scattering into the mn diffraction peak, which we denote by $|S_{oo \to mn}(k_z)|^2$ depends on m,n and on the incident momentum in the z direction k_z. Gerber and Yinnon[23] treated the inversion problem for this system under the condition (the Sudden condition):

$$k_z^2 \gg |\underline{G}_{mn}|^2 \quad \text{(for all important peaks m,n).}$$

Using the Sudden approximation for surface scattering [6] it was possible to show that if $|S_{oo \to mn}(k_z)|^2$ is available from experiments for all peaks (mn) over a wide range of incident momenta k_z then the data can be inverted to yield $V_1(z)$ and $Q(x,y)$ uniquely. Fortunately, there are other methods for finding $V_0(z)$. Moreover, the inversion procedure is computationally simple and quite accurate. The key to further progress on this is in obtaining experimental results of sufficient resolution and over a large enough range of k_z values. There is, however, solid ground for expecting rapid developments in this respect. Fig. 5 shows results of an inversion carried out on <u>simulated</u> data for He scattering from the W(110) surface[6] . The results seem most encouraging.

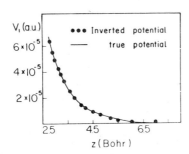

Figure 5. Inversion Results for $V_1(z)$;
Case of He/W(110) surface.

REFERENCES

1. Buck, U., 1974, Rev. Mod. Phys. 46, pp. 369-389.
2. Buck, U., 1975, Advan. Chem. Phys. 30, pp. 314-388.
3. Gerber, R.B., Shapiro, M. Buck, U. and Schleusener, J., 1978, Phys. Rev. Letters 41, pp. 236-239.

4. Child, M.S., and Gerber, R.B., 1979, Mol. Phys. 38, pp. 421-432.
5. Gerber, R.B., Buch, V. and Buck, U., 1980, J. Chem. Phys. 72, pp. 3596-3603
6. Gerber, R.B. and Yinnon, A.T., 1980, J. Chem. Phys. 73, pp. 3232-3238.
7. Gerber, R.B., 1980, in "Molecular Structure and Dynamics" (Lectures in honor of S. Lifson), edited by M. Balaban (International Science Services, Philadelphia) pp. 81-101.
8. Newton, R.G., 1966, "Scattering Theory of Waves and Particles" (McGraw Hill, New York) p. 493.
9. Newton, R.G., 1968, J. Math. Phys. 9, pp. 2050-2058.
10. Gerber, R.B. and Karplus, M., 1970, Phys. Rev. D1, pp. 998-1012.
11. Gerber, R.B. and Shapiro, M., 1976, Chem. Phys. 13, pp. 227-234.
12. Shapiro, M. and Gerber, R.B., 1976, Chem. Phys. 13, pp. 235-242.
13. Sabatier, P.C., 1965, Nuovo Cimento 37, pp. 1180-1227; Miller, W.H., 1969, J. Chem. Phys. 51, pp. 3631-3638.
14. Child, M.S., 1974, "Molecular Collision Theory" (Academic Press, London) p.
15. Child, M.S. and Gerber, R.B., 1979, Mol. Phys. 38, pp. 421-432.
16. Bandrauk, A.D. and Child, M.S., 1970, Mol. Phys. 19, pp. 95-111.
17. Buck, U., Huisken, F., Schleusener, J. and Pauly, H., 1977, Phys. Rev. Letters 38, pp. 680-683. Gentry, W.R., Giese, C.F., 1977, J. Chem. Phys. 67, pp. 5389-5391.
18. Bergmann, K., Engelhardt, R., Hefter, U. and Witt, J., 1979, J. Chem. Phys. 71, pp. 2726-2728.
19. Gerber, R.B., Buch, V. and Buck, U., 1980, J. Chem. Phys. 72, pp. 3596-3603, and to be published.
20. Bosanac, S. and Balint-Kurti, G.G., 1975, Mol. Phys. 29, pp. 1797-1811.
21. McGuire, P. and Kouri, D.J., 1974, J. Chem. Phys. 60, pp. 2488-2499.
22. Gerber, R.B., Buch, V., Buck, U., Maneke, G. and Schleusener, J., 1980, Phys. Rev. Letters 44, pp. 1397-1400.

SELECTIVE VIBRATIONAL INELASTICITY IN PROTON-MOLECULE COLLISIONS

F.A. Gianturco[+] and V. Staemmler[x]

+ Gruppo di Chimica Teorica, Nuovo Ed. Chimico, Città
  Universitaria - 00185 Rome, Italy.

x Theoretische Chemie, Ruhr-Universität, ˋ4630 Bochum 1, FRG.

ABSTRACT

The scattering of slow protons from simple molecules constitutes
an attractive model for the study of ion-molecule reactions. In fact,
the presence of long-range polarization forces and the strong coupling of the motion of the impinging H^+ with the internal degrees of
freedom of the targets are primarily responsible for the existence of
many compound states found by experiments and theory.

The present paper examines the nature of some vibrational resonances at low energies that appear in H^+-H_2 collisions and finds results that confirms the model behaviour already shown by the H^++ CO
case. The unusually high vibrational inelasticity of the O_2-H^+ system
is also analysed in detail from the point of view of the full potential
energy surface and a model explanation is provided for the experimental
observations on the basis of that surface characteristics.

INTRODUCTION

The inelastic collisions that involve small molecules and 'free'
protons in the gas phase constitute at first sight a fairly simple
class of phenomena that can provide useful models for ion-molecule
reactions. For these reactions, in fact, it is well known that the
corresponding rate constant remains as large as $10^{-9} \cdot cm^3 \cdot s^{-1}$ down to
very low temperatures, a fact which is usually explained in terms of
the Langevin theory[1], where the polarization potential:

$$V_{Pol} = - \frac{\alpha e^2}{2R^4}$$

is assumed to be the dominant interaction at large R distances. In the

B. Pullman (ed.), Intermolecular Forces, 79–99.

above equation α is the polarisability of the neutral molecule and
e is the elementary charge. The effective potential is then given by
the sum of V_{pol} and the centrifugal potential, the latter being   a
function of the impact parameter b. If the collision energy is lower
than the barrier height, the trajectory will be pushed away from the
barrier and the collision will be a distant one. If the collision
energy is higher than the barrier height, the interacting pair will
approach each other with acceleration and come into close contact.
For each collision energy one can then obtain a critical impact parameter b_c that gives in turn the corresponding Langevin cross section,
which is sometimes called the 'orbiting' cross section or, somewhat
misleadingly, the 'capture' cross section. The essential point here
is that such a collision leads to a close contact between interacting
partners, hence it causes the necessary condition for large energy
exchange (reactive and/or subreactive) to take place. For polar molecules, moreover, the observed reaction rate is sometimes larger than
the Langevin value[2,3] and calls for a more detailed inclusion of
static interactions.

The main assumption of the above theory is that the existence of
a deep well in the potential energy surface (PES) is very likely
to cause, at low collision energies, a long-lived and strongly-coupled
complex which will decay statistically into the products. The latter
assumption has been recently analysed in detail for the H^+-D_2 processes
in the kinetic energy range from 0.1 to 4 eV[4]. It was shown there
that the complex formation criteria need to be carefully defined and
that Langevin cross sections agree with experiments only over rather
narrow ranges of collision energies, while overestimating them at
higher E_{cm} values.

Collisional experiments with molecular beams can provide further
evidence n the 'structure' of such complex systems formed during
orbiting collisions, whether they break up along reactive or simply
inelastic exit channels, since the corresponding energy-sharing and
energy-deposition between internal degrees of freedom could in
principle be analysed under single-collision conditions[5].

Proton projectiles therefore become very interesting in model
studies where the roto-vibrational inelastic cross sections, integral
and differential, provide indication of strong collisions and large
energy exchanges during complex formation or, in general, during
close-contact encounters even at relatively low velocities or small
collision angles.

In both situations one finds that the nature of the strong interaction between species mirrors the changes in 'chemical' structure of the bound target orbitals as caused by the impinging H^+ atom. This could therefore imply that *specific* degrees of freedom, or specific modes within them, are preferentially activated depending on the particular molecular orbitals that are involved at a given collision energy or for a given impact parameter.

In the following Sections we are thus presenting theoretical models and results that point out this dynamical selectivity in proton-molecule collisions.

Due to the meagre amount of information available on the corresponding potential energy hypersurfaces, we have begun with simple targets and employed, when possible, a model treatment of the dynamics. Section II is thus discussing internal energy re-distribution during complex formation, when only the vibrational coordinate is considered, for the H^+-H_2 case which has been extensively studied in the recent literature. Section III is examining the more complex case of O_2-H^+ and various 'cuts' of the potential surface are studied *vis à vis* the experimental findings for the differential cross sections, also vibrationally inelastic, measured in molecular beams.

SELECTIVE 'CORE' EXCITATIONS

a) <u>The Theoretical model</u>

When the system under study is constituted by a diatomic target that interacts with the proton beam, the full adiabatic (Born-Oppenheimer) interaction potential is usually cast in the familiar form of a multipolar expansion:

$$V(\underline{R},\underline{r}) = \sum_{\lambda=0}^{\lambda_{max}} V_\lambda (R,r) P_\lambda (\hat{r} \cdot \hat{R}) \qquad (1)$$

where $\underline{R}$ is the proton distance from the molecular centre of mass and $\underline{r}$ is the internal molecular coordinate.

For strongly interacting partners several values of λ are usually needed to fit the computed PES points at all $\underline{R}$ and for various geometries $\underline{r}$ of the target[6]. In practice, it is commonly found that the spherical component in eq. (1) is essentially the only one which shows a deep potential well and that therefore plays a major role in studying orbiting resonances and compound-state resonances.

This is·particularly important when studying the latter type of resonance as a model for transition-state structures, since the usual conversion of kinetic energy into internal excitation energy of the partners (that which takes place in an inelastic collision) may make it possible sometime to excite states not otherwise energetically accessible. If this occurs, a complex of temporary stability may be formed and the resulting adduct can be described as occupying one of the bound states supported by the PES of eq. (1), i.e. supported most-ly by its V_o component.

When several internal states are involved, then there will be a competition between kinetic energy redistribution among these internal states and energy back-flow into the translational mode, thus creating a predissociation state with finite lifetime[7].

The competing processes of above may also show up in some of the inelastic channels that are open at the considered energy, hence the corresponding cross sections may exhibit unusually large values at specific energies. On the other hand, they may also exhibit rapid drops at some other energy values thereby implying that the correspond-ing S-matrix element goes rapidly through zero during complex formation. In the latter instance, the lengthened lifetime of the adduct during the resonances allows the forces at play to favour instead energy 'back dumping' into the translational mode without the molecular ex-citation that takes place in the former case.

All the above possibilities indicate here the existence of strong relationships between the molecular structure, the interaction po-tential and the energy flows during the dynamical process and suggest that model studies of such flow could teach us something of value about the internal structures of the possible adducts that are tem-porarily formed[8].

By approximating the relevant interaction only via the spherical component of eq. (1), we can introduce a very straight-forward model that describes quantum-mechanically a structureless particle of mass μ colliding with a diatomic target which exhibits in turn only oscillatory motion under a potential $v(r)$, along the r coordinate and has reduced mass M:

$$\{[-\frac{1}{2M} \nabla^2_r + v(r)] - \frac{1}{2\mu} \nabla^2_R + V_o(r,R) - E\} \Psi(r,\underline{R}) = o \qquad (2)$$

One can then study in a transparent form the influence of the coupling potential between translational modes and internal vibrational modes at energies where resonances occur. The positions of the latter may be approximated by considering the quasi-bound levels of the adduct

to be the eigenvalues of the two-particle (ion-molecule) system in
an equivalent or effective potential. At the simplest level of approximation, such a potential may be defined as the rigid rotor
potential for the isotropic mutual interaction between the excited
molecular vibrator and the incoming proton, i.e. one can simply write:

$$V_o(r,R) \simeq V_o(r_{eq},R) + <f|v|f> \qquad (3)$$

where both potentials $V_o(r,R)$ and $v(r)$ are defined as in eq. (2).
The last term in eq. (3) is independent of R and simply gives the
bound states of the asymptotic target structure as 'level shifting'
factors.

The potential $V_o(r_{eq},R)$ represents now the complex as free to
rotate in space, with the internal vibrational mode totally decoupled.
Its bound states are straight-forwardly obtained by solving the 1D
Schrödinger equation and provide a zeroth order approximation for the
pseudo-bound states of the real systems. The potential $v(r)$ describes
the situation of vanishing coupling and could be given by a simple
harmonic oscillator form.

The description (3) can be now tested against a more realistic
situation where the r-motion and the R-motion *are* coupled and where
the asymptotic conditions imposed on the total wavefunction are those
of a scattering problem. The corresponding equations to be solved
thus become:

$$\left\{\frac{d^2}{dR^2} + \frac{\ell(\ell+1)}{R^2} + k_n^2\right\} F_\ell^n(R) = 2\mu \sum_{n'} <n'|V_o|n> F_\ell^n(R), \qquad (4)$$

where $V_o(r,R)$ is now the full spherical component of eq. (1) and where
the radial, unknown functions originate from the more correct expansion of the w.f. in (2):

$$\Psi(r,\underline{R}) = \sum_{n,m,\ell} \chi_n(r) F_{\ell m}^n(R) R^{-1} Y_\ell^m(\hat{R}). \qquad (5)$$

The above w.f. in turn exhibits the familiar asymptotic form for each
of its radial components[9]. The functions $\chi_n(r)$ now describe the asymptotic eigenfunctions of the target diatom.

2. Inelastic cross sections for H^+-H_2

We examined first the collinear scattering situation, i.e. for
$\ell = o$ in eq. 4, so that orbiting resonances coming from the changes

in shape of the effective potential are not allowed and one can therefore test more clearly the isolated effect of purely compound-state
or 'core-excited' resonances. The latter are in fact a crucial step
in modelling for simple situations the relevant stable, predissociat-
ing structures of a transition complex in ion-molecule reactions.

In the above instance, the corresponding cross sections associated
with the scattering amplitudes generated by each coupled solution of
eq. (4) are given by:

$$\sigma_{n \to n'}(E_{coll}) = \frac{\pi}{k_n^2} \left| S_{n \to n'} - \delta_{nn'} \right|^2 \tag{6}$$

with $S_{n \to n'}$, being the S-matrix element associated to the process under
consideration.

A similar study carried out by us for the CO-H$^+$ system[9] had
provided an interesting insight into the structural effects that alter
the forms of resonances and their corresponding widths, the latter
being ultimately related to the life times. The case here reported
involves the simpler H$^+$-H$_2$ interaction for which many collisional
studies where done before [10,12] and whose potential energy surface has
been recently computed with great accuracy[13] and tested successfully
against experimental findings[14].

The spherical component of that PES presents a well depth of
∿0.15 hartrees when the H$_2$ distance is 1.4 a_0 and the H$^+$-H$_2$ distance
is ∿ 1.8 a_0. The next anisotropic contribution, V_2, is nearly always
repulsive with a small well of ∿ 10^{-3} hartrees and therefore one can
associate the gross freatures of the resonances with the bound states
supported by the V_0 component only. By numerically solving the corresponding Schrödinger equation a total of 17 bound states was found
within the well thus indicating the strong likelihood of Feshbach resonances whenever a sizeable coupling exists between translational
motion and vibrator states.

The latter coupling in eq. (4) was introduced by linearizing the
V_0 dependence on r and by then defining a spherical stretching parameter β(R) that was responsible for the coupling:

$$\beta(R) = \frac{1}{V_0(r,R)} \frac{\partial}{\partial r} V_0(r,R) \bigg|_{r=r_{eq}} \tag{7}$$

Since the above parameter exhibited oscillatory dependence on
the R-variable[15], we decided to start out by taking its average,
asymptotic value for the rapid generation of the coupling matrix

elements needed in eq. (4). The latter were found to be much larger than in the previous CO-H$^+$ system and therefore yielded a much stronger pattern of resonant behaviour.

Figure 1 presents an example of the computed inelastic cross sections as function of total collision energy in units of oscillator spacing ($\hbar\omega$. $\simeq 0.5$ eV for the H$_2$ case). The case reported treats a situation where only the $n = 1$ excitation channel is open, while the n=2 vibrational threshold is reached at $E_{coll} = 2.5$ $\hbar\omega$. One clearly sees there the appearance of strong 'dips' associated with zeros in eq. (6) and related to virtual excitations during complex formation, as discussed before.

By taking into consideration the bound states for each of the potentials on the r.h.s. of eq. (3), one can classify the resonances with two separate indices (n,i), the first being the target vibrator level and the second the bound level supported by $V_o(r_{eq},R)$, were i=o labels the highest of them[9]. The latter are the resonant levels for each complex formed during 'core' vibrational excitation.

Such an analysis immediately shows that at least three closed channels interact with the open ones, as opposed to the weaker CO-H$^+$ case where only at the most a couple where needed to reach convergence of results[9].

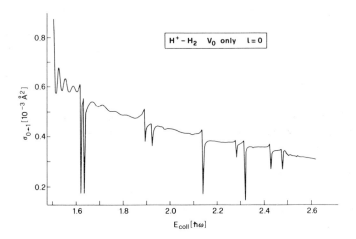

Fig. 1 - Vibrationally inelastic integral cross section for the (0➤1) excitation process in H$^+$-H$_2$ collision. The energy units are H.O. level spacings ($\hbar\omega$ $\simeq$ 0.5 eV)

Moreover one finds that:

(i) the strength of coupling remains nearly constant for the levels involved but does *not* progress uniformly along each sequence of i values for a fixed n value.

(ii) the pseudo-bound states are strongly localized and show rather little residual coupling that shifts the resonance positions from their zeroth order values[16]. On the contrary, the CO case exhibited larger shifts and more diffuse wavefunctions for its 'core-excited' resonant states[9].

(iii) the larger spacings between the bound states of the H_2-H^+ adduct, as opposed to the more densely packed levels of the $CO-H^+$ compound states, causes the resonances to be fewer in number but more marked in character. The corresponding linewidths are here of the order of $\sim 10^{-3}$ $\hbar\omega$. With additional rotational broadening they might thus become measurable quantities in real life.

In conclusion, the break-up pattern of such special class of resonances appears to be strongly dependent on the particular structure of the formed adduct rather than exhibiting statistical behaviour, as suggested for complex transition states[17], in phase space. When such a limiting behaviour can be reached in detailed model computations is still an open question for which the present model treatment provides an easy to implement computational approach.

VIBRATIONAL INELASTICITY IN BEAMS

1. The O_2-H^+ case: the experimental evidence

A large variety of experiments have looked into the rotational and vibrational excitations of several simple molecules by collision with collimated beams of Li^+ and H^+ projectiles[18]. Even more complicated systems[19] have been studied under beam conditions in order to elucidate the selectivity of specific vibrational inelasticity by collision with H^+ and Li^+. In this latter instance it was indeed found for SF_6 targets that the differences in relative velocity of the two ions selected different vibrational modes to be excited, provided that the major coupling mechanisms was assumed to be the gradient of the transition dipole moment. For the methane molecule, however, other experiments[20] suggest instead the possibility of 'localized' charge-exchange processes as an explanation for enhanced inelasticity of certain vibrational modes.

The interesting implications of the above results, both from a fundamental and an applied point of view, have therefore spurred a detailed investigation of the behaviour of several simple molecules, diatomics and polyatomics, when colliding with slow protons that excite their vibrational and rotational modes[21].

Figure 2 presents as an example the measured results recently obtained[21] for the total, average energy transfer $<\Delta E_{av}>$ as a function of scattering angle and for a proton beam with an energy of about 10 eV. The latter quantity is computed by direct integration of the time-of-flight spectrum $S(t)$[22]

$$<\Delta E_{av}> = \left\{\int_{-\infty}^{\infty} S(t)\ \Delta E(t)\ \frac{d}{dt}\ \Delta E\ dt\right\} \cdot \left\{\int_{-\infty}^{\infty} S(t)\ \frac{d}{dt}\ \Delta E\ dt\right\}^{-1} \qquad (8)$$

where $\Delta E(t)$ is the energy transfer associated with a time delay τ and measured with respect to the elastic peak. The vibrationally inelastic distributions can also be fitted in some cases by a modified Poisson distribution[19] and, in any event, allow to define the vibrational energy transfer from the deconvoluted transition probabilities P_n[22]:

$$<\Delta E_{vib}>_k = \hbar\omega_k \sum_n n\ P_n^k \qquad (9)$$

where n labels, as before, the target vibrational levels of a given mode k with oscillator spacing $\hbar\omega_k$. Since this result is independent of the integration performed in eq. (8), one can further define a way to estimate the rotational energy transfer:

$$<\Delta E_{rot}> = <\Delta E_{av}> - <\Delta E_{vib}> \qquad (10)$$

The above equality should be, in principle, quite reliable since it does not depend on any specific, and often uncertain, interpretation of the broadenings and shifts in observed peaks.

It clearly appears from the above Figure that sulphur hexafluoride is by far the most efficiently excited molecular target, as one would expect from its closely spaced vibrational levels. On the other hand, CH_4 and CO_2 are less efficiently excited than the O_2 target which, within the series examined, indeed appears to present the second largest inelasticity, after the $S F_6$ system.

If one also compares rotational inelasticity, as obtained via equation (10), for N_2, CO and NO a decreasing trend is observed along the series apparently indicating that the permanent dipoles of the targets have no noticeable effect on energy transfer[21]. Moreover they

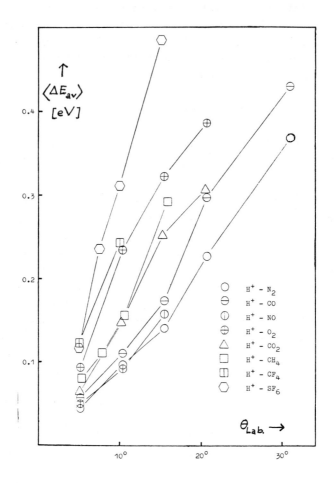

Fig. 2 -- Average total energy transfers as functions of scattering
        angle and for several molecular targets. The proton energy
        was about 10 eV in the centre of mass (from ref. 21).

all show vibrational inelasticity that is much smaller than for the
O_2 system. The latter molecule, in fact, presents $<\Delta E_{vib}>$ values that
are about three times larger than in NO and 10 times larger than in
N_2 and CO. An opposite, albeit slightly smaller, increase is shown by
the $<\Delta E_{rot}>$ values[23] when going from O_2 to N_2, through NO and CO.

    The above results far diatomic targets thus tend to group the
molecules according to the presence or absence of unpaired electrons:
the stronger coupling between open-shell bound electrons and the incoming protons apparently leads to weaker rotational energy transfers

but to some sort of 'chemical' interaction that favours vibrational energy transfer. It is with the aim of elucidating this mechanism that the following paragraphs are dedicated to the study of several 'cuts' for the various PES involved in O_2-H^+ interaction and to a simple, but effective, modelling of the dynamics.

2. O_2-H^+: the potential energy surfaces

The four lowest states of the O_2 molecule have all the orbital configuration $(1\sigma_g^2 \ 1\sigma_u^2 \ 2\sigma_g^2 \ 2\sigma_u^2 \ 3\sigma_g^2 \ 1\pi_{ux}^2 \ 1\pi_{uy}^2 \ 1\pi_{gx} \ 1\pi_{gy})$ and are usual-ly classified as the $^3\Sigma$ (coupled triplet), the $^1\Delta_g$ state with two components and the $^1\Sigma_g^+$ state. The corresponding positive ion, O_2^+, exhibits then one singly occupied π_g-orbital and has the overall elec-tronic symmetry of a $^2\Pi_g$ state. Upon interaction with a proton one can then presume that, depending on the relative collision energy, the asymptotic channels after the encounters will either contain O_2 or O_2^+, each in one of its electronic states and in one level of its vibration-al manifold. Figure 3 therefore shows the energy dependence on the in-ternal molecular coordinate r for both systems in their lowest states and when the proton-molecule coordinate R goes to infinity. This ob-viously represents a particular cut of a PES which, in the present instance, is given by a functional $V_\Gamma(\underline{r}, \underline{R})$ for each of the electronic states Γ considered.

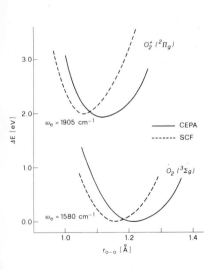

One sees from the Figure that cal-culations at the SCF level, employing a GTO basis set of 38 functions (TZD) quality)[24], come already fairly close to the experimental equilibrium geo-metries for O_2 (1.208 Å) and O_2^+(1.117 Å). The solid curves, on the other hand, exhibit the results for corre-lated calculations of the CEPA-type already employed with success for other ion-molecule potential energy surfaces[25,26]. They indicate here a very good agreement with both geo-metries ($r_{O_2}^{CEPA}$ = 1.212 Å and $r_{O_2^+}^{CEPA}$ = = 1.115 Å) and a fairly good re-production of the oscillator experi-

Fig. 3 — Potential energy curves for the O_2 and O_2^+ lowest electronic states. See text for meaning of symbols.

mental frequencies reported in the Figure: the computed values of ω_e are, in fact, 1,565 cm^{-1} for O_2 and 1,950 cm^{-1} for O_2^+. One can then attach a reasonable degree of confidence to the quality of the target description in both instances, thereby progressing with the actual calculations to other regions of the relevant PES's. It is here worth mentioning that previous calculations for the isolated neutral molecule[27,28] found SCF results of similar level of accuracy.

Upon addition of a proton in the xz plane the $D_{\infty h}$ symmetry of the O_2 molecule is distorted, the σ and π_x orbitals are mixed to form a' orbitals, while the π_y become a" orbitals. One component of the $^1\Delta_g$ state can then interact with the $^1\Sigma_g^+$ to form two mixtures that give rise to two $^1A'$ states. On the other hand, the O_2 $^3\Sigma_g^-$ will go into a $^3A"$ configuration while the other component of the $^1\Delta_g$ will give rise to a $^1A"$ state. The relative ordering of these PES, for large R values, will thus presumably follow their corresponding asymptotes. The $^3A"$ state will therefore be the lowest and will be described by a single configuration (...7a' 2a"), the same that pertains to the $^1A"$ state, while the $^1A'$ states will be one lower than the latter configuration and the other the highest, in analogy with the $^1\Sigma_g^+$ case. This fourth state will not be considered explicitly here because of its large excitation energy with respect to the collision velocities employed in the experiments.

The above discussion applies to a general configuration of the three atoms that belongs to the C_S symmetry, the more general 'cut' of the PES for each r value. One should, however, also consider the special geometries that have the incoming proton along the molecular bond ($C_{\infty v}$ symmetry) or the projectile that impinges perpendicularly to the O_2 band (C_{2v} symmetry). In the former case, the two lowest configurations of the O_2-H^+ adduct ($^3A"$ and $^1A'$) go into the $^3\Sigma^-$ and into one component of the $^1\Delta$ configurations, respectively. In the latter situation, on the other hand, they become instead $^3B_1$ symmetries,while the next higher configuration ($^1A"$) goes into the $^1B_1$ representation. One should therefore expect that, by keeping the O_2 geometry 'frozen' at its equilibrium value, several different orbital mixing schemes will apply when varying the ($\hat{r}_{eq}$, $\hat{R}$) angle hence will cause rather strong changes in thestructure of the overall PES of eq. (1), for each Γ electronic state. Moreover, by applying the same approach to those electronic configurations that asymptotically relate to the O_2^++H system one can also study in some detail the spatial regions where interaction between the two different sets of channels can occur.

Figure 4 reports the computed results for the collinear approach with the molecular bond kept fixed at the equilibrium value of the neutral species. One clearly sees the crossings of the adiabatic curves

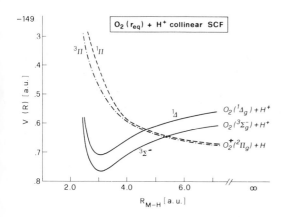

Fig. 4 - Computed 'cuts' of the potential energy surfaces for the O_2-H^+ and O_2^+-H systems. The molecular coordinate is kept constant at its equilibrium value for O_2. See text for meaning of symbols.

the one between lowest-lying states occurring around 5.5 a.u.; because of the symmetries involved, transitions can only occur via non-adiabatic angular momentum coupling between $^3\Sigma$ and $^3\Pi$ curves. Although no calculations are available for it in the present system, the occurrence of such coupling usually requires strong torques applied to the electronic structure of the system during fast collisions[33], much faster than in the cases examined here.

The corresponding results for the C_{2v} geometries are reported in Figure 5 for the three lowest curves of the (O_2-H^+) system and the corresponding two lowest curves of the (O_2^+-H) case. In the latter instance the collinear, $^{3,1}\Pi$ configurations go into the $^{3,1}(A_2, B_2)$ representations. Here again angular momentum coupling is required to cause non-adiabatic transitions between the lowest triplet states, a not very likely event at the collision energies of the experimental data.

In both cases, however, the computed cuts of the PES reveal substantial potential wells in the O_2 entrance channels with the consequent likely formation of adducts when the collision energies become smaller. Although the latter instance corresponds to situations still outside the current experimental conditions, the existence of 'core excited" resonances obviously could play a non-negligible role even in this

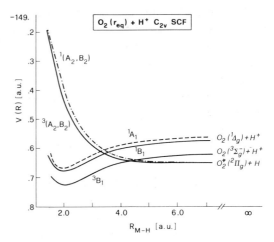

Fig. 5 - Same as in Fig. 4 but for the atomic projectile impinging perpendicularly to the bond. The latter is again held fixed at its eq. value.

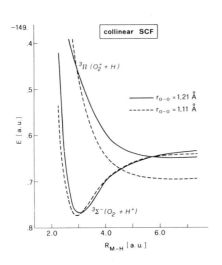
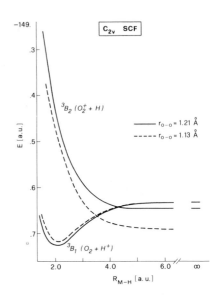

Fig. 6 - Potential energy curves of the (O_2-H^+) and (O_2^+-H) systems for collinear (left) and perpendicular (right) encounters and at different molecular geometries. The dashed lines correspond to O_2^+ equilibrium values.

case, as will be investigated in the future.

Correlation effects at various R values have also been studied for the above situations[29] and found to generally cause a lowering of the PES of an average 0.2 a.u. They do not, however, alter the general shapes of the relative curves reported in the above Figures[29].

What is likely, however, to alter the relative shapes of the various cuts discussed above is of course the motion along the other radial coordinate, i.e. the changes of O_2 and O_2^+ geometries. As an example of this effect, calculations are reported in Figure 6 for the same collisional arrangements of before but with two different values of the molecular bond distances. Only the lowest surfaces that adiabatically cross are reported in that Figure.

In both cases the SCF calculations show a marked lowering of the repulsive curves that asymptotically correlate with the molecular ion, while the corresponding attractive potentials are left essentially unchanged. The ground state, neutral target interaction with the incoming proton is controlled, at large R values, by the polarization of the outer orbitals which change little during bond compression, while in the other case the outer, single electron of π-symmetry rearranges after charge transfer to reach the more stable configuration

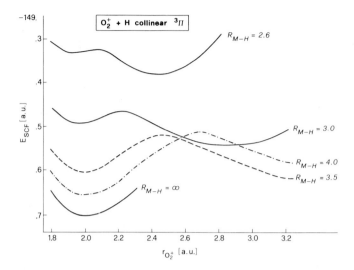

Fig. 7 - Potential energy curves, as function of internal molecular
coordinate, for various distances of the incoming projectile.
See text far meaning of symbols.

of O_2^+, thus moving the adiabatic curve crossing to smaller impact
parameters. This means, therefore, that the necessary non-adiabatic
coupling will have to act between partners that come even closer to
each other, a fact which requires then even higher collision energies
than postulated before.

An interesting result, however, is obtained when the full dependence on the molecular internal coordinate is examined for the
above geometries of the surfaces[29]. Most of the previous curves in
fact exhibit a sort of 'valley' behaviour along the R-coordinate,
with their side walls being steeper as functions of r for small R
values, while becoming shallower when the proton (hydrogen) moves
further away[29]. The only notable exception is given by the collinear
approach in the repulsive, $^3\Pi$, configuration that asymptotically correlates with the O_2^+-H channel. The latter behaviour is shown in
Figure 7 where one clearly sees that, as the projectile comes in, a
second valley appears and a lower equilibrium position is reached
with the central O atom sitting roughly midway between the other two
particles: the sharing of the hydrogen charge with the ionic molecule
tends to make more stable a configuration that essentially contains
an O_2 diatom. Whenever such a symmetry can be reached in the reaction
region with incoming channel ($O_2 + H^+$) it will thus tend to produce
back again the latter products, possibly with the O_2 molecule in its
ground vibrational state.

3. O_2 vs NO: a dynamical model

The previous discussion of the $(O_2 + H^+)$ system at various geometries has given us some insight into the forces at play along the
reaction coordinate for some of the lowest-lying configurations. If
one carries out, even at a qualitative level, a similar analysis for
the molecular targets reported in Fig. 2, it is fairly easy to see
that most states and geometries do not involve the crossing of adiabatic configurations of the same symmetry. Hence some sort of non-
adiabatic coupling via electronic effects (curve-hopping) needs to
be invoked to explain strong vibration inelasticity in collisions[20].
Moreover, some of the above systems have ionization potentials which
are larger than that of H, thereby excluding any direct curve crossing.

A rather marked exception to the above situation is constitued
by the O_2 and NO molecules interacting with protons in the more
general C_S symmetry. The corresponding experimental findings show
that oxygen exhibits a factor of three greater vibrational transfer
and roughly a factor of two less rotational energy transfer than the
NO system[23]. The adiabatic potential curves for both systems in the
C_S geometry are shown in Figure 8, where the occurrence of avoided
crossings is reported at different impact parameter values. Only the
PES 'cuts' that one obtains with $r=r_{eq}$ are presented in the Figure.

Both systems have ionisation potentials smaller than 1.0 Rydberg
and exhibit unoccupied π-orbitals as their outer, open-shell Molecular
Orbitals. In the case of the $(NOH)^+$ system, previous calculations[30,31]
indicate that the outer 7a' and 8a' orbitals change their general
character as the projectile approaches the molecule. At large distances, in fact, the 7a' is mostly described by 1s functions on the H
atom and therefore correlates asymptotically with the $NO^+(^1\Sigma) + H(^2S)$
channel. On the other hand the 8a' orbital, in the same radial region,
is mostly located on the molecule thereby asymptotically correlating
with the $NO(^2\Pi) + H^+$ channel.

As the proton approaches the target, however, the 7a' orbital
increasingly acquires the character of a π-orbital of the NO molecule
while the 8a' begins to take up more and more the character of an
atomic 1s orbital on the H. Because of the overall $^2A'$ symmetry for
both triatomic situations then one can deduce that, as the proton
comes into the small-R region, an avoided crossing must occur since
the incoming proton meets an NO target in the entrance channel and
the latter in turn changes its structure to an NO^+ molecular ion as
R becomes smaller.

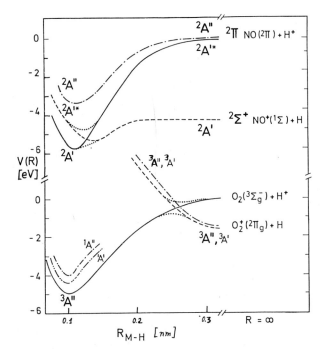

Fig. 8 - Potential energy curves of C_S symmetry for the $(NOH)^+$ system
(upper curves) and the $(O_2H)^+$ case (lower curves) at fixed
molecular geometries. See text for meaning of symbols (from
ref. 23).

Such a behaviour is reported in Fig. 8, where the $^2A'^*$ curve refers to the orbital configuration with NO^+ character at small distances,
while this character holds only at large R values for the curve
labelled $^2A'$.

The behaviour of the adiabatic potential curves with C_S symmetry
that pertain to the $(O_2H)^+$ case are shown in the lower part of the
same Figure 8. Here the orbitals that change character are now $7a'$
and $2a''$ because of the extra π-electron of the O_2 molecule. Moreover,
because of the smaller energy gap between asymptotic channels, this
change now takes place at larger R values with smaller slopes between
branches of the adiabatic curves near the avoided crossing. The incoming proton therefore interacts substantially longer with the O_2
target since at the considered collision energies and scattering
angles[23] the curve crossing region is most probably reached for O_2
but not for NO targets.

In order to model, in at least a crude way, the dynamics of these
collisions we could now turn back to the asymptotic cuts of the real

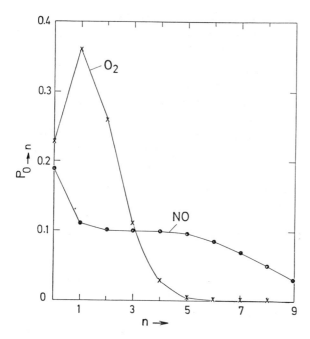

Fig. 9 – Excitation probabilities computed for the O_2 molecule (eq. 12) and the NO molecule (eq. 13) treated as harmonic oscillators. The abscissae report the vibrational quantum numbers for the final states.

potential hypersurfaces shown in Fig. 3. Due to the difference in geometry between the M molecular target and the M^+ ionic target, in fact, the probability of the latter being formed in a vibrationally excited state $|K\rangle$ could be simply associated to the corresponding Franck–Condon factor between vibrational wave functions ϕ_M and ϕ_{M^+} at $R = \infty$:

$$P_1(o \rightarrow k) \alpha \, | \langle \phi_{M^+}(k) | \phi_M(o) \rangle |^2 \tag{11}$$

where the initial, neutral target is considered in its vibrational ground state and we assume the proton interaction with M to be short in comparison with its vibrational period. The ionic species formed temporarily during close encounters therefore do not have time in such sudden conditions to relax to their specific ground states (with different r_{eq}!).

If one now remembers that at the collision energies considered the non-adiabatic charge-exchange process is not very important, as also shown by the lack of proton flux losses, when the proton leaves the system the target will in turn rapidly acquire again the neutral molecular configuration it had before. One can then define a second probability $P_2(k \rightarrow n)$ by writing:

$$P_2(k \rightarrow n) \alpha \, | \langle \phi_M(n) | \, \phi_{M^+}(k) \rangle |^2 \tag{12}$$

The final total probability of observing the molecule M in the nth vibrational level, is then qualitatively given by the summed products:

$$P_{o \rightarrow n} = \sum_k P_1(o \rightarrow k) \cdot P_2(k \rightarrow n) \tag{13}$$

At the considered collision energies the flight-time is here of the order of $2.3 \cdot 10^{-14}$ s/nm, while the harmonic oscillator vibrational times are $\approx 2.7 \cdot 10^{-14}$s for NO and $\approx 3.2 \cdot 10^{-14}$ s for O_2. From the crossing positions of Fig. 8 one expects that O_2^+ exists for several vibrational units of time during the interaction with the proton, while the NO^+ species exists for a much shorter time. It follows then that the NO could conceivably obey the probability law of eq. (13), a law which has been derived under sudden conditions whereby no time is given to the vibrating target to relax during the rapid, close interaction with the perturbing proton. On the other hand, the O_2^+ species should be allowed to relax to its ground vibrational state during the much longer interaction time that occurs with the impinging projectile. It then follows that the final distribution of the vibrational states in the neutral O_2 could be modelled by only using eq. (12), but this time with k=o.

By employing the harmonic oscillator model discussed by Katriel[32] and the molecular data of refs. (34,35) we can compare the outcomes of the two equations for the above two systems. The results are displayed in Fig. 9. The relaxation of temporarily formed O_2^+ provides final neutral O_2 molecules in vibrationally excited states with a relative probability that is three times larger than in the NO case, where no relaxation can take place during the shorter time of NO^+-formation. This result, based on a very crude picture, is also in keeping with the present experimental findings. It suggests the importance of electronic structure differences in driving interacting partners into different exit channels. This aspect could be further explored with the curve-hopping model treatment, within classical trajectory calculations, once a more extended portion of the full PES for all the symmetries involved becomes available from bound state calculations of the adiabatic electronic energies.

ACKNOWLEDGEMENTS

This work was begun when one of us (F.A.G.) was visiting the Max-Planck-Institut für Strömungsforschung in Göttingen, FRG, and several illuminating discussions that occured there with Professor J. Peter Toennies are warmly acknowledged. The von-Humboldt-Stiftung is also thanked by F.A.G. for the award of a Senior Fellowship. Finally the help of Dr. F. Battaglia and Mr. G. Drolshagen with some of the computations discussed here is gratefully remembered. The ab initio calculations of the PES for O_2-H^+ have been performed on the interdata 8/32 minicomputer sponsored by the Deutsche Forschungsgemeinschaft to the Lehrstuhl für Theoretische Chemie at the Ruhr-Universität Bochum.

REFERENCES

1. Gioumousis G. and Stevenson D.P.: 1958, J. Chem. Phys., 29, pp. 294-301.

2. Dugan Jr., J.V. and Magee, J.L.: 1967, J. Chem. Phys. 47, pp. 3103-16.

3. Sakimoto,K. and Takayanagi, K.: 1980, J. Phys. Soc. Japan, 48, pp. 2076-83.

4. Gerlich, D., Nowotny U., Schlier Ch. and Teloy E.: 1980, Chem. Phys., 47, 245-55.

5. Toennies Peter J.: 1979, Comments Atom. Mol. Phys., 8, pp. 137-53.

6. Gianturco, F.A., Lamanna, U.T. and Ignazzi D.: 1980, Chem. Phys., 48, pp. 387-98.

7. see: Feshbach, H.: 1967, Ann. Phys. (N.Y.), 43, pp. 410-24.

8. Eastes, W. and Marcus, R.A.: 1973, J. Chem. Phys., 59, pp. 4751-54.

9. Drolshagen, G., Gianturco, F.A. and Toennies, J.P.: 1980, J. Chem. Phys., 73, pp. 5013-24.

10. Gentry, W.R. and Giese, C.F.: 1977, J. Chem. Phys., 67, pp. 5389-98, .

11. Krutein, J. and Linder, F.: 1977, J. Phys. B, 10, pp. 1363-71.

12. Gianturco, F.A. and Tritella, P.: 1977, Phys. Rev., A16, pp. 542-51.

13. Schinke, R., Dupuis, M. and Lester, W.A. Jr.: 1980, J. Chem. Phys., 72, pp. 3909-15.

14. Schinke, R.: 1980, J. Chem. Phys., 72, 3916-22.

15. Schinke, R.: private communication.

16. Gianturco, F.A. and Drolshagen, G.: unpublished results.

17. Pechukas, P. and Light, J.C.: 1965, J. Chem. Phys., 42, pp. 3281-95.

18. For a good review see: Faubel, M. and Toennies, J.P.: 1977, Adv. At. Mol. Phys., 13, pp. 229-62.

19. Ellenbroeck T., Gierz, U. and Toennies, J.P.: 1980, Chem. Phys. Lett., 70, pp. 459-64.

20. Gentry, W.R., Udseth H. and Giese C.F.: 1975, Chem. Phys. Lett., 36, 671-3.

21. Gierz, U.: 1980,Doctoral Thesis, University of Göttingen, FRG.

22. Eastes, W., Ross, U. and Toennies, J.P.: 1979, J. Chem. Phys., 70, 1652-61.

23. Gianturco, F.A., Gierz U. and Toennies, J.P.: 1981, J. Phys. B, 14, 667-77.

24. Staemmler, V.: 1980, unpublished results.

25. Staemmler, V.: 1975, Chem. Phys., 7, pp. 17-26.

26. Staemmler, V.: 1976, Chem. Phys., 17, pp. 187-96.

27. Peyerimhoff, S.D., and Buenker, R.J., 1972, Chem. Phys. Lett., 16, pp. 235-43.

28. Van Lenthe, J.H. and Ruttink, P.J.A., 1978, Chem. Phys. Lett., 56, pp. 20-24.

29. Staemmler, V.: 1981, unpublished results.

30. Marian C., Bruna, P.J., Buenker, R.J. and Peyerimhoff S.D.: 1977, Mol. Phys., 33, pp. 63-74.

31. Bruna, P.J., and Marian, C.M.: 1979, Chem. Phys., 37, pp. 425-44.

32. Katriel, J.: 1970, J. Phys. B, 3, pp. 1315-22.

33. Bottcher, C.: 1980, Adv. Chem. Phys., 42, pp. 169-206.

34. Field, R.W.: 1973, J. Mol. Spectr. 47, pp. 194-203.

35. Herzberg, G.: 1950, Spectra of Diatomic Molecules, D. Van Nostrand Publ. Co. (N.Y.).

ACCURATE MOLECULAR PROPERTIES, THEIR ADDITIVITY, AND THEIR USE IN
CONSTRUCTING INTERMOLECULAR POTENTIALS

William J. Meath, Daniel J. Margoliash, B.L. Jhanwar,
A. Koide, and G.D. Zeiss
Department of Chemistry, University of Western Ontario,
London, Canada

ABSTRACT. The use of dipole oscillator strength distributions to
accurately evaluate atomic and molecular isotropic dipole properties,
and to study their additivity, is discussed. The results suggest that
this technique, together with additivity formalisms, offers a general
approach for the reliable estimation of the isotropic dipole properties
of large species. The use of accurate dipole-dipole dispersion energy
coefficients in constructing reliable potential models is also discussed
and used as a vehicle for emphasizing the need for reliable higher
multipole dispersion energies as well.

1. INTRODUCTION

Over the past few years globally reliable ground state isotropic
dipole oscillator strength distributions (DOSDs) have been constructed
for a variety of atoms and molecules and used to accurately evaluate
(often with estimated error $\lesssim$ 1%) their isotropic dipole properties [1-5].
Here some of this work will be discussed with particular emphasis on the
use of this approach as a general means of studying the additivity of
molecular dipole properties and of obtaining accurate results for the
properties of large molecules from those of small molecules. Both the
properties of individual molecules and interactive properties involving
two or more molecules (e.g. long range dispersion energies) will be
considered.

The construction of the molecular DOSDs and the related evaluation
of the isotropic molecular dipole properties are reviewed briefly in
Sec. 2. From a practical point of view, the DOSD method is apparently
the only available approach for the accurate evaluation of all relevant
dipole molecular properties. Formal expressions that relate the
properties of a molecule to those of its constituent parts are given in
Sec. 3 and used to discuss the additivity of the isotropic dipole
properties, as a function of the nature of the property, using the
following atoms and molecules as models: $H, N, O, H_2, N_2, O_2, NO, NH_3, H_2O,$
N_2O, the normal alkanes through octane, the primary alcohols through

101

propanol, and the 1-alkenes through butene. The use of additivity techniques, in conjunction with the DOSD approach for smaller molecules, to predict the properties of large molecules is illustrated in Sec. 4. In general the study of the additivity of the properties for a given molecule gives insight into the nature of the intramolecular forces operating within that molecule.

The discussion in this paper is largely limited to isotropic dipole molecular properties. The accurate evaluation of isotropic higher multipole properties, and of anisotropic molecular properties, is also of considerable importance. This is illustrated in Sec. 5 where it is shown that reliable values of the higher multipole long range dispersion energies, as well as accurate results for the dipole-dipole dispersion energy, are needed in order to fully understand several recent, successful models for representing the isotropic part of the potential energy curve for the interaction of closed shell systems. In what follows atomic units are used unless indicated otherwise.

2. DIPOLE OSCILLATOR STRENGTH DISTRIBUTIONS AND PROPERTIES

The isotropic dipole oscillator strength distribution, or DOSD, of a molecule is a distribution consisting of the differential dipole oscillator strength (df/dE) as a function of molecular excitation energy E. Once the DOSD for an atom or molecule is constructed many important dipole properties can be evaluated [1-8]. Some properties of individual atoms and molecules that are of interest include the dipole oscillator strength sums S_k and L_k, mean excitation energies I_k, and molar refractivity R_λ for wavelength λ:

$$S_k = \int_{E_0}^{\infty} dE \left(\frac{df}{dE}\right) E^k \tag{1}$$

$$L_k = \int_{E_0}^{\infty} dE \left(\frac{df}{dE}\right) E^k \ln E \tag{2}$$

$$I_k = \exp \left[L_k/S_k\right] \qquad , \quad 2.5 > k > -\infty \tag{3}$$

$$R_\lambda = \frac{4}{3} \pi N \int_{E_0}^{\infty} dE \frac{(df/dE)}{[E^2 - \varepsilon^2]} \tag{4}$$

where E_0 is taken to be the ultraviolet absorption threshold for the molecule of interest, N is Avogadro's number and ε is the energy of a photon with wavelength λ. The accurate evaluation of the properties (1) - (4) provides useful information for a variety of research areas: for example (1) the straggling (I_1,S_1), and the stopping (I_0) and total (I_{-1},S_{-1}) inelastic scattering cross sections, of fast charged particles in matter; (2) dipole polarizabilities (S_{-2}); (3) charge densities at nuclei (S_2); and (4) Lamb shifts (I_2).

Interactive properties (i.e. properties involving two or more molecules) that are of particular interest include the dipole-dipole and triple-dipole dispersion energy coefficients. The isotropic dipole-dipole dispersion energy for the interaction of molecule A with molecule B is given by $-C_6(A,B)R^{-6}$, where R is the distance between the two molecules and

$$C_6(A,B) = \frac{3}{2} \int_{E_0(A)}^{\infty} dE_A \int_{E_0(B)}^{\infty} dE_B \frac{(df/dE)_A (df/dE)_B}{E_A E_B (E_A + E_B)} \qquad (5)$$

Both dispersion energies, especially the dipole-dipole energy, play important roles in determining atomic and molecular interactions in various states of matter [9,10].

The methods used to construct the DOSDs have been discussed in detail previously [1-5]. The atoms and molecules analyzed so far include those listed in the introduction. Their DOSDs were constructed from extensive experimental and theoretical information, including discrete oscillator strength, photoabsorption, and high energy inelastic scattering data, and were constrained to satisfy the Thomas-Reiche-Kuhn sum rule $S_0 = N$, where N is the number of electrons in the species under consideration, and to reproduce available accurate refractivity and dispersion measurements for the relevant dilute gases. Once the DOSDs are obtained the corresponding results for the molecular properties can be obtained by numerically integrating (1) - (5). A much more efficient way of evaluating interactive properties, without loss of accuracy, is to use pseudo-DOSD techniques [11,12] which permit the dipole properties to be evaluated in pseudo-spectral form by using discrete representations of the recommended DOSDs.

The isotropic dipole properties obtained by using these DOSDs have been carefully analyzed. For the molecules studied the errors associated with the results for S_k, L_k and I_k, $2 \gtrsim k \gtrsim -8$, and $C_6(A,B)$, have been estimated to be generally less than one to two percent (assuming the molar refractivity data used to constrain the DOSDs are reliable to a few tenths of a percent); for the N and O atoms the errors are larger than this particularly for $k \lesssim -2$.

The various dipole properties can depend markedly on different excitation energy regions of the DOSDs [5,13]. Properties corresponding to $k \geq 0$ emphasize the "additive" higher energy portions of the DOSDs while those for $k < 0$ emphasize the "chemical", non-additive low energy part of the DOSDs, moreso as k becomes more negative. These trends follow from the nature of the excitation energy weighting factors E^k multiplying the DOSD in (1) - (5) and are illustrated for H_2O in Figure 1. This figures shows $(df/dE)E^k$ versus E for $k = 0$, -2 and -6; the areas under the curves represent $S_0 = 10$, S_{-2}, and S_{-6}. Clearly the dipole properties become additive for $k \gtrsim 0$ and become increasingly non-additive as $k < 0$ decreases.

The DOSD method outlined here is the only practical approach for obtaining reliable results for a wide variety of isotropic dipole

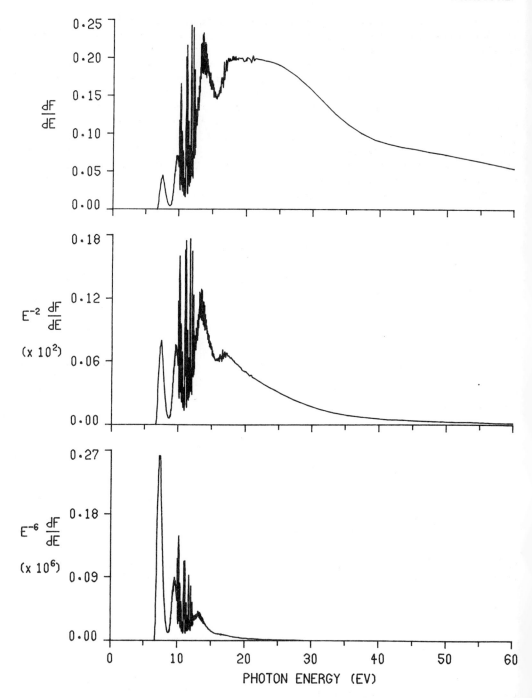

Fig. 1. Graphical representations of S_0, S_{-2}, and S_{-6} for the H_2O molecule; $E^{-n}(df/dE)$ is in units of $(eV)^{-n-1}$.

properties. For example, the Padé approximant technique [14,15] for obtaining bounds for $C_6(A,B)$ can be unreliable [2,11,16] because of the sensitive dependence of the results to relatively small errors in the input $S_{-2\ell}$ values for $\ell > 1$ (the latter are usually determined by fitting molar refractivity data to Cauchy moment expansions of (4)). In principle ab initio or non-empirical methods [17] can be used to evaluate the dipole properties but in practice these methods apparently yield accurate results only for relatively small systems or for rather special cases [18]. Purely theoretical approaches are most viable for multipole properties that can be related, through sum rule techniques, to expectation values since this avoids the difficult problem of obtaining reliable representations of the excitation spectrum for the molecule of interest. For the dipole properties of ground state species sum rules [7] are available only for k = 2, 1, 0, -1.

3. ADDITIVITY EXPRESSIONS FOR MOLECULAR PROPERTIES

Additivity relations are often used to provide easy to obtain estimates for certain properties of molecules. These expressions can be derived for all the dipole properties discussed in Sec. 2 by applying what amounts to the zeroth order theory of intermolecular forces to the problem [13].

Consider a molecule $A_\ell B_m C_n$ which can be considered to be made up of the elemental substances A, B, and C (A, B, and C may be atoms or groups of atoms). One can then show that to "zeroth order" the properties of the molecule are given by [13]

$$\frac{df}{dE}(add.) = \ell \left(\frac{df}{dE}\right)_A + m \left(\frac{df}{dE}\right)_B + n \left(\frac{df}{dE}\right)_C \tag{6}$$

$$S_k(add.) = \ell S_k(A) + m S_k(B) + n S_k(C) \tag{7}$$

$$L_k(add.) = \ell L_k(A) + m L_k(B) + n L_k(C) \tag{8}$$

$$I_k(add.) = \exp[L_k(add.)/S_k(add.)] \tag{9}$$

while the additivity expression for the isotropic dipole-dipole dispersion energy coefficient for the interaction of $A_x B_y C_z$ with $A_\ell B_m C_n$ is

$$C_6(add.) = x\ell \, C_6(A,A) + ym \, C_6(B,B) + zn \, C_6(C,C) + (xm + \ell y) \, C_6(A,B)$$

$$+ (xn + \ell z) \, C_6(A,C) + (yn + mz) \, C_6(B,C) \tag{10}$$

Some of the additivity rules (6) - (10) are well known. For example (6) is the often used mixture rule [1,19] which is used to estimate photoabsorption cross sections for high photon energies. Equation (9) with k = 0 is a reformulation of Bragg's additivity rule [20,21] for the stopping cross section of fast charged particles in matter, in terms of the stopping mean excitation energies. The additivity expression for dipole polarizabilities, (7) with k = -2, is related to Landolt's rule [22] which relates the molar refractivity of a compound to those of its constituent parts. The additivity of the dispersion energy constant C_6 has also been studied frequently.

The success of additivity rules depends in a critical way on the choice of the elemental substances used to compute the results. Appropriate choices yield effective zeroth order approaches, relative to the use of isolated atoms, that can very effectively account for the intramolecular forces operating within the molecules. The correct choice of the effective zeroth order problem becomes progressively more crucial as the index k associated with the properties S_k, L_k and I_k decreases below zero (cf. the discussion in Sec. 2 based on Figure 1).

These ideas can be nicely illustrated [13] by using the reliable DOSD results for the properties of $H, N, O, H_2, N_2, O_2, NO, NH_3, H_2O$ and N_2O as models. Additivity results for S_k, I_k, L_k and C_6 have been obtained for the diatomic and triatomic molecules by using the atoms as elemental substances, and for the triatomic molecules and NO by using the homonuclear diatomic molecules as elemental substances. The results for S_k, I_k, and $C_6(A,A)$ are compared with the recommended values in Tables 1,2, and 3 respectively.

Table 1. Some results for S_k(add.) for some simple (H,N,O) – containing molecules expressed as a percent deviation from the recommended DOSD results. Note that S_0(add.) = N.

Molecule	H, N, O as atoms				H, N, O, from H_2, N_2, O_2			
	S_1	S_{-1}	S_{-2}	S_{-4}	S_1	S_{-1}	S_{-2}	S_{-4}
NH_3	-1	16	43	49	-0.1	0.8	-4	-37
H_2O	-1	18	45	83	-0.3	6	11	4
N_2O	-0.2	0.7	-0.9	-8	0.1	-5	-14	-34
NO	-0.2	3	6	0.9	0.2	-1	-3	-16
H_2	-20	29	66	166				
N_2	-0.4	8	24	81				
O_2	-0.3	-0.01	-6	-34				

The dipole sums S_k clearly become progressively less additive as k becomes more negative; the two cases in Table 1 where this is not the case are misleading since the percent deviation of S_k(add.) from the accurate results changes sign at $k \simeq -5$. For $k \geq 0$ the additivity of the S_k is very good and improves with increasing k, with the exception of H_2 when atomic additivity is used; the additivity results for k = 2 are essentially exact, with the exception of $S_2(H_2)$ which is in error by ~29%. The additivity results based on using the homonuclear diatomic molecules as elemental substances yield results for the S_k that are in much better agreement with the reliable results for the hydrogen containing molecules than those based on atomic additivity, whereas no overall improvement is found for the molecules which contain no hydrogen. This is in agreement with the fact that molecule formation causes larger modifications in the behaviour of electrons near protons, relative to the isolated atom, than is the case in most heavier atoms (N in N_2 is another exceptional case) [13,23].

The discussion of the additivity results for the average energies I_k and the dipole-dipole dispersion energy coefficients C_6 is analogous to that for the dipole sums S_k; see Tables 2 and 3 and [13]. It is worth noting that the I_k are more additive than the S_k for k < -2. Furthermore, I_0, unlike S_0, is not trivially additive so that the test of the additivity of I_0 provides a test of the validity of Bragg's rule that is independent of the errors associated with the experimental measurements of the stopping cross sections [21].

In general it appears that additivity rules based on atoms or diatomic molecules as elemental substances yield unreliable estimates for the molecular properties corresponding to k < -2. Generally for k ≥ -2 and for C_6, the additivity relations can be used to obtain results that are reliable to within ≤20 percent with the estimates for the S_k and I_k improving substantially as k increases. Clearly some important properties are additive to a very good approximation at the atomic or diatomic molecule reference level. The list of additive properties can be extended markedly if the elemental substances are chosen carefully. In general these should be obtained from atoms or molecules that reflect the electronic structural features of the molecule of interest. Examples using group additivity are considered next.

4. RELIABLE EVALUATION OF PROPERTIES USING GROUP ADDITIVITY

The object is to reliably evaluate all the interesting dipole properties of a "large" molecule by using additivity techniques. The required group properties, which will serve as elemental properties to accomplish this goal, generally should be constructed by considering several members of the appropriate "homologous" series of molecules. Models to illustrate this are furnished by some recent results for the isotropic dipole properties of the n-alkanes [5,12], through octane, the primary alcohols, through propanol, and the 1-alkenes, through 1-butene. Only a portion of this extensive collection of data has been carefully analyzed and the following should be regarded as a reasonably firm summary of the analysis, the details of which will be published elsewhere [24].

Table 2. Some results for I_k(add.) <u>expressed as a percent deviation</u> from the recommended values.

Molecule	H, N, O as atoms				H, N, O from H_2, N_2, O_2			
	I_1	I_0	I_{-1}	I_{-2}	I_1	I_0	I_{-1}	I_{-2}
NH_3	3	-11	-18	-13	0.4	-1	1	10
H_2O	2	-9	-18	-17	0.9	-4	-6	-3
N_2O	0.4	-1	0.4	3	-0.3	2	8	13
NO	0.4	-2	-3	-2	-0.3	0.9	1	3
H_2	-17	-22	-22	-22				
N_2	0.8	-4	-11	-15				
O_2	0.5	-1	3	11				

Table 3. Some results for C_6(add.), for some like-molecule interactions <u>expressed as a percent deviation</u> from the recommended values.

Molecule	Recommended values	H, N, O as atoms	H, N, O from H_2, N_2, O_2	N, O as atoms H from H_2
NH_3	89.1	75	1	15
H_2O	45.4	72	20	17
N_2O	184.9	1	-16	
NO	69.8	10	-4	
H_2	12.1	115		
N_2	73.4	31		
O_2	62.0	-4		

4.1   The n-alkanes

The required CH_3 and CH_2 elemental properties can be obtained from any pair of alkanes. As examples we choose the pairs $[C_2H_6, C_4H_{10}]$, $[C_2H_6, C_6H_{14}]$, and $[C_4H_{10}, C_6H_{14}]$ since the input data needed for the construction of the DOSDs is most extensive for these molecules. The group properties have been used to evaluate S_k, L_k, I_k, $2 \geq k \geq -17$, and C_6 for all the alkane molecules except CH_4. The comparison of the additivity results with the recommended values for S_k and C_6 is summarized in Table 4; the results for the I_k are similar to the S_k.

Table 4.   A summary of the properties of the n-alkanes obtained by using group additivity.

CH_3 and CH_2 elemental properties taken from	Maximum percent deviation from recommended results for the normal alkanes C_nH_{2n+2}, $n=2(1)8$						
	S_2	S_1	S_{-2}	S_{-4}	S_{-6}	S_{-8}	C_6
C_2H_6, C_4H_{10}	0.5	1.3	1.0	1.1	−0.7	−4.8	1.6
C_2H_6, C_6H_{14}	−0.3	−0.5	−0.9	−0.9	+0.9	3.9	−1.4
C_4H_{10}, C_6H_{14}	1.1	2.1	1.7	2.5	−1.8	−12.8	2.1
	0.5	1.1	0.01	0.5	+0.4	−2.7	−0.5
	[C_2H_6 deleted from consideration]						

It is clear that the additivity scheme works exceedingly well. Even the potentially very non-additive properties like S_{-4} and S_{-6} are generally predicted to within the errors inherent in the globally reliable DOSDs. The slightly higher percentage deviations obtained when $[C_4H_{10}, C_6H_{14}]$ are used to provide the CH_3 and CH_2 group properties are removed when the properties of C_2H_6 are not considered for comparison purposes (see the last entry in Table 4); chemically C_2H_6 is an exception relative to the other n-alkanes since it has no CH_2 groups.

4.2   The 1-alkenes and primary alcohols

Tests of the validity of group additivity for the 1-alkenes and primary alcohols can be obtained by using the recommended DOSD properties for propene and ethanol, and CH_2 group results obtained from any two alkanes, in conjunction with the additivity relations

$$H_2C=CHCH_3 + CH_2 \rightarrow H_2C=CHCH_2CH_3 \tag{11}$$

and

$$HO-CH_2CH_3 + CH_2 \rightarrow HO-CH_2CH_2CH_3 , \tag{12}$$

to predict the dipole properties of 1-butene and n-propanol. Similarly the additivity results for the (1-butene)-(1-butene) or (n-propanol)-(n-propanol) dipole-dipole dispersion energy constants can be evaluated by using the group interaction constants obtained from the accurate C_6 values arising from the six interactions involving the two alkanes and propene or ethanol respectively. A comparison of the additivity and recommended values for S_k and C_6 is given in Table 5; [C_2H_6, C_4H_{10}], which is not the "best" choice, has been used to help provide the required elemental properties.

Table 5. A comparison of the group additivity results for some of the properties of 1-butene and n-propanol with the recommended values; C_2H_6 and C_4H_{10} are used to help provide the group elemental properties.

Molecule	Percent deviation from recommended values							
	S_2	S_1	S_{-2}	S_{-3}	S_{-4}	S_{-6}	S_{-8}	C_6
1-butene	0.1	0.2	0.3	-0.4	-1.6	-4.9	-8.5	1.0
n-propanol	0.1	0.2	-0.5	-0.8	-0.7	0.5	2.7	-0.7

The additivity estimates are again in excellent agreement with the recommended values, with those for n-propanol being more reliable than those for 1-butene as $k \leq -4$ decreases, in agreement with chemical intuition.

4.3 Conclusions

It appears reasonable that DOSD results like those discussed here will permit the reliable prediction of a wide range of isotropic dipole properties of large molecules for which little dipole oscillator strengt or other data is available. For example the predictions of the dipole properties of the higher 1-alkenes and primary alcohols, based on 1-butene and n-propanol, should be of comparable reliability to those obtained in Table 5 for 1-butene and n-propanol since the additional CH_2 groups become more alkane-like as the molecules become elongated. The properties of many molecules containing the groups CH_3, CH_2, $H_2C=CH$, and OH can be reliably estimated by using the results now available. The extension to other bonding situations, groups, atoms, and molecules is underway and it is clear that, at present, there is a lack of relevant input data for the construction of reliable DOSDs for many important molecules.

5. HIGHER MULTIPOLE PROPERTIES

The accurate evaluation of anisotropic and higher multipole isotropic properties is of considerable importance. To date, due to a paucity of symmetry analyzed or higher multipole oscillator strength data, ab initio or non-empirical quantal techniques are the methods of choice for determining all but isotropic dipole properties. As in the isotropic dipole case, see Sec. 2, the purely theoretical methods are not generally reliable [18,25]. In general ab initio calculations must be extended to well beyond the SCF limit to obtain results having an accuracy comparable to that associated with the isotropic dipole properties determined from reliable DOSDs.

An example of the importance of reliable results for higher multipole molecular properties (e.g. dispersion energy coefficients) is provided by a consideration of some recent models for the isotropic part of the interaction energy between closed shell systems. An important feature of these models is the use of damped dispersion energy representations of the attractive part of the potential which require as input C_6, C_8 and C_{10}, the dipole-dipole, dipole-quadrupole, and the sum of the quadrupole-quadrupole and dipole-octopole, dispersion energy coefficients. The HFD [26] and the Tang-Toennies [27] models represent the repulsive part of the potential by SCF results for the dimer and use the damped dispersion energy as a representation of the correlation energy. Another approach [28,29] is to use a damped dispersion energy to help represent the coulomb inter-action energy which is coupled with a model for the exchange energy. The object of these approaches is to reduce the number of adjustable parameters in the potential to a minimum and, in principle, eventually to zero.

The success of these model potentials in a variety of applications has been reviewed recently by Scoles [30]. The point to be made here is that the errors in the input values of C_8 and C_{10}, as well as C_6, can substantially affect the results obtained from the models. This arises because the damping functions used are derived from results that assume the knowledge of the exact values of C_6, C_8 and C_{10}. For many inter-actions the uncertainty in C_8, C_{10}, and sometimes C_6, are such that the values used for the dispersion coefficients will be inconsistent with the damping functions unless additional flexibility is incorporated into the potential models. Reliable DOSD results for C_6 are accurate to ~1% while non-empirical results [17] for C_8 and C_{10}, if scaled using reliable results for C_6, appear [18] to be often in error by 20 and 40% respect-ively. Padé approximant results for C_6, C_8 and C_{10} can also have large errors, or error bounds.

As a specific example we consider the potential model

$$E_{int} = [1 - \gamma(1 + 0.1R)]E_c^{(1)}$$

$$-[C_6F_6(SR)R^{-6} + C_8F_8(SR)R^{-8} + C_{10}F_{10}(SR)R^{-10}]G(SR) \qquad (13)$$

where

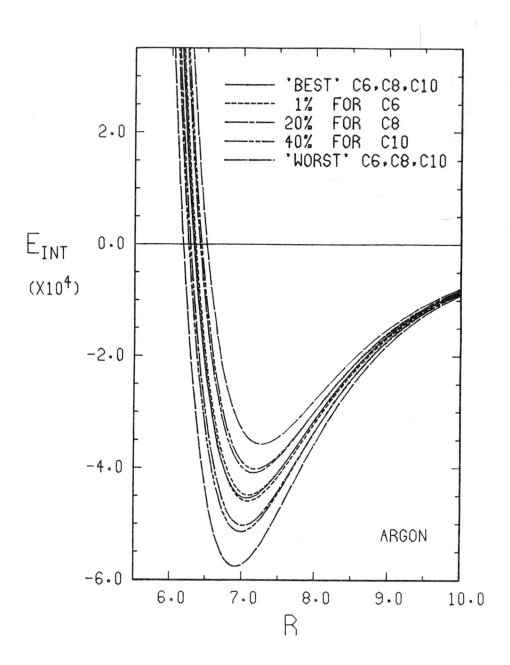

Fig. 2.   Some model Ar-Ar interaction potentials based on (13) and (14)
and the input values $C_6 = 66(\pm 1\%)$, $C_8 = 1508(\pm 20\%)$, and $C_{10} = 52648(\pm 40\%$

$$G(SR) = 1 + 41.34 \exp(-0.8588 \; SR). \tag{14}$$

Here Υ is an adjustable parameter which controls the balance between the exchange and the coulomb components of E_{int}, $E_c^{(1)}$ is the first order coulomb energy, and $S = 7.82 \; R_m^{-1}$ is a scaling factor where R_m is the internuclear distance associated with the minimum in E_{int}. The damping functions F_n, which correct for the divergent nature of the R^{-1} multipolar results as R becomes small, and the function G, which accounts for other neglected terms, are obtained by analyzing the H-H $^3\Sigma_u^+$ interaction energy. This potential model has not yet been fully optimized and details of its construction and final form will be given elsewhere [31]. It is a revised version of a single damping function model [28] which has the form of (13) with the $F_n = 1$. The Tang-Toennies model and a revised version [32] of the HFD potential also have damping functions associated with the individual multipole terms $C_n R^{-n}$.

The difficulties that can arise due to errors in the C_n can be nicely illustrated by using the potential model (13) to discuss the Ar-Ar interaction. For all practical purposes there are two equivalent reference potentials for this interaction that reproduce accurately all of the bulk properties of Ar gas and the uv-spectrum of the dimer [29,33]. In both cases flexibility was provided in the potential by allowing $C_6 = 67.2$ ($\pm 5.4\%$), $C_8 = 1480$ ($\pm 20\%$), and $C_{10} = 42800$ ($\pm 29\%$) to vary between the bounds specified by Tang, Norbeck and Certain [34] and by adjusting a parameter associated with the repulsive part of the model. The parameter Υ and the values of the C_n in (13) were determined by fitting second virial data for argon and by using the reference potential [29] as a discriminator between the almost equivalent potentials thereby obtained.

The "best" resulting potential is essentially identical to the reference potential and corresponds to $\Upsilon = 1.9878$, $C_6 = 66.0$, $C_8 = 1508$, $C_{10} = 52648$, $R_m = 7.090$ and $E_{int}(R_m) = 4.538(-4)$. It is illustrated in Figure 2 together with potentials with the same value of Υ obtained by assuming errors of $\pm 1\%$, $\pm 20\%$, and $\pm 40\%$ in C_6, C_8 and C_{10} respectively. Clearly the effects of the "errors" in C_8 and C_{10} are huge, illustrating the point at hand. A redetermination of Υ compensates for all of the shift in the potential caused by the error in C_6 and for a good deal of that caused by the individual errors in C_8 or C_{10}. We have also performed a similar set of calculations assuming the errors [34] associated with the original Padé approximant input data for C_6, C_8 and C_{10}. The Padé error bounds of $\pm 5\%$ for C_6 cause the same change in the potential as that caused by the 20% error in C_8, see Figure 2. Errors of $\pm 5\%$ in C_6 are too large to be compensated for adequately by adjustments in the parameter Υ.

In summary, flexibility is still required in model potential energy functions for isotropic closed shell interactions. With the errors inherent in the current values of C_8, C_{10}, and sometimes C_6, it can be feasible to use carefully constructed potential models that do not have individual damping functions F_n if the C_n are adjusted within their

estimated errors; the effects of individual damping functions are
generally smaller than those due to the errors in the C_n until R becomes
quite small. Individual damping functions are of course theoretically
correct [35] and can be important in practical ways as well. If they are
indeed universal, the values of C_n, obtained by fitting experimental
data with potentials involving them, should be closer to the correct
quantal results than those obtained otherwise. Individual damping
functions will also be required for reliable potential models if accurate
values for the relevant C_n are available as input. Finally, as pointed
out by Tang and Toennies [27] and by Scoles [32], they may well be
important (even with inaccurate values of the C_n) when considering
interactions whose R^{-1} multipole dispersion energy series are strongly
divergent as a function of decreasing R. In general reliable determinations of the higher multipole dispersion energy coefficients, as well
as C_6, are needed to develop a better understanding of the interesting
finer details of interaction potential energies.

ACKNOWLEDGEMENT. This research was supported by a grant from the
Natural Sciences and Engineering Research Council of Canada.

REFERENCES

[1]   Zeiss, G.D., Meath, W.J., MacDonald, J.C.F., and Dawson, D.J.:
       1975, Radiat. Res. 63, 64.
[2]   Zeiss, G.D., and Meath, W.J.: 1977, Molec. Phys. 33, 1155.
[3]   Zeiss, G.D., and Meath, W.J., MacDonald, J.C.F., and Dawson, D.J.:
       1977, Can. J. Phys. 55, 2080.
[4]   Thomas, G.F., and Meath, W.J.: 1977, Molec. Phys. 34, 113.
[5]   Jhanwar, B.L., Meath, W.J., and MacDonald, J.C.F.: 1980,
       Can. J. Phys. 59, 185.
[6]   Fano, U., and Cooper, J.W.: 1968, Rev. Mod. Phys. 40, 441 and
       1969, Rev. Mod. Phys. 41, 724.
[7]   Hirschfelder, J.O., Byers Brown, W., Epstein, S.T.: 1964, Adv.
       Quant. Chem. 1, 255.
[8]   Dalgarno, A.: 1966, In "Perturbation theory and its applications
       in quantum mechanics". Edited by C.H. Wilcox (John Wiley and Sons)
       pp. 145-183.
[9]   Hirschfelder, J.O., Curtiss, C.F., and Bird, R.B.: 1954,
       "Molecular Theory of Gases and Liquids" (John Wiley and Sons).
[10]   Barker, J.A., Watts, R.O., Lee, J.K., Schafer, T.P., and Lee, Y.T.
       1974, J. Chem. Phys. 61, 3081.
[11]   Margoliash, D.J., and Meath, W.J.: 1978, J. Chem. Phys. 68, 1426;
       Margoliash, D.J., Proctor, T.R., Zeiss, G.D., and Meath, W.J.:
       1978, Molec. Phys. 35, 747.
[12]   Jhanwar, B.L., and Meath, W.J.: 1980, Molec. Phys. 41, 1061.
[13]   Zeiss, G.D., Meath, W.J., MacDonald, J.C.F., and Dawson, D.J.:
       1980, Molec. Phys. 39, 1055.
[14]   Langhoff, P.W., Karplus, M.: 1970, J. Chem. Phys. 53, 233.
[15]   Starkschall, G., and Gordon, R.G.: 1971, J. Chem. Phys. 54, 663.
[16]   Jhanwar, B.L., and Meath, W.J., to be published.

[17]   Mulder, F., and Huiszoon, C.: 1977, Molec. Phys. 34, 1215;
       Mulder, F., Van Hemert, M., Wormer, P.E.S., and Van der Avoird,
       A.: 1977, Theoret. chim. Acta 46, 39; Lekkerkerker, H.N.W.,
       Coulon, Ph., and Luyckx, R.: 1977, Physica A88,375.
[18]   Mulder, F., Thomas, G.F., and Meath, W.J.: 1980, Molec. Phys. 41,
       249; Thomas, G.F., Mulder, F., and Meath, W.J.: 1980, Chem.
       Phys. 54, 45.
[19]   Henke, B.L., and Elgin, R.L.: 1970, Adv. X-ray Analysis 13, 639.
[20]   Bragg, W.H., and Kleeman, R.: 1905, Lond. Edinb. Dubl. Phil. Mag.
       10, 318.
[21]   Zeiss, G.D., Meath, W.J., MacDonald J.C.F., and Dawson, D.J.:1977,
       Radiat. Res. 70, 284.
[22]   Landolt, H.H.: 1866, Annln. Phys. 117, 353.
[23]   Cooper, J.W.: 1974, Phys. Rev. A9, 2236.
[24]   Jhanwar, B.L., Margoliash, D.J., and Meath, W.J., to be published.
[25]   Mulder, F., and Meath, W.J.: 1981, Molec. Phys., in press.
[26]   Hepburn, J., Penco, R., and Scoles, G.: 1975, Chem. Phys. Lett.
       36, 451; Ahlrichs, R., Penco, R., and Scoles,G.: 1977, Chem. Phys.
       19, 119.
[27]   Tang, K.T., and Toennies, J.P.: 1977, J. Chem Phys. 66, 1496 and
       1978, J. Chem. Phys. 68, 5501.
[28]   Ng, K.C., Meath, W.J. and Allnatt, A.R.: 1978, Chem. Phys. 32,
       175 and 1979, Molec. Phys. 37, 237.
[29]   Koide, A., Meath, W.J., and Allnatt, A.R.: 1980, Molec. Phys. 39,
       895.
[30]   Scoles, G.: 1980, Ann. Rev. Phys. Chem. 31, 81.
[31]   Koide, A., Meath, W.J., and Allnatt, A.R., to be published.
[32]   Douketis, C., Scoles, G., Marchetti, S., Zen, M., and Thakkar, A.J.:
       1981, J. Chem. Phys., in press.
[33]   Aziz, R.A., and Chen, H.H.: 1977, J. Chem. Phys. 67, 5719.
[34]   Tang, K.T., Norbeck, J.M., and Certain, P.R.: 1976, J. Chem. Phys.
       64, 3063.
[35]   Kreek, H., and Meath, W.J.: 1969, J. Chem. Phys. 50, 2289;
       Murrell, J.N., and Shaw, G.: 1968, J. Chem. Phys. 49, 4731.

CORRELATED STATES IN POLYENES AND ION-RADICAL ORGANIC SOLIDS

Thomas E. Miller and Zoltán G. Soos
Department of Chemistry, Princeton University
Princeton, New Jersey 08544

Without extensive configuration interaction (CI), molecular orbital or band theories are not even qualitatively adequate for some properties of weakly-overlapping, open-shell systems. Site representations and valence bond (VB) methods are required for such correlated electronic states. The complete CI problem in models based on one valence state per site is treated by diagrammatic VB theory by expanding correlated states in terms of conventional VB structures. The interpretation and manipulation of correlated states is discussed in both ion-radical organic solids and the Pariser-Parr-Pople model of polyenes. The efficient exact analysis of all-trans octatetraene illustrates correlations in excited states and in π bond orders. The electrostatic energy and partial ionicity of organic charge-transfer complexes involve superpositions of correlated states, rather than a self-consistent single-determinantal MO treatment. Such superpositions also occur in mixed-valent organic conductors, whose enormous orbital degeneracy reflects about one valence electron per two equivalent sites in segregated stacks.

I. VALENCE BOND REPRESENTATION OF CORRELATED STATES

The convergence of molecular orbital (MO) and valence bond (VB) methods is traditionally illustrated with 1s atomic orbitals (AOs) in H_2. MO is preferable for large overlaps and small internuclear separations, VB for small overlaps and large separations. Neither a single-determinantal MO function nor a single VB structure is usually adequate. The generalization to N' electrons in N orbitals yields far larger, but still finite, configuration-interaction (CI) matrices for either MO or VB basis states. We have found site representations[1] and

B. Pullman (ed.), Intermolecular Forces, 117–132.
Copyright © 1981 by D. Reidel Publishing Company.

diagrammatic VB methods to be particularly convenient for modeling
ion-radical organic conductors[2], semiconductors[3], charge-transfer complexes[4], and magnetic insulators[5] with random exchanges. The relevant
solid-state models are formally similar to the Pariser-Parr-Pople (PPP)
model[6] for π-electrons. We consider below several manifestations of
electron correlations in both polyenes and ion-radical organic solids.
We naturally focus on features that are qualitatively incorrect in the
best, or Hartree-Fock, single-determinantal MO approach.

Ion-radical organic solids have open electronic shells. The currently known organic conductors are mixed-valent systems, with different numbers N' of electrons and N of equivalent sites. MO or band
theory yields a half-filled, metallic band for N' = N and equivalent
sites. The semiconducting nature of such ion-radical organic solids
immediately points to strong correlations. Open-shell molecular crystals form a distinctly new class of solids. They resemble metals in
having partly-filled bands, at least for N' ≠ N. They resemble closed-
shell inorganic salts in having substantial electrostatic (Madelung)
energies. Their structures, vibrational frequencies, and single site
excitations, on the other hand, indicate weakly-perturbed molecules
reminiscent of typical closed-shell organic solids. Their unusual
electric and magnetic properties, as well as characteristic charge-
transfer (CT) excitations, reflect the partly-filled MO retained in
the site representation.

The related PPP approach to aromatic molecules is based on a
carbon $2p_z$ AO at each atomic site. Spin-independent Coulomb interactions
are again responsible for correlations. We present here the complete
PPP analysis of several electronic states of all-trans octatetraene,
a linear N' = N = 8 system shown in Fig. 1. This conventional VB
representation, the Kekulé structure, is immediately seen to have
four π-bonds and one π-electron in every $2p_z$ AO. It represents
one basis vector in the VB analysis. Previous complete MO[7] and VB[8]
treatments stopped at N' = N = 6, although some fourth-order, MO-based
CI results have recently been reported[9] for octatetraene. Coulomb
correlations place the 2^1A_g excited state below the 1^1B_u level, in
agreement with experiment[10,11] and contrary to single-determinantal
MO results. The cis-trans photoisomerization of such conjugated molecules, an important early step in the visual process,[12,13] probably
involves the low-lying 2^1A_g state.

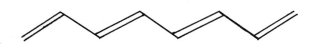

Fig. 1. All-trans octatetraene

Diagrammatic VB methods are convenient whenever there is one valence state per site, which can be empty, doubly occupied, or singly-occupied with spin α or β. The number operator for site p,

$$n_p = a_{p\alpha}^+ a_{p\alpha} + a_{p\beta}^+ a_{p\beta} \tag{1}$$

is consequently integral (0, 1, or 2) for every VB diagram like Fig. 1. For carbon AOs, empty C^+ sites correspond to carbonium ions, doubly-occupied C^- sites to carbanions, while lines denote covalent bonds with paired spins. A pictorial representation of parallel spins, or antibonding pairs, is straighforward.[2,8] Every VB diagram represents a perfectly-correlated state in which electrons occupy localized orbitals. These diagrams form the VB basis and the entire CI computation is carried out in terms of transformations induced among diagrams by electron transfers. Both electron transfers and intermolecular interactions conserve spin in organic solids, since spin-orbit coupling is weak. Intermolecular forces are consequently fully diagonal and go as $V_{pp'} n_p n_{p'}$. The coefficient $V_{pp'}$ is a parameter representing all intermolecular interactions. Its evaluation or measurement has been an important and longstanding goal. We pursue here the consequences of linear combinations of VB diagrams involving different n_p at each site.

The singlet VB diagrams in Fig. 2 are now readily interpreted in terms of covalent bonds, long bonds, dots for C^+ and crosses for C^-. A similar description holds for open-shell organic solids, for any degree of filling determined by $N' < 2N$. The emphasis is naturally quite different, since there is negligible energy gain in the solid for singlet correlations among distant sites. The VB basis is then primarily a convenience for preserving the total spin, S. Single MO configurations, by contrast, involve Slater determinants that are eigenfunctions of S_z only. Any VB diagram with p lines thus automatically represents 2^p Slater determinants with fixed relative phases. This economy, as well as the simple chemical interpretation of the diagrams in Fig. 2, are important advantages of diagrammatic VB methods.

We then represent all molecular or solid-state wavefunctions as linear combinations of VB diagrams. Spatial symmetries are readily included as linear combinations of diagrams. The finite dimensions of the space for one valence state per site ensures that various operators produce linear combinations of the VB diagrams. As summarized for octatetraene, the entire CI analysis is then efficiently done diagrammatically, thereby completely avoiding previous difficulties[14] with the direct manipulation of VB functions. We also emphasize the necessary equivalence of MO and VB approaches to finite-dimensional CI problems. Several beautiful and elegant formal developments[15,16] focus on the group-theoretical aspects of many-electron states. They provide alternatives to, and justification for, the present method.

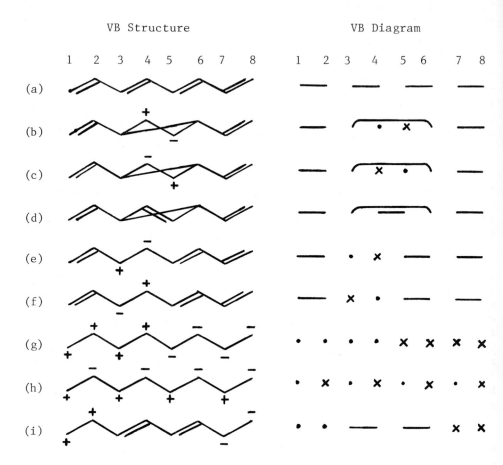

Fig. 2. Representative singlet VB diagrams for octatetraene

II. DIAGRAMMATIC VB TREATMENT OF OCTATETRAENE

The second-quantized PPP hamiltonian for a linear polyene of N sites and N electrons is

$$\mathcal{H} = \sum_{p=1,\sigma}^{N-1} |t_p| (a^+_{p\sigma} a_{p+1\sigma} + a^+_{p+1\sigma} a_{p\sigma}) + \sum_{pk}^{N} V_{pk} (n_p - 1)(n_k - 1)/2 \quad (2)$$

Here $a^+_{p\sigma}$, $a_{p\sigma}$ create, annihilate an electron with spin σ in the valence state of the pth carbon. The first term of Eq. 2 describes nearest-neighbor electron transfers, with spin conservation, that are retained in Huckel theory with the identification $t = \beta$. The inversion symmetry of all-trans octatetraene in Fig. 1 leads to

$t_1 = t_7$, $t_2 = t_6$, $t_3 = t_5$. The intersite Coulomb interactions V_{pk} are given in the Ohno parametrization[17] as

$$V_{pk} = 14.397eV \ (1.67 + r_{pk}^2)^{-\frac{1}{2}} \qquad (3)$$

were r_{pk} is the distance between carbons p and k in Angstroms. The number operator n_p in Eq. 1 gives the number (0, 1, or 2) of electrons in the valence state. The on-site (p=k, r=0) repulsions in Eq. 3 are U = 11.13eV and correspond to the only correlation retained in Hubbard models.[18] The distance dependence of t is taken to be[17,6]

$$t_{pp+1} = -2.6 \ eV + 3.21 \ (r_{pp+1} - 1.397)eV \qquad (4)$$

The choice of r = 1.35A and 1.46A for double and single bonds in the ground state leads to $t_1 = t_3 = -2.75eV$ and $t_2 = t_4 = -2.40eV$, respectively. All angles in Fig. 1 are fixed at 120°. This completely defines the PPP hamiltonian for all-trans octatetraene.

The Kekulé structure in Fig. 1 and other VB diagrams in Fig. 2 are normalized basis states. The general result[2] gives 1764 linearly independent singlet diagrams for octatetraene. These expand to over 6000 Slater determinants with $S_z = 0$, thereby spoiling a direct CI approach. Since n_p is diagonal for each site of each VB diagram, the interaction terms in Eq. 2 are diagonal for all 1764 singlets. They are simply found by setting n = 0 for C^+, n = 2 for C^-, and n = 1 for sites connected by lines. Purely covalent structures like (a) or (d) in Fig. 2 have vanishing Coulomb interactions, while the highly-ionized diagram (g) is over 60eV higher and is hardly expected to contribute to low-lying states.

The t terms of Eq. 2 conserve S, move electrons to adjacent sites, and thus connect different VB diagrams. A transfer from 4 to 5 in diagram (a) leads to (b); returning the electron to 4 yields (d). The magnitudes and phases of the transformation coefficients are known[2] in general. The transfer terms consequently generate a matrix of coefficients that interrelate the VB diagrams. The matrix is not symmetric because some VB diagrams are not orthogonal, a fact that hindered early work,[14] but direct analysis of nonsymmetric matrices completely circumvents[2-5] the nonorthogonality problem.

The singlet subspace of octatetraene can be reduced through the reflection symmetry σ and the electron-hole symmetry I. The plus and minus linear combinations of (b) and (c) in Fig. 2 are evidently even and odd under σ and thus belongs to A and B representations, respectively. As discussed elsewhere[19], electron-hole symmetry is closely related to the Pariser alternancy symmetry. In terms of VB diagrams[3], I merely interchanges C^+ and C^- sites, with a phase factor $(-1)^{p+p'}$ for interchanging p dots and crosses in diagrams with p' singly-occupied odd numbered sites. Thus (e) $\pm$ (f) in Fig. 2 transform as

$I = \pm 1$. Linear combinations of the 1764 singlet diagrams yield four subspaces. The ^1A_g state, with $\sigma = I = 1$, has 485 basis vectors; the ^1B_u state, with $\sigma = I = -1$, has 460; the $\sigma = -1$, $I = 1$ and $\sigma = 1$, $I = -1$ subspaces have 404 and 415, respectively.

Computer generation[20] of unnormalized, symmetry-adapted linear combinations of singlet VB diagrams is straightforward for linear systems. Each 8-site VB diagram in Fig. 2 is uniquely encoded as an 8-digit number. Each site is assigned one of four numbers, depending on whether it is C^+, C^-, or the right or left end of a line. Reflection symmetry amounts to reading backwards and relabeling right and left ends of lines. Electron-hole symmetry involves a phase factor and interchanging of C^+ and C^- sites.

The transformation coefficients induced by the $|t|$ terms of Eq. 2 are then computer generated. The resulting coefficient matrices are sparse since transferring a single electron interconnects few diagrams. The lowest eigenvalue in ^1A_g or ^1B_u was found[20] by iteration[21], with 300 cycles yielding 5 significant figures. Special care was taken to multiply only the nonvanishing elements of the sparse matrices, thereby reducing the computation time by an order of magnitude. Deflating[21] the ^1A_g subspace and further iteration yields the 2^1A_g energy. The unnormalized eigenstates are given in each case as the coefficients of the symmetry-adapted VB diagrams. The entire procedure takes less than one minute of computer time. The direct solution of 485 x 485 or 460 x 460 matrices is far more expensive and the additional high-energy states are of little current interest.

Table 1 contains results for the parameter values above. The order of the states agrees with experiment, but not with single-determinantal MO results. The excitation energies are about 0.5eV too high. They are close to the fourth-order CI values plotted, but not listed, in Ref. 9. Their $2^1A_g - 1^1B_u$ splitting of 0.56eV is very close to our 0.58eV result. The conventional Ohno parameters in the PPP hamiltonian are clearly adequate, even with parameters that preceded extensive CI computations, and smaller values of t improve the excitation energies.

The choice $V_{pk} = U\delta_{pk}$ along the diagonal yields Hubbard-model results, while neglecting all correlations in Eq. 2 regains Huckel results for the sum of the occupied MOs. Quite aside from providing convenient numerical checks,[20] these limits show that a common off-diagonal VB representation of the transfer terms simultaneously describes both uncorrelated electrons and complete CI. It is consequently very easy to change models or parameter values in Eq. 2. The resulting exact wavefunctions represent all orders of CI. They are conveniently given, probably for the first time, in terms of understandable VB diagrams. Although somewhat tedious for large subspaces, their normalization is also straightforward.

Table 1. Exact PPP results for all-trans octatetraene

State	$1\,^1A_g$	$2\,^1A_g$	$1\,^1B_u$
Energy (eV)	-18.533	-14.477	-13.895
Excitation Energy	-	4.056	4.638
Experiment (ref. 10)	-	3.55	3.98
(ref. 11)	-	3.54	3.96
Bond Order (Eq. 5)			
p_1	0.892	0.554	0.769
p_2	0.342	0.660	0.478
p_3	0.831	0.358	0.574
p_4	0.360	0.688	0.549

III. PROPERTIES OF CORRELATED STATES

The π bond order p_{nm} between carbons n and m was introducted by Coulson[22] in his classic Huckel-MO analysis of polyenes. The bond order p_k between carbons k and k+1 is

$$p_k = \frac{1}{2} \frac{\partial E}{\partial t_k} \qquad (5)$$

where t_k is the transfer (or β) integral connecting the two sites and E is the π-electron energy. Although originally defined for MO coefficients in a single-determinantal ground state, p_k generalizes naturally to excited states and to total π-electron energies of correlated systems. The Hellman-Feynman theorem relates p_k to changing t_k in the PPP hamiltonian. There are various other possible definitions[23] of bond orders, none of them entirely satisfactory.

We comment on the one peculiarity of Eq. 5. The bonding $(\sigma_b)^2$, or MO, ground state of H_2 has unit bond order for any internuclear separation R. But the Heitler-London, or purely covalent, state has vanishing bond order due to equal admixtures of bonding $(\sigma_b)^2$ and antibonding $(\sigma_a)^2$ configurations. This illustrates nicely the fact that the covalent limit is only achieved for large R, or vanishing t, where the partial derivative in Eq. 5 vanishes. Thus Huckel theory tends to give too large π bond orders, while the correlations in more realistic PPP models decrease the sensitivity of E to t_k variations. Thus changes in p_k, rather than its absolute magnitude, are important.

The derivatives in Eq. 5 are readily found numerically[20] by evaluating E for different t_k. In order to preserve the octatetraene symmetry, we varied $t_1 = t_7$, $t_2 = t_6$, and $t_3 = t_5$ together. Bond orders for the ground state values of $t_1 = t_3 = -2.75eV$ and $t_2 = t_4 = -2.40eV$ are listed in Table 1. They do not change significantly with 10% changes in $|t|$ values. The clear evidence for comparable partial double and single bonds in the 1^1A_g ground state is expected, since the standard parameters reflect various self-consistent treatments[6,24] for correlating the bond length r, bond order p(r), and t(r). The excited-state bond orders in Table 1, on the other hand, are clearly inconsistent with the ground-state values of t_k and point to different backbone geometry in the excited states. The excited state bond orders are hardly changed[20], however, on adjusting t in Eq. 4 to reflect the new geometry.

Previous MO results for octatetraene bond orders[25] parallel the 1^1A_g and 1^1B_u values. As already noted, Huckel theory is quite inadequate for the 2^1A_g energy, and no bond orders have been reported. The most striking feature of the 2^1A_g bond order in Table 1 is the small value of p_3. This erstwhile double bond, which becomes comparable to the ground-state single bonds, is an obvious candidate for facile cis-trans isomerization in the 2^1A_g manifold. Such photo-isomerization is observed[26] even at 4.2K in a solid matrix. The correlated-state description clearly implicates the 2^1A_g state.

The π-electron contribution to force constants Q_k for C_k-C_{k+1} stretches are simply

$$Q_k = 2p_k\left(\frac{d^2t_k}{dr_k^2}\right) + \left(\frac{\partial^2E}{\partial t_k^2}\right)\left(\frac{dt_k}{dr_k}\right)^2 + \left(\frac{\partial^2E}{\partial r_k^2}\right) \qquad (6)$$

The last term does not arise in Huckel theory, but must be included in PPP or other models with interactions that explicitly depend on r_k. The other two terms have been discussed previously.[24,6] The usual association[27] of large force constants with high bond order is embodied in the first term. The $(\partial^2E/\partial t_k^2)$, or bond polarization, contribution is relatively small[25] in octatetraene. Our complete CI results also yield small self polarizabilities as well as small r_k second derivatives. The behavior of $(\partial^2 t_k/\partial r_k^2)$ consequently dominates Q_k, but requires going beyond the PPP model. The linear t(r) dependence in Eq. 4 yields t"(r) = 0 and is clearly inadequate, while an exponential[24] t(r) yields a negative contribution. The original choice[27] of large positive t"(r) assumes both σ and π contributions.

The expansion of the complete CI wavefunctions in terms of VB diagrams also provides new insights about selection rules and transition moments[3] for CT excitations, the degree of ionicity[4] in CT solids, the role of nearest-neighbor interactions,[2] or any other property that depends primarily on the valence states described in Eq. 2. It is usually straightforward to find the second-quantized representation of the relevant operator. Since the VB basis is complete, all matrix elements necessarily reduce to coefficients of VB diagrams, although possibly not within a single subspace. Thus dipole transitions change both σ and I, while spin relaxation mechanisms change S. Explicit computations with correlated states are still fairly primitive, as are their interpretation.

For example, the Kekulé structure (a) in Fig. 2 has the largest coefficient in the 1^1A_g state. But the PPP parameters give the largest admixture for CT states like (e) or (f), with an adjacent electron-hole pair, rather than purely covalent states like (d) whose diagonal energy is lower. VB diagrams that are directly linked by a single $|t|$ term to the Kekulé structure are favored, not too surprisingly. All 1^1B_u diagrams necessarily contain one or more electron-hole pairs, with the lowest state again dominated by adjacent C^+C^- pairs. The 2^1A_g state, by contrast, becomes one of 42 purely covalent states in the limit of large intrasite correlations U. Their energy spread of the order of $|t|^2/U$ is then related[1] to the antiferromagnetic exchange of a linear Heisenberg chain. The crossover of 2^1A_g and 1^1B_u with increasing correlations is thus guaranteed. The small $2^1A_g-1^1B_u$ gap in octatetraene points to intermediate correlations, slightly on the covalent side. The far larger difference between optical and magnetic gaps[1] in many organic ion-radical solids indicates more correlated behavior, with a smaller ratio of $|t|/U$.

IV. MADELUNG ENERGY OF PARTLY-IONIC SOLIDS

Organic CT complexes usually crystallize in mixed $...D^{+\gamma}A^{-\gamma}D^{+\gamma}A^{-\gamma}...$ stacks. The degree of ionicity, γ, is zero in largely neutral complexes of weak donors (D) and acceptors (A). The ionicity approaches unity in several recently prepared very strong complexes.[28] The valence state now involves the half-filled HOMO of the ion-radicals D^+ and A^-. The relation between the charge q_p and the number operator n_p in Eq. 1 is

$$q_p = 1 - n_p \qquad p = C \text{ site}$$

$$q_p = \ - n_p \qquad p = A \text{ site} \qquad (7)$$

$$q_p = 2 - n_p \qquad p = D \text{ site}$$

The Kekulé diagram (a) in Fig. 2 now describes a spin state of the ionic

lattice, while diagram (h) corresponds to a neutral lattice with As at
odd sites and Ds at even sites. The Coulomb interactions in Eq. 2 are
suitably modified, with $V_{pp'}$ evaluated[29] for delocalized HOMOs and the
crystalline geometry.

We consider a DA solid of N equivalent A sites, N equivalent D
sites, and $2\gamma N$ charges. The ground-state ionicity γ is necessarily
the average charge of either $D^{+\gamma}$ or $A^{-\gamma}$ sites. The Madelung operator
per DA pair is

$$M(\gamma) = \sum_{p'} V_{pp'} q_p q_{p'} \qquad (8)$$

The self-consistent Madelung energy $\langle M(\gamma) \rangle$ is given by setting
$\langle q_p \rangle = \pm \gamma$ for D and A sites in Eq. 8. Coulomb exchange corrections[30]
go as $\langle q_p q_{p'} \rangle - \langle q_p \rangle \langle q_{p'} \rangle$ and are small for the large, almost
Van der Waals, separations in organic CT complexes. Metzger[31] has
reviewed $\langle M(\gamma) \rangle$ computations in mixed-valent segregated organic stacks
(M $\sim$ 2-3eV) and ionic CT complexes (M $\sim$ 4-5eV). Theory does not support
the formation or stability of partly-ionic lattices unless additional
postulates are introduced[32] for fractional ionization potentials and
electron affinities.

Setting $\langle q_p \rangle = \pm \gamma$ in Eq. 8 reduces the Madelung operator to a
sum in which electron correlations are neglected. This is a Hartree
approximation if γ is found self-consistently. We present an improved
treatment based on correlated states. The $|t| \to 0$ limit of zero over-
lap, for example, clearly requires integral charges (electrons) at all
sites. The Madelung sum is then minimized[32] by a Wigner lattice with
$2\gamma N$ charges and $2(1-\gamma)N$ neutral molecules. The Hartree sum scales as
γ^2, the Wigner sum as approximately γ. For $\gamma = \frac{1}{2}$, the latter gives
$-M/2$ for a lattice in which every other sites is ionic. The large
difference between $-M/2$ and $-M/4$ is 1eV for M = 4eV, stabilizes the
correlated state, and exceeds any contribution from $|t| \sim 0.1-0.2$eV.
This pure correlation effect is largest around the interesting
$\gamma \sim \frac{1}{2}$ regime of good organic conductors.

The static picture of correlated ionic and neutral sites is not
consistent with magnetic, optical, or structural properties of open-
shell organic solids.[1,2] Finite $|t|$ along at least one crystal axis
admixes VB diagrams with different charge distributions thereby giving
an average, fractional charge at each site. In contrast to previous
Hartree results, however, the charges remain correlated. Fortunately,
most intermolecular contributions[31] are actually evaluated at $\gamma = 0$
or 1 and remain useful parameters in a correlated-state description.

Exact VB analysis of 10-site mixed stacks[2] already yield over
15,000 VB diagrams. This suffices for modeling the neutral-ionic
interface,[2] the structure of magnetic excitations,[34] or the effects

of nearest-neighbor interactions.[4] But direct computations of three-dimensional lattices remain prohibitive, even with computer generation, although their main features can be anticipated.

We first note that the charge operators q_p in Eqs. 7 and 8 are diagonal and that VB diagrams with different charge distributions are rigorously orthogonal.[2] A normalized crystal state $|S,\gamma\rangle$ with spin S and ionicity γ thus involves linear combinations of diagrams with $N\gamma$ cationic D^+ sites and $N\gamma$ anionic A^- sites. Diagrams with the same or symmetry-related electron distributions, like (a), (d) or (b), (c) or (e), (f) in Fig. 2, have the same electrostatic energy and are not orthogonal in general. We collect all such diagrams and construct a normalized state $|m,S,\gamma\rangle$ for the mth charge distribution. The crystal state can now be expanded as

$$|S,\gamma\rangle = \sum_m c_m |m,S,\gamma\rangle \qquad (9)$$

The orthonormality of $|m,S,\gamma\rangle$ shows $|c_m|^2$ to be the weight of a particular charge distribution. We finally let $|S,\gamma\rangle$ be the ground state and evaluate the expectation value of Eq. 8 to obtain various Madelung sums weighted according to their coefficients $|c_m|^2$. Finite $|t|$ thus produces admixtures of correlated states with different coefficients, the largest presumably corresponding to electrostatic energies close to the Wigner lattice.

The relative magnitudes of the c_m in Eq. 9 can be estimated from small systems. For example, four electrons on seven equivalent A sites in a ring yield a segregated $...A^-\,{}^\gamma A\,{}^\gamma A^-\,{}^\gamma...$ stack, with $\gamma = 4/7$, close to the filling in tetrathiofulvalene-tetracyanoquinodimethane (TTF-TCNQ). Exact VB results[2] suggest that the magnetic susceptibility and CT transitions can be approximately fit with $|t| \sim 0.13ev$, $U = V_{pp} = 1.40ev$, and $V = V_{pp+1} = 0.40ev$. These parameters are in the difficult regime of intermediate correlations, with U-V comparable to the bandwidth $4|t|$. Table 2 shows the relative weights $|c_m|^2$ of the S = 0 and S = 1 ground states for three charge distributions. Class I contains all diagrams, like a Wigner lattice, with minimum Coulomb energy. Class II involves all other diagrams without a doubly-occupied A^{-2} site. Class III diagrams contain one or more A^{-2} site. Their small weight belies their crucial role in producing anti-ferromagnetic exchange interactions that dominate the magnetic susceptibility. The electrostatic interactions in the Madelung sum, by contrast, clearly depend primarily on Class I diagrams.

The VB analysis of electrostatic interactions in enormous linear combinations of correlated states thus combines equivalent sites with partial ionicity γ, magnetic properties due to small admixtures of doubly-occupied sites, and crystal energies closer to the Wigner

Table 2.   Decomposition of the S = 0 and S = 1 VB ground states for
4 electrons on a segregated, 7-site ring, from ref. 2b.
U and V are the on-site repulsion and nearest-neighbor
Coulomb interactions, respectively, in Eq. 2.   Setting
V = 0 gives the Hubbard model, with no preference for
Class I diagrams.

Correlations		Total Spin	Type of VB Diagram		
$U/\sqrt{2}\lvert t\rvert$	$V/\sqrt{2}\lvert t\rvert$	S	% Class I	% Class II	% Class III
6	2	0	70.44	26.23	3.33
		1	64.88	27.87	7.25
10	3	0	81.17	17.64	1.19
		1	78.58	17.97	2.70
6	0	0	43.13	54.69	2.18
		1	39.09	55.56	5.35

lattice than to the Hartree result of $M\gamma^2$.   Correlated states lead to
large corrections around $\gamma \sim \frac{1}{2}$, where there is maximum orbital degeneracy, and reduce to the Hartree result for $\gamma = 0$ or $\gamma = 1$,   when
there is no orbital degeneracy if doubly-charged sites are excluded.

V.   DISCUSSION

Although organic molecules are routinely represented by VB
structures, their electronic properties are almost exclusively discussed in MO language.   Solid-state or molecular models based on
one valence state per site, by contrast, are conveniently treated
using a VB basis.   Second-quantized and diagrammatic methods yield
the complete CI for some $10^4- 10^5$ linearly independent VB diagrams,
and further efficiencies can safely be anticipated.   Direct VB computations[15,35] via Slater determinants are tedious even for 10-50
states.   Diagrammatic VB methods thus put many new problems within
reach.   On the other hand, the many successes and great efficiency
of modern MO methods makes it prudent to seek situations with significant correlations, for example in weakly-overlapping open-shell
molecular solids.   We comment briefly on the scope, advantages, and
difficulties of VB methods.

The most important initial problem is deceptively simple: to find
the correspondence between correlated many-electron VB functions and
the usual one-electron MO description.   The manipulation and interpretation of correlated states is still primitive.   For example,
suppressing correlations in Eq. 2 regains the Huckel limit and
necessarily connects the one- and many-electron densities of states.
The effects of correlations on the density of states will be needed

for many solid-state applications. One-electron language is so deeply
ingrained that it is often the only available description.

The second point is that PPP, Hubbard, or other models are at
best approximations. Reparametrization to obtain some exact excitation energies would still fall short of a quantitative description.
All electron treatments of molecules, with judicious combinations[36]
of both MO and VB ideas, are the way to quantitative results. Models
highlight trends and are applicable to more complicated cases. Diagrammatic VB methods give a clear picture of correlated states.
For example, the different length dependences[6] and spectral
characteristics of polyenes and cyanine dyes have led us to model
the latter as N+1 electrons on N sites, with N odd, with modified
PPP parameters for the terminal nitrogens. Various ionic VB diagrams
are favored[20] at the expense of the expected covalent structures.
The excited state bond orders do not interchange, in contrast to the
octatetraene results in Table 1, and there is again a low-lying A
state. Some reparametrization is needed to establish the sensitivity
of the results, but this is secondary to the task of interpreting the
correlated states and contrasting them to polyenes.

The high energies of unfavorable VB diagrams like (g) or (i)
in Fig. 2 suggest that they can safely be neglected, just as high
energy D^{+2} and A^{-2} states are neglected[4] in CT solids. Our third
point is that approximations in VB treatments generally involve
truncating the basis. Either unfavorable individual sites, or
repulsive states involving adjacent doubly-occupied sites, or total
electrostatic energies can be used. An electrostatic, or diagonal,
cutoff of 15eV for the PPP parameters of octatetraene excludes[20]
about half of the VB diagrams, but only leads to 0.1eV errors for
the energies in Table 1. Since the number of states increases[2]
rapidly with the number of sites, such approximations are of
limited utility. For example, modeling the cis-trans isomeration
of octatetraene by varying t_3 in Eq. 2, as discussed[37] for isomeriz-
ing stilbene, destroys the σ symmetry and yields 889 diagrams with
I = 1. Since 1^1B_u transforms as I = -1, the two lowest I = 1 roots
give the desired energy gap. Truncating at 15eV gives 525
states and a reduced[20] gap of 2.3eV. We cannot test directly the
hypothesis,[13] in rhodopsin, that the $2A_g$ barrier is attractive,
however, since PPP models neglect nonbonded interactions and
overestimate the ground-state barrier. The "$2A_g$" excited state
is either slightly repulsive or attractive at $t_3 = 0$, which is
consistent with facile rotation about this bond.

We note in closing that polyacetylene, $(CH)_x$, is a polymer
nominally based on infinite polyenes, although the chain lengths
are limited by cross links or CH_2 sites. Either involves sp^3
carbons whose ^{13}C resonance can be resolved[38] from the majority of

sp^2 hybrids in the chain. Some 4% sp^3 carbons clearly lead to rather short conjugated segments. Doping $(CH)_x$ with a few percent of either donors or acceptors produces a semiconductor to metal transition and increases the conductivity by many orders of magnitude.[39] The physical properties of neat and doped $(CH)_x$ are consequently of intense current interest and even controversy. None of the different proposed models,[39,40] however, have so far required going beyond one-electron ideas, possibly because several different properties must be modeled accurately to reveal inconsistencies. The clear evidence for correlations in octatetraene and other polyenes suggests that rather different models for $(CH)_x$ will emerge.

We gratefully acknowledge support of this work through NSF-DMR-7727418 A01.

REFERENCES

1. Z. G. Soos and D. J. Klein, in Molecular Association, vol. 1 (ed. R. Foster, Academic, New York, 1975) pp. 1-119; D. J. Klein and Z. G. Soos, Mol. Phys. 20, 1013 (1971); Z. G. Soos, J. Chem. Ed. 55, 546, (1978).

2. S. Mazumdar and Z. G. Soos, Synthetic Metals 1, 77 (1979); Phys. Rev. B23, 2810 (1981).

3. S. R. Bondeson and Z. G. Soos, Chem. Phys. 44, 403 (1979).

4. Z. G. Soos and S. Mazumdar, Phys. Rev. B18, 1991 (1978).

5. S. R. Bondeson and Z. G. Soos, Phys. Rev. B22, 1793 (1980).

6. L. Salem, The Molecular Orbital Theory of Conjugated Systems (Benjamin, New York, 1966). Chs. 1, 3, 7 and 8; R. Pariser and R. G. Parr, J. Chem. Phys. 21, 446, 767 (1953); J. A. Pople, Trans. Far. Soc. 42, 1375 (1953).

7. K. Schulten, J. Ohmine, and M. Karplus, J. Chem. Phys. 64, 4422 (1976).

8. S. R. Bondeson and Z. G. Soos, J. Chem. Phys. 71, 3807 (1979); 73, 598 (1980).

9. P. Tavan and K. Schulten, J. Chem. Phys. 70, 5407 (1979).

10. M. F. Granville, G. R. Holtom, and B. E. Kohler, J. Chem. Phys. 72, 4671 (1980); M. F. Granville, G. R. Holtom, B. E. Kohler, R. L. Christensen, and K. D'Amico, J. Chem. Phys. 70, 593 (1978).

11. R. M. Gavin, Jr., C. Weisman, J. K. McVey, and S. A. Rice, J. Chem. Phys. 68, 522 (1978).

12. B. Honig, Ann. Rev. Phys. Chem. 29, 31 (1978); B. Honig, T. Ebrey, R. H. Callender, U. Dinur, and M. Ottolenghi, Proc. Natl. Acad. Sci. USA 76, 2503 (1979).

13. R. R. Birges and L. M. Hubbard, J. Amer. Chem. Soc. 102, 2195 (1980).

14. L. Pauling, J. Chem. Phys. 1, 280 (1933); H. Eyring, T. Walter, and G. E. Kimball, Quantum Chemistry (Wiley, New York, 1944) Ch. 13.

15. J. Paldus, Theor. Chem: Advs. and Perspectives, 2, 131 (1976); J. Paldus and M. J. Boyle, Physica Scripta, (Nobel Symposium #46 Special Issue) 1979.

16. F. A. Matsen, Accts. Chem. Res. 11, 387 (1978); Advances in Quantum Chemistry, Vol. II (ed. P.O. Lowdin, Academic, New York, 1978) p. 223 and references therein.

17. K. Ohno, Theor. Chim. Acta 2, 219 (1964). These parameters are also used in Ref. 9.

18. J. Hubbard, Proc. Roy. Soc. Ser. A276, 238 (1963); A277, 237 (1964); A281, 401 (1964).

19. J. Cizek, J. Paldus, and I. Hubac, Int. J. Quant. Chem. 8, 951 (1974).

20. T. E. Miller, Senior Thesis, Princeton University, 1981 (unpublished).

21. A. Jennings, Matrix Computation for Engineers and Scientists (Wiley, New York, 1977) Ch. 10.

22. C. A. Coulson, Proc. Roy. Soc. (London) A169, 413 (1939). See also ref. 6.

23. K. Jug, J. Amer. Chem. Soc. 99, 7800 (1977).

24. H. C. Longuet-Higgins and L. Salem, Proc. Roy. Soc. (London) A251, 172 (1959). See also Ch. 8 in ref. 6.

25. R. M. Gavin, Jr. and S. A. Rice, J. Chem. Phys. 55, 2675 (1971).

26. M. F. Granville, G. R. Holtom, and B. E. Kohler, Proc. Natl. Acad. Sci. USA 77, 31 (1980).

27. C. A. Coulson and H. C. Longuet-Higgins, Proc. Roy. Soc. (London) A193, 457 (1948); J. E. Lennard-Jones and J. Turkevich, Proc. Roy. Soc. (London) A158, 297 (1937).

28. Z. G. Soos, H. J. Keller, K. Ludolf, J. Queckbörner, D. Wehe, and S. Flandrois, J. Chem. Phys. (in press).

29. R. M. Metzger, J. Chem. Phys. $\underline{57}$, 1870, 1876, 2218 (1972).

30. Z. G. Soos and A. J. Silverstein, Mol. Phys. $\underline{23}$, 775 (1972).

31. R. M. Metzger, Ann. N.Y. Acad. Sci. $\underline{313}$, 145 (1978). R.M. Metzger, ed., Cohesive and Conformational Energies, Springer Topics in Current Physics, (Springer, Berlin-Heidelberg, in press).

32. Z. G. Soos, Chem. Phys. Letters $\underline{63}$, 179 (1979); A. N. Bloch, Bloch, Bull. Amer. Phys. Soc. $\underline{25}$, 255 (1980) and private communication.

33. J. Hubbard, Phys. Rev. $\underline{B17}$, 494 (1978). B.D. Silverman, Phys. Rev. $\underline{B16}$, 5153 (1977), discusses correlated states and finite t, but retains the Hartree γ^2 scaling.

34. Z. G. Soos and S. R. Bondeson, in Extended Linear Chain Compounds, Vol. 3 (ed. J. S. Miller, Plenum, in Press).

35. R. Daudel, R. Lefèbre, and C. Moser, Quantum Chemistry (Interscience, New York, 1959) pp. 99-110.

36. W. A. Goddard, III, T. H. Dunning, Jr., W. J. Hunt, and P.J. Hay, Acct. Chem. Res. $\underline{6}$, 368 (1973).

37. C. H. Ting and D. S. McClure, J. Chinese Chem. Soc. $\underline{18}$, 94 (1971).

38. M. M. Maricq, J. S. Waugh, A. G. MacDiarmid, H. Shirakawa, and and A. J. Heeger, J. Amer. Chem. Soc. $\underline{100}$, 7729 (1978).

39. A. J. Heeger and A. G. MacDiarmid, in the Physics and Chemistry of Low Dimensional Solids, NATO ASI Series C56 (ed. L. Alcacer, Reidel, Dordrecht, Holland, 1980) p. 353.

40. W. P. Su, J. R. Schrieffer, and A. J. Heeger, Phys. Rev. Lett. $\underline{42}$, 1698 (1979); Y. Tomkiewicz, T. D. Schultz, H. B. Broom, T. C. Clark and G. B. Street, Phys. Rev. Lett. $\underline{43}$, 1532 (1979).

THEORETICAL STUDY OF THE INTERMOLECULAR HCL POTENTIAL

Christian Votava and Reinhart Ahlrichs
Lehrstuhl für Theoretische Chemie
Universität Karlsruhe, D-7500 Karlsruhe

The HCl pair-potential is investigated by means of
extended quantum chemical computations. These involve highly correlated wave functions obtained
from large basis sets within the CEPA. It turns
out that repulsive, electrostatic and dispersion
interactions are all of great importance. The computed interaction potential is fitted to a simple
analytical expression. This is then used to compute
the second virial coefficient B(T) and structure
factors $S_{\alpha\beta}$ by means of MD calculations. Comparison
with previous potentials and corresponding experiments indicates the accuracy of the present potential.

1. MOTIVATION

The theoretical first principles treatment of thermodynamic
properties of gases or condensed matter requires in the
first place the quantum-mechanical computation of potential
hypersurfaces which are the prerequisit for the subsequent
application of statistical mechanics to obtain e.g. the
virial coefficient for a gas or the pair correlation function for a liquid.
In this article we describe the results of such a study for
the case of HCl. This example was selected for the following reasons:
(a) it appeared feasible, without being trivial,
(b) results could be checked against recent measurements
    (1,2),
(c) which in turn can be complemented by theory,
(d) and last not least we wanted to demonstrate the present
    state of the art.

The accurate computation of the intermolecular HCl

133

B. Pullman (ed.), Intermolecular Forces, 133–147.
Copyright © 1981 by D. Reidel Publishing Company.

potential is by no means an easy task since repulsive,
electrostatic and dispersion forces are of great importance
with smaller contributions arising from polarization
effects. Matters are somewhat facilitated in the subse-·
quent MD treatment since electrostatic forces are not too
strong, hydrogen bonding is very weak, and - in part as a
consequence of these facts - three-body and higher forces
are expected to be rather small too.
Previously published HCl pair potentials were based on
small basis set SCF-computations (which neglect dispersion
forces) (3) or derived by fitting a reasonable Ansatz to
experimental data (4). A comparison with these potentials
and e.g. the virial coefficient derived from them in an
extended and careful study by Mc Donald et al. (4) can
further serve to establish the merits of corresponding
procedures and the accuracy required for these potentials.

2. METHODS OF COMPUTATION

The electronic structure calculations were performed with
the CEPA-1(SD) method (5), which may be briefly described
in the following way. One starts from an SCF computation,
then includes explicitly all single and double replace-
ments from valence shell MOs and accounts for the bulk of
higher then doubly substituted configurations by means of
the CEPA-1 procedure which guarantees size "extensivity"
(6). Our programm is technically a matrix formulated
direct CI variant (7) formally without a complete integral
transformation, and resembles closely methods recently
developed by Meyer (8) and Pople et al. (9).
Since inner shell Cl-AO's are of virtually no importance
for the present problem they have been simulated by the
Habitz-Schwartz pseudopotential method (10). This helps to
reduce computational expences without inferring a loss in
accuracy as has been verified in extensive tests on systems
such as NH_3, H_2O, PH_3, H_2S, and HCl (11).

For the Cl-atom we employed an uncontracted (4s,4p,2d)
(η_{1d}=1.4 and η_{2d}=0.35) GTO basis set derived from
Huzinaga's (12s,9p) (12) set by leaving out the corresponding steep "inner shell" functions which are superfluous
in the pseudopotential approach used here.
The hydrogen basis was of (5s,1p) type contracted to
(3s,1p) with η_p=0.3. For the $(HCl)_2$ we thus have an extended valence shell basis consisting of 64 CGTOs giving rise
to up to 101024 singly and doubly excited configuration
state functions (in the case of no symmetry) which are explicitly accounted for in the CEPA-1(SD), whereas about
10^8 quadruple excitations are treated in an approximate
way.

These large scale computations can nowadays be performed on small computers (e.g. mini computers or a UNIVAC 1108) with modest core storage ($\sim$ 300 KB) and IO requirements.

The methods used to evaluate virial coefficients and to perform MD calculations employed standard procedures and need not be discussed here further.

3. POTENTIAL SURFACE CALCULATIONS: RESULTS AND DISCUSSION

3.1. The HCl molecule

In the first place we want to establish the accuracy achieved in the description of the HCl molecule by the present approach (CEPA-1(SD) with the basis set given above). For this purpose we list our results for the equilibrium distance r_e, the dipole moment μ, and the quadrupole moment Θ (at r_e) in comparison with experimental results which are given in parantheses (13):

$$
\begin{aligned}
r_e &= 2.415 & \text{a.u.} &= 1.278 & &\text{Å} & (1.275 \text{ Å}) \\
\mu &= 0.4675 & \text{a.u.} &= 1.198 & &\text{D} & (1.094 \text{ D}) \\
\Theta &= 2.818 & \text{a.u.} &= 3.79 & &\text{DÅ} & (3.75 \text{ DÅ})
\end{aligned}
$$

Our value for μ is 0.09 D too small but the relatively large quadrupole moment is reproduced with an error of 1%. This indicates strongly that we get a proper electron density distribution for HCl, which is important for the description of the HCl pair-potential. Werner and Rosmus (14) have recently demonstrated that the CEPA-1 method yields the correct $\mu(r)$ if a very large CGTO basis set (14s, 11p, 3d, 1f; 7s, 2p, 1dσ)/(11s,9p,3d,1f; 7s,2p,1dσ) is used. We note that the $\mu(r)$ obtained in the present work displays a constant shift of 0.09 D for $2 < r < 3$ as compared to Werner and Rosmus' very accurate result.

3.2. The Ar...HCl potential

As a further test of our approach as described in section 2., we have computed the Ar...HCl interaction for which a number of gas phase data are available (15). For the most stable linear arrangement Ar...HCl we find the equilibrium distance
$$R(Ar...Cl) = 7.74 \text{ a.u.} = 4.10 \text{ Å}$$
with an almost insignificant increase in r(HCl) by 0.01 a.u. and a binding energy $E_b = E(Ar) + E(HCl) - E(Ar...HCl)$
$$E_b = 565 \text{ μH}$$
corresponding experimental results (15) range from 7.48 a.u. - 7.81 a.u. for R(Ar-Cl) and 729 μH - 836 μH for E_b. An ab initio study of Diercksen (16) (CI(SD) plus Davidson correc-

tion using a (13s, 10p, 1d; 5s, 1p)/(8s, 6p, 1d; 3s, 1p)
CGTO basis) yields R(ArCl)∼ 8 a.u. and E_b∼ 501 µH. This is
in line with our result since a second d set on Ar and Cl
- included in our work but not in Diercksen's - gives an
improved description of dispersion forces which increases
E_b by ∼ 60 µH. The discrepancy between our and experimental
estimates for E_b of about 200 µH appears to be mainly due
to the slow convergence of the dispersion interaction towards the basis set limit. This point of view is supported
by additional computations which indicate an increase of E_b
by about 60 µH if an f set is included on Ar and Cl.

As a summary of section 3.1 and 3.2 we may conclude
that the present approach seems to give a proper description of repulsive and polarization interactions but it
underestimates dispersion (by about 200 µH near the equilibrium Ar...HCl).

3.3. The $(HCl)_2$ pair-potential V.

We considered only the interaction of rigid HCl subunits
with intramolecular distance of 2.41 a.u., the experimental
r_e. The inclusion of intramolecular geometry relaxation
would have been to expensive and was not expected to lead
to considerable changes in V, in Ar...HCl it contributes a
few µH only.

For the determination of V we started with a series of
computations for the following six relative arrangements of
HCl molecules: The three linear structures with H...Cl,
Cl...Cl, and H...H approaches, and three planar geometries
with 0°, 90°, and 180° angles between HCl axes, the 90°
case with collinear ClH...Cl alignement.
For either of these six curves we considered intermolecular
distances R(ClCl) from ∼ 15 a.u. down to R values where
V > 25 kcal/mol. between 10 and 18 points were computed on
the SCF level for each curve. Since the correlation energy
contribution to V shows a much smoother behaviour than the
SCF part, it turned out to be sufficient to perform CEPA-1
calculations for only half the number of points. The remaining correlation energies were obtained from one dimensional fits. Additional computations were done near the
equilibrium geometry and for three-dimensional structures.
The total number of discrete points is 100, of which 62
were treated on CEPA level. The gross features of V are
apparent from V(R) curves for the just mentioned six
arrangements which are shown in Fig. 1.

The dominant features of these curves may be qualitatively understood as resulting from the structure of electrostatic interactions. This is due to the small anisotropy

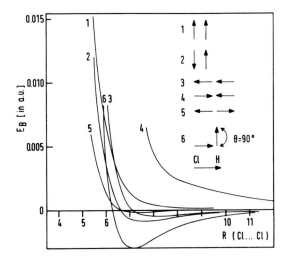

Figure 1. Computed HCl...HCl interaction energy for some typical arrangements.

in the HCl polarizability ($\alpha_\perp$ = 2.50 $\overset{o}{A}{}^3$; $\alpha_\parallel$ = 2.81 $\overset{o}{A}{}^3$) (14) which leads to a rather isotropic dispersion interaction. The deepest minimum is found for the rectangular arrange-ment where dipole-quadrupole ($\mu\theta$) and $\theta\theta$ are attractive, and $\mu\mu$ is zero or almost zero, depending on the choice of molecular centers. Next we find the planar structure with antiparallel HCl ($\mu\mu$ attractive, $\mu\theta{\sim}0$, $\theta\theta$ repulsive) and the linear "hydrogen bonded" case ($\mu\mu$ twice as attractive as before, $\mu\theta{\sim}0$, $\theta\theta$ about 2.7 times as repulsive as before). It is certainly the large quadrupole moment of HCl that determines the relative order of minima in this case. A rather shallow minimum is found for the linear structure with Cl...Cl approach ($\mu\mu$ repulsive, $\mu\theta$ attractive, $\theta\theta$ re-pulsive). No minima occur for the parallel arrangement of HCl molecules ($\mu\mu$ repulsive, $\mu\theta$ = 0, $\theta\theta$ repulsive) and for the linear H...H approach ($\mu\mu$, $\mu\theta$ and $\theta\theta$ all repulsive).

In order to get better insight into the relative im-portance of the various intermolecular forces we consider two cases in more detail.

In Fig. 2. we have plottet the SCF contribution to V (accounting for Pauli-Principle repulsion, polarization and electrostatic effects), the correlation energy contri-bution (mainly dispersion) and the total V(R). The SCF interaction energy is rather small in absolute value and shows only a shallow minimum at R(ClCl)$\sim$ 12 a_o, with V $\sim$ - 120 μH. This behaviour my be rationalized as a cancella-

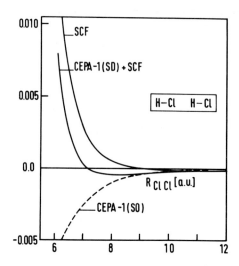

Figure 2. Energy contributions to the interaction
energy as a function of R(ClCl).

tion of attractive ($\mu\mu$ and polarization) and repulsive forces (mainly $\theta\theta$). Addition of the attractive- and smoothly
varying-correlation energy contribution leads to a markedly
different V(R). We now find a much deeper minimum (V $\sim$
-442 μH) at a much smaller distance, R(ClCl)$\sim$8.2 a_o. If
ever necessary, this example demonstrates again the importance of correlation energy for an accurate description of
intermolecular interaction even for polar molecules such as
HCl.

Fig. 3 shows V and various contributions to V as a
function of HCl rotation which transforms the linear hydrogen bonded structure into the rectangular one. The behaviour
of V is dominated by the electrostatic interaction (as obtained from our model explained in the next section), which
is contained in the SCF energy, of course. There is also a
slight variation of the correlation energy favouring the
linear structure, which is in line with the angular dependence of dispersion forces as given by Buckingham's asymptotic formula (18). This anisotropy results from the fact
that $\alpha_\parallel > \alpha_\perp$ for HCl, as noted above.

3.4. Analytic representation of the HCl pair-potential V

The most difficult problem occuring in the course of the
present study was the fit of computed interaction energies
to an analytical form which was an almost endless procedure of trial and error.

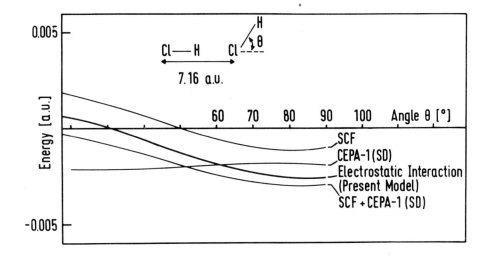

Figure 3. Energy contributions to the interaction
energy as a function of θ.

Since electrostatic forces play an important role we first
of all have to find an accurate representation of these
effects. For this purpose we computed the electrostatic
potential of an isolated HCl and fitted a point charge model to reproduce this potential. Our best representation
was obtained for a model where charges are located on the
atoms H, Cl and a dummy center D, with the following charges and distances:

```
                    1.0              2.41
               D--------Cl-----------H
charges q:     0.506    -0.909       0.403
```

Since HCl has a large quadrupole moment it is quite expected that a three charge model is necessary. In our experience it is furthermore important to put the third center D
as shown here, i.e. near chlorine but outside the bond.
Since our point charge model was fitted to reproduce the
global electrostatic potential of HCl, we are confident to
get a reliable description of the latter and not only of
effects arising solely from dipole or quadrupole moment
interactions, see also the discussion in section 3.5.

Our final fit is of the form
$$V = V_1 + V_2 \tag{1}$$
where V_1 denotes the electrostatic interaction

$$V_1 = \sum_{i,j} q_i \, q_j \, / \, R_{ij} \tag{2}$$

and i,j run over the centers D, H, Cl of the two HCl molecules, respectively, and R_{ij} denotes the corresponding intermolecular distances. V_2^{ij} reads (in a.u.)

$$V_2 = 0.1322 \quad \exp \, (-1.9 \, R_{HH}) +$$

$$+ 113.1362 \quad \exp \, (-1.6 \, R_{ClCl}) \, -7219.288/R_{ClCl}^9 \tag{3}$$

$$+ \, 15.1689 \quad \exp \, (-1.9 \, R_{HCl'}) \, - \quad 7.883/R_{HCl'}^5$$

$$+ \, 15.1689 \quad \exp \, (-1.9 \, R_{H'Cl}) \, - \quad 7.883/R_{H'Cl}^5$$

$$- \, 1.0889 \quad \exp \, (-1.3 \, R_{DD'})$$

in an obvious notation. We did not succeed to represent V_2
by an atom-atom potential involving H and Cl atoms only.
The fact is simply that HCl has not only a relatively large
quadrupole moment but is a rather "quadrupolar molecule"
which requires to include at least $R_{DD'}$ in the intermolecular potential.
Although we cannot discuss the details of eqs.(1-3) here,
we note two gratifying feathures of V:
(a) it is of a simple form and
(b) it conforms with physical intuition.

     In the development of V_2 by means of a least squares
fit it was of great help to use $|V|^{-1}$ as a weight function.
Least squares fits usually give a good representation near
end points (in the present case small R's, with a $|V|^{-1}$
weighting we enforce in addition a good description of
asymptotes and nodes of V, and it was no serious problem
to get the minima right too.

     The analytical representation (1-3), reproduces the
location of minima of the one dimensional curves discussed
above with an error of 0.1 a_0, and the corresponding binding
energies with an error of 100 - 300 µH.

     Fig. 4 gives the contour lines of V for a collinear
ClH...Cl arrangement with R_{ClCl} and θ ($\theta = \sphericalangle$ H...ClH) as
variables. We find only a single minimum in this plane at
$R_{ClCl} = 7.10 \, a_0$, $\theta = 90.8°$, $V_0 = -3310$ µH. Further calculations showed that a variation of the angle = ClH...Cl lowe-
red V_0 in the µH range only.

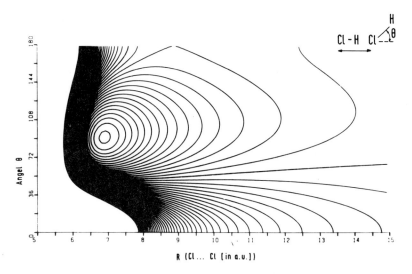

Figure 4. Contour lines of V in the r,θ-plane.

3.5. Comparison with other (HCl)₂ pair-potentials

Mc Donald et al. (4) have recently compared and investigated four pair-potentials, denoted model A - D, which were
either taken from previous publications, model A (19), or
developed by these authors, model B, C, D.

For a comparison of model A - D with the present V we
have plotted for two typical arrangements the corresponding
potentials as a function of intermolecular distances, see
Fig. 5a and 5b. Agreement with our result is best for model
D, also favoured by Mc Donald et al. However, it should be
noted that all curves A - D show marked deviations from
extended quantum mechanical calculations, such as a shift
of repulsive parts by as much as 1 - 2 a_o, see Fig. 5a.

The crucial point of the whole business seems to be
the location of the dummy center, the presence of which
seems to be indispensable. McDonald et al. have put the
third center between the two atoms, but we did not succeed
to get good representations of V_1 and V_2 unless D was located as described above. In the following table we give
for comparison dipole to hexadecapole moments resulting
from the two electrostatic models in question.

	Mc Donald et al.	present
μ (D)	1.04	1.18
Θ (DÅ)	3.37	3.83
Ω (DÅ²)	5.00	3.65
φ (DÅ³)	6.54	5.31

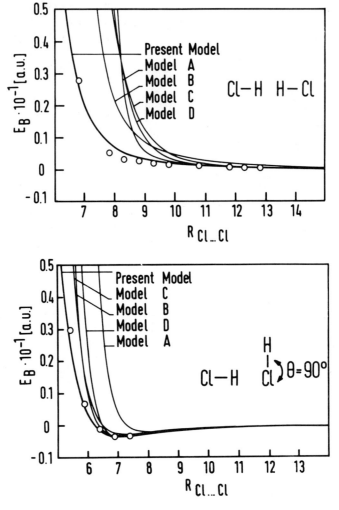

Figure 5a,b. Comparison of the present V with
            models A - D of ref. (4).

3.6. A comparison with (HF)$_2$

Extended basis set ab initio computations have been performed by Lischka (20) for the linear hydrogen bonded (HF)$_2$
configuration. An even superficial consideration reveals
marked differences between (HF)$_2$ and (HCl)$_2$.
HF has a much larger dipole moment, $\mu=1.82$ D (21), and a
smaller quadrupole moment, $\theta=2.6$ DÅ (22), than HCl. This
and the much more pronounced effects of hydrogen bonding in
HF lead to quite different features of intermolecular interactions, the most important of which is the minor role of

correlation (dispersion) effects in $(HF)_2$, which are of
great importance for $(HCl)_2$. One should thus proceed with
extreme care in carrying over results from one system to
the other, which is in agreement with all chemical experience.

4. THE SECOND VIRIAL COEFFICIENT B(T) of HCl

The second virial coefficient B(T)

$$B(T) = N_L \int (1-\exp (-V/RT)) d\tau \qquad (4)$$

is determined by the corresponding intermolecular pair
potential and provides a sensitive test for V. In Fig. 6
we give the reliable experimental data of Schramm et al.(1)
together with theoretical results obtained from the present
V, eq. (1 - 3), and those published by Mc Donald et al. In
a comparison of these curves one has to keep in mind that
the present V describes the interaction of rigid HCl molecules, i.e. we neglect effects of intramolecular relaxation
and vibrational averiging and excitation on B(T), which are
not easily estimated in a quantitative way. However, we
note the much better agreement of our computed B(T) with
experiment as compared to models A - C of Mc Donald et al.
Their model D is in almost perfect agreement with experiment, but these authors note: "... the excellent fit ...
is not particularly significant in view of this ambiguity
in handling the "switching function" ...".

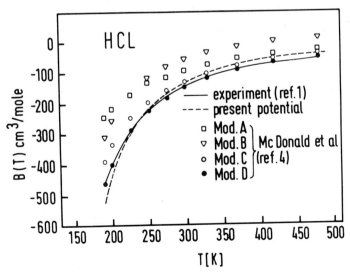

Figure 6. Comparison of experimental and theoretical B(T)

Eq. (4) for B(T) shows a crucial dependence on V, and
a detailed investigation reveals in fact that only minor
errors in V -in the order of a few percent or 100 - 200 µH
near the equilibrium geometry- are required to explain the
deviation of the present B(T) from experiment. If one is
aiming for higher accuracy than achieved for B(T) now, one
clearly has to employ an even more sophisticated method of
computation and a more accurate fit for V.

5. RESULTS OF MD CALCULATIONS

Let us finally describe an application of the pair-potential V developed in this work in a study of the structure
of dense HCl gas. Dr. A. Geiger from our institute has performed MD computations (at $\rho = 0.176$ g/cm^3, T = 351.5 K,
216 molecules in the periodicity cell), since experimental
neutron diffraction results by Soper and Egelstaff (2) are
available for comparison in this case. In Fig. 7a, 7b, 7c
we give our theoretical intermolecular partial structure
factors $(S_{\alpha\beta}(Q)-1)$ with the corresponding experimental ones.
A detailed comparative discussion of these results is not
easy, mainly due to the large error bars in the neutron
diffraction results. These comprise, as far as we understand it, statistical errors only -evaluation of 5 experiments for 3 curves- but do not account for systematic
effects.

The good agreement between theory and experiment is
obvious for the Cl...Cl structure factor -which shows the
best resolution- whereas there could be a deviation in the
$S_{HCl}(Q)$, allthough it is hard to tell whether this is significant. We further note that the theoretical evaluation
of $S_{\alpha\beta}(Q)$ has to cope with increasing difficulties for
$Q < 1\text{Å}^{-1}$.
Further MD simulations for higher densities are in
progress.

6. CONCLUSIONS

Our investigations first of all show that an accurate computation of intermolecular potentials requires a proper
description of all interaction mechanisms, in the case of
(HCl)$_2$ especially the Pauli-Principle repulsion, electrostatic and dispersion interactions. The relative importance
of corresponding contributions changes from case to case
-e.g. relative orientations and/or properties of contributing molecules- as discussed in sec. 3.3.
For the fit to an analytical potential it turned out
to be important to start from a good description of the

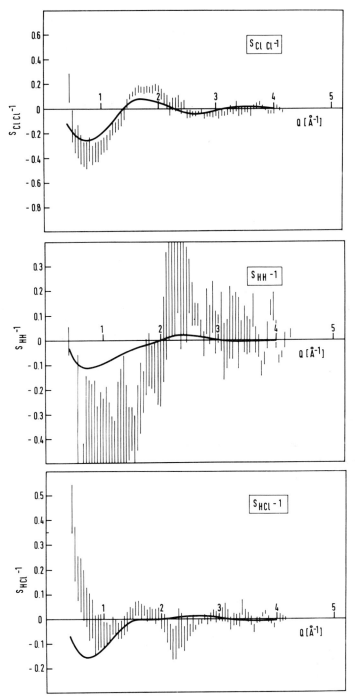

Fig. 7. Comparison of experimental and computed structure factors $S_{\alpha\beta}(Q)-1$.

electrostatic interaction. For this purpose the procedure described in sec. 3.4. - we used a point charge model - was successful. The location of the point charges provides useful hints to develop the final potential.

Our extended quantum chemical calculation provides a good insight into the structure of intermolecular forces, and the comparison of results obtained for the virial coefficient and structure factors with experiment and/or previous theoretical work is encouraging.

ACKNOWLEDGEMENTS

We are indebted to Dr. A. Geiger for making his MD results available before publication and also for valuable discussions. The Ar...HCl computations have been performed by R. Becherer. All computations were performed at the "Rechenzentrum der Universität Karlsruhe". This work was partly supported by the "Fonds der Chemischen Industrie".

REFERENCES

(1) Schramm, W., Leuchs, U.: 1979, Ber.d.Bunsengesellschaft
     83, 847
(2) Soper, A.K., Egelstaff, P.A.: 1980, Mol.Phys. 39,1201
(3) Kollman, P.: 1977, J.Am.Chem.Soc. 99, 4875
(4) Mc Donald, I.R., O'Shea, S.F., Bounds, D.G., Klein,
     M.L.: 1980, J.Chem.Phys. 72, 5710
(5) The present method is described in (5a), for recent
     review and further articles see (5b)
     (5a) Zirz, C., Ahlrichs, R.: "recent developments in
          coupled pair theories" in Electron Correlation,
          proceedings of the Dares Lury Study Weekend
          17 - 18 November 1979, Ed. Guest, M.F., Wilson,S.
     (5b) Meyer, W.: 1972, J.Chem.Phys. 58, 1017
     (5c) Kutzelnigg, W.: "Pair Correlation Theories" in
          Modern Theoretical Chemistry Vol. III, Ed.
          Schaefer III, H.F. (New York, Plenum Press 1977)
     (5d) Ahlrichs, R.: 1979, Computer Phys. Comm. 17, 31
(6) March, N.H., Young, W.H., Sampanther, S.: New York
     1977, "The Many Body Problem in Quantum Mechanics",
     Cambridge University Press
(7) Roos, P.O., Siegbahn, P.E.M.: New York, Plenum Press
     1977, "The Direct Configuration Interaction Method
     from Molecular Integrals" in Modern Theoretical
     Chemistry Vol. III, Ed. Schaefer III, H.F.
(8) Meyer, W.: 1975, J.Chem.Phys. 64, 2901
(9) Pople, J.A., Seeger, R., Krishnan, R.: 1977, Int.J.
     Quant.Chem. 511, 149

(10) Chang, T.C., Habitz, P., Pittel, B., Schwarz, W.H.E.: 1974, Theor. Chim. Acta 34, 263
(11) Ahlrichs, R., Votava, C. unpublished results
(12) Huzinaga, S.: 1971, "Approximate Atomic Functions I/II", Department of Chemistry of the University of Alberta, Canada
(13) Rank, D.H., Rao, B.S., Wiggins, T.H.: 1965, J.Mol. Spectrosc. 17, 122
     Smith, F.G.: 1973, J. Quant. Spectrosc. Radiat. Transfer 13, 717
     De Leeuw, F.H., Dymanus, A.:1973, J.Mol.Spectrosc. 48, 427 see also ref. 22
(14) Werner, H.J., Rosmus, P.: 1980, J.Chem.Phys. 73, 2319
(15) Holmgren, S.L., Waldman, H., Klemperer, W.: 1978, J.Chem.Phys. 69, 1661
     Danker, A.M., Gordon, R.U.: 1976, J.Chem.Phys. 64, 354
     Miziolek, A.W., Pimentel, G.C.: 1976, J.Chem.Phys. 65, 4462
     Farrar, J.M., Lee, Y.T.: 1974, Chem.Phys. Letters 26, 428
     Novick, S.E., Davies, P., Harris, S.J., Klemperer, W.: 1973, J.Chem.Phys. 59, 2273
     Neilsen, W.B., Gordon, R.G.: 1973, J.Chem.Phys. 58, 4149
     Rank, D.H., Sitaram, P., Glickman, W.A., Wiggins, T.A.: 1963, J.Chem.Phys. 39, 2673
(16) Diercksen, G.H.F., private communication
(17) Bridge, H.J., Buckingham, A.D.:1966, Proc.R.Soc. London Ser. A 295, 334
(18) Buckingham, A.D.: New York 1978, "Basic Theory of Intermolecular Forces: Applications to Small Molecules", in Intermolecular Interactions: From Diatomics to Biopolymers, ed. Pullmann, B., Wiley, J.
(19) Powles, J.G., Evans, W.A.B., McGrath, E., Gubbins, K.E., Murad, S.: 1979, Mol.Phys. 38, 893
(20) Lischka, H.: 1974, J.Am.Chem.Soc. 96, 4761
(21) Weiss, R.: 1963, Phys. Rev. 131, 659
(22) Stogryn, D.E., Stogryn, A.P.: 1966, Mol.Phys. 11, 371

"NEW" MOLECULAR BOUND AND RESONANCE STATES

Phillip R. Certain
Theoretical Chemistry Institute
Department of Chemistry
University of Wisconsin
Madison, Wisconsin 53706 U.S.A.

Nimrod Moiseyev
Department of Chemistry
Technion-Israel Institute
 of Technology
Haifa, Israel

It has been shown previously that a new type of molecular bound state can result from corrections to the Born-Oppenheimer approximation. It is shown here that "new" bound and resonance states can occur in a variety of problems involving coupled wave equations. Several examples are discussed, the most important one involving the bound and resonance states of van der Waals complexes.

A. INTRODUCTION

Some years ago, it was pointed out that non-adiabatic coupling between two diatomic electronic states can lead to "new" bound vibrational states (1,2). If a purely repulsive potential energy curve which supports no bound states is coupled in a particular way to an excited potential which does support bound states, a new bound state can occur with an energy below the dissociation limit of the repulsive curve. Apparently there have been no experimental observations of such a state, and indeed, the conditions for observing and recognizing its effects in spectral data are very stringent (2,3).

New bound states may arise not only in systems which violate the Born-Oppenheimer separation, but also in scattering processes involving a single Born-Oppenheimer potential surface. We discuss this situation in this paper, and show that concomitant with the new bound states are "new" resonance states in the continuum. In addition, we discuss the possibility of new resonance states due to non-adiabatic coupling of two repulsive Born-Oppenheimer potentials, neither of which supports bound states. "New" resonance may occur, particularly when the interaction between them is localized and strong. The coupling of continuum states resulting in a bound state is well-known in the theory of superconductivity (4).

Of course, a state of a system can be designated as "new" only with reference to an approximate description of the system in which the state in question is absent in the first approximation. When it

B. Pullman (ed.), Intermolecular Forces, 149–160.

appears at a higher level of approximation, it is termed "new".

The descriptions we are considering are based on sets of coupled wave equations which result from averaging a complete wave equation over all degrees of freedom but one of special interest. Denoting this special degree of freedom by R and all others by r, the assumed form of the exact wavefunction is

$$\Psi_\alpha(R,r) = \sum_n \chi_{n\alpha}(R) \, \phi_n(r,R) \qquad (1)$$

where the ϕ_n are some fixed basis functions which are complete and orthonormal in the variables r, and the $\chi_{n\alpha}$ are functions to be determined from the wave equation. For simplicity, we shall assume that the ϕ_n are real functions. In the Born-Oppenheimer approximation for a diatomic molecule, the r are the electronic coordinates and R is the internuclear separation. In the scattering problem on a single Born-Oppenheimer potential, the R is the intermolecular separation, and the r are the internal degrees of freedom (rotations, vibrations) of the scattering molecules plus the overall rotational degrees of freedom of the complex.

Substitution of the wavefunction Ψ_α into the wave equation leads to a set of coupled equations for the functions $\chi_{n\alpha}$,

$$(\hat{T}_R + V_{nn}(R) - E_\alpha)\chi_{n\alpha}(R) + \sum_m V_{nm}(R) \, \chi_{m\alpha}(R) = 0 \qquad (2)$$

where $\hat{T}_R$ is the kinetic energy operator for R and $V_{nm}(R)$ is the matrix element of the Hamiltonian in the basis $\phi_n(r,R)$.

If all of the off-diagonal elements V_{nm} are small, the essential features of the problem are revealed by neglecting them completely and solving the resulting uncoupled equations. The "ordinary" bound states are obtained at this level of approximation. "New" bound states may result when the off-diagonal elements are considered, as discussed previously (1-3) and in the next section of the present paper.

The form of the wavefunction, Eq. (1), is not unique (5-7), however, since it is invariant to an R-dependent orthogonal transformation of the ϕ_n. Such a transformation leads to coupled equations of the same form as Eq. (2), but with redefined potentials V_{nm} and hence a different definition of what are "ordinary" and what are "new" bound states. Thus, "new" bound states are new only with respect to a particular choice of basis functions ϕ_n, and may be recast as "ordinary" bound states by a proper choice of basis.

B. "NEW" BOUND STATES

In this section we review the analysis (2,3) which provides the condition for a "new" bound state to appear at a particular stage in a sequence of approximate solutions to Eq. (2).

We assume that the first level of approximation is to neglect all off-diagonal potentials in Eq. (2). This gives a set of uncoupled equations,

$$(\hat{T}_R + V_{nn}(R) - E^0_{n\alpha})\chi^0_{n\alpha}(R) \; = \; 0 \qquad\qquad (3)$$

which may be solved for a set of eigensolutions $(E^0_{n\alpha}, \chi^0_{n\alpha})$ for each potential $V_{nn}(R)$. For convenience of notation, we shall assume a box normalization $[0 \leq R \leq R_{max}, \chi_{n\alpha}(R_{max}) = 0]$, so that the continuous spectrum of $(\hat{T}_R + V_{nn}(R))$ is represented by a closely-spaced sequence of eigenvalues. At the end of the discussion we may consider the limit $R_{max} \to \infty$. The case we wish to consider is when the lowest potential $V_{00}(R)$ [i.e. $V_{00}(R_{max}) \leq V_{nn}(R_{max})$, n > 0] supports no bound states, $E^0_{n\alpha} \geq V(R_{max})$, all α, as shown in Figure 1.

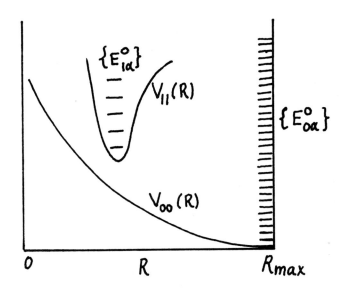

Figure 1.   Potential Curves and Eigenenergies for Discussion of New Bound States.

Each of the sets of eigenfunctions $\{\chi^0_{n\alpha}(R)$, n fixed, $\alpha = 0,1,2,...\}$ is assumed to provide a complete basis, so that in higher levels of approximation where the off-diagonal potentials in

Eq. (2) are not neglected, each $\chi_{n\alpha}$ may be expanded in terms of the corresponding set $\{\chi_{n\alpha}^0\}$,

$$\chi_{n\alpha}(R) \;=\; \sum_{\beta} c_{n\beta}^{\alpha}\, \chi_{n\beta}^{0}(R) \; . \qquad (4)$$

These expansions transform the coupled wave equations (2) into a set of coupled algebraic equations,

$$(E_{n\beta}^{0} - E_{\alpha})c_{n\beta}^{\alpha} \;=\; -\sum_{m\gamma} V_{n\beta,m\gamma}\, c_{m\gamma}^{\alpha} \qquad (5)$$

where

$$V_{n\beta,m\gamma} \;=\; <\chi_{n\beta}^{0}\,|\,V_{nm}\,|\,\chi_{m\gamma}^{0}> \; . \qquad (6)$$

A sequence of approximations may be defined in which more and more terms are included in the summation in Eq. (5).

At any level of approximation, the solution of the set of Eqs. (5) is conveniently discussed in terms of familiar partitioning techniques (8). At the second level of approximation, only two sets, n = 0 and 1, are included in Eqs. (5),

$$(E_{0\beta}^{0} - E_{\alpha})c_{0\beta}^{\alpha} \;=\; -\sum_{\delta} V_{0\beta,1\delta}\, c_{1\delta}^{\alpha} \qquad (7a)$$

$$(E_{1\gamma}^{0} - E_{\alpha})c_{1\gamma}^{\alpha} \;=\; -\sum_{\beta} V_{1\gamma,0\beta}\, c_{0\beta}^{\alpha} \; . \qquad (7b)$$

Solving the first equation for $c_{0\beta}^{\alpha}$ and substituting the result into the second yields

$$(E_{1\gamma}^{0} - E_{\alpha})c_{1\gamma}^{\alpha} - \sum_{\beta\delta} V_{1\gamma,0\beta}(E_{0\beta}^{0} - E_{\alpha})^{-1}\, V_{0\beta,1\delta}\, c_{1\delta}^{\alpha} \;=\; 0 \; . \qquad (8)$$

The eigenvalue condition for E_{α} is that the determinant of coefficients of the $c_{1\gamma}^{\alpha}$ vanishes, which may be written

$$\det\{\delta_{\gamma\delta}(E_{1\gamma}^{0} - E_{\alpha}) - \sum_{\beta} V_{1\gamma,0\beta}(E_{0\beta}^{0} - \mathcal{E})^{-1}\, V_{0\beta,1\delta}\} \;=\; 0 \qquad (9a)$$

and

$$E_\alpha(\mathcal{E}) = \mathcal{E} . \qquad (9b)$$

The general properties of the functions $E_\alpha(\mathcal{E})$ are well-known (8). For $\mathcal{E}$ large and negative each E_α approaches one of the eigenvalues $E^0_{1\gamma}$. In addition each E_α is a monotonically decreasing function of $\mathcal{E}$ with poles at the eigenvalues $E^0_{0\beta}$, which in the present application are closely spaced points on the positive energy axis representing the continuum of eigenvalues of $(\hat{T}_R + V_{00}(R))$. Hence a graphical solution of Eq. (9) has the qualitative appearance shown in Figure 2

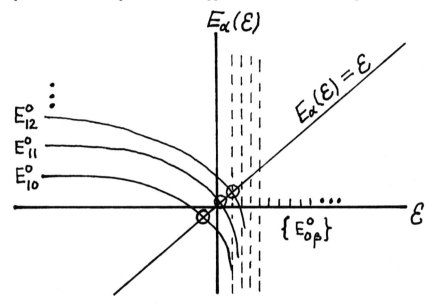

Figure 2.   Qualitative behavior of $E_\alpha(\mathcal{E})$. The circles denote points at which the eigenvalue condition, Eq. (9), is satisfied.

"New" bound states correspond to eigenvalues $E_\alpha < 0$. It is clear from Figure 2 that for any finite R_{max} the number of such states is given by the number of negative eigenvalues $E_\alpha(0)$; that is, the number of negative roots of the secular equation,

$$\det\{\delta_{\gamma\delta}(E^0_{1\gamma} - E_\alpha(0)) - \sum_\beta V_{1\gamma,0\beta} V_{0\beta,1\delta} / E^0_{0\beta}\} = 0 , \qquad (10a)$$

$$E_\alpha(0) < 0 . \qquad (10b)$$

These two equations are the general conditions for "new" bound states to appear at the second level of approximation. In principle we should consider the limit $R_{max} \to \infty$; however, since each eigenvalue $E_\alpha(\mathcal{E})$ is expected to be a decreasing function of R_{max}, the number of "new" bound states for a finite R_{max} is an upper bound to the true number.

Further analysis depends on the particular nature of the problem under consideration, as this determines the magnitudes of the matrix elements appearing in Eq. (10). Previous analysis (2,3,7) has focussed on problems involving the Born-Oppenheimer separation and will not be repeated here. In the following sections, examples such as arise in scattering problems will be discussed.

General qualitative conditions for "new" bound states can be given from an examination of Eq. (10). If the coupling matrices $V_{1\alpha,0\beta}$ are small, it is clear that E_{10}^0 must also be small (i.e. close to threshold). If there are a number of small eigenvalues $E_{1\gamma}^0$, or if the coupling elements are large, then there may be a number of "new" bound states. This apparently does not happen in Born-Oppenheimer problems (2), but it may in scattering problems. Finally, if the set of eigenvalues $E_{1\gamma}^0$ also are a discrete representation of a continuum, "new" bound states may appear which persist in the limit $R_{max} \to \infty$.

We turn now to a discussion of a few examples. Although these are model problems, they exhibit the qualitative behavior found in real scattering problems.

C. EXAMPLES

a. A Spin Problem

It is instructive to consider first an exactly soluble model problem corresponding to the idealized Stern-Gerlach experiment illustrated in Figure 3.

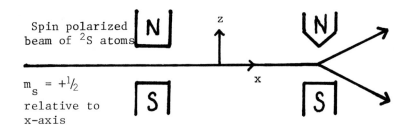

Figure 3. Modified Stern-Gerlach Experiment. The left magnetic field is homogeneous, while the second (Stern-Gerlach) field is inhomogeneous.

A polarized beam of paramagnetic ^2S atoms whose spins are quantized (m_s = +1/2) with respect to the laboratory x-axis enters a region of homogeneous magnetic field oriented in the z-direction. With respect to the z-axis of quantization, the beam contains equal components of α and β spin. For the α-component, the homogeneous magnetic field presents a square barrier and the scattering of the wave packet will show reflection, transmission, and for certain energies, resonances. For the transmitted α-component there will be a time-delay which may be long in the vicinity of resonances. For the β-component, the homogeneous field presents a square well which may support a number of bound states, and the transmitted component will be accelerated, particularly in the vicinity of resonances. Thus, as the beam emerges from the homogeneous field region, it will have unequal α and β spin components with respect to the z-axis of quantization, so the Stern-Gerlach magnet will split the beam into two, with the relative intensities of the two beams depending on the kinetic energy of the beam and the strength of the homogeneous magnetic field.

The quantum mechanical description of this spin scattering experiment involves solving the wave equation for the stationary state of the Hamiltonian for the system in the region of homogeneous magnetic field.

$$\hat{H} = (p^2/2m) + \lambda \vec{B} \cdot \vec{S} \tag{11}$$

where the first term is the atomic kinetic energy and the second the magnetic interaction of the spin with the homogeneous field. In the central region

$$\vec{B} = B(x)\hat{z} \tag{12}$$

where $B(x)$ vanishes except between the pole faces of the central magnet, where it has a constant value. If the stationary state eigenfunctions of H are expressed in terms of spin-components quantized along the x-axis, the wave equation is equivalent to two coupled equations

$$(\frac{p^2}{2m} - E)\psi_\alpha^x = \lambda B \ \psi_\beta^x \tag{13a}$$

$$(\frac{p^2}{2m} - E)\psi_\beta^x = \lambda B \ \psi_\alpha^x . \tag{13b}$$

With this quantization, ψ_α^x and ψ_β^x may be viewed as two free particle scattering states which are coupled by a localized interaction $\lambda B(x)$. The fact that this coupling can give rise to "new" bound and resonance states is obvious if instead we express the wavefunction in terms of spin-components quantized along the z-axis. Then the two spin-components are uncoupled and satisfy the equations

$$(\frac{p^2}{2m} + 1/2 \ \lambda B(x) - E)\psi_\alpha^z = 0 \qquad (14a)$$

$$(\frac{p^2}{2m} - 1/2 \ \lambda B(x) - E)\psi_\beta^z = 0 \qquad (14b)$$

the solutions of which are discussed in many textbooks (9). The second equation may possess a number of bound states (10) depending on the strength and width of the magnetic field, while both equations exhibit resonances for positive energies.

Thus, the existence of "new" bound and resonance states is revealed in this spin example by a unitary transformation of the hamiltonian which diagonalizes the potential matrix. A special feature of this model is the fact that the unitary transformation commutes with the kinetic energy operator so that the transformed equations are completely uncoupled. In the general case, the transformed equations, which exhibit the "new" bound states explicitly, are a better first approximation than the original set.

b. Atom-Diatomic Molecule Scattering

A common class of problems where "new" bound states can occur involves atom-diatom scattering on a single Born-Oppenheimer potential surface. It is conventional in such problems to represent the surface in terms of Legendre polynomials,

$$V(R,\theta) = \sum_{\ell=0} V_\ell(R) \ P_\ell(\cos\theta) \qquad (15)$$

where the variables are defined in Figure 4.

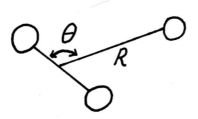

Figure 4.   Variables for Atom-Diatom Scattering.

The scattering wavefunction is obtained as a solution to a set of coupled equations of the form of Eq. (2), where the variables r which have been averaged correspond to the vibration and rotation of the diatomic, and the rotation of the triatomic complex as a whole. The potential terms $V_{nm}(R)$ are linear combinations of the $V_\ell(R)$ which result from requirement that the wavefunction be an eigenfunction of total angular momentum.

For example, for a collision of a ^1S atom with a polar $^1\Sigma$ molecule, the simplest potential has only two terms,

$$V(R,\theta) = V_0(R) + V_1(R)\cos\theta \qquad (17)$$

and the coupled equations for zero total angular momentum have the form

$$\left(\frac{P_R^2}{2\mu} + V_{jj}(R) - E\right)\chi_j(R) + a_j V_1(R)\chi_{j+1} + b_j V_1(R)\chi_{j-1} = 0 , \qquad (18)$$

$$j = 0,1,2,\ldots$$

where

$$V_{jj}(R) = V_0(R) + j(j+1)\left[\frac{1}{2\mu R^2} + \frac{1}{2I_d}\right] \qquad (19a)$$

$$a_j = (j+1) / \sqrt{(2j+1)(2j+3)} \qquad (19b)$$

$$b_j = a_{j-1} . \qquad (19c)$$

In these equations, μ is the reduced mass of the atom-diatomic pair and I_d is the moment of inertia of the diatomic (assumed to be rigid).

If the spherical potential $V_0(R)$ is purely repulsive, all of the potentials $V_{jj}(R)$ will be purely repulsive, so that if the anisotropy $V_1(R)$ is neglected, there will only be scattering solutions to Eq. (18). On the other hand, even if $V_0(R)$ is repulsive it is quite possible for the total potential Eq. (17) to have a well at one end of the diatomic which is deep enough to support bound states. These are the vibrational states of the triatomic van der Waals complex. In this case the bound state solutions are obtained as a result of the terms in Eq. (18) which couple purely repulsive potentials.

There is a broad class of van der Waals molecules described by intermolecular potentials similar to Eq. (17). The present analysis suggests that it is inappropriate to use a scattering-like basis $\phi_n(r)$ in the treatment of the bound states of these molecules, and that angular momentum decoupling procedures appropriate to scattering problems should be used with care.

c.   "New" Resonance States

So far we have discussed mainly the appearance of "new" bound states. We now give a simple example to show that "new" resonance states are expected to accompany the occurrence of "new" bound states. Consider the pair of coupled equations

$$[\frac{p^2}{2m} + V_{00}(R) - E]\chi_0(R) + V_{01}(R)\chi_1(R) = 0 \tag{20a}$$

$$[\frac{p^2}{2m} + V_{11}(R) - E]\chi_1(R) + V_{01}(R)\chi_0(R) = 0 \tag{20b}$$

where $V_{00}(R)$, $V_{01}(R)$, and $V_{11}(R)$ have the qualitative appearance shown in Figure 5a

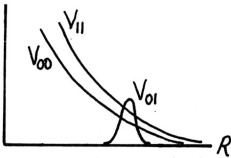

Figure 5a.   Potential Functions appearing in Eq. (20).

These equations may arise, for example, by truncating the equations for our atom-diatom scattering problem such as was discussed previously.

If we transform the basis, for example by the R-independent transformation,

$$\chi_0' = (\chi_0' + \chi_1')/\sqrt{2} \tag{21a}$$

$$\chi_1' = (\chi_0' - \chi_1')/\sqrt{2} \tag{21b}$$

then we obtain new potentials

$$V'_{00} = 1/2(V_{00} + V_{11}) - V_{01} \qquad (22a)$$

$$V'_{11} = 1/2(V_{00} + V_{11}) + V_{01} \qquad (22b)$$

$$V'_{01} = 1/2(V_{00} - V_{11}) \qquad (22c)$$

which have the qualitative behavior shown in Figure 5b

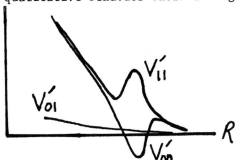

Figure 5b.   Transformed Potentials Resulting from the Transformation
Eq. (21).

If the equations are decoupled by neglecting V'_{01}, the potential V'_{11} can have shape resonances which persist even when V'_{01} is subsequently included. When these equations are appropriate for treating van der Waals molecules, the "new" resonance states are presumably associated with the temporary trapping of the scattering atom in the potential well of the diatomic.

D.   SUMMARY

We have seen that "new" bound and resonance states may arise in a variety of problems involving sets of coupled wave equations. The identification of a state as "new" results from a particular choice of approximate wavefunction, and a "new" state may be transformed into an "ordinary" one by the proper change of basis. Although "new" states resulting from deviations from the Born-Oppenheimer separation are expected to be extremely rare, they may be more common in molecular scattering events.

Acknowledgement: Support of this research by the National Science Foundation (US) and the University of Wisconsin Graduate School is gratefully acknowledged.

References

1. Riess, I.: 1970, J. Chem. Phys. 52, p. 871.

2. Rosenfield, J., Voigt, B., Mead, C. A.: 1970, J. Chem. Phys. 53, p. 1960.

3. Gelbart, W. M., Jortner, J.: 1971, J. Chem. Phys. 54, p. 2070.

4. Cooper, J. N.: 1956, Phys. Rev. 104, p. 1189.

5. Smith, F. T.: 1969, Phys. Rev. 179, p. 111.

6. Levine, R. D., Johnson, B. R., Bernstein, R. B.: 1969, J. Chem. Phys. 50, p. 1694.

7. Atabeck, O., Lefebve, R.: 1973, J. Chem. Phys. 59, p. 4145.

8. Löwdin, P. O., in "Perturbation Theory and Its Applications in Quantum Mechanics", Wilcox, C. H., ed. (John Wiley & Sons, New York, 1965), pp. 255-294.

9. Messiah, A., "Quantum Mechanics", Vol. I (John Wiley & Sons, New York, 1961), Chapter III.

10. Of course, these states are unbound in the suppressed y-direction.

COMPARISON BETWEEN ACCURATE AB INITIO AND ELECTRON GAS POTENTIAL ENERGY SURFACES

M.C. van Hemert
Gorlaeus Laboratories, department of Physical Chemistry,
University of Leiden, P.O.Box 9502, 2300 RA Leiden,
The Netherlands

INTRODUCTION

The ab initio calculation of the interaction energy with an accuracy sufficient to reproduce many experimental observations appears to be possible only for systems with very few electrons. The recent potential energy surfaces for the He-H_2[1] and H_2-H_2 [2] systems obtained by Meyer and coworkers, clearly form a landmark. For systems with more electrons there are many difficulties to overcome. A commom problem in all ab initio calculations is the computer time, needed for integral evaluation, that goes up with the fourth power of the size of the basis set, which is itself more or less proportional to the number of electrons in the system.

The need for potential surfaces for systems containing many electrons and having many internal and external degrees of freedom, has therefore resulted in the use of simpler methods. Among these methods the electrongas approximation (EGA) as introduced by Gordon and Kim [3], has gained popularity. The attractive aspects of this method are that it is 100 to 1000 times faster than ab initio methods and that it does (at least in principle) not contain adjustable or empirical parameters. The EGA potential energy surfaces did however not always compare well with experimental or ab initio surfaces. Part of the discrepancies could be removed by making improvements in the formalism. Furthermore, also the generally somewhat older experimental or ab initio surface were not always reliable. Therefore it is useful to analyze once more the results of the EGA, by making comparisons with the most recent accurate ab initio potential energy surfaces, necessariy still for systems with a limited number of electrons.

In the next section we will briefly review the theory of the original EGA and discuss some of the improvements. Since in its present form the EGA uses the undeformed charge distribution for the monomers, as obtained from ab initio SCF wavefunctions, we expect that the EGA potential without its (statistical) correlation contribution should equal the potential obtained from first order exchange perturbation theory, also using SCF wavefunctions for the monomers. We will give a series of examples where this proves to be the case, both for the isotropic and for the anisotropic part of the potential. We will show that useful complete potentials can be obtained be replacing the (statisti-

B. Pullman (ed.), Intermolecular Forces, 161–174.

cal) correlation energy contribution by a dispersion term obtained from
a multipole expansion. This is, of course, limited to the intermolecular distance region where the multipole expansion is valid.

THEORY

     In the statistical theory of atoms [4](i.e. the electron gas
approach) the energy of a system is expressed as the sum of Coulomb,
exchange, kinetic and correlation energy contributions. Each of those
terms is completely determined by the charge distribution $\rho(\vec{r})$ for that
system. The Coulomb energy expression can be easily derived from the
assumption that in every part of the charge distribution the behaviour
of the electrons is represented by plane waves for which only the lowest k vectors are allowed. The exchange term stems from the requirement
that also the Pauli principle should be satisfied.

$$E_{coul} = \tfrac{1}{2} \int \rho(\vec{r}) \frac{1}{|\vec{r}-\vec{r}'|} \rho(\vec{r}') d\vec{r} d\vec{r}' - \int \sum_k \frac{Z_k}{|\vec{R}_k-\vec{r}|} \rho(\vec{r}) d\vec{r} + \sum_{k<\ell} \sum \frac{Z_k Z_\ell}{|\vec{R}_k - \vec{R}_L|} \tag{1}$$

and

$$E_{exch} = - C_{ex} \int \rho^{4/3}(\vec{r}) d\vec{r} \quad , \quad C_{ex} = \tfrac{3}{4} (\tfrac{3}{\pi})^{1/3} \tag{2}$$

As usual $\vec{R}_k$ indicates the position of a nucleus with charge Z_k. It
should be noticed that (1) and (2) contain the so-called self Coulomb
and self exchange contributions respectively. These two terms have
equal absolute values but opposite sign, so in the sum of (1) and (2)
the self interactions are properly discarded. The use of the relation
between the value of the k vector and the momentum of the electron
($\vec{p}=\vec{k}$ h/2π) leads under the assumptions stated above to the expression
for the kinetic energy

$$E_{kin} = C_{kin} \int \rho^{5/3}(\vec{r}) d\vec{r} \quad , \quad C_{kin} = \tfrac{3}{10} (2\pi^2)^{2/3} \tag{3}$$

For the correlation energy a rather general expression is used

$$E_{corr} = \int \rho(r) \, g_{corr}(\rho(\vec{r})) d\vec{r} \tag{4}$$

Here $g_{corr}(\rho)$ can be considered as the correlation energy density, for
which different forms have been suggested for low and high density regions (see [4]).

     In order to find the density and thus the energy, the energy is
minimized as a function of $\rho(\vec{r})$. The resulting equation for $\rho(\vec{r})$ has
been known for a long time under the name of the Thomas-Fermi-Direc(TFD)
equation.

     A supermolecular approach for the determination of the interaction
energies using the TFD solutions, seems not well suited, since especial-

ly the tail of $\rho(\vec{r})$ appears to be badly represented in the solution of the TFD equation. Gordon and Kim [3] therefore used known Hartree-Fock densities instead of the TFD densities in the statistical expression for the energy. They defined the interaction energy by

$$\Delta E^{EG} = E^{EG}_{AB}(\rho_{A,SCF} + \rho_{B,SCF}) - E^{EG}_A(\rho_{A,SCF}) - E^{EG}_B(\rho_{B,SCF}) \tag{5}$$

So it is assumed that the charge density distribution in the supermolecule AB can be found from the sum of the underformed charge densities of the monomers A and B. For the various contributions to the interaction energy one thus has (we will drop the subscript SCF in the following)

$$\Delta E^{EG}_{coul} = \int [\rho_A(\vec{r}) - \sum_{k=A} \delta(\vec{r} - \vec{R}_k)] V_B(\vec{r}) d\vec{r} \tag{6}$$

$$\Delta E^{EG}_{exch} = -C_{ex} \int [(\rho_A + \rho_B)^{4/2} - \rho_A^{4/3} - \rho_B^{4/3}] d\vec{r} \tag{7}$$

$$\Delta E^{EG}_{kin} = C_{kin} \int [(\rho_A + \rho_B)^{5/3} - \rho_A^{5/3} - \rho_B^{5/3}] d\vec{r} \tag{8}$$

$$\Delta E^{EG}_{corr} = \int [(\rho_A + \rho_B) g_{corr}(\rho_A + \rho_B) - \rho_A g_{corr}(\rho_A) - \rho_B g_{corr}(\rho_B)] d\vec{r} \tag{9}$$

In (6) V_B stands for the potential resulting from the electrons and nuclei of molecule B. The speed of this electron gas approach for the interaction energies of molecular systems, which is called the Gordon-Kim (GK) method, results from the possibility to compute the integrals (6)-(9) in a highly efficient way using 3 dimensional quadrature [5].

In figures 2 and 3 we compare the interaction energy as obtained from the GK method with the experimental curves for He_2 and Ne_2. It is obvious that the potential is too little repulsive at short range and too little attractive at long range.

Of the terms that contribute predominantly at short range the Coulomb part is, of course, correct, since it is exactly equal to the equivalent term resulting from perturbation theory. Rae [6], however, stated that in the sum of Coulomb and exchange terms as given by equations (6) and (7) no proper cancelation of self interaction occurs. According to Rae the exchange contribution of each species (A, B and AB) should be scaled by a factor γ that depends only on the number of electrons N in that species. Rae obtained this factor by a comparison of the exact (statistical) expression for the exchange energy of an electron gas, given as a discrete sum over k vector pairs and the approximate integral expression appearing in (2). For the exchange contribution to the interaction energy one thus writes

$$\Delta E^{EG}_{exch} = -C_{ex} \int [\gamma(N_A + N_B)(\rho_A + \rho_B)^{4/3} - \gamma(N_A)\rho_A^{4/3} - \gamma(N_B)\rho_B^{4/3}] d\vec{r} \tag{10}$$

In order to avoid problems associated with incorrect asymptotic behaviour of this expression, Rae originally used the same correction

factor $\gamma(N_A+N_B)$ for all terms in the integral of (10). This has thus
the effect of multiplying the exchange energy from the GK method by a
factor depending only on the total number of electrons in the supermolecule. The curve of the correction factor as a function of the
number of electrons is given in figure 1. Later Rae [7] has argued,
using the theory of plasma oscillations, that it would be better to
choose the value of $\gamma(N_A)$ for all γ's in (10) whenever $N_A=N_B$ and replace $\gamma(N_A+N_B)$ by a sort of geometric mean of $\gamma(N_A)$ and $\gamma(N_B)$ elsewhere.
    Later also Waldman and Gordon modified the expression for the ex-
change contribution[8]. As Rae, they used a scaling that depends on the
number of electrons in the supermolecule. The expression for the
scaling factor, however, was found by determining the fraction of self
exchange (per valence electron) in the exchange energy as obtained
from Hartree-Fock calculations for a series of atoms and ions. For a few
systems the correction factors are indicated in figure 1. It has to be
remarked that Waldman and Gordon used the exchange correction in conjunction with a gradient correction for the kinetic energy.
    Quite recently Gazquez and Ortiz [9] derived a refined expression
for the exchange potential. This potential has the correct asymptotic
behaviour since it explicitly deals with the separate exchange and
self exchange portions. Furthermore it accounts for the finite number
of electrons. The derivation was based on the properties of the one and
two particle density matrix. In the resulting expression for the exchange energy there not only appears a self exchange term, but a scaled
electron gas term as well. In order to obtain a practical formula for
the exchange contribution to the interaction energy, when starting from
this expression, we still have to assume the same scaling factor for
A, B and AB. At this level of approximation the self exchange corrections cancel and the remaining scaling of the GK exchange energy (7) is
completely due to the finiteness of the charge distribution. The scaling
factor is given by

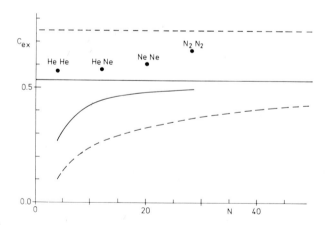

Figure 1. Exchange factors C_{ex} as function of the total number of electrons
in the system (N). ---- according to Rae [6], —— according to eq.(11).
Horizontal lines are the limits for an infinite number of electrons.

$$\gamma = \frac{110}{243} \cdot 2^{2/3} (1 - \frac{2}{N}) \tag{11}$$

where N is the total number of electrons in the supermolecule (see figure 1).

The correlation density functionals $g(\rho(r))$ have in the past been derived with the intention to improve the single determinantal wavefunction picture underlying the derivation of expressions (1)-(3). In practice it appears that non of the suggested functionals is able to describe the intermolecular correlation, i.e. the dispersion energy, correctly. We therefore decided to drop this contribution completely.

Since the Hartree Fock densities are used for the monomers and no deformation is allowed for, when forming the supermolecule (so there is no equivalent of induction), we assumed that the sum of the Coulomb exchange and kinetic energy terms ((6)-(8)) should equal the first order energy from (exchange) perturbation theory. In order to obtain a compete potential, of course, a second order energy has to be added.

In the next section we will analyze the results that we have obtained when making use of the modifications to the GK method as outlined above. We have used for all our calculations a slightly modified version of the Parker Pack program [10]. The SCF charge distributions were always derived from the best LCAO-SCF wavefunctions with very large Slater type basis sets.

POTENTIAL ENERGY SURFACE

Atom Atom interactions; He_2 and Ne_2.

In figures 2 and 3 we compare potential curves for the He_2 and Ne_2 systems respectively. In the left hand sides we focus on the comparison between various first order values. The data for the first order (exchange) perturbation theory curve for He_2 have been taken from refs. [11,12], where the zeroth order wave function was obtained as the product of highly accurate Hartree-Fock wavefunctions for the atoms. The curve resulting from the original GK expression (without correlation contributions) is completely incorrect since it is negative for R smaller than 4.8 Bohr. The modifications in the exchange energy contribution, as suggested by Rae, make the curve R somewhat too repulsive, it is overcorrecting. The Waldman and Gordon corrections (curve W) are clearly insufficient for R greater than 5 Bohr. Only the use of the exchange correction factor as defined by equation (11) gives good results over the whole distance range considered (curve M(modified)).

For Ne_2 we used the data of ref. [13] for the first order perturbation theory curve. The first order electron gas curve (GK) is again insufficiently repulsive when only the original GK terms are summed. On the average the correction factor from eq.(11) (curve M) is more effective than the Rae correction (curve R). The corrections from Waldman and Gordon lead to a curve (not drawn) between the GK and M curves. In the case of He_2 the complete potential (figure 2 right hand side) is found by adding to the first order energy a second order energy term, approximated by a multipole expansion using the ab initio

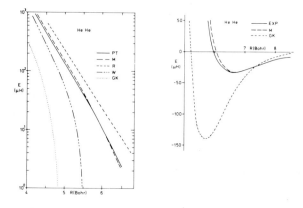

Figure 2. First order (a) and total (b) interaction energy for He$_2$. PT:
ab initio perturbation theory; M: electrongas using eq.(11); R: electrongas using Rae correction; W: electrongas with Waldman and Gordon correction;
GK: unmodified electrongas.

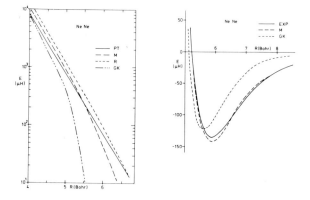

Figure 3. First order (a) and total (b) interaction energy for Ne$_2$.
Abbreviations as in figure 2.

(correlated) C_6 and C_8 coefficients from ref. [14]. At R=5.6 Bohr the
multipole expansion still accounts for 92% of the (correlated) dispersion energy. The neglect of correlation in the first order energy does
not matter that much in the case of He$_2$ [14], so it is not surprising
that the curve, computed as described above, agrees very well with the
experimental curve [15]. For Ne$_2$ empirical C_6, C_8 and C_{10} coefficients
have been taken from ref.[16]. Here also the complete potential energy
curve obtained from the modified electron gas method (curve M) agrees
nicely with the experimental curve [17]. The remaining differences are
largely caused by the use of the (severely truncated) multipole expansion at distances where the expansion is clearly no longer permitted.

Atom diatomic molecule interactions; He H$_2$, Ne H$_2$, He CO.

In order to be able to analyze in a systematic way both the distance (R) and the orientation (θ) dependence (the anisotropy) in the
two dimensional potential energy surfaces, we expand the interaction

energy in the usual form

$$E(R,\theta) = \sum_\lambda V_\lambda (R) P_\lambda (\cos\theta) \tag{12}$$

where P_λ is a Legendre polynomial. Thus our comparison will be concentrated on the expansion coefficients V_λ, which we obtain from a numerical integration procedure [18], rather than from the usual least squares fit. Since no extremely accurate first order perturbation theory data are available for these systems, we assume that a very good approximation is given by the results from SCF calculations, provided that they use extensive basis sets and take the basis set superposition error into account. For all 3 systems the induction contribution will probably be negligibly small.

In figures 4 through 6 we compare only the first order energy results. The V_0 and V_2 SCF curves for He H_2 and for Ne H_2 have been obtained from data in refs. [1] and [19] respectively, under the assumption that the potential is completely determined by these two coefficients. Again the modified electron gas curves (M) agree best with the ab initio results, especially for the anisotropy. In order to get an impression of the importance of the remaining discrepancies, we have included in the figures the absolute value of the dispersion energy. For He H_2 we used the unexpanded (correlated) dispersion energy from ref.[1]. From the combination of the modified electron gas term and the dispersion energy the following characteristic values for the

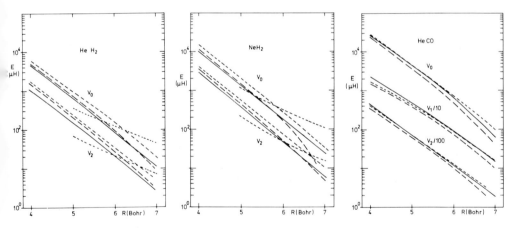

Figure 4, 5, and 6. Expansion coefficients for the first order energy.
——— ab initio SCF   - - - - electrongas using eq.(11); —————— electrongas using Rae correction; ..... absolute value of the dispersion energy. Figure 4 for He H_2, figure 5 for Ne H_2 and figure 6 for He CO.

well of V_0 are found: R_0=5.73 Bohr, R_m=6.48 Bohr, ε=44 µH, all within
1% of the suggested experimental values in ref.[1]. For V_2:R_0=6.2
Bohr, R_m=6.8 Bohr, ε=4.4 µH, also very close to the values from ref.
[1]. In passing we mention that the use of the (correlated) multipole
expansion coefficients from ref.[20], would reduce the characteristic
values by only a few percent, thereby making the well also a few percent deeper. Only for R values below 6 Bohr, the multipole expansion
results differ more than 10% from the unexpanded dispersion energy
term. For Ne H_2 we used the C_6, C_8 and C_{10} coefficients derived from
frequency dependent polarizabilities[21]. For the combined potential
(using the first order contribution M) we find for V_0:R_0=5.35 Bohr,
R_m=6.0 Bohr and ε=149 µH. Compared with the values from ref.[21], R_m
is too small and the well is too deep. This is partially caused by
a somewhat too little repulsive electron gas potential at long range,
partially by a too large dispersion contribution due to too large
$C_{6,8,10}$ coefficients and the neglect of penetration effects. For V_2
the combined potential curve gives characteristic values (R_0=6.05 Bohr,
R_m=6.80 Bohr and ε=10.5 µH) that are in very good agreement with those
listed in ref.[21]. Clearly penetration effects in the dispersion term
start to be important for R values below 6.5 Bohr.

For He CO the SCF curves for V_0, V_1 and V_2 from ref.[22] are
clearly best reproduced when the Rae correction,curve, R, is used.
The use of the exchange correction factor from eq.(11) results in
molecules that are about 0.1 Bohr softer than the SCF molecules. However, when an isotropic dispersion term, derived from C_6 and C_8 coefficients (based on the frequency dependent polarizability[23]) is added
to the first order energy, it is only for the M curve that one finds
the V_0 curve referred to as experimental in ref.[22]. (we find R_0=6.3
Bohr, R_m=7.1 Bohr, ε=85 µH). For the V_2 with λ=3 to 6 the agreement
between the R and the SCF values is similar to that for the lower λ
values.

Interactions between diatomic molecules: H_2 - H_2, N_2 - N_2, H_2 - CO.

A very suitable way to describe the 4-dimensional (or 6-dimen-
sional if the internal degrees of freedom are also considered) potential energy surface is given by the use of a spherical expansion in
terms of the angles describing the molecular orientation (θ_A,θ_B and ϕ.
see figure 7).
According to ref.[18] we can write

$$\Delta E(R,\theta_A,\theta_B,\phi) = (4\pi)^{3/2} \sum_{L_A} \sum_{L_B} \sum_{L} V_{L_A L_B L}(R) A_{L_A L_B L}(\theta_A,\theta_B,\phi) \qquad (13)$$

The orthonormal functions $A_{L_A L_B L}$ are defined by

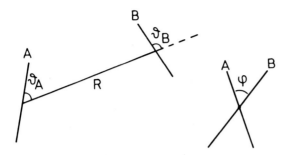

Figure 7.   Definition of internal coordinates.

$$A_{L_A L_B L}(\theta_A, \theta_B, \phi) = \sum_{M=0}^{\min(L_A, L_B)} X_{L_A L_B LM} P_{L_A}^M(\cos\theta_A) P_{L_B}^M(\cos\theta_B) \cos M\phi \quad (14)$$

The expression for the numerical constants $X_{L_A L_B LM}$ is given in ref.[24].

When the potential has been calculated at a set of well-chosen orientations, the expansion coefficients can be obtained by numerical integration methods [18].

For H_2-H_2 interactions a number of very accurate $V_{L_A L_B L}(R)$ curves are published in ref.[2]. It is, however, hard to make a comparison with our electron gas results, since in [2] only the CI values are given, and no decomposition into first and second order contributions has (yet) been published. The long range part of the SCF curve for V_{000} in figure 8 has therefore been approximated by the difference between the CI values from ref.[2] and an estimate of the dispersion energy obtained from a multipole expansion with correlated C_6, C_8 and C_{10} values from ref.[20]. Of course, this estimate will, in the case of H_2-H_2 be good only for R values larger than 7 Bohr, so only the asymptotic behaviour of the approximated long range SCF curve in figure 8 will be correct. Somewhat less accurate SCF and CI calculations (a smaller basis set and less configurations) had been published before, only valid for the short range [25]. We have used these SCF data for V_{000} in the 2.5 to 5 Bohr range. The two SCF curves in figure 8 cannot be connected by a straight line and probably the short range part is already insufficiently repulsive at R=5 Bohr.

Our electron gas results were obtained from calculations at 12 R values ranging from 2.5 to 8 Bohr. The first 14 unique (non zero) expansion coefficients at each R were determined from the computed energies at 30 (unique) orientations. As for the He-He and He-H_2 interactions, use of the exchange correction factor from eq.(11) gives a V_{000} curve that agrees well with the ab initio results for R<7 Bohr. Although a guess could also have been made for the other ab initio SCF expansion coefficients, the results would not have been very accurate. In particular the anisotropic dispersion coefficients are only available from accurate ab initio calculations which neglect correlation [26], and it is

well known that correlation may significantly alter the dispersion ani-
anisotropy. The remaining modified electron gas expansion coefficient
curves in figure 8 are given as a prediction. In order to obtain a
curve for V_{224} that has an approximately exponential distance dependence, we have, as in previous studies [24], removed the electrostatic
quadrupole-quadrupole interaction in it. Our electron gas calculations

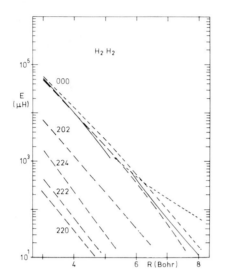

Figure 8.   Dominant expansion coefficients for the first order energy for
H_2-H_2. For explanation see figure 4.

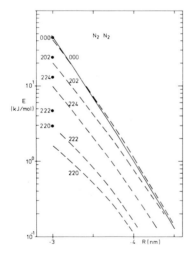

Figure 9. Dominant expansion coefficients for the first order energy for
N_2-N_2. ● : ab initio first order values at .3nm, full line: distance dependence of ab initio results, dashed line: results of electrongas method with
Rae correction.

showed that expansion coefficients with L_A (or L_B) $\geqslant 4$ are unimportant. Our first order potential is determined to within 5% accuracy by the three coefficients V_{000}, V_{202} and V_{224}. This appears also to be the case for the ab initio potential from ref.[2].

For N_2-N_2 extensive ab initio first order calculations have recently been performed. The complete first order potential energy surface was represented in the form of the spherical expansion of eq.(13). Since the detailed anisotropy of the potential was only determined at one intermolecular distance, it was assumed in [18] that all expansion coefficients had the same distance dependence, which was determined from the energies for a few selected orientations at other distances. In figure 9 we have indicated the distance dependence only for the ab initio first order V_{000} coefficient. For a few other coefficients we have given the ab initio value at R=.3nm (5.67 Bohr). Electron gas calculations have been made for 6 distances and 105 orientations. Details are given in ref.[24]. We found that when we used the Rae correction the ab initio V_{000} curve was very well reproduced. Furthermore we noticed that the distance dependence of the various expansion coefficients in the modified electron gas method is not at all the same. At the time we performed these calculations, eq.(11) was not yet available; its use would have resulted in a somewhat insufficiently repulsive curve. In ref.[18] a complete potential was obtained by adding second order contributions from a scaled ab initio multipole expansion[27] to the ab initio first order energy. In the case of

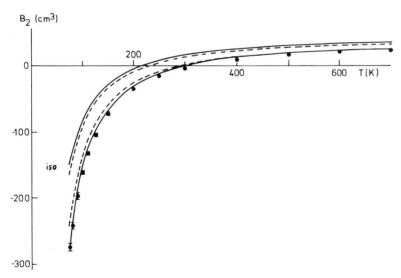

Figure 10.   Temperature dependence of the second virial coefficient. ⧫ experimental points. Solid line from electrongas potential, dashed line from ab initio potential. Curves marked iso are obtained using the V_{000} part only.

N_2-N_2 interactions, penetration effects in the dispersion energy show
up only at relatively short distances[18]. This complete ab initio
potential worked very well in describing crystal properties [18,28].
This potential and the potential found by adding the dispersion to the
electron gas potential (including the Rae correction) were used in the
calculation of the temperature dependence of the second virial coefficient $B_2(T)$[24]. As is seen from figure 10, the agreement with experiment is very good, especially for the curve based on the electron gas
potential.

For the interaction between CO and H_2 we have recently completed
the calculation of the 6 dimensional potential energy surface using the
electron gas formalism [29]. R values ranged from 3.5 to 8.5 Bohr and
a series of intermolecular distance of CO and H_2 (r_A and r_B) was chosen
in such a way that the potential can be used in theoretical studies of
vibrational energy transfer. We found that the inter- and intra-molecular distance dependence of the expansion coefficients can be accurately
represented by

$$V_{L_A L_B L}(R,r_A,r_B)=a_{L_A L_B L}(r_A,r_B)\exp\left[b_{L_A L_B L}(r_A,r_B)R+c_{L_A L_B L}(r_A,r_B)R^2\right]$$

with

$$a_{L_A L_B L}(r_A,r_B)=a^o_{L_A L_B L}+a^A_{L_A L_B L}r_A+a^B_{L_A L_B L}r_B+a^{AB}_{L_A L_B L}r_A r_B \qquad (15)$$

and similar expressions for b and c. Since there is no accurate ab
initio information in the form of a spherical expansion of first order
or SCF energies, we have given in figure 11 a comparison between SCF
results for a few representative cuts of the potential surface and the
electron gas results with the Rae correction, using in the last case
the expansion (13) and fit (15). The electron gas results agree reasonably well with the most recent SCF values [30]. The earlier SCF results
[31] clearly suffer from basis set deficiencies at the larger R values.

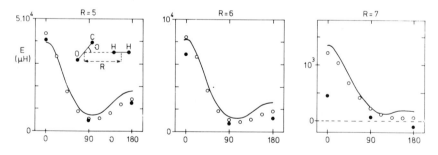

Figure 11. Cuts through the first order potential energy surface for CO
Full line electrongas results using expansion (13) and fit (15).
Ab initio SCF results: 0 [30] and ●[31].

CONCLUDING REMARKS

From the examples shown, we gain the impression that the electron gas method provides a means to obtain rather accurate first order interaction energies. A prerequisite is a modification of the original expression for the exchange contribution. For systems with few electrons the new correction factor (eq.11) seems to be appropriate. For other systems the Rae correction works better. Possibly slightly more complicated corrections, announced in ref.[9], will merge the two approaches. The question as how to obtain the additional accurate dispersion (and sometimes also induction) contributions is left unanswered. The existing electron gas formulae for the correlation[3] and the Drude model [32]do not yet provide acceptable values for the dispersion contribution.

REFERENCES

1. W.Meyer, P.C.Hariharan and W.Kutzelnigg, J.Chem.Phys. 73, 1880 (1980)
2. W.Meyer, and J.Schaefer (to be published), cited in L.Monchick and J.Schaefer, J.Chem.Phys. 73, 6153 (1980)
3. R.G.Gordon and Y.S.Kim, J.Chem.Phys. 56, 3122 (1972)
4. P.Gombas, Die statistische Theorie des Atoms, Springer Verlag, Wien, 1949
5. G.A.Parker, R.L.Snow and R.T Pack, Chem.Phys.Lett. 33, 399 (1975)
6. A.I.M.Rae, Chem.Phys.Lett. 18, 574 (1973)
7. A.I.M.Rae, Mol.Phys. 29, 467 (1975)
8. M.Waldman and R.G.Gordon, J.Chem.Phys. 71, 1325 (1979)
9. J.L.Gázquez and E.Ortiz, Chem.Phys.Lett. 77, 186 (1981)
10. G.A.Parker and R.T Pack, Program 305, Quantum Chemistry Program Exchange, Indiana University, Bloomington, IN 47401
11. B.Jeziorski, M.Bulski and L.Piela, Int.J.Quant.Chem. 10, 281 (1976)
12. W.Kołos and E.Radzio, Int.J.Quant.Chem. 13, 627 (1978)
13. M.Bulski, G.Chałasinsky and B.Jeziorski, Theor.Chim.Acta (Berl.) 52, 93 (1979)
14 K.Szalewicz and B.Jeziorski, Mol.Phys. 38, 191 (1979)
15. A.L.J.Burgmans, J.M.Farrar and Y.T.Lee, J.Chem.Phys. 64, 134 (1976)
16. G.Starkshall and R.G.Gordon, J.Chem.Phys. 56, 2801 (1972)
17. J.M.Farrar, Y.T.Lee, V.V.Goldman and M.L.Klein, Chem.Phys.Lett. 19, 359 (1973)
18. R.M.Berns and A.van der Avoird, J.Chem.Phys. 72, 6107 (1980)
19. P.C.Hariharan, cited as private communication in K.T.Tang and J.P.Toennies, J.Chem.Phys. 68, 5501 (1978)
20. W.Meyer, Chem.Phys. 17, 27 (1976)
21. K.T.Tang and J.P.Toennies, J.Chem.Phys. 68, 5501 (1978)
22. L.D.Thomas, W.P.Kraemer and G.H.F.Diercksen, Chem.Phys. 51, 131 (1980)
23. G.A.Parker and R.T Pack, J.Chem.Phys. 69, 3268 (1978)
24. M.C.van Hemert and R.M.Berns, to be published in J.Chem.Phys. (15 januar 1982)

25. F.H.Ree and C.F.Bender, J.Chem.Phys. 71, 5362 (1979)
26. F.Mulder, A.van der Avoird and P.E.S.Wormer, Mol.Phys. 37, 159 (1979)
27. F.Mulder, G.van Dijk and A.van der Avoird, Mol.Phys. 39, 407 (1980)
28. T.Luty, A.van der Avoird and R.M.Berns, J.Chem.Phys. 73, 5305 (1980)
29. M.C.van Hemert (to be published)
30. G.H.F.Diercksen and W.P.Kraemer, results presented at the meeting
    "New methods for intermolecular forces" Orsay, France, June 1980
31. J.Prisette, E.Kochanski and D.R. Flower, Chem.Phys. 27, 373 (1978)
32. M.Waldman and R.G.Gordon, J.Chem.Phys. 71, 1340 (1979)

Solute-Solute Interactions in Dilute Solutions of Gases in Liquids

Robert A. Pierotti, Stephen L. Parrott* and Mark A. Tallent
School of Chemistry, Georgia Institute of Technology, Atlanta, Georgia 30332
*Phillips Petroleum Company, Bartlesville, Oklahoma

Abstract

Equations are developed to extract in-solution solute-solute virial coefficients from gas-liquid phase equilibrium data. Experimental data for the solubility of helium, neon and hydrogen in liquid argon are analyzed and second virial coefficients for these solutes, B_2^*, in liquid argon are determined over the entire liquid range of argon from 84° to 148°K. The values of B_2^* are negative at the lowest temperature and approach zero or positive values as the critical temperature of argon is approached. The magnitude and temperature dependence of the virial coefficients are considered in terms of the reversible work of cavity formation and an effective local density.

Introduction

During the past twenty years, the development of statistical mechanical theories of rigid sphere fluids[1-3] along with a well-defined perturbation theory of rigid sphere fluids[4-6] has lead to notable advances in our understanding of infinitely dilute solutions[7-10]. This success is manifest by the growing literature dealing with Henry Law behavior in solution by theorists and experimentalists and particularly by those interested in the properties of aqueous solutions[11-13] Interestingly much of the recent interest in dilute solution behavior is concerned with hydrophobic interactions, i.e. the "real" or "apparent" interaction between solute molecules in liquid water[14-17]. A number of researchers have attempted to use the success of statistical mechanical theories of <u>infinitely dilute</u> aqueous solutions to speculate on the nature of hydrophobic or solvophobic interactions[18-19]. In an infinitely dilute solution (the Henry Law region), only solute-solvent and solvent-solvent interaction can be important and there can be no information concerning solute-solute interactions - not even a second order effect. If any information about solute-solute interactions is to be obtained, one must consider systems which show deviations from the Henry Law region. A clever attempt to circum-

B. Pullman (ed.), Intermolecular Forces, 175–197.

scribe the lack of studies and data outside the Henry Law region, is
that by Ben-Naim in which the Henry Law solubility of essentially
monomers and dimers in water are compared and their difference related
to the hydrophobic effect[20].

The present study attempts to develop the technique for studying
dilute solutions of gases in simple liquids in which deviations from
Henry Law behavior takes place and from which solute-solute interactions can be interpreted in molecular terms. The results will be to
obtain solute-solute virial coefficients in solution from gas-liquid
equilibrium data.

Theory

The theory for multicomponent systems developed by McMillan and
Mayer[21,22] indicated that for systems in osmotic equilibrium the
osmotic pressure of the system is given by a power series in the
activity of the solute. The coefficients of the series were shown to
be analogous to the virial coefficients of an imperfect gas in that
they involved the same form of configuration and cluster integrals.
The important difference in these integrals was the replacement of
the interaction energy by the potential of the average or mean force
existing among the set of solute molecules as a result of the presence
of the solvent. The statistical thermodynamics basis of the McMillan-
Mayer theory is related to the grand canonical ensemble whose representative thermodynamic system is a member of the $(\{\mu\}, V, T)$ ensemble.
While this is the natural ensemble to use for osmotic equilibrium
where at least one component is non-diffusible, it is not practical
for normal gas-liquid phase equilibria where all components are dif-
fusible.

The System of Interest

We consider a thermodynamic system in which a given number of
moles of one component called the solvent is introduced into a equilibration vessel at temperature T and volume V. If the conditions
are suitable, the solvent will exhibit typical liquid-vapor equilibrium properties in which it will exert a vapor pressure $P_1^*(T)$. If a
second gaseous component is added, it partitions itself between the
liquid and vapor phase, such that the total pressure is P, the gas
phase composition is (y_1, y_2) and the liquid phase composition is
(x_1, x_2), where y_i is the gas phase mole fraction of i and x_i is the
liquid phase mole fraction of i. The vapor pressure of the solvent
changes as a result of the presence of dissolved solute along with
effect of increased pressure (the Poynting effect)[23]. The convenient measure of the activity of a component in the gas is its fugacity.
We will be interested in properties of the solution entrapolated to
infinite dilution or zero fugacity of the solute in a specified
amount of solvent hence the representative system will be one containing sets of solute molecules $\{N_2\}$, at a given activity (λ_2) in N_1

solvent molecules at a pressure $P > P_1^*$ and temperature T. The natural ensemble for this system is the $(\{N_2\}, N_1,P,T)$ ensemble or the semi-grand isobaric-isothermal ensemble. This ensemble has been discussed by several authors[24-26] and we follow the notation of Hill. Since we are interested at present in binary solutions and preserving notational simplicity, we develop the theory for a two component system although it can be generalized for many components without difficulty.

Statistical Treatment

The appropriate ensemble is given by[24]

$$\Gamma(\lambda_2,N_1,P,T) = \sum_{N_2 \geqslant 0} \Delta_{N_2} (N_2,P,T)\lambda_2^{N_2} \tag{1}$$

where Γ is the semi-grand isobaric-isothermal partition function open with respect to solute but at fixed N_1, λ_2 is the absolute activity of solute given by $\exp(\mu_2/kT)$ and Δ_{N_2} is the isothermal-isobaric partition function for a set of N_2 solute molecules and is given by

$$\Delta_{N_2} (N_1,P,T) = \sum_V Q_{N_2}(N_1,V,T)e^{-\beta PV} \tag{2a}$$

$$= \sum_V \frac{Z_{N_2}(N_1,V,T)e^{-\beta PV}}{(\Lambda_1^3/j_1)^{N_1}(\Lambda_2^3/j_2)^{N_2}N_1!N_2!}e \tag{2b}$$

where $Q_{N_2}(N_1,V,T)$ is the canonical partition function for N_i molecules of i in fixed volume V and temperature T. P is the total hydrostatic pressure, β is (1/kT), k is the Boltzman constant, (Λ_i^3/j_i) is the ratio of the deBroglie thermal wavelength and the molecular partition function for the internal degrees of freedom which are assumed to remain unchanged in this treatment. Z_{N_2} is the usually configuration integral given by

$$Z_{N_2} = \int \cdots \int_V \exp\{-\beta U_{N_1+N_2} (\bar{r}_1\cdots\bar{r}_{N_1+N_2})\} d\bar{r}_1\cdots d\bar{r}_{N_1+N_2} \tag{3}$$

where $\bar{r}_i$ is the generalized position of the i-th molecule.

Since we are interested in the properties of the solution relative to the properties of the pure solvent, we consider the ratio of Γ to the isobaric-isothermal partition function for N_1 molecules of pure solvent, $\Delta_0^*(N_1,P,T)$. The characteristic function of the logarithm of the ratio Γ/Δ_0^* is

$$-N_1(\mu_1-\mu_1^*)/kT = \ln(\Gamma/\Delta_0^*) \tag{4}$$

where $\mu_1(N_1/N_2,P,T)$ and $\mu_1^*(P,T)$ are the chemical potentials of the solvent in the solution and in the pure liquid state.

The ratio of the partition functions (Γ/Δ_0^*) can be written

$$(\Gamma/\Delta_0^*) = 1 + \sum_{N_2 \geqslant 1} (\Delta'_{N_2}/\Delta_0^*)(\lambda_2^{N_2}/N_2!)(\Lambda_2^3/j_2)^{-N_2} \tag{5}$$

where $\Delta'_{N_2} = (\Lambda_2^3/j_2)^{N_2} N_2!\Delta_{N_2}$ and for classical systems can be written

$$\Delta'_{N_2} = \int dv \int \cdots \int \exp\{-\beta U_{N_1+N_2}(\bar{r}_1 \cdots r_{N_1+N_2})-\beta PV\} \, d\bar{r}_1 \cdots d\bar{r}_{N_1+N_2} \tag{6}$$

If we define a relative activity $z_2 = \lambda_2/(\Lambda_2^3/j_2)$ then

$$\ln(\Gamma/\Delta_0^*) = \ln(1 + \sum_{i \geqslant 1} \delta_i z_2^i) \tag{7}$$

where δ_i is the ratio $(\Delta'_{N_2}/\Delta_0^*)$ and hence

$$(\mu_1-\mu_1^*) = (-kT/N_1) \sum_{i \geqslant 1} \beta_i^* z_2^i \tag{8}$$

where β_i^* are sets of reduced cluster integrals. The first two of which are

$$\beta_1^* = \delta_1 = \Delta_1'/\Delta_1^* \tag{9a}$$

$$2\beta_2^* = 2\delta_2 - \delta_1^2 = 2(\Delta_2'/\Delta_0^*) - (\Delta_1'/\Delta_0^*)^2 \tag{9b}$$

Data Treatment

We consider two methods of data treatment equivalent to each other theoretically, but of somewhat different experimental practicality. The first method is gas phase based and we proceed directly to the properties of the partition function (Γ/Δ^*). The average number of molecules in a system of the ensemble is given by[24],[25]

$$\bar{N}_2 = z_2[\partial \ln(\Gamma/\Delta_0^*/\partial z_2]_{N_1,P,T} \tag{10a}$$

$$= \sum_{i \geqslant 1} i\beta^*_i \cdot z_{22}^i \tag{10b}$$

$$= \sum_{i \geqslant 1} i\beta^*_i (f_2/kT)^i \tag{10c}$$

where the relative activity z_2 has been replaced by the activity of an ideal gas (f_2/kT) where f_2 is the fugacity of the solute in the gas phase. Equation (10) can be rewritten in terms of the mole ratio $\bar{m}_2$ as

$$\bar{m}_2 = \bar{N}_2/N_1 = \sum_{i \geqslant 1} i(\beta^*_i/N_1)(f_2/kT)^i \tag{11a}$$

or

$$\bar{m}_2/f_2 = \beta^*_1(P,T)/N_1 kT + 2\beta^*_2(P,T)f_2/N_1(kT)^2 + \ldots \tag{11b}$$

The Henry Law constant is by definition

$$\lim_{f_2 \to 0} (f_2/m_2) = K^0_H(P,T) = N_1 kT/\beta^*_1(P,T) \tag{12}$$

The coefficient of the f_2 term is related to the solute-solute interaction and will be discussed later.

The second method of data treatment is based upon the liquid phase composition. Here we proceed by using the Gibbs-Duhem relationship along with equation (8) to obtain the activity of the solute

$a_2 = \gamma_2 m_2$ in the solution where γ_2 is the mole ratio activity coefficient. For this purpose it is useful to express the relative activity z_2 in terms of a_2 and the one particle partition function ratio β_1^*. The relationship is $a_2 = (\beta_1^*/N_1)z_2$ and in these terms

$$\mu_1 - \mu_1^* = -kT/N_1 \sum_{i \geqslant 1} \beta_i^* (N_1/\beta_1^*)^i a_2^i \tag{13}$$

and from the Gibbs-Duhem equation

$$(N_2/a_2) = -N_1 (\partial \ln(\mu_1 - \mu_1^*)/\partial a_2)_{P,T} \tag{14}$$

or

$$m_2 = \sum_{i \geqslant 1} i(\beta_i^*/N_1)(N_1/\beta_1^*)^i a_2^i = \sum_{i \geqslant 1} i\theta_i a_2^i \tag{15}$$

Reversion and rearrangement of the series yields

$$\ln(a_2/m_2) = \sum_{i \geqslant 1} \gamma_{i+1} m_2^i \tag{16}$$

where the first two coefficients in this series are

$$\gamma_2 = -2\theta_2 = -2\beta_2^* N_1/ (\beta^*)^2 \tag{17a}$$

$$\gamma_3 = 6\theta_2^2 - 3\theta_3 \tag{17b}$$

and where

$$\theta_i = N_1^{i-1}(\beta_i^*/\beta_1^*)^i \tag{18}$$

Since the gas and solution phase are in equilibrium

$$\mu_1^0(T) + RT\ln f_2 = \mu_1^*(T,P) + RT\ln a_2 \tag{19}$$

or

$$\ln f_2/m_2 = [\mu_1^*(T,P) - \mu_i^0(T)]/RT + \sum_{i \geq 1} \gamma_{i+1} \, m_2^i \tag{20a}$$

$$= \ln K_H^0(P,T) + \sum \gamma_{i+1}(P,T)m_2^i \tag{20b}$$

The coefficient γ_2 is the first term which contains information concerning the solute-solute interaction.

Although the isobaric ensemble was used above, it is clear that the total pressure varies as the solute fugacity increases hence one cannot obtain β_2^* or γ_2 directly from equations (11) or (20). Rather it is necessary to obtain the pressure derivatives of these equations and extrapolate the results to P_1^*. The results are

$$\lim_{P \to P_1^*} (\partial(m_2/f_2/\partial P)_T = \overline{V}_2^0/\alpha K_H^{0\,2}RT + 2\beta_2^*/N_1(kT)^2 \tag{21a}$$

or

$$\lim_{P \to P_1^*} (\partial \ln(m_2/f_2)/\partial P)_T = \overline{V}_2^0/\alpha RT + \gamma_1 \tag{21b}$$

where $\alpha = (\partial m_2/\partial P)_T$ and $\overline{V}_2^0/RT = (\partial \ln K_H^0/\partial P)_T$ where $\overline{V}_2^0$ is the partial molar volume of solute at infinite dilution.

The Virial Coefficients of Mean Force

The reduced cluster integrals β_1^* and β_2^* in the isobaric-isothermal representation can be related to the McMillan-Mayer virial coefficients. We consider first the term β_1^* which is given by

$$\beta_1^*(P,T) = \Delta_1'/\Delta_0^* = \frac{\int dv \int \cdots \int \exp\{-\beta U_{N_1+1}(\bar{r}_1 \cdots \bar{r}_{N_1}, \bar{r}_a) - \beta PV\} d\bar{r}_1 \cdots d\bar{r}_{N_1} d\bar{r}_a}{\int dv \int \cdots \int \exp\{-\beta U_{N_1}(\bar{r}_1 \cdots \bar{r}_{N_1}) - \beta PV\} d\bar{r}_1 \cdots d\bar{r}_{N_1}}$$

(22a)

$$= \int dv \int d\bar{r}_a \int \cdots \int \frac{\exp\{-\beta U_{N_1} \bar{r}_1 \cdots r_{N_1}) - \beta PV\}\exp\{-\beta\Delta U^{(1)}(\bar{r}_a)\}}{\int dv \int \cdots \int \exp-\{\beta U_{N_1}(\bar{r}_i \cdots \bar{r}_{N_1}) - \beta PV\} d\bar{r}_1 \cdots d\bar{r}_{N_1}} d\bar{r}_1 \cdots d\bar{r}_{N_1}$$

(22b)

$$= \int dv \int d\bar{r}_a \int \cdots \int P_{NV_mT} \ \delta(V-V_m)\exp\{-\beta\Delta U^{(1)}(\bar{r}_a)\} d\bar{r}_1 \cdots d\bar{r}_{N_1} \tag{22c}$$

$$= \int \delta(V-V_m)dv \int d\bar{r}_a \int \cdots \int P_{NV_mT} \ \exp\{-\beta\Delta U^{(1)}\bar{r}_a)\} d\bar{r}_1 \cdots d\bar{r}_{N_1} \tag{22d}$$

$$= \int_{V_m} \ <\exp\{-\beta\Delta U^{(1)}(\bar{r}_a)\}> d\bar{r}_a \tag{22e}$$

In the above equation (22b) the right hand integral is the conditional probability of observing the configuration $\bar{r}_1 \cdots \bar{r}_2$ in a system containing N_1+1 molecules at a pressure P, temperature T, regardless of the volume. This is reflected in equation (22c) where we also note that for macroscopic systems this probability (P_{NV_mT}) is concentrate about the maximum value of the volume V_m. For this reason a Dirac delta function is included whose argument is $\delta(V-V_m)$. Note also that the interaction energy has been written in such a manner as to isolate the coordinates of the solute molecule ($\bar{r}_a$). This could be handled more judiciously by including a coupling parameter and an appropriate charging integral. We avoid this for simplicity and note that $\Delta U^{(1)}(\bar{r}_a) = U_{N_1}(\bar{r}_1 \cdots \bar{r}_a) - U_{N_1}(\bar{r}_1 \cdots \bar{r}_{N_1})$ is the binding energy of a single solute molecule in solution referred to by Ben-Naim[27]. In (22e) it can be shown that if one defines $W^{(1)}(\bar{r}_a, P,T)$ by the following equation

$$\exp-\{W^{(1)}(\bar{r}_a,P,T)\} = \langle\exp\{-\beta\Delta U^{(1)}(\bar{r}_a)\}\rangle \qquad (23)$$

that $-(\partial W^{(1)}(\bar{r}_a)/\partial\bar{r}_a)_{P,T} = \langle-\partial U_{N+1}(\bar{r}_1\cdots\bar{r}_{N_1^2},\bar{r}_a)/\partial\bar{r}_a\rangle$ and hence $W^{(1)}(\bar{r}_a)$ is the one particle potential of mean force as defined at infinite dilution by the McMillan-Mayer theory of solutions. Equation (23) is also equal to the singlet distribution function $g^{(1)}(\bar{r}_1\cdots\bar{r}_N,\bar{r}_a)$. Finally we define a relative n-particle potential of mean force $\omega^{(n)}$ by

$$\omega^{(n)}(\bar{r}_1\cdots\bar{r}_n) = W^{(n)}(\bar{r}_1\cdots\bar{r}_n) + kT\ln(kT/\nu_1^* K_H^0)^n \qquad (24)$$

where ν_1^* is the molecular volume of pure solvent. When this is substituted in equation (23) and (22e) and noting that the relative value of $\omega^{(1)}$ is zero, we obtain the integral for β_1^* indicated above.

The second cluster integral β_2^* can be shown in a similar manner to be related to the two particle potential of mean force $W^{(2)}(\bar{r}_a,\bar{r}_b)$ or more usefully $\omega^{(2)}(\bar{r}_a,\bar{R})$ the pair potential of mean force. The relationship is

$$\beta_2^*(T,P) = (\bar{V}_1^*/2N_1)(kT/\nu_1^* K_H^0)^2(\nu_2^0-B_2^*) \qquad (25)$$

where

$$B_2^*(T,P_1^*) = -1/2 \int_0^\infty (e^{-\omega^{(2)}(\bar{R})}-1)d\bar{R} \qquad (26)$$

is the second virial coefficient of the solute in the solution and is the same as the second osmotic virial coefficient in the McMillan-Mayer theory.

Results

We are interested in the nature of the virial coefficient of solutes in simple liquid mixtures, hence we consider data for the solubility of helium, neon and hydrogen in liquid argon. Although data on such systems are limited due to the experimental difficulty associated with cryoscopic high pressure gas-liquid systems, there have been some important measurements made as a function of temperature over the entire liquid range of argon. These measurements include those reported by Mullins and Ziegler[27] for helium in argon from 85°

to 110°K, by Sinor and Kurata for helium in argon from 93° to 148°K,[28] by Streett for neon in argon from 84° to 130°K[29] and by Volk and Halsey[30] for hydrogen in argon from 87° to 140°K. As an example of the quality of the data and the mode of data presentation, Figures 1 and 2 show $\ln(f_2/m_2)$ plotted versus m_2 for the helium data, Figure 3 displays m_2/f_2 versus f_2 for hydrogen in argon data and Figure 4 shows m_2 versus $(P-P_1^*)$ for the neon data. Table I gives the values of

---

TABLE I. Henry Law Constant in Liquid Argon

He-Ar		Ne-Ar		H_2-Ar	
T°K	$K_H^0 \times 10^{-3}$(atm)	T°K	$K_H^0 \times 10^{-3}$(atm)	T°K	$K_H^0 \times 10^{-3}$(atm)
85.0	13.7	84.42	1.11	87.0	.855
90.0	10.3	87.42	1.05	94.0	.754
95.0	7.86	95.82	.978	100.0	.699
100.0	6.07	101.94	.714	110.0	.601
105.0	4.70	110.78	.578	120.0	.525
110.0	3.66	121.36	.418	130.0	.448
93.2[+]	9.82	129.32		140.0	.371
113.2[+]	3.58				
133.2[+]	1.25				
148.2[+]	.396				

[+]Data of Sinor-Kurata

---

of K_H^0 for the various systems while Figure 5 displays the temperature dependence of $\ln K_H^0$.

Intercepts of graphs of $(\ln f_2/m_2 - \ln K_H^0)/m_2$ versus m_2 and of $(m_2/f_2 - 1/K_H^0)f_2$ versus f_2 are related to $\overline{V}_2^0$ and B_2^* through equations (21a) and (21b). In order to calculate B_2^*, values of $\overline{V}_2^0$ are required. Volk and Halsey determined $\overline{V}_2^0$ for their H_2 in argon studies. Figure 6 compares $\overline{V}_2^0$ for hydrogen in argon with the molar volume $\overline{V}_1^*$ of liquid argon. In lieu of anything better we have used a quadratic representation of the combined hydrogen and liquid argon data to represent $\overline{V}_2^0$ for all of the solutes. With this assumption values of B_2^* were calculated and are given in Table II and are displayed as a function of temperature in Figure 7. The values of B_2^* are the mean values determined by using the two methods of data analysis indicated by equation (21).

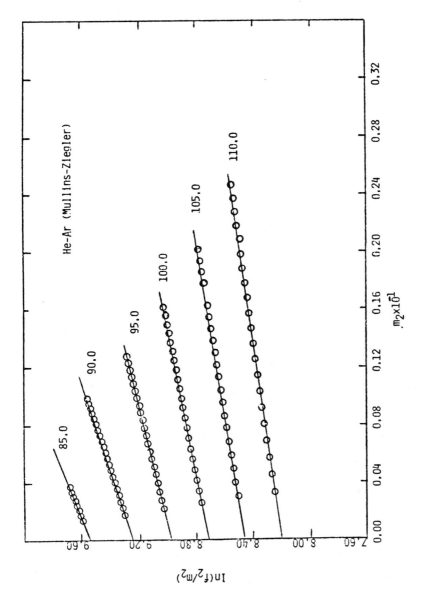

Figure 1. Helium-Argon Data of Mullins and Ziegler

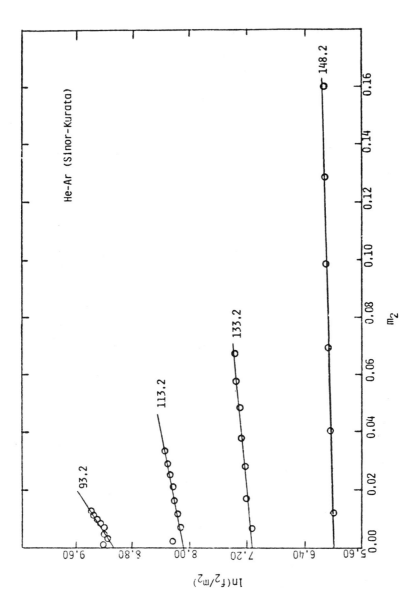

Figure 2.   Helium–Argon Data of Sinor and Kurata

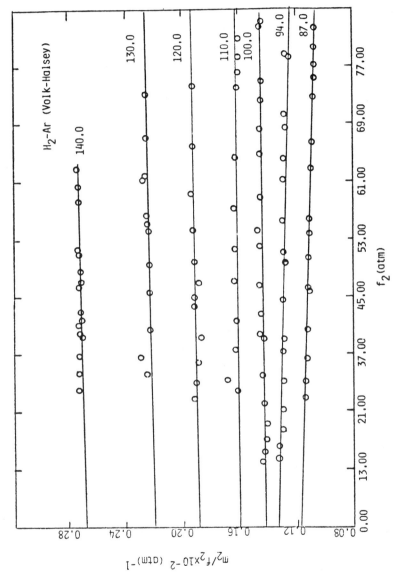

Figure 3. Hydrogen-Argon Data of Volk and Halsey

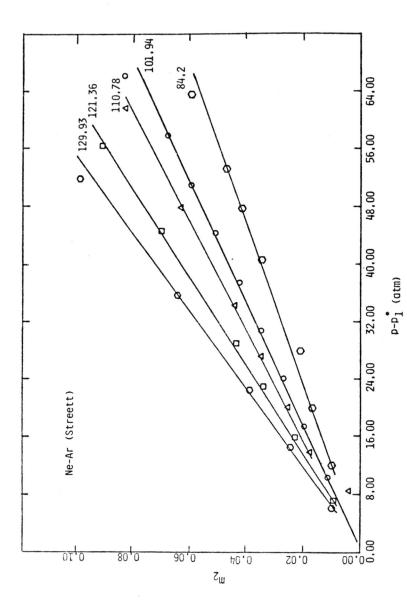

Figure 4. Neon-Argon Data of Streett

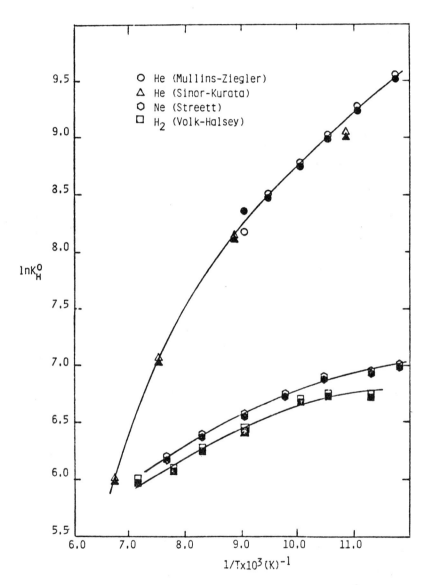

Figure 5. Temperature Dependence of $\ln K_H^O$

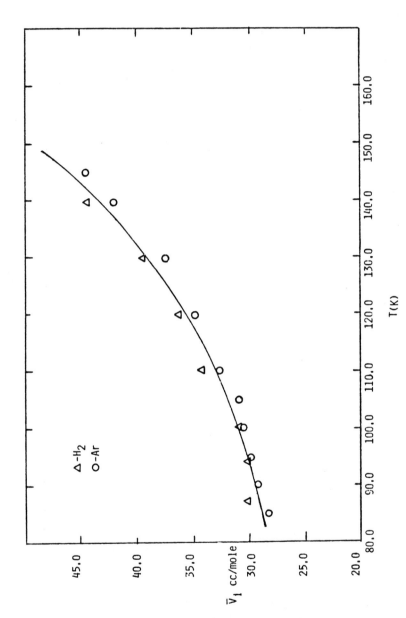

Figure 6.   Comparison of the Partial Molar Volume of H_2 in Argon with the
Molar Volume of Liquid Argon.

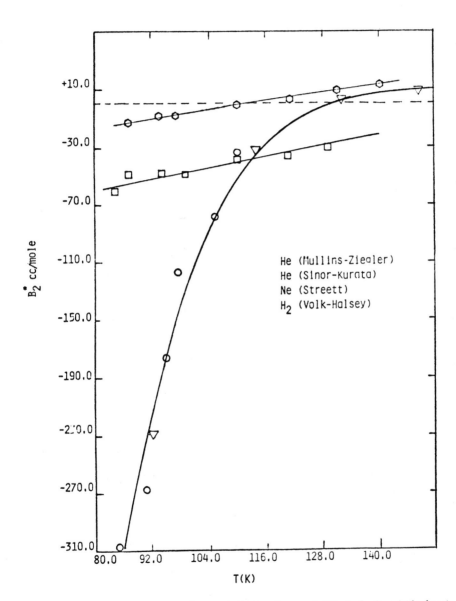

Figure 7. Solute-Solute Second Virial Coefficients in Liquid Argon

TABLE II.   Solute-Solute Second Virial Coefficients
in Liquid Argon

He-Ar		Ne-Ar		H_2-Ar	
T°K	B_2^*(m^3/atm)	T°K	B_2^*(m^3/mole)	T°K	B_2^*(m^3/mole)
85.0	-307.	84.42	-62.3	87.0	-15.0
90.0	-265.	87.42	-50.9	94.0	- 9.66
95.0	-177.	95.82	-52.7	100.0	- 9.50
100.0	-119.	101.94	-53.0	110.0	- 2.30
105.0	- 92.8	110.78	-57.6	120.0	+ 2.88
110.0[+]	- 34.3	121.36	-40.6	130.0	+ 6.64
93.2[+]	-231.	129.32	-32.2	140.0	+10.1
113.2[+]	- 32.5				
133.2[+]	- 6.79				
148.2[+]	+ 25.2				

[+]Data of Sinor-Kurata

Discussion

The most notable features of Figure 7 are the magnitudes of the
second-virial coefficients and the order of helium having the largest
negative value. Figure 8 is a plot of the gas phase values of the
second-virial coefficients of helium, neon and argon. Clearly the
potential of mean force between helium molecules is dominated by
solvent effects. It is particularly interesting that the large
negative virial coefficient is indicative of a strong apparent interaction between the helium atoms (an effect which if observed in a
water solution would be ascribed to a hydrophobic effect). In essence
all three solutes display solvophobic interactions in liquid argon
with helium being the most pronounced. One must be concerned with the
possibility of quantum effects playing some role with these solutes
especially helium in liquid argon, but the large negative virial
coefficients for helium imply that solvent effects must be dominant
and quantum effects can only be a perturbation to the solvent effect.

The question arises as to what form of the potential of mean force
can yield the trends of the curves in Figure 7. To explore this we
consider the following process: 1) First we start with two solute
particles immersed in the solvent but infinitely far apart where the
pair potential of mean force is zero and we imagine one of the
particles, say at position $\bar{r}_a$ to be fixed and the particle at $\bar{r}_b$ a
distance R = ∞ from $\bar{r}_a$ to be reversibly collapsed in a manner such as

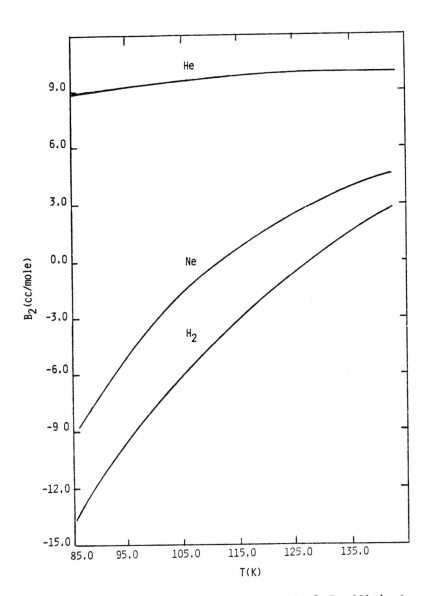

Figure 8. Gas Phase Second-Virial Coefficients

to completely uncouple it from the solvent. The reversible work to do this is given by $-\bar{G}_c(\infty)-\bar{G}_i(\infty)$ where $\bar{G}_c$ is cavity work and $\bar{G}_i$ is the interaction or charging work. The sum of these terms could be estimated using a form of the scaled particle theory or the perturbation theory approach used to treat infinitely dilute solutions. 2) We consider next a process by which a second solute particle is coupled to the solvent at a distance R from particle one. The reversible work here is $\bar{G}_c(R) + \bar{G}_i(R)$.

The pair potential of mean force is then given as a function of R as

$$\omega^{(2)}(R) = \bar{G}_c(R)-\bar{G}_c(\infty) + \bar{G}_i(R)-\bar{G}(\infty) \tag{27}$$

Efforts to obtain the radial distribution function for hard sphere fluids in terms of the scaled particle theory can be useful in estimating $\Delta\bar{G}_c(R)$[31,32]. These efforts have been based upon the notion of an "effective" local chemical potential or local density[33-34]. A particle scaling procedure directly relates $\omega^2(R)$ to integrals involving an effective local density. This can be written as[31]

$$\omega^2(R) = u(R) - 24 \int_0^{\xi'} \xi^2 yG(y,\xi)d\xi - 12 \int_0^{t^*} y(t)dt \int_0^{\xi'} \xi^2 G(y(t),\xi)\, d\xi \tag{28}$$

where ξ is a coupling parameter with $\xi' = (\sigma_{11} + \sigma_{12})/2$, y is the reduced density $\pi\rho\sigma_{22}^3/6$, σ_{11} and σ_{22} are the collisional diameters of the solute and solvent respectively, $t = \cos\theta$ (see Figure 9), $yG(y,\xi)$ is the conditional probability that a molecular center is located in spherical shell of thickness ξ to $\xi + d\xi$ and $y(t)G(y(t),\xi)$ is the conditional probability based on an effective local reduced density $y(t)$ and ρ is the number density of the solvent. It should be noted that for rigid sphere $G(\xi,y)$ is the probability function that contributes only at particle contact and hence $G[\xi,y(t)]$ is zero for those configurations for which contact is impossible. Reference to Figure 9 along with the restriction that u(R) is a rigid sphere potential and that the local density $y(t) = y$ for all $\theta \leq \theta^*$ and $y(t) = 0$ for all $\theta > \theta^*$ yields after integration.

$$\omega^{(2)}(R) = \begin{cases} \infty & R < \sigma_{11} \\ \tfrac{1}{2}W_c(1-R/2\alpha_{12}) & \sigma_{11} \leq R \leq \sigma_{11} + \sigma_{22} \\ 0 & R > \sigma_{11} + \sigma_{22} \end{cases}$$

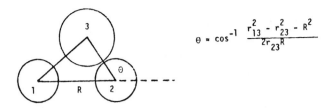

$$\theta = \cos^{-1} \frac{r_{13}^2 - r_{23}^2 - R^2}{2r_{23}R}$$

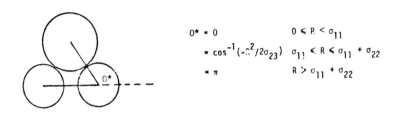

$$\theta^* = 0 \qquad\qquad 0 \leqslant R < \sigma_{11}$$
$$= \cos^{-1}(-R^2/2\sigma_{23}) \quad \sigma_{11} \leqslant R \leqslant \sigma_{11} + \sigma_{22}$$
$$= \pi \qquad\qquad R > \sigma_{11} + \sigma_{22}$$

Figure 9. Geometry of Solute-Solvent Triplets

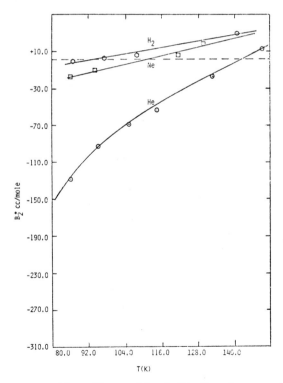

Figure 10. Theoretical B_2^* for Apparent Hard-Sphere Solutes in Argon

where W_c is reversible work of creating a cavity of dimeter σ_{12} for a single solute molecule in pure solvent and $\sigma_{12} = (\sigma_{11} + \sigma_{12})/2$. We could calculate W_c using say the scaled particle theory of a perturbation theory, but instead we use the relationship of W_c to the experimental Henry Law constant thereby including solvent-solvent interactions in W_c. For this purpose we note that

$$W_c = W^{(1)} = kT\ln(K_H^O \overline{V}_1^*/RT) \tag{30}$$

where the symbols have been previously defined. When equation (30) is substituted in (29) and (26) the resulting B_2^*'s (see Figure 10) are qualitatively and semiquantitatively similar to the experimental B_2^*'s for helium, neon and hydrogen are as previously assumed dominated by solvent forces. If $u(R)$, the actual solute-solute interaction potential, had been more realistic than that for a rigid sphere the change would have been in a direction to improve the results.

It is clear that a treatment of $\omega^{(2)}$ based upon experimental liquid argon radial distribution function similar to that or Pratt and Chandler[35] or based upon computer simulation techniques similar to Pangali, Rao and Berne[36] would be interesting and a worthwhile endeavor. The availability of temperature dependent B_2^*'s for simple systems can be a valuable test of theoretical treatments of solution phenomena.

References

1. H. Reiss, H. L. Frisch and J. L. Lebowitz, J. Chem. Phys., 31, 369 (1979).
2. H. L. Frisch, Adv. Chem. Phys., 6, 229 (1963).
3. H. Reiss, Adv. Chem. Phys., 9, 1 (1966).
4. J. A. Barker and D. Henderson, J. Chem. Phys., 47, 4714 (1967).
5. P. J. Leonard, D. Henderson, and J. A. Barker, Trans. Faraday Soc., 66, 2439 (1979).
6. J. D. Weeks, D. Chandler, and H. C. Anderson, J. Chem. Phys., 54, 5237 (1971).
7. R. A. Pierotti, Chem. Rev., 76, 717 (1976).
8. R. O. Neff and D. A. McQuarrie, J. Phys. Chem., 77, 413 (1973).
9. S. Goldman, Acc. Chem. Res., 11, 409 (1979).
10. F. H. Stillinger, J. Solution Chem., 2, 141 (1973).
11. R. A. Pierotti, J. Phys. Chem., 69, 281 (1965).
12. A. Ben-Naim, Water and Aqueous Solutions, Plenum Press, New York, 1974.
13. M. Lucus, J. Phys. Chem., 76, 4030 (1972).
14. S. Marcelja, D. J. Mitchell, B. W. Ninham, and M. J. Sculley, J. Chem. Soc. Faraday Trans. 2, 73, 630 (1977).
15. A. H. Clark, F. Franks, M. D. Pedley, and D. S. Reid, J. Chem. Soc. Faraday Trans. 1, 73, 290 (1977).

16. L. R. Pratt and D. Chandler, J. Chem. Phys., $\underline{67}$, 3683 (1977); J. Solution Chem., $\underline{9}$, 1 (1980).
17. A. Ben-Naim, Hydrophobic Interactions, Plenum Press, New York, 1980.
18. M. Lucas, J. Phys. Chem., $\underline{80}$, 999 (1976).
19. E. Wilhelm and R. Battino, J. Chem. Phys., $\underline{56}$, 563 (1972).
20. A. Ben-Naim, J. Chem. Phys., $\underline{54}$, 1387 (1971).
21. W. G. McMillan and J. E. Mayer, J. Chem. Phys., $\underline{13}$, 276 (1954).
22. T. L. Hill, Statistical Mechanics, McGraw-Hill, New York, 1956.
23. T. L. Hill, Introduction to Statistical Thermodynamics, Addison-Wesley, Reading, Mass., 1960.
24. T. L. Hill, J. Am. Chem. Soc., $\underline{79}$, 4885 (1957).
25. W. B. Brown, Mol. Phys., $\underline{1}$, $\underline{68}$ (1958) and R. A. Sack, Mol. Phys., $\underline{2}$, 8 (1959).
26. A. Ben-Naim, J. Chem. Phys., $\underline{59}$, 6535 (1973).
27. J. C. Mullins and W. T. Zigler, Tech. Report #3, Contract No. CST-1154, National Standard Reference Data Program, National Bureau of Standards, Washington, D. C. (1965).
28. J. E. Sinor and F. Kurata, J. Chem. Eng. Data, $\underline{11}$ 537 (1966).
29. W. B. Streett, J. Chem. Phys. $\underline{42}$, 500 (1965).
30. H. Volk and G. D. Halsey, Jr., J. Chem. Phys., $\underline{33}$, 1132 (1960).
31. S. J. Harris and D. M. Tully-Smith, J. Chem. Phys., $\underline{55}$, 1104 (1971).
32. H. Reiss and R. V. Casberg, J. Chem. Phys., $\underline{61}$, 1107 (1974).
33. T. L. Hill, J. Chem. Phys., $\underline{30}$, 1521 (1959).
34. E. Helfand and F. H. Stillinger, Jr., J. Chem. Phys., $\underline{37}$, 2646 (1962).
35. L. R. Pratt and D. Chandler, J. Chem. Phys., $\underline{73}$, 3434 (1980).
36. C. Pangali, M. Rao, and B. J. Berne, J. Chem. Phys., $\underline{71}$, 2975; 2982 (1979).

STUDIES OF INTERMOLECULAR FORCES BY VIBRATIONAL SPECTROSCOPY

Werner A.P. Luck
Physikalische Chemie, Universitaet Marburg, Germany

Abstract: A review is given on the possibilities to study intermolecular
effects by vibrational spectroscopy. Van der Waals forces can be studied
by the matrix spectroscopy at low temperatures, their small frequency
shift $\Delta\nu$ needs small half width of bands. The sign of $\Delta\nu$ of bending
modes of XH groups is able to differ between van der Waals forces and
H-bonds. The Badger-Bauer rule relation between $\Delta\nu$ and H-bond energy
is discussed as well as the relation between half width and $\Delta\nu$ including their reasons of H-bond distances and angles. The differences between experiments of fundamental and overtone spectroscopy are demonstrated and first theoretical aspects to understand the different intensity
influence by H-bonds are given. - Possibilities to apply the efficient
solutions methods of H-bond spectroscopy on pure liquids are discussed
and applied on water and alcohols. Applied experiments to study H-bonds
in aqueous systems like electrolyte solutions, membranes or biologic
samples are referred.

1.   INTRODUCTION

The language of atoms and molecules is spectroscopy. Questions of intermolecular forces can be answered too with this method, as well as by
disturbances of electron excitations or by disturbances of vibrations.
Examples of the influence of intermolecular forces on electron excitation
spectroscopy is the enhance of the absorbance of iodine vapour in the
visible band by foreign gases (1-2). By shorting of the life time of
electron states by collisions the half width of the rotation fine
structure is changed. The optical collision diameter determined by the
electron excitation band of I_2 is by a factor 1.2 till 1.5 larger than
the gaskinetic diameter, depending on the intermolecular potential (2).
Similar effect has been observed by the pressure broadening of rota-
tional-vibration bands of vapours in the I.R. (2-3) or by the pressure
induced I.R. absorption of HD by rare gase collisions (4-6).
A number of interesting intermolecular details can be observed by disturbances of vibration frequencies:

B. Pullman (ed.), Intermolecular Forces, 199–215.
Copyright © 1981 by D. Reidel Publishing Company.

2.   INFLUENCE OF VAN DER WAALS FORCES ON VIBRATION BANDS

Normal van der Waals forces give only small disturbances of IR bands.
But the so called matrix technique at low temperatures (7) can be used
to study intermolecular effects because the small half width got by this
technique. Small amounts of CH_3OH or H_2O in matrices of rare gases or
other vapours at 4K show the OH stretching band of monomers and at higher
concentrations of H-bonded dimers or higher aggregates. All these bands
are a little shifted by the intermolecular potential of the matrix molecules. The pair potential E_{11} of intermolecular van der Waals forces can
be estimated by RT_C (T_C: critical temperature) (8-9). The pair potential
E_{12} between two unlike molecules 1 and 2 can be estimated by the known
rule given by Berthelot (10) and theoretically founded by London:
$E_{12} = \sqrt{E_{11}E_{22}}$, which we could assume as: $\sim \sqrt{T_{C1}T_{C2}}$. Fig. 1 demonstrates: the small shifts of OH stretching vibration frequency induced by
the intermolecular forces of the matrices are straight lines by plotting
the maxima of absorption bands as function of the square root of T_C of
the matrices $\sqrt{T_{C2}}$. The upper line give the matrix effects on the monomeric species of CH_3OH and the lower line of the H-bonded dimer complex
band, which splits in a fine structure of three lines (measurements by
Schrems (11)). Full lines give the effect of rare gas matrixes and dotted
one of D_2, N_2 or CH_4 matrices.
The influences of different matrixes on the stretching modes of water
are very similar (12) to the demonstrated effects on alcohols. Only the
coupling of the two water modes ν_1 and ν_3 has to taken in account in
the case of so called 1:1 complexes (1 H_2O + 1 H-bond acceptor molecule)
(13-14).
The CH vibrations of alcohols are less sensitive on matrix effects compared with the OH stretching bands (Table 1).

Table 1

Influence of different matrix molecules on OH or CH stretching bands of
CH_3OH

Matrix	Neon	Argon	Krypton	Xenon
$\nu_{OH}(cm^{-1})$	3689	3667	3656	3641.5
$\nu_{CH}(cm^{-1})$	2994	2992	2982	2982

In the series Neon to Xenon the OH vibration has a frequency shift of
$\Delta\nu$ = 47.5 cm^{-1} but the CH only of 12 cm^{-1}. The bigger effect on the OH
band is still existing comparing the quotient $\Delta\nu$ /ν. This effect demonstrates the higher interactions induced by the OH dipoles.
On the other side the effect of van der Waals forces on OH stretching
frequencies is small compared with effects by H-bonds. In both cases
one observes a reduction of the frequencies by interactions (red shift).
The story is different studying the bending mode of OH. Van der Waals
forces induces a red shift too (12) but H-bonds a blue shift of the
bending mode, because the H-bond field is perpendicular to the vibration
amplitudes. This experiment allows to differ between van der Waals force
and small H-bond interactions.

3.   HYDROGEN BONDS

The shifts by intermolecular forces between the OH groups and matrix forming gases are small compared with the shifts induced by H-bonds. Empirically this H-bond shifts are proportional to the H-bond energy (so called Badger-Bauer rule (15-17)).

3.1. Badger-Bauer rule

This relation can be studied with a solution technique: for instance with solutions of small amounts of CH_3OH in inert solvents like CCl_4 and an excess of different bases B. We can assume an equilibrium:

$$OH_{free} \ + \ B_{free} \ \rightleftharpoons \ OH \ bonded \tag{1}$$

$$K = \frac{OH_b}{OH_F \ B_F} = \frac{(C_O - C_F)}{C_F \ (C_{OB} - C_b)} = \frac{(C_O - C_F)}{C_F \ [C_{OB} - (C_O - C_F)]} \tag{2}$$

(C_O: molar concentration of alcohol; C_F: concentration of non H-bonded alcohol (called free alcohol); C_{OB}: molar concentration of base; C_B: non H-bonded base concentration).
The concentration of "free" alcohol can be determined quantitatively with high accuracy by the intensity (absorbance - coefficient) of the undisturbed OH stretching vibration. Its half width is very small, this effect enlarges the accuracy of this spectroscopic method. Equilibrium constants determined with this method belongs to the best which physical chemistry knows (18). For instance we found real constant values of the dimerisation of lactams by H-bonds in a concentration region of lactam varying by a factor of 1000 (18). This high accuracy is especially garantueed in the overtone spectroscopy. The high superiority of overtone spectroscopy to determine H-bonds with high accuracy is not generally known. Therefore the reputation if the I.R. methods to study intermolecular effects is not as high as it should be.
By temperature studies of H-bond equilibrium constants analogic to equation (2) we could determine H-bond energies ΔH_H. In the literature are given straight lines for plots ΔH_H as f ($\Delta \nu$) (Badger-Bauer plots). For instance this plot given for different alcohols as H-bond donors and N = C-CH_3 as acceptor is a good straight line ΔH_H = a$\Delta \nu$ starting at the zero point in the series: starting with $(CH_3)_3$ OH as weakest donor   (ΔH_H = - 1.96 kcal/mol and $\Delta \nu$ = 64 cm^{-1}), over different alcohols like CH_3OH with - 2.55 kcal/mol and 92 cm^{-1} , or CF_3CH_2OH with - 4.2 kcal/mol and 130 cm^{-1} to $(CHCl_2)_2CHOH$ with - 4.66 kcal/mol and 175 cm^{-1} to $(CF_3)_2CHOH$ with - 5.83 kcal/mol and 209 cm^{-1} (datas from (19); such results are in agreement with the simple theory of H-bonds given by Lippincot and Schroeder (20)).

Own experiments with CH_3OH as donor and different oxy-bases as acceptors give a proportionality:

$$\Delta H_H = a\Delta\nu + b \qquad (3)$$

a straight line for stronger acceptors which is not starting at the zero point. Own experiments with pyrrolidon as donor and different bases (16) has been given a result $\Delta H_H = a\Delta\nu + b$ with $b\neq0$ too. Similar results with $b\neq0$ have been published by Rao and al. (17) and Drago et al. (21-23). There are not distinct reasons known if a linear relationship with $b=0$ is real or the value b in equation (3) has additional resons. The plot given by H.Kleeberg during this conference ν of the water combination band $\nu_3 + \nu_2$ against the heat of vaporisation of water in presence of different H-bond acceptors indicates a similar property: a linear relationship for higher values but a lower slope at low ones. Still without clearing the meaning of the value b in eq. (3) Badger-Bauer plots give a useful possibility to study H-bond interactions in well studied systems like alcohols or water.

In solution experiments we can assume that the H-bond complexes can free adjust the minimum of the H-bond potential. The base strength of the acceptors or the acidity of protons give the parameter for the observed ΔH_H or $\Delta\nu$ -values.

3.2. H-Bond parameters in condensed phases

In condensed phases there are in addition other parameters for H-bonds. For instance as consequence of different H-bond strength in crystals are observed different oxygen...oxygen distances $R_{0...0}$ between the proton donor OH and its acceptor. There are different plots of $\Delta\nu = f(R_{0...0})$ known (24). In such relations too we observe two regions with different slopes. For instance one of the first the known Wall and Horning plot (24) gives a low slope at high $R_{0...0} > 2.9$ Å a proportionality $\Delta\nu \sim (R_{0...0})^{-3}$ and a higher slope $\Delta\nu \sim (R_{0...0})^{-14}$ for $R_{0...0} < 2.8$ Å (16).

In addition in steric hindered condensed media we have to taken in account the angle dependance of the H-bonds (17, 25). We have tried to study this by coordination of the distinct different concentration dependant H-bond bands in the matrix technique of lower aggregates of CH_3OH or water (16, 17, 25). There are a lot of papers discussing if the first observed aggregates in the matrix technique are linear or cyclic dimers. This discussion has been dominated by a paper by Tursi and Nixon (26), they have discussed by symmetry rules that the dimer has to be a linear one. But we could demonstrate that this paper has been disturbed by a correlation of different bands to the same dimer which has different concentration dependencies (27, 28), meaning these bands belongs to different aggregates. This question is still not solved completely. It seems that both the linear and cyclic dimer are present. Some papers are starting the question if a cooperativity, that means induction effects in a longer chain of H-bonded molecules $(ROH)_n$ strengthens the H-bonds in higher aggregates.- H-bond bands can also observed in the vapour phase (29-31).

Indications given that the $\Delta\nu$ values observed by H-bonded aggregates in the vapour phase are generally lower than in condensed media like CCl_4-solutions or in matrices. For instance measurements in my working group (30, 31) have shown that in the example of $RO^{\prime H}$ ...HCl complexes the ratios of the $\Delta\nu$ values are the following ones:

Table 2
Ratio of frequency shifts of ROH-HCl complexes in argon matrixes to the gase phase and ratios in CCl_4-solutions to vapour phase (30-31).

Acceptor	$\Delta\nu$ Argon Matrix: vapour	$\Delta\nu$ CCl_4-solution:vapour
CH_3OH	1.36	1.36
CH_3CH_2OH	1.33	1.34
$(CH_3)_2CHOH$	1.30	1.32

New experiments are in work to prove if this effect is caused by a induced shorter $RO^{\prime H}$...HCl distance in condensed phases or by other additive secondary effects.

3.3. Discussion of the Half Width $\Delta\nu_{1/2}$ of H-Bond Bands

A series of different studies (see Fig. 2) (32) in our Marburg lab have established the known fact that the half width $\Delta\nu_{1/2}$ of H-bond bands are:

$$\Delta\nu_{1/2} \sim \Delta\nu \tag{4}$$

The observation that the half width at same $\Delta\nu$-values are different in solutions or liquids and in the solid phase of crystalline hydrate demonstrates that the distribution of $R_{0...0}$-distances and H-bond angles contribute to the relation (4). This could be the consequence (32) of the relation

$$\Delta\nu \sim (R_{0...0})^{-n}$$
$$\Delta\nu_{1/2} = |d(\Delta\nu)(dR)| \sim (R_{0...0})^{-n-1} \sim \Delta\nu\, R_{0...0} \sim \Delta\nu^{(1+\frac{1}{n})} \tag{5}$$

and a similar relation may be valid (32) for the angle dependance $\Delta\nu = f(\beta^m)$. This would mean that the half widths in condensed media content informations on the distance and angle distribution function, which are generally assumed as typically for the properties of liquids. It may be difficult to differ both partition functions on distances and angles which are of importance for H-bonded liquids. At a simpler point of view we could stress that the contour of H-bond bands content the information of the H-bond energy partition function. To determine it the Badger-Bauer plots could be helpful.

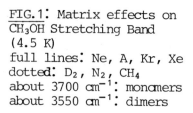

FIG.1: Matrix effects on
CH₃OH Stretching Band
(4.5 K)
full lines: Ne, A, Kr, Xe
dotted: D₂, N₂, CH₄
about 3700 cm⁻¹: monomers
about 3550 cm⁻¹: dimers

abscissa: critical T of
the Matrix gas.

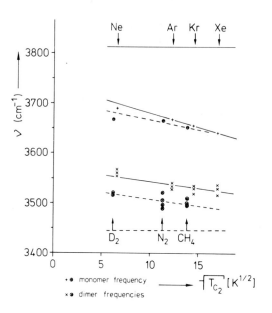

Band Maxima CH₃OH (Matrix) 4K

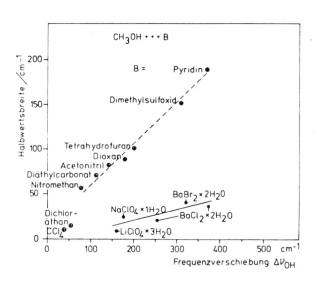

FIG.2: Half-width as
function of Frequency
Shift Δν of OH Band.

Dotted: CH₃OH solutions
in presence of different
bases B.

Full lines: H₂O in crys-
tal-line hydrates.

4. COMPARISON FUNDAMENTAL AND OVERTONE REGION, INTENSITY PROBLEMS

Fig. 3 and 4 demonstrate the differences between the fundamental OH
stretching property (Fig. 3) and its first overtone (Fig. 4). In both
figures we see the small band of undisturbed OH groups with its small
half width and   red shifted the absorption region of H-bonded OH
groups with broader half width. In the fundamental region the intensity
$\int \varepsilon d\nu$ increases strongly by H-bonds ($\int \varepsilon d\nu \sim \Delta \nu$),but in the overtone
region the area of OH bands $\int \varepsilon d\nu$ is nearly constant during the H-bond
formation (33). As reason we could assume:
The transition moments matrix elements for the fundamental band $(0 \rightarrow 1)$
R_{10} and for the first overtone $(0 \rightarrow 2) R_{20}$ can be given in approximation
(34-37) as:

$$R_{10} = \frac{\Theta}{\sqrt{2}} \; p_1 - \frac{5}{\sqrt{2}} \; b\Theta^2 p_2 \tag{6}$$

$$R_{20} = b \frac{\Theta}{\sqrt{2}} \; p_1 + \frac{\Theta^2}{\sqrt{2}} \; p_2 \tag{7}$$

(p_1: first derivation dp_0 / dq of the dipole moment p_0; p_2: second
dipole moment derivation $d^2 p_0 / dq^2$; Θ and b constants depending on
molecular properties, b contains the anharmonicity constant).
The discussion of eq. (6) and (7) shows: if we treat the OH vibration
of methanol as diatomic oszillator in approximation (35-37) and because
the constant b is negative, in cases in which the dipole moment derivations are increasing with stronger H-bonds, equations (6) and (7)
demonstrate easy why the intensity of the fundamental band increases
and the overtone intensity changes much less. Unfortunately the integre-
ted band intensities $\int \varepsilon d\nu$ · are proportional to R_{10}^2. Therefore there
are two possibilities to calculate bond moments from band intensities.
By our measurement of methanol...base H-bonds in overtone and fundamental measurements we would estimate the bond derivations of CH_3OH
with the two possible sets of Table 3.

Table 3
Dipole moment p_0 and its first p_1 and second differential quotient p_2
of CH_3OH in presence of CCl_4 or acetonitrile: AN; dioxane: DIO; tetra-
hydrofuran: THF or dimethylsulfoxide: DMSO (units Debyes, 10^{-18} esucm).

CH_3OH in	+ +			+ −		
	$-p_0$	p_1	p_2	p_0	p_1	p_2
CCl_4	1.83	1.00	2.83	2.32	1.16	−0.16
AN	0.94	2.18	3.12	2.15	2.29	0.14
DIO	0.83	2.52	3.35	2.11	2.62	0.51
THF	0.73	2.71	3.44	2.17	2.82	0.65
DMSO-d_6	1.10	3.26	4.36	1.78	3.38	1.60

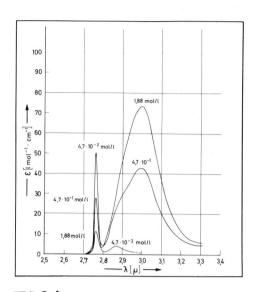

FIG. 3 :

Fundamental OH-Band of cyclohexanol-
solutions (CCl$_4$)

(Room-Temperature)

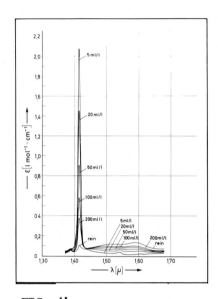

FIG. 4:

First Overtone OH-Band of
Cyclohexanol-solutions

(CCl$_4$)

(20°C)

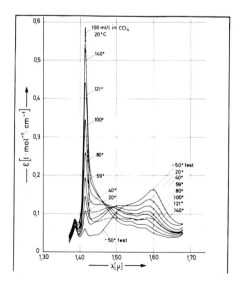

FIG. 5:

First overtone OH-Band of
liquid cyclohexanol (at
different temperatures)

dotted: CCl$_4$-solution 20°C.

We get the result that the H-bonds increase the differential quotients
of the dipole moment. Generally the overtone measurements have a much
higher precision compared with fundamental ones. But in the case of the
areas of bands belonging to H-bonded OH groups we have to taken in
account that in solutions or liquids exist simultaneous excitation of
vibrations in neighbour molecules are H-bonded (38). This effect
disturbes the intensity measurement in the wavelength region in fig.4
from 1.55 till 1.65 µm.
The measurements of the content of non H-bonded so called "free" OH
groups give same results done with overtone or fundamental bands (16,
39). The precision in this case is much higher in the overtone method.

5.   STUDIES OF THE H-BOND STATE OF PURE LIQUIDS

The I.R. method to study the H-bond state in solution is well established.
Attempts to extend this method on pure liquids has been successful too
(16, 40, 41, 42). Fig. 5 shows as example the overtone band of cyclo-
hexanole in liquid state (29). By comparison with fig. 4 we could re-
cognize the break of H-bonds in liquid cyclohexanol with increasing
temperature. At high T the sharp OH vibration of free OH appears in the
liquid at the same frequency like in dilute solutions. Extended studies
of liquid water or alcohols (33, 40-43) have shown, that the areas
$\int \varepsilon d\nu$ of the free OH overtone band is insensitive on its environment.
Therefore we can estimate the concentration of "free" OH in H-bonded
liquids.
It could be shown (40-43) that this information is very useful to describe
the anomal caloric properties of H-bonded liquids quantitatively by the
simple assumption of the equilibrium between open and closed H-bond
similar to equation (2). These results are in agreement with computer
simulation calculations of the water structure. Which have shown that
the H-bonds are oscillating around the optimum H-bond angle ß. This
angle is a fundamental parameter too for the secondary and tertiary
structure of H-bonded molecules in biochemistry (18, 44).
The H-bond energy ΔH_H in liquid water could be determined too by an
equilibrium similar to eq. (1) and (2):

$$OH_{free} \;+\; \Theta_{free} \;\rightleftharpoons\; OH_{bonded} \tag{8}$$

(Θ: lone pair electrons)

We took two different methods:
1. Measuring the T-dependance of the concentration of OH_{free} at the 1.
overtone band of HOD (see fig. 6) at 7100 cm^{-1} or 1.4 µm (41) with
the result: $\Delta H_H = -3.8$ kcal/mol.
2. The HOD band 6200 till 5800 cm^{-1} corresponds to a simultaneous excitation of the OH and OD stretching vibrations in two neighboured H-
bonded molecules (38). This band is a measure of the content of OH
(bonded). The H-bond equilibrium (8) can be determined therefore too by
measurements of the T-dependance of OH (free) at 7100 cm^{-1} and the OH
(bonded) at the simultaneous band. This method gave $\Delta H_H = -3.7$ kcal/mol
in good agreement with the first one (38).

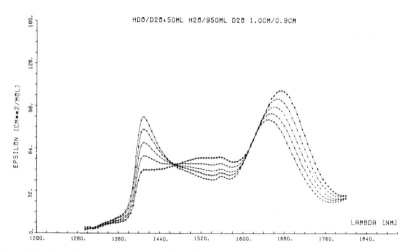

FIG. 6: First overtone-band of liquid HOD at different T.

(left band from top to bottom: 90°C; 70°C; 50°C; 30°C; 10°C).

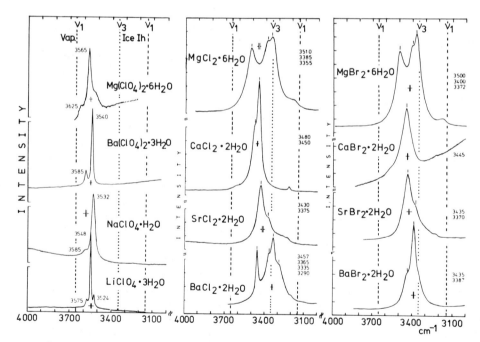

FIG. 7: Raman-Water Spectra in Crystalline Hydrates.

dotted: water-vapour and ice-bands.

A fairly good control of the method to study pure liquids is the determination of ΔH_H in the well estabilished solution technique and in pure
liquids. Both done with CH_3OH gave for the self-association: ΔH_H - 4.2
kcal/mol in CCl_4 (medium concentrations) and -4.1 kcal/mol in pure
CH_3OH.
The anomal properties of CH_3OH and C_2H_5OH could be calculated too by
the knowledge of the content of OH_{free} (43). Estimations of the H-bond
state in liquid pyrrolidone, N-ethylacetamide or in 6-polyamids samples
are done too (29).

6.   APPLICATIONS ON AQUEOUS SYSTEMS

The big success to study H-bonds in solutions and to extend this spectroscopic method on pure liquids induces the attempt to apply it on
more complicated systems too. The big importance of aqueous systems
compels to look for first informations.

6.1. Electrolyte solutions

In first approximation salt additives change the I.R. spectra of water
similar like a temperature change (45, 46). Therefore we could determine
the so called structure temperature T_{str} as a T at which pure water
would have a similar I.R. overtone spectra, that means having a similar
H-bond structure like the solution. The series of salts orders with its
T_{str} gives the so called Hofmeister or lyotropic ion series (46). This
would mean that the Hofmeister series is the series of the H-bond
structure of solutions. The first group of salts has a $T_{str} < T_{solution}$,
having a stronger H-bond system as water at the same T. Such salt induces salt-out effects on third solutes (solution is less hydrophilic).
This effect can be described by the orientation of water dipoles in the
Coulomb field of the ions (Debye, 47). But there is a big salt group
with $T_{str} > T_{solution}$, having a weaker H-bond system like pure water at
the same T. These salts can induces salt-in effects (being more hydrophilic) (46), they are called structure breaker. The breaker mechanism
has been unclear.
Therefore we have measured the frequency shifts of water in crystalline
hydrates, remembering that in hydrates the distances and orientations
of water are in nearly optimum conditions. The question of these experiments has been could we learn the water...ion interactions by this
technique (fig.7)? We took Raman spectra of hydrates (32, 48), because
the asymetric OH stretching vibration is Raman inactive and the spectra
are therefore easier to interprete. Unexpected the most Raman stretching
band of hydrate water have a smaller $\Delta\nu$ as ice (32, 48, 49). Only some
salts known as strong electrolytes or inducing hydrolysis give higher
$\Delta\nu$ -values. How to interprete this result? In a first step we could try
to assume that similar Badger-Bauer plots as we know from ROH or H_2O
experiments with different organic bases could be applied on aqueous
electrolyte solutions too. With this assumption we have to ask could the
interaction energy HOH...anion per mole OH groups be smaller than the
HOH...OH_2 interaction? This contradicts the big discussed hydration

energies ΔH_{hyd} in the seize of 100 till 200 kcal/mole. But firstly we
could remember that the Debye-Hueckel theory assumes too that the
Coulombic energy in a electrolyte solution is small compared with RT.
Secondly we could now understand why salts have finite solubility in
spite of its high hydration energies - which are higher than the normal
covalent chemical bonds. Hydration energies are given per mole salt.
If we assume that only a group of water molecules could induce the hydration energy we could understand both: 1. we need a finite solubility,
we need in minimum the group of water which induce ΔH_{hyd}. 2. Rememberin
that every ion has 4 active interacting groups: 2 protons interacting
with anions and 2 lone pair electrons interacting with cations. A rough
estimation of the seize of the hydration sphere by the solubility of a
salts and multiply the number of H_2O molecules got in this way by the
factor of 4, we get an estimation of the interaction energies ion...per
active water group (50). Indeed for many salts we get in this estimation
at 25°C energies not bigger than 5 kcal/mole, a seize which may be the
optimum of H-bond energies in a water dimer. Exceptions are: CsF (- 20
kcal/mol); LiCl (- 19.2); LiBr (- 18.6); KF (- 15.5); RbF (- 10.9);
NaI (- 9.4); CsCl (- 8.2) or LiF (- 10 kcal/mol).
In addition we have to taken in account that the vibrational spectroscop
of OH stretching modes is mainly sensitive to anion...OH interactions
which may be smaller than cation...OH_2 interactions.
Secondly we have to discuss ion pair formation in solutions which influence the heat of solutions strongly. But last not least we have to
look if the vibrational methods could expect similar effect by ions on
the OH vibrations like normal H-bonds. The computer simulations simulate
H-bonds as Coulomb forces between point charges. This favored our spectroscopic attempts. But we are not sure if ΔH_H influence $\Delta \nu$ values or
the gradients of the interaction fields and so on.
As first experiments we have determined the H-bond interactions in HOD
solutions with the technique of the simultaneous bands (see chapter 5).
We got reduction of the so calculated average H-bond energies in solutions (48) (fig. 8). In fig. 8 is plotted the average ΔH_H energies of
aqueous solutions (15 water per ionpair) against the frequency shift
$\Delta \nu$ of the fundamental band in the respected crystalline hydrate (48).
This method establishes a reduction of H-bond interactions and is in
agreement with the assumed consequences of our observed Raman shifts
of hydrate water. Further experiments have to be done to look, if our
ideas of electrolyte solutions are useful.
Our success to calculate the anomal properties of liquid water based on
the spectroscopic content of "free" OH has been successful too calculating the specific heat. As result we have to assume that the change of
H-bond content per degree is the main factor of the intermolecular part
of the specific heat of water.
Measurements of aqueous electrolyte solutions have established the that
this part is reduced by normal salts (48). On this base the negative
partial molar heats of ions could be understood.

6.2. Membrane Mechanism

The combination band of water $\nu_3 + \nu_2$ (stretching + bending) is useful t

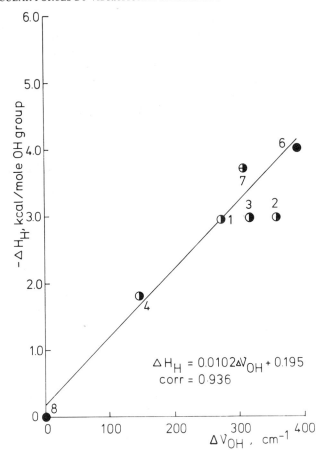

FIG. 8: Maximum of the Fundamental Water Band as Function of the Average H-Bond Energy of HOD in Solutions (H_2O : salt = 15 : 1)

1. LiCl · $1H_2O$

2. KF · $2H_2O$

3. LiBr · $2H_2O$

4. $NaClO_4$ · $1H_2O$

6. Ice IR (HOD, $-4^{\circ}C$)

7. Liquid HOD $25^{\circ}C$

8. H_2O vapour

study water properties in mixtures with organic solutes, because CH or alcoholic OH vibrations are absent in this frequency region (see too lecture Kleeberg on this conference).

The I.R. spectra of water inside desalination membranes (51, 52) are different from spectra of liquid like water. We conclude that the water structure inside efficient membranes is different from liquid like water. This hydration water does not solve salts and is preventing the ion flux through desalination membranes.

The conclusion from chapter 6.1., that we need a group of water per ionpair, is important for the membrane mechanism too.

6.3. Structure of Water in Biologic Samples

In mixtures water/biochemicals the band maximum of the $\nu_3 + \nu_2$ water combination band is shifted in a small amount (53, 54). In the last fig. 9 is demonstrated in the middle part the position of this band maximum as function of relative humidity in tendon or gelatine. We observe a plateau till the first hydration shell is completed and than a slow shift in the direction of pure water. Reverse we observe a small shift of a biopolymer I.R. band, during forming the hydrate and after a plateau at high relative humidities. There are a lot of discussions

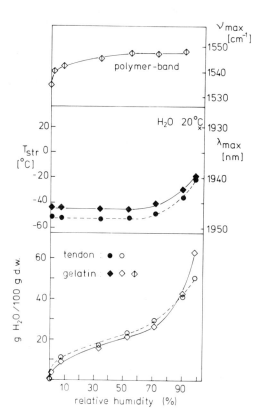

FIG. 9 a)
Shift of 1535 cm^{-1} Gelatine Band by Water.

FIG. 9 b)
Shift of the Water Combination band (right scale)
Left scale: Structure temperature of the hydrate water.

FIG. 9 c)
Desorption Isotherms.

if we could differ different types of water in living cells, called
"bound" and "free" water. We would establish this difference by our
difficile technique, but have to stress the differences are small. In
addition we would reject the nomenclature of these two types. We would
prefer the nomenclature: hydrate and liquid like water.
The lect scale in fig.9 (middle) is given the structure T, T of pure
water with the same band maximum like the water inside tendon. This
scale demonstrates why the small differences in the water structure
observed by the band shift at different relative humidities could induce big effects in biologic samples (for instance seed does not grow
at low relative humidities, or small reduction of rel. humidities can
stop bacteria grows on food).

I thank the Deutsche Forschungsgemeinschaft for funds to visit this
conference and funding my work; many thanks to my coworkers for their
excellent cooperation.

REFERENCES

1. Luck, W.A.P.: 1951, Z.Naturforsch. 6a, p. 313.
2. Luck, W.A.P.: 1952, Z.Elektrochem. 56, p. 870.
3. Penner, S.S., and Weber, D.: 1951, J.Chem.Phys., 82, p. 451.
4. Van Kranendonk, J.: 1980, "Intermolecular Spectroscopy", North
    Holland Co., Amsterdam, p. 77.
5. Poll. J.D., Tipping, R.H., and Prasad, R.D.G.: 1976, Phys.Rev.Lett.
    36, p. 248.
6. Prasad, R.D.G., and Reddy, S.P.: 1975, J.Chem.Phys., 62, p. 3582;
    1977, J.Chem.Phys., 66,p. 707.
7. Hallam H.: 1973, "Vibrational Spectroscopy of Trapped Species",
    John Wiley + Sons, London.
8. Luck, W.A.P.: 1979, Angew.Chem. 91, p. 408; 1979, Angew.Chem.Int.
    Ed.Engl. 18, p. 350.
9. Luck, W.A.P.: 1972, "Empirische Regeln zur Abschaetzung physikal.-
    chemischer Eigenschaften von Gasen und Fluessigkeiten", in: Ull-
    manns Enzyklopaedie der techn. Chemie, Verlag Chemie, Weinheim
    Bd. I 1/1, p. 55.
10. Berthelot, D.: 1856/1898, C.R. 126, p.1713.
11. Schrems, O.: 1981, Dissertation, Marburg.
12. Behrens, A., Luck, W.A.P., and Schrems, O.: in preparation
13. Luck, W.A.P., and Schioeberg, D.: 1979, Advan.Mol.Relaxation
    Processes, 14, p. 277.
14. Schioeberg, D., and Luck, W.A.P.: 1979, J.Chem.Soc.Faraday Trans.I,
    75, p. 762.
15. Badger, R.M., and Bauer, S.H.: 1937, J.Chem.Phys. 5, p. 839.
16. Luck, W.A.P.: 1973, "Infrared Studies of Hydrogen Bonding in Pure
    Liquids and Solutions", in: The Hydrogen Bond in Water: A compre-
    hensive Treatise, Plenum Publishing Corp., ed.F.Franks, New York,
    Chapter 4, Vol.II, p. 225.

17. Rao, C.N.R., Dwived, P.C.: .Ratajczak, H., and Orville-Thomas, W.J.: 1975, Farad.Trans.II, 71, p. 955.

18. Luck, W.A.P.: 1965, Naturwissenschaften 52, p. 25, 49; 1967, Naturwissenschaften 54, p.601.

19. Joesten, M.D., and Schaad, L.J.: 1974, "Hydrogen Bonding", Marcel Dekker, Inc. New York; Murthy, A.S.W., and Rao, C.N.R.: 1969, "Spectroscopic Studies of the Hydrogen Bond", in Appl. Spectroscopy Rev., Marcel Dekker, New York, Vol. 2.

20. Lippincott, E.R., and Schroeder, R.: 1955, J.Chem.Phys., 23, p.1099.

21. Joesten, M.D., and Drago, R.S.: 1962, J.Am.Chem.Soc. 84, p.3817.

22. Purcell, K.F., and Drago, R.S.: 1967, J.Am.Chem.Soc. 89, p. 2874.

23. Drago, R.S., O'Bryan, and Vogel, G.C., 1970, J.Am.Chem.Soc. 92, p. 3924.

24. Wall, T.T., and Hornig, D.F.: 1965, J.Chem.Phys. 43, p. 2079.

25. Luck, W.A.P.: 1976, "The Angle Dependence of Hydrogen-Bond-Interactions" in: The Hydrogen Bond, ed.Schuster-Zundel-Sandorfy, Verlag North Holland Publ., Vol.II, Kap.11, p. 527.

26. Tursi, A.J., and Nixon, E.R.: 1970, J.Chem.Phys. 52, p. 1521.

27. Behrens, A., and Luck, W.A.P.: 1980, J.Mol.Struct., 60, p.337.

28. Luck, W.A.P., and Schrems, O.: 1980, J.Mol.Struct., 60, p.333.

29. Luck, W.A.P., and Ditter, W.: 1967/68, J.Mol.Struct. 1, p. 261.

30. Luck, W.A.P., and Schrems, O.: 1980, Spectroscopy Letters, 13, p. 719.

31. Luck, W.A.P., and Schrems, O.: 1980, Chem.Phys.Letters 76, p. 75.

32. Buanam-Om, C., Luck, W.A.P., and Schioeberg, D.: 1979, Z.Phys. Chem.N.F. 117, p. 19.

33. Luck, W.A.P., and Ditter, W.: 1968, Ber.d.Bunsenges.Phys.Chem. 72, p. 365.

34. Yao, S.L., and Overend, J.: 1976, Spectrochim.Acta 32A, p. 1059.

35. Singh, S., and Luck, W.A.P.: 1981, Chem.Phys.Letters 78, p. 117.

36. Singh, S., and Luck, W.A.P.: J.Mol.Struct., in press.

37. Singh, S., Schioeberg, D., and Luck, W.A.P.: Spectroscopy Letters, in press.

38. Schioeberg, D., Buanam-Om, C., and Luck, W.A.P.: 1979, Spectroscopy Letters 12, p. 83.

39. Mecke, R.: 1948, Z.Elektrochem. 52, p. 269.

40. Luck, W.A.P.: 1980, Angew.Chem. 92, p. 29; 1980, Angew.Chem.Int. Ed.Engl. 19, p. 28.

41. Luck, W.A.P.: 1974, "Structure of Water and Aqueous Solutions", ed.W.A.P.Luck, Verlag Chemie/Physik Verlag, Weinheim.

42. Luck, W.A.P.: 1976, "Hydrogen Bond in Liquid Water", in: The Hydrogen Bond, ed.Schuster-Zundel-Sandorfy, Verlag North Holland Publ. Vol.III, Kap. 28, p. 1369.

43. Luck, W.A.P.: 1967, Discuss.Faraday Soc. 43, p. 115.

44. Luck, W.A.P.: 1968, Naturwiss.Rundschau, 21, p. 236.

45. Luck, W.A.P.: 1965, Ber.Bunsenges.Phys.Chem. 69, p. 69.

46. Luck, W.A.P.: 1964, Fortschr.Chem.Forsch. 4, p. 653.

47. Debye, P.: 1929, "Polare Molekeln", Leipzig, Hirzel; 1927, Phys.Ztschr. 28, p. 199.

48. Buanam-Om, C.: 1981, Dissertation Marburg.
49. Luck, W.A.P., and Shah, S.S: 1978, Progr.Coll.and Polymer Sci.
     65, p. 53.
50. Luck, W.A.P.: 1980, "Structure of Electrolyte Solutions and
     Membrane Mechanism", Proceedings of the 7th Intern.Symposium
     on Fresh Water from the Sea, Amsterdam, Vol. 1, p. 77.
51. Luck, W.A.P., Schioeberg, D., and Siemann, U.: 1980, J.C.S.Faraday
     II, 76, p. 136.
52. Luck, W.A.P., Schioeberg, D., and Siemann, U.: 1979, Ber.Bunsenges.
     Phys.Chem. 83, p. 1085.
53. Kleeberg, H., and Luck, W.A.P.: 1977, Naturwissenschaften, 64,
     p. 223.
54. Luck, W.A.P., and Kleeberg, H.: 1978, "Structure of Water and
     Aqueous Systems", in: Photosynthetic Oxygen Evolution, ed.H.
     Metzner, Academic Press, London-New York-San Francisco, p. 1.

INTERMOLECULAR FORCES AND SPECTRA IN WEAK CHARGE TRANSFER INTERACTIONS

Robert L. Strong
Rensselaer Polytechnic Institute
Troy, New York   12181

INTRODUCTION

Electron donor-acceptor (EDA) complexes have been known for many years. Although they manifest themselves in many ways, such as solubility and chemical reactivity, probably the most characteristic property of EDA complexes is the appearance of an intense absorption band that is not present in the separated donor and acceptor species. This band, often broad and structureless, usually is in the visible and/or near ultraviolet regions, giving rise in many cases to large color changes when the individual components are mixed. Many spectroscopic studies generating various thermodynamic parameters have been carried out on EDA complexes over the past thirty years following the classic papers by Benesi and Hildebrand (1) on 1:1 iodine complexes with aromatic hydrocarbons and Mulliken's (2,3) subsequent explanation of the absorption band based on a charge transfer (CT) resonance model. This CT theory has been widely covered in many review articles and monographs, including those of Mulliken and Person (4,5).

It is now quite well established that the Mulliken CT resonance model (summarized below) adequately represents EDA spectra, and that in fact it is necessary to explain some unique properties of these complexes. Associated with this model is the concept of (attractive) CT resonance forces that supplement the classical intermolecular electrostatic contributions to ground-state stability of EDA complexes. It should be pointed out, however, that the presence of a CT absorption band alone says nothing at all about the source, or sources, of ground-state stability; in fact, very strong CT absorption spectra are observed in many cases where there is virtually no attraction between the two components in a simple encounter -- i.e., contact charge-transfer spectra (6,7) -- which is the limiting example of CT interaction. Much of the early work on EDA complexes tended to discount classical electrostatic attractions because such strong CT absorption bands and dipole moments were obtained even for non-polar donor and acceptor species. It is now apparent, however, that other interactions such as quadrupole-induced dipole and quadrupole-quadrupole must be considered, and the question has

B. Pullman (ed.), Intermolecular Forces, 217–231.

developed in recent years as to the origin of EDA ground-state stability and the importance of CT resonance forces relative to the general van de Waals attraction when all types of interactions are taken into account (8).

MULLIKEN'S CT RESONANCE MODEL

Following Mulliken's treatment, the ground-state wave function of a 1:1 EDA complex between even-electron D and A systems may be taken to be a linear combination of two resonance structures $\Psi_0(D,A)$ and $\Psi_1(D^+-A^-)$

$$\Psi_N \simeq a\Psi_0(D,A) + b\Psi_1(D^+-A^-) \tag{1}$$

Here, $\Psi_0(D,A)$ is the wave function for the "no-bond" structure, and is an antisymmetrized product of the ground-state wave functions Ψ_D and Ψ_A. The dative structure, $\Psi_1(D^+-A^-)$, represents transfer of an electron from D to A, and may be an important contribution to Ψ_N depending on the mixing coefficients a and b. Assuming Ψ_0 and Ψ_1 are normalized but not orthogonal, the fractional contributions of "no-bond" and "dative" characters in normalized Ψ_N are given respectively by $(a^2 + abS_{01})$ and $b^2 + abS_{01})$, where $S_{01} = \int\Psi_0\Psi_1 d\tau$; for loose complexes S_{01} is very small, $a^2 + b^2 \sim 1$, $a^2 \gg b^2$, and hence b^2 gives an approximate measure of the fractional transfer of charge from D to A.

Equation 1 implies quantum mechanically an excited CT electronic state involving the mixing of the same wave functions Ψ_0 and Ψ_1:

$$\Psi_V \simeq a^*\Psi_1(D^+-A^-) - b^*\Psi_0(D,A) \tag{2}$$

The excited state wave function Ψ_V is normalized and orthogonal to Ψ_N, and as a first approximation $a \sim a^*$ and $b \sim b^*$. The characteristic absorption band of EDA complexes results from transition from the ground Ψ_N state to the excited Ψ_V state. More precisely, however, additional terms must be included to account for back charge transfer and electronically excited local (D* and A*) and CT structures. Equations 1 and 2 then become respectively

$$\Psi_N = a\Psi_0(D,A) + \sum_i b_i\Psi_i(D^+-A^-) + \sum_j c_j\Psi_j(D^--A^+) + \sum_k d_k\Psi_k(D^*,A)$$

$$+ \sum_\ell e_\ell\Psi_\ell(D,A^*) \tag{3}$$

and

$$\Psi_V = \sum_i a_i^*\Psi_i(D^+-A^-) - b^*\Psi_0(D,A) + \sum_j c_j^*\Psi_j(D^--A^+) + \sum_k d_k^*\Psi_k(D^*,A)$$

$$+ \sum_\ell e_\ell^*\Psi_\ell(D,A^*) \tag{4}$$

where the summations are over all CT and locally excited state manifold

In principal Ψ_D and Ψ_A as combined in Ψ_0 are corrected for geometrical changes of D and A in the complex reorganization energy and for any polarization effects from electrostatic, dispersion, and exchange forces, although in practice they are often approximated by the undistorted wave functions of isolated D and A. An additional stability in Ψ_N is then implied by the inclusion of Ψ_1 in Equation 1; it is the extent of this contribution over and above the classical electrostatic interactions and exchange repulsions that has been strongly questioned over the past several years.

If the energy between neutral D and A at infinite separation is $W_0(\infty)$, then the "no-bond" energy at the equilibrium intermolecular separation d_{DA} associated with the wave function $\Psi_0(D,A)$ is

$$W_0 = W_0(\infty) + G_0 \tag{5}$$

where G_0 (which may be positive or negative) includes all of the geometrical changes and electrostatic attractions and repulsions between D and A. The ground-state energy W_N then includes the (negative) resonance interaction energy X_0:

$$W_N = W_0 + X_0 \tag{6}$$

The total ground-state stability is $\Delta U = G_0 + X_0$. Similarly the energy of the CT state, W_V, is

$$W_V = W_1 + X_1 \tag{7}$$

where W_1 is associated with $\Psi_1(D^+-A^-)$ and represents the net ionic energy between D^+ and A^-:

$$W_1 = W_1(\infty) + G_1 \tag{8}$$

G_1 is primarily coulombic, and is negative.

Assuming only one CT state and only wave functions Ψ_0 and Ψ_1, it can be shown (4) that solution of the Schrödinger wave equation for Equation 1 or 2 gives

$$W_{N\ or\ V} = \frac{1}{1-S_{01}^2}\left[\frac{W_0+W_1}{2} - S_{01}H_{01} \pm \sqrt{(\Delta/2)^2 + \beta_0\beta_1}\right] \tag{9}$$

where β_0 and β_1 are the resonance integrals $(H_{01}-S_{01}W_0)$ and $(H_{01}-S_{01}W_1)$, respectively, $\Delta = W_1-W_0$, and S_{01} is the overlap integral $\int\Psi_0\Psi_1 d\tau$. The lower energy root is W_N, the upper is W_V, and the difference between W_V and W_N is the energy of the CT band maximum:

$$h\nu_{CT} = W_V - W_N = \frac{2\sqrt{(\Delta/2)^2 + \beta_0\beta_1}}{1-S_{01}^2} = \frac{\Delta}{1-S_{01}^2}\sqrt{1 + \frac{4\beta_0\beta_1}{\Delta^2}} \tag{10}$$

For weak complexes $(\Delta/2)^2 \gg \beta_0\beta_1$, $S_{01}^2 \ll 1$, and $W_0 \sim W_N$; second-order perturbation theory then leads to

$$h\nu_{CT} = \Delta + \frac{\beta_0{}^2 + \beta_1{}^2}{\Delta} + \cdots \simeq \Delta + \frac{\beta_0{}^2 + \beta_1{}^2}{\Delta} \qquad (11)$$

These energy relationships are shown in Figure 1.

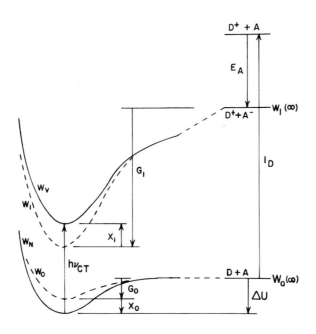

Figure 1.   Schematic EDA potential energy diagram.

It is seen that $\Delta = W_0 - W_1 = I_D^V - E_A^V + (G_0 + G_1)$, where I_D^V is the vertical ionization energy of D and E_A^V is the vertical electron affinity of A.   Letting $G_0 + G_1 = C$ and substituting into Equation 11 gives

$$h\nu_{CT} = I_D^V - E_A^V + C + \frac{\beta_0{}^2 + \beta_1{}^2}{I_D^V - E_A^V + C} \qquad (12)$$

or, for stronger complexes where Equation 11 is not valid, Equation 10 leads to

$$h\nu_{CT} = \frac{I_D^V - E_A^V + C}{1 - S_{01}{}^2} \sqrt{1 + \frac{4\beta_0\beta_1}{(I_D^V - E_A^V + C)}} \qquad (13)$$

In most cases data are reported for a series of donor species with a single acceptor, so that $E_A^V - C = C_1$ is taken to be constant.   For example, a wide variety of donors with molecular iodine as acceptor

follow Equation 12 quite well (9), although the relatively strong ali-
phatic amine-I_2 complexes are better fitted to the parabolic Equation
13 (10). Similar $h\nu_{CT}$-I_D correlations for other acceptors are extensive
and well documented (9, 11, 12), and the concept of CT spectra is
firmly established.

CT CONTRIBUTIONS TO GROUND-STATE STABILITY

Experimental

Spectroscopic demonstration of the importance of CT forces in
ground-state EDA complex stabilization is much less extensive than
experimental justification of CT spectra described above. Dewar and
Thompson (13) argued that for π-π^* complexes, such as tetracyanoethylene
(TCNE) acceptor with a wide range of unsubstituted aromatic hydrocarbons
(a series strongly favoring charge transfer), the relative variation
in complex stability constant K_C with the wavelength of the CT transition λ_{CT} did not suggest any major CT contribution to ΔU. However, for
π-σ^* complexes involving iodine monochloride acceptor with aromatic
hydrocarbons in non-polar solvents the same correlation between K_C and
λ_{CT} suggests dominant CT contributions to ground-state stability (14),
as do the data for TCNE with methylbenzene (15). It should be pointed
out, however, that the correlation used in References 13 and 14 is based
on variations in K_C being due to energy rather than entropy changes;
this may not be valid even over a homologous series of donors (16).

One problem in determining the extent of CT stabilization in the
ground state is in ascertaining the role of the solvent. Spectral and
thermodynamic data for EDA complexes are for the most part on liquid-
phase systems involving relatively non-volatile components, whereas the
Mulliken's CT resonance model theory is based on isolated donor, acceptor, and complex species that would be present only in a low-pressure
vapor phase. On the other hand, most vapor-phase thermodynamic results
based on spectroscopic measurements are subject to rather large error
limits because of the low D and A concentrations and concomittant small
transmittance changes (17). In most cases substances conveniently
studied in the liquid phase inherently have low vapor pressures at
reasonably accessible temperatures.

Solvent effects on EDA complexes have been reviewed (18). These
may be separated into two broad categories: specific interactions (such
as hydrogen bonding), which in fact correspond to additional complex
formation; and non-specific interactions, associated with random encounters. The difference in CT transition energies between the vapor
and liquid phases is given by (17):

$$\Delta h\nu_{CT} = h\nu_{CT}(\text{vapor}) - h\nu_{CT}(\text{liquid}) = \Delta X_N - \Delta X_V \qquad (14)$$

where ΔX_N and ΔX_V are the solvation stabilization energies of the complex in its ground and excited states, respectively. If ΔX_V is greater
than ΔX_N, as expected for weak complexes where $b^2 < a^2$ (very little ground-
state charge-transfer), there will be a red shift (lower energy) in

going from vapor to liquid. The magnitude of this shift will depend
on the strength of the complex, and in fact the transition may even be
blue-shifted for strong complexes. ΔX_N and ΔX_V are difficult to evaluate, however, because the CT transition occurs to the Franck-Condon
excited state having the geometry of the ground state and existing in
a ground-state solvation sphere, which is followed by rapid relaxation
to the lower-energy equilibrium excited state. The difference between
the Franck-Condon and equilibrium states would be expected to be large
for CT complexes. Also, although trends in the red shift as a function
of complex stability appear to follow Equation 14 for closely related
donors with a common acceptor, there is no correlation among widely
different donors. Activity coefficients and the specific concentration
scales appropriate for non-ideal behavior in solutions (18) are generally neglected. These and other factors discussed in Reference 18 make
the experimental evaluation of CT <u>vs</u> electrostatic forces based on
liquid-phase spectroscopic data very questionable. It is worth pointing
out that the conclusion of dominant CT stability of the ground state
reported in Reference 14 was based on non-polar solvent results; the
effect of electrostatic forces could not be ruled out in the case of
polar solvents.

One area that is very sensitive to ground-state stability which
has been extensively studied in the vapor phase is the three-body recombination of halogen atoms. For the most part (except at high pressures)
the recombination is kinetically third order:

$$X + X \ (+ \ M) \rightarrow X_2 \ (+ \ M); \quad \frac{d[X_2]}{dt} = k_{recomb}[X]^2[M] \tag{15}$$

This recombination occurs either via an energy-transfer (ET) mechanism
involving an unbound quasidimer X_2^* (19)

$$2X \ \rightleftarrows \ X_2^* \tag{16}$$

$$X_2^* + M \rightarrow X_2 + M \tag{17}$$

or through an M-X intermediate (the radical-molecule, RMC, mechanism):

$$X + M \ \rightleftarrows \ M\text{-}X \tag{18}$$

$$M\text{-}X + X \rightarrow X \ + M \tag{19}$$

The efficiency of the recombination, as manifested in k_{recomb}, is determined by the stability of M-X and hence to some extent on the nature of
M. (In the liquid phase, of course, there is no explicit need for the
third-body M even though interactions represented by Equation 17 still
exist, and recombinations in "inert" solvents are kinetically second-
order in halogen atom concentration.) Transient charge-transfer spectra
associated with M-X in both liquid and vapor phases have been detected
in many photochemical or radiological systems containing halogen atoms
(7).

Russell and Simons (20) showed a correlation between log k_{recomb}
and the boiling point or the critical constant of M, as well as

a regular inverse variation between k_{recomb} and the ionization energy of M. It was concluded that general van der Waals forces are involved rather than more specific types of interaction. This inverse relationship between k_{recomb} and I_D of M does not in itself allow a distinction between electrostatic and CT stabilization of M-X, however, since both are generally dependent on ionization energies and polarizabilities in the same way (3,8,21). On the other hand, it has been shown (22) that the non-CT dipole-quadrupole, quadrupole-quadrupole, polarizability-polarizability, and polarizability-dipole interactions are greater for I atoms than for Br atoms with alkyl halides, whereas the electron affinity of I is less than that for Br. Thus, opposite effects should be observed depending on the dominant nature of the intermolecular interaction when different halogen atoms recombining are compared with a common third body M: k_{recomb} will be greater for Br recombination than for I recombination if M-X stability is predominantly charge-transfer, whereas the opposite will be the case for primarily electro-static inductive effects. Neglected in this evaluation is any ET contribution to the recombination (Equations 16 and 17), which will become more significant with high I_D third bodies.

Results on vapor-phase recombination of Br and I atom recombination following flash photodissociation of the parent diatomic species are given in Table I for several third bodies covering over a 2 eV range in I_D (22). Except for Br in benzene, strong CT absorption bands are observed in all cases (7). Although Br atoms in benzene vapor recombine much faster than I atoms, for both halogens k_{recomb} decreases with

Table I

Third-order Rate Constants for Vapor-phase Recombination
of Bromine or Iodine Atoms at Room Temperature

Third Body	I_D/e_V	$k_{recomb}/10^{-32}$ cm^{6}molecule^{-2}s^{-1}	
		Br	I
C_6H_6	9.24	145	23
C_2H_5Br	10.29	69	22
CH_3Br	10.53	33	15
C_2H_5Cl	10.97	15	16
CH_3Cl	11.33	9.6	11

increasing I_D, consistent with either CT or electrostatic interactions. However, the decrease is greater for Br than for I, with a reversal in recombination efficiency at the highest donor ionization energy. These results suggest that both CT and electrostatic interactions are involved in stability of the intermediate M-X. For low-I_D third bodies, CT interactions dominate, whereas at high-I_D, the non-CT forces become relatively more important. In similar recombination studies with different third bodies, DeGraff and Lang (23) concluded that CT forces are important in vapor-phase halogen atom recombinations but that factors

other than E_A of the halogen atom must be the dominant influence. The
third bodies used by them, however, had for the most part ionization
energies that would put their results primarily in the electrostatic
region. Since the halogen atoms have very large electron affinities
compared to most acceptors involved in EDA studies and therefore strongly
favor CT interactions, these results suggest that in most cases of
lesser E_A the classical electrostatic forces may be the major contributor to ground-state stability.

Theoretical

Several theoretical studies have been carried out on the origin
of intermolecular interactions in EDA ground-state complexes, and it is
clear that each specific type of interaction -- CT, dipole-dipole,
quadrupole-induced dipole, etc. -- must be considered. Hanna and coworkers have shown, for example, that quadrupole-quadrupole interactions
may be the dominant contributions to ground states of weak $\pi-\pi^*$ and
$\pi-\sigma^*$ complexes involving non-polar components (8,23).

Morokuma and coworkers have carried out extensive analyses of
interaction energies and geometries of EDA complexes using energy and
charge distribution decomposition (ECDD) analyses, as summarized in (24)
Ab initio SCF calculations are performed in which the ground-state
interaction energy of the complex ΔU is decomposed into five components
electrostatic (ES), exchange repulsion (EX), polarization (PL), charge
transfer (CT), and the coupling term (MIX) -- with the first four being
calculated separately. ES is the classical electrostatic interaction
between the undistorted charge distributions on D and A, which are
assumed to have the same geometries whether as isolated molecules or
complexed. This contribution includes the interactions of all permanent
charges and multipoles, and may be positive or negative; it does not
allow any electron transfer between D and A. The polarization contribution PL is the energy effect from the mutual distortion of electron
clouds and higher order coupling terms, and is always negative (attractive). EX, a consequence of the Pauli exclusion principle, is a short-
range repulsion due to electron cloud overlap. The CT term is as discussed above, and MIX is the difference between the total interaction
energy ΔU and the sum of the other four components; it is in fact the
sum of various coupling terms between the other four energy components.
Not included in this energy decomposition scheme within the SCF framework is the dispersion energy term DISP. For most polar A and D molecul
the five terms included in the calculations adequately describe the
observed interaction. For relatively large non-polar components, however, dispersion effects may become important (as seen, for example, in
the earlier calculations of Hanna (8)), and in these cases the contribution of DISP has been estimated by a sum-of-states second-order perturbation calculation (25).

Calculations were carried out within the closed shell LCAO-SCF
approximation using the 4-31G basis set with standard exponents, contraction coefficients, and scale factors; details of the calculation

are given by Morokuma et al (26). In the 4-31G basis set (27), each atomic orbital is represented by four Gaussian functions, but each valuence orbital is split into inner and outer parts described by three and one Gaussians, respectively. This basis set gives a reasonable estimate of the component energies but it does exaggerate molecular polarities, thereby overemphasizing the electrostatic interaction ES. Apparently this tendency has been reduced by using a modified basis set (4-31G**) that includes a set of polarization functions on each atom (28).

Selected results from Reference 24 are summarized in Table II for several strong n-σ* and σ-σ*, intermediate π-π*, and weak halogen π-π* and π-σ* EDA complexes.

Table II

Energy Components and Total Stabilization Energy ΔU
at Equilibrium Intermolecular Separation[a]

D	A	R_e/nm	Type	ES[b]	EX	PL	CT	MIX	DISP	ΔU
NH_3	BH_3	0.170	n-σ*	-92.9	86.9	-17.2	-27.1	5.6		-44.7[d]
CO	BH_3	0.163	σ-σ*	-60.9	98.9	-61.8	-68.3	63.6		-28.5[d]
C_6H_6	$OC(CN)_2$	0.36[c]	π-π*	-2.8	1.8	-1.7	-1.6	0.1	-2.6	-6.8
NH_3	F_2	0.300	n-σ*	-0.8	0.6	-0.3	-0.6	0.0		-1.1[d]
C_6H_6	Cl_2	0.36	π-σ*	-0.5	0.7	-0.1	-0.8	0.0	-0.7	-1.4

[a]From Reference 24

[b]Energies in kcal mol^{-1} (1 kcal mol^{-1} = 4.184 kJmol^{-1})

[c]Not optimized

[d]Dispersion not evaluated, but estimated to be a minor component

Some interesting conclusions are to be drawn from these calculations of Morokuma and coworkers. In all cases it is apparent that the electrostatic interaction ES is a major contributor to the total ground-state stabilization energy, although no single component becomes the dominant origin. As expected, the equilibrium intermolecular separation R_e is smallest for the polar components that lead to large electrostatic interactions and strong complexes, and is largest for the very weak complexes generated from non-polar D and A. A rather large proportion of ΔU is made up of electrostatic and dispersion contributions in the carbonyl cyanide-benzene complex, whereas CT and dispersion effects are major contributors for the weak benzene-molecular chlorine complex. These relative contributions are quite sensitive to intermolecular separation and choice of orbital basis set. Thus, in an earlier paper

(25) using the STO-3G rather than the 4-31G basis set, the results
shown in Figure 2 were obtained. Although the total stabilization

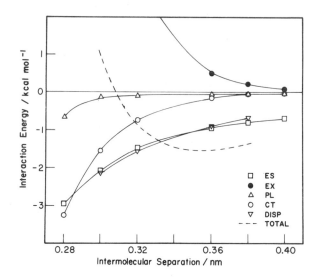

Figure 2.   Energy components and total stabilization energy for carbonyl
cyanide-benzene π-π* complex.   Data from Reference 25.

energy is appreciably smaller than that obtained with the 4-31G basis
set and the distance between the two molecular planes is larger than
that for comparable π-π* complexes, the general trends and relative
values may be compared.   Increasing intermolecular separations has the
result of reducing the CT contribution relative to the total electrostatic and dispersion components, which are of equal magnitude, and
this may justify the conclusion that CT contributions to EDA ground-
state stability are significant mainly in the strong (i.e., small intermolecular separation) complexes.

CONTACT CHARGE TRANSFER

     For the strong complexes (large binding energies) listed in Table
II the donor-acceptor intermolecular separations are close to those of
covalent bonds.   A well-defined structure results, and presumably the
same configuration exists in solution and in a crystalline form made
up of individual 1:1 complexes.   With decreasing magnitude of the interaction energy the complex becomes less stable and its geometry becomes less defined, so that its properties become the statistical
average over all thermally equilibrated configurations.   As ΔU approaches
thermal energy (RT) there can only be at best a brief association
between D and A with almost free rotation of one molecule relative to
the other.   In the limiting case of $\Delta U=0$ (in which case $W_N = W_O$) there
can be no thermodynamically preferred orientation, and the only interaction between D and A are repulsive effects at short internuclear

distances. This limiting situation is possible in the liquid phase where solvent-donor, solvent-acceptor, and solvent-solvent interactions may cancel completely the D-A interactions, but it cannot occur in the strictest sense in the vapor phase for real condensible gases. As a practical matter we have established as one operational criterion for contact charge transfer (CCT) that $\Delta U > -1.0$ kcal mole^{-1}, or $\sim -RT$ (7).

If D and A are so far apart that no overlap occurs between the donor and acceptor orbitals ($S_{01} = 0$) then the states are orthogonal and no charge transfer transitions are possible. Such a transition can occur, however, even though there is negligible stabilization in the ground-state (6). Mulliken has pointed out (29), for example, that the acceptor molecular orbital for a molecule such as I_2 is the strongly antibonding σ^*, which should be appreciably larger than the outer orbitals of the isolated molecule leading to overlap with donor orbitals even at large (contact) separations. It has also been shown by Murrell (30) that in fact this CCT transition may be very intense due to mixing of the CT states with one or more locally excited donor states, so that the CT band borrows intensity from the strongly allowed donor transitions; this may be partially forbidden in more stable complexes where an unfavorable geometry exists. (Thus, the intensity of the CT bands for a homologous series of methylated benzenes with molecular iodine decreases with increasing complex stability as the proportion of the intense CT contribution from the random higher energy configurations to the intensity decreases with increasing encroachment of the weaker stable complex CT intensity (6).)

Equation 12 takes on a particularly simple form for CCT. The resonance integrals β_0 and β_1 are negligible, $G_0 \sim 0$, and G_1 is the Coulomb attraction between the free ions D^+ and A^-. Thus, in the vapor phase

$$h\nu_{CCT} \simeq I_D^V - E_A^V - \frac{e^2}{4\pi\varepsilon_o d_{DA}} \qquad (20)$$

where e is the electronic charge, ε_o is the permittivity of vacuum, and d_{DA} is the contact pair size, or the diameter of the so-called "electron-acceptation" volume (29).

In view of their ionization energies, vapor pressures, reactivities, and spectral absorption properties, the alkyl halides represent one of the most convenient homologous donor series for spectroscopic studies and correlations of halogen acceptor EDA interactions in vapor and solution systems. Morita and Tamres (31) have shown that the liquid-phase UV band for alkyl iodides with I_2 (n-heptane solvent) can be resolved into three bands: two are CT bands separated by ca 0.5 eV and arise from the two components of the first ionization energy of the alkyl iodides; the third is a blue-shifted and enhanced n-σ^* local excitation in the alkyl iodide molecule. Plots of CT absorbance vs. RX concentration (fixed I_2), or absorbance against I_2 concentration at fixed [RX] for several alkyl iodides, bromides, and chlorides (31) were linear passing through the origin over quite large concentration changes, indicative of CCT interactions (7). Although these I_2 results were

obtained in the liquid phase, Morita and Tamres (31) estimated the
red shift in going from vapor to liquid (which according to Equation 14
should be the largest possible for contacts) to be $\sim$ 0.5 eV, and on
this basis calculated from Equation 20 the d_{DA} values given in Table
III. The general trend shown in d_{DA} of increasing size with increasing

Table III

Contact Pair Size for Alkyl Halide-I_2 Contacts (29)

	Donor					
	CH_3I	C_2H_5I	$n-C_3H_7I$	$i-C_3H_7I$	$n-C_3H_7Br$	$i-C_3H_7Cl$
d_{DA}/nm	0.639	0.664	0.679	0.691	0.671	0.660

complexity of D is qualitatively reasonable. The magnitudes are considerably larger than normal van der Waals separations, again consistent
with contacts involving I_2 (29).

Using the flash photolysis technique with both spectral and tem-
poral resolution, we have measured the transient CT spectra of I atoms
interacting with a series of alkyl iodides and bromides in the vapor
phase (32), and with the same alkyl iodides in hexane solution (33),
following flash photodissociation of I_2. As with the alkyl iodide-I_2
spectra, two CT bands are resolved in each case with the splitting
being approximately constant for each RX series: 0.53 eV (vapor) and
0.56 eV (liquid) for RI, 0.28 eV for RBr. On Figure 3 the spectral
data $h\nu_{CT}$ for the vapor and liquid phase systems are plotted as a
function of I_D according to Equation 12; also included are the data
from Reference 31 for the comparable RI-I_2 species in n-heptane solution, showing the very similar behavior among the two I atoms and one
molecular iodine-alkyl iodide series.

The reason for the failure to correlate $h\nu_{CT}$ data between the lower
and upper ionization energy regions is not resolved. One explanation
is based on predominantly contact interactions. Equation 20 implies
that for contacts the $h\nu_{CT}$-I_D plots should have unit slope if d_{DA} is
unaffected by variations in the RI donors. In fact, however, the
coulombic term in Equation 20 becomes a major contributor to the CT
energy for the systems reported here in which E_A is large, so that it
is not surprising that slopes in either ionization energy region are
approximately linear yet are less than unity if d_{DA} increases gradually
over the homologous donor series.

Whether or not d_{DA} will be different for the two ionization energy
components for a given DA pair depends on the isomeric orientation
resulting from the different transitions; since the spin-orbit splitting
in RI that leads to the two ionization energies does not produce
orbitals with different nodal planes, different orientations would
not be expected for the RI-I contacts and d_{DA} from Equation 20 should

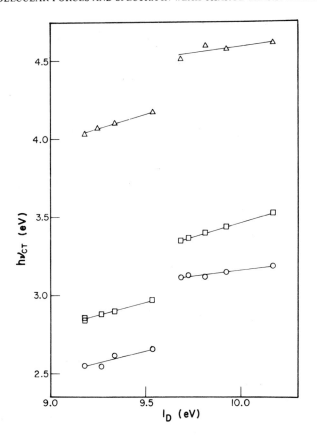

Figure 3. CT transition energies for alkyl iodide-iodine interactions. Circles: RI-I in hexane; squares: RI-I in the vapor phase; triangles: RI-I$_2$ in n-heptane.

be the same for a given DA pair. That this is approximately the case for all the RI-I vapor-phase species is shown in Table IV.

Table IV

Intermolecular Separation for Vapor-phase RI-I DA Pairs (32)

Donor	d_{DA}/nm	
	Lower I_D Component	Upper I_D Component
CH$_3$I	0.41	0.40
C$_2$H$_5$I	0.43	0.42
n-C$_3$H$_7$I	0.43	0.43
i-C$_3$H$_7$I	0.44	0.44
i-C$_4$H$_9$I	0.44	0.44

A second explanation involves modification of the spin-orbit splitting of the alkyl iodide ionization energy by a strong CT interaction such that the decoupling of spin and orbital precession in the donor ion does not occur. Qualitatively, for a strong complex there will be appreciable mixing in the (D,A) and (D$^+$-A$^-$) states in both the ground and excited states thus decreasing the local electric field from that associated with the ionic $^2\Pi$ or ^2E state of the free halide. In the limiting case this requires that the doublet splitting for the "ionized" donor in the complex be that of the strongly spin-orbit coupled free iodine atom -- 0.94 eV -- rather than that for the isolated alkyl iodide ($\sim$ 0.6 eV). When the upper ionization energy points in Figure 3 are shifted upwards by $\sim$ 0.34 eV, the two regions become in general agreement for each of the three series (32). Of course Equation 20, applicable only to contacts, is no longer valid in this case; reasonable agreement is also possible, however, with Equation 13 (32), which is applicable to strong complexes.

As pointed out above, Morita and Tamres conclude that the alkyl halide - I$_2$ interactions at least in the liquid phase are primarily contact (31). On the other hand the vapor-phase recombination of I atoms is sensitive to the nature of the alkyl halide and is over an order of magnitude greater with alkyl halides than with inert-gas third bodies, implying interactions beyond solely contact (22). We are now carrying out ab initio calculations based in part on Morokuma's energy and charge distribution decomposition analyses (24) to clarify the magnitude and specific intermolecular interactions involved in these systems.

REFERENCES

1. Benesi, H.A. and Hildebrand, J.H.: 1949, J. Am. Chem. Soc. 71, pp. 2703-2707.
2. Mulliken, R.S.: 1950, J. Am. Chem. Soc. 72, pp. 600-608.
3. Mulliken, R.S.: 1952, J. Am. Chem. Soc. 74, pp. 811-824.
4. Mulliken, R.S. and Person, W.B. "Molecular Complexes. A Lecture and Reprint Volume", Wiley-Interscience, New York, 1969.
5. Mulliken, R.S. and Person, W.B.: in H. Eyring, D. Henderson and W. Jost (eds.), "Physical Chemistry, an Advanced Treatise", Vol. 3, Academic, New York, 1969, pp. 537-595.
6. Orgel, L.E. and Mulliken, R.S.: 1957, J. Am. Chem. Soc. 79, pp. 4839-4846.
7. Tamres, M. and Strong, R.L.: in R. Foster (ed.), "Molecular Association", Vol. 2, Academic Press, London, 1979, pp. 331-456.
8. Hanna, M.W. and Lippert, J.L.: in R. Foster (ed.), "Molecular Complexes", Vol. 1, Crane, Russak, New York, 1973, pp. 1-48.
9. Briegleb, G.: "Elektronen-Donator-Acceptor-Komplexe", Springer-Verlag, Berlin, 1961.
10. Yada, H., Tanaka, J., and Nagakura, S.: 1960, Bull. Chem. Soc. Japan 33, pp. 1660-1667.
11. Andrews, L. and Keefer, R.M.: "Molecular Complexes in Organic Chemistry", Holden-Day, San Francisco, 1964.

12. Foster, R.: "Organic Charge-Transfer Complexes", Academic, London, 1969.

13. Dewar, M.J.S. and Thompson, Jr., C.C.: 1966, Tetrahedron Supp. 7, pp. 97-114.

14. Ghosh, B.D. and Basu, R.: 1980, J. Phys. Chem. 84, pp. 1887-1889.

15. Merrifield, R.E. and Phillips, W.D.: 1958, J. Am. Chem. Soc. 80, pp. 2778-2782; data plotted in Reference 13.

16. B. Nagy, O. and B. Nagy, J.: in B. Pullman (ed.), "Environmental Effects on Molecular Structure and Properties" (Proceedings of the Eighth Jerusalem Symposium), Reidel, Dordrecht, 1976, pp. 179-203.

17. Tamres, M.: in R. Foster (ed.), "Molecular Complexes", Vol. 1, Crane, Russak, New York, 1973, pp. 49-116.

18. Davis, K.M.C.: in R. Foster (ed.), "Molecular Association", Vol. 1, Academic, London, 1975, pp. 151-213.

19. Kim, S.K. and Ross, J.: 1965, J. Chem. Phys. 42, pp. 263-271.

20. Russell, K.E. and Simons, J.: 1953, Proc. Roy. Soc., Ser. A 217, pp. 271-279.

21. Slater, J.C. and Kirkwood, J.G.: 1931, Phys. Rev. 37, pp. 682-697.

22. Basila, J.R. and Strong, R.L.: 1974, J. Am. Chem. Soc. 96, pp. 701-706.

23. Lippert, J.L., Hanna, M.W., and Trotter, P.J.: 1969, J. Am. Chem. Soc. 91, pp. 4035-4044.

24. Morokuma, K.: 1977, Acct. Chem. Research 10, pp. 294-300.

25. Lathan, W.A., Pack, G.R., and Morokuma, K.: 1975, J. Am. Chem. Soc. 97, pp. 6624-6628.

26. Morokuma, K., Iwata, S., and Lathan, W.A.: in R. Daudel and B. Pullman (eds.), "The World of Quantum Chemistry", Reidel, Dordrecht, 1974, pp. 277-316.

27. Ditchfield, R., Hehre, W.J., and Pople, J.A.: 1971, J. Chem. Phys. 54, pp. 724-728.

28. Umeyama, H., Morokuma, K., and Yamabe, S.: 1977, J. Am. Chem. Soc. 99, pp. 330-343.

29. Mulliken, R.S.: 1956, Recl. Trav. Chim. Pays-Bas 75, pp. 845-852.

30. Murrell, J.N.: 1959, J. Am. Chem. Soc. 81, pp. 5037-5043.

31. Morita, H. and Tamres, M.: 1976, J. Phys. Chem. 80, pp. 891-898.

32. Strong, R.L. and Kaye, J.A.: 1976, J. Am. Chem. Soc. 98, pp. 5460-5464.

33. Strong, R.L. and Venditti, Jr., F.: 1977, Chem. Phys. Lett. 46, pp. 546-550.

STRUCTURE DETERMINATION OF COLLISION COMPLEXES BY NMR METHODS

P. Cristinziano[+], F. Lelj[+], T. Tancredi[°] and P.A. Temussi[+]

+ Laboratorio di spettroscopia, Istituto Chimico, Università di Napoli, Via Mezzocannone 4, I-80134 Napoli, Italy

° ICMIB del C.N.R., Via Toiano 2, I-80072 Arco Felice, Italy

ABSTRACT

The possibility of a rigorous determination of the structure of collision complexes has been evaluated using the systems dimethylacetamide-benzene and caffeine-benzene. Simulation of the experimental chemical shifts by means of ring-current contributions is very good for protons but is only rough for all carbon shifts, thus yielding only approximate equilibrium structures for the complexes. The possible causes of the discrepancies of the carbon shifts are critically discussed in terms of specific changes in electron density induced by complexation and/or of possible contributions from several 1:1 complexes.

INTRODUCTION

Collision complexes have been studied from many points of view in the last three decades (1) but there is little doubt that one of the most interesting aspects of these studies is the possibility of elucidating the role of intermolecular forces in solution.

Although indirect information can be drawn from any of the various physicochemical techniques employed in these studies, a crucial step in the elucidation of the forces involved in complexation is the determination of the structure of the complex since the relative orientation of the molecules taking part in the complex is often incompatible with some of the possible forces (2).

NMR spectroscopy (3) is best suited among all techniques so far employed (1) to give structural information, since it may give independent spectral parameters for atoms located in different parts of the complex. Simulation of these parameters, e.g. by attributing the chemical shift

B. Pullman (ed.), Intermolecular Forces, 233–256.

changes mainly to ring current effects, may in turn yield the average
geometrical arrangement of the molecules forming the complex.

Qualitative estimates of the geometry of the complexes have in
fact been given in all NMR studies of collision complexes (4-8), but
quantitative interpretations have been precluded by two fundamental
difficulties: the number of NMR parameters (of a given nucleus) that
can be used for structural determinations is generally lower than the
number of degrees of freedom and secondly, the complexity of the equilibria has been often underestimated in studies at low magnetic fields.
Only in the last few years in fact it has been shown clearly that many
of the systems studied consist of mixtures of 1:1 and 1:2 complexes
(9-12) (at least) rather than a simple 1:1 complex as assumed in all
older studies. The number of degrees of freedom necessary to define the
geometry of a 1:2 complex is prohibitively high even recording resonances of more than one type of nucleus, but it is essential when dealing
with complicated multiple equilibria to obtain reliable equilibrium parameters for the 1:1 complex, i.e. to disentangle the contributions of
say 1:1 and 1:2 complexes. These problems may be exemplified by systems
in which the so-called "acceptor" molecule is a purine and the "donor"
molecule is an aromatic hydrocarbon. The intermolecular forces responsible for the formation of these complexes have been identified as charge-
tranfer by many authors, mainly on the basis of solid state studies (13-
15), but other forces, such as dipole-induced dipole or dispersion forces have also been proposed (8).

The study of these systems in solution has attracted our interest
(as well as that of many others) also because of their potential biological relevance (8). Two limiting literature cases will be briefly mentioned hereafter: that of tetramethyluric acid (TMU) in which (7) it
was possible to determine the structure of the complex in a semirigorous
way, and that of caffeine (11) in which such a determination was not
even attempted. The linear behaviour of the shifts of the methyl groups
of TMU as a function of benzene concentration was interpreted on the basis of a single 1:1 complex to yield four widely different limiting
chemical shifts for the pure complex. Although this number is smaller
than the minimum required (that is three Euler's angles plus the coordinates of the origin (16)) a semirigorous determination was possible (7)
by imposing that the minimum distance of approach be of the order of
the sum of van der Waals radii, and by making use of the cylindrical
symmetry of the magnetic field of benzene (17). The degrees of freedom
are thus reduced to four and it was possible to show that the two molecules are not parallel but form an angle of ca. 45° in the average geometrical arrangement in solution. Such a geometry is not consistent
with a significant contribution of charge-tranfer forces.

A more complex situation is presented by the system caffeine-benzene, the first complex between a purine and an aromatic hydrocarbon to be studied by NMR spectroscopy (18). This complex was studied by Hanna and Sandoval in carbon tetrachloride solution at 60 MHz; owing to the superposition of the aromatic resonance with that of benzene and to the low field employed (1.4 T) only the resonance of one methyl group could be used for the measurement of the association constant, on the assumption of the existence of only a 1:1 complex. The corresponding limiting chemical shift was interpreted as a proof of the superposition of the benzene ring and of the five membered ring of the purine. A subsequent study (11) at higher field revealed that the system is more complex since the NMR data can be interpreted only by assuming an equilibrium between 1:1 and 1:2 complexes. Rather than attempting a semirigorous approach for the determination of the geometry of the 1:1 complex, as in the case of TMU, we chose to increase the number of NMR parameters by measuring also carbon chemical shifts. The resulting structural determination is one of the topics of this paper but, since the preliminary goal is the discussion of the possibility of structure determination in itself, most of the data reported hereafter will deal with the simpler system composed by benzene and N,N'-dimethyl-acetamide (DMA). The latter compound,although, superficially, quite different from purines, may be regarded as a good model for the forces under investigation since it contains but one feature of the mentioned purines, i.e. the N-methyl amide bond.

This system was considered ideal for our purposes of total structure determination because it can furnish a large enough number of limiting chemical shifts (i.e. seven, considering protons and carbons simultaneously) in spite of the small molecular weight of the "acceptor" molecule and also because the structural features of DMA itself have been studied extensively in various solvent systems with a variety of techniques (4,5). The actual possibility of an accurate structural determination was based on the following general considerations derived from our previous studies on collision complexes (11,12).

Spectrometers operating at fields of the order of 6 T (or higher), which are commonly available nowadays, enable routine measurements for complexes with limiting proton chemical shifts as low as 0.1 ppm and in very dilute solutions. Accordingly it is now fairly easy to measure formation quotients using spectral data from all the protons of an acceptor molecule whereas most of the older studies were based only on the most favourable chemical shifts. The simultaneous measurement of carbon shifts will generally yield a total number of limiting shifts larger than six, the minimum required for a quantitative geometrical characterization (16).

The interpretation of experimental chemical shifts is now possible even in systems showing marked non linear changes owing to the availability of more sophisticated methods of analysis of multiple equilibria, either based on graphical procedures or on numerical approaches (19).

It may be mentioned that a fairly rigorous geometrical characterization of caffeine dimers has been published (20) while our work was in progress; however the procedure for the geometrical fitting of the NMR shifts is not as general as the one described in this paper.

EXPERIMENTAL SECTION

Caffeine was purchased from Merck (Milano, Italy) and used without further purification. Spectrograde carbon tetrachloride was purchased from Fluka AG (Buchs, Switzerland); benzene-d_6, cyclohexane-d_{12} and chloroform-d were purchased from Wilmad Glass Co. Inc. (Buena, N.J., USA). Dimethylacetamide and cyclohexane were purchased from Carlo Erba (Milano, Italy) and used without further purification. All caffeine solutions were prepared in such a way that the molar concentration of caffeine remained approximately constant (0.025 M) along the whole series of both ^1H and ^{13}C spectra. The cyclohexane and chloroform solutions of DMA for both ^1H and ^{13}C spectra were prepared in such a way that the molar concentration of DMA remained approximately constant (0.03 M). The same concentration was used for ^1H spectra in carbon tetrachloride solutions, while ^{13}C spectra were carried out using more concentrated solutions (0.05 M). Although it is well known (21) that both caffeine and DMA may give rise to some association effects, the low concentration used and its constancy along the whole series of spectra should prevent both equilibrium ratios and $\triangle$'s to be affected by errors due to self association.

All the spectra were carried out on a Bruker WH-270 spectrometer operating at 270 MHz and 67.88 MHz for ^1H and ^{13}C respectively, in the FT mode, with internal deuterium lock. In the case of carbon tetrachloride solutions the first spectrum of the series was performed using deuterated benzene for external lock. The chemical shift measurements were performed using both tetramethylsilane (TMS) and cyclohexane as internal references.

Energy calculations served mainly the purpose of checking the reliability of the final geometries obtained from the experimental chemical shifts, i.e. to ascertain whether the relative arrangement of donor and acceptor molecules gives rise to severely destabilising interactions Semiempirical methods were excluded since it has been shown convincingly (22) that even in their most sophisticated forms they fail to predict reasonable equilibrium distances and energies of intermolecular complexes. Ab initio computations were also excluded because of their prohibitive cost when dealing with systems of this complexity.

Accordingly we employed empirical methods (23,24) based on 6-12 potentials (25-28) with several different sets of parameters and including a monopole term for electrostatic interactions. It had in fact already been shown (29-32) that this kind of approach can give satisfactory results in dealing with systems of complexity comparable to ours.

Molecular models of the heavy atoms skeleton of DMA and caffeine, with the numbering scheme used throughout the work, are shown below:

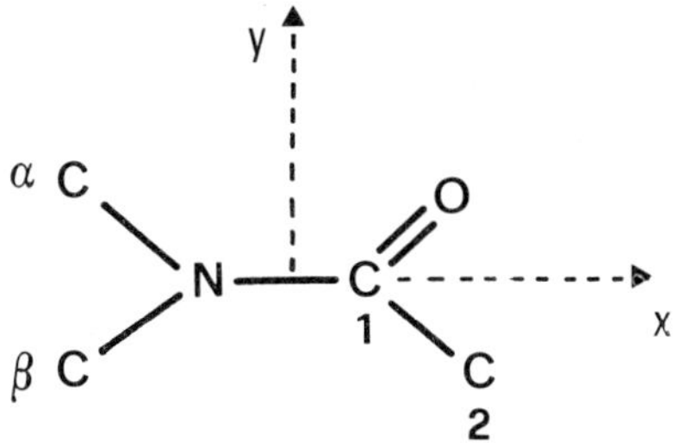

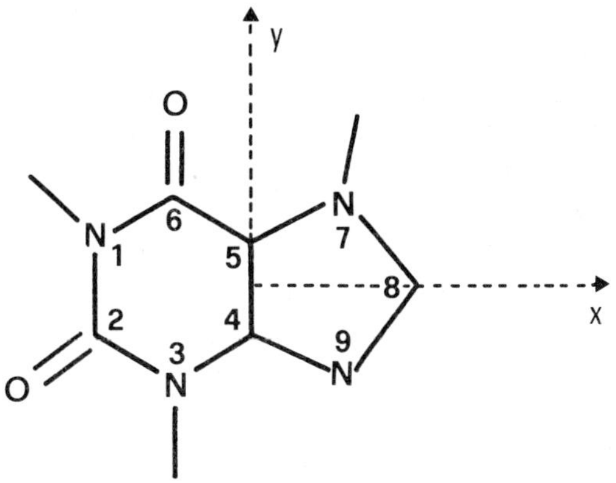

The left-handed mobile coordinate systems used in the simulation of the chemical shift are also shown, superimposed to the models. The fixed coordinate system, in both cases, was obviously centered on the benzene molecule.

NMR RESULTS

Both ^1H and ^{13}C chemical shifts of DMA were recorded as a function

of benzene concentration in three different solvents. The use of solvent
systems with a different potential ability of interacting with DMA and/or
benzene served the specific purpose of discriminating between solvent
effects and intermolecular forces between the two molecules forming the
complex. Another important reason for employing various solvent systems
is that complexation equilibria may be simplified (8) by a given solvent
in a way that is difficult to foresee.

The assignments of all resonances could be safely taken from numerous literature sources (33); besides the peaks are enough spaced to assume that the relative positions do not change in different solvents. A
preliminar graphical treatment of the data showed that only the shifts
of cyclohexane solutions could be fitted satisfactorily by means of linear equations typical of systems containing only 1:1 complexes (34),
such as

$$\frac{1}{\Delta_{obs}} = \frac{1}{Q_1 \Delta_1^\circ} \frac{1}{m} + \frac{1}{\Delta_1^\circ} \tag{1}$$

where Δ_{obs} is the chemical shift change induced by a benzene molality
equal to m, Δ_1° is the corresponding shift of the pure complex and Q_1
is the formation quotient. The shifts of chloroform and carbon tetrachloride solutions, on the other hand, show a marked non linear behaviour
when introduced into equation (1) or into equivalent formulation based
on the existence of 1:1 complexes, but can be fitted rather well by means
of equations that take explicitly into account the existence of at least
a second equilibrium in solution involving a 1:2 complex with a formation
quotient Q_2

$$A + 2D = AD_2$$

and a limiting chemical shift Δ_2° for the acceptor resonances of the pure AD_2 complex.

The numerical values of the equilibrium parameters (Δ's and Q's)
for each system were obtained with the aid of an automatic minimization
procedure, previously described (11) by some of us. All shifts, i.e.
both proton and carbon chemical shifts, were used simultaneously to
yield the equilibrium parameters. The fit was generally very good as
shown by the graphs of figures 1 trough 6, with one significant exception; the chemical shifts of the carbonyl group of DMA cannot be fitted
by the equilibrium parameters that fit all other chemical shifts in chloroform solutions. The carbonyl shifts were in fact excluded in the final
minimization procedure for this system, and in figure 6 they are shown

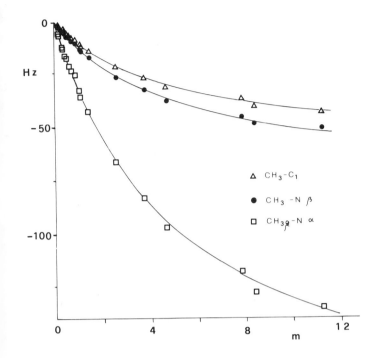

Figure 1. Plots of ^{1}H shifts of DMA induced by increasing molalities of C_6D_6 in C_6D_{12}. The curves, in this and all subsequent plots, are interpolations drawn by the minimization program.

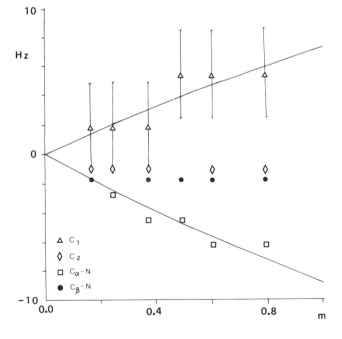

Figure 2. Plots of ^{13}C shifts of DMA induced by increasing molalities of C_6D_6 in C_6D_{12}.

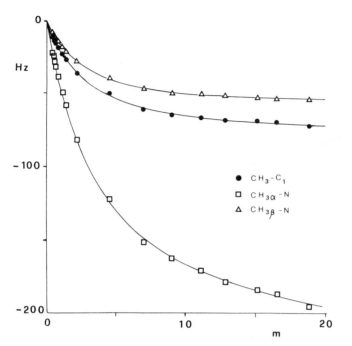

Figure 3.   Plots of ^1H shifts of DMA induced by increasing molalities of C_6D_6 in CCl_4.

● CH_3-C_1
□ $CH_{3\alpha}-N$
△ $CH_{3\beta}-N$

Figure 4.   Plots of ^{13}C shifts of DMA induced by increasing molalities of C_6D_6 in CCl_4.

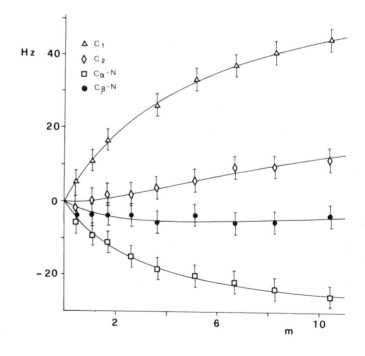

△ C_1
◇ C_2
□ $C_\alpha-N$
● $C_\beta-N$

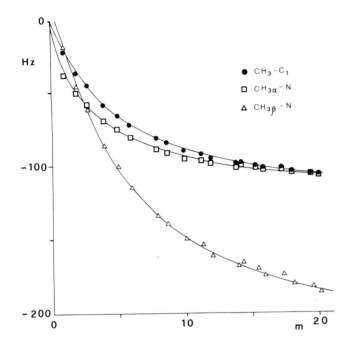

Figure 5. Plots of ^1H shifts of DMA induced by increasing molalities of C_6D_6 in $CDCl_3$.

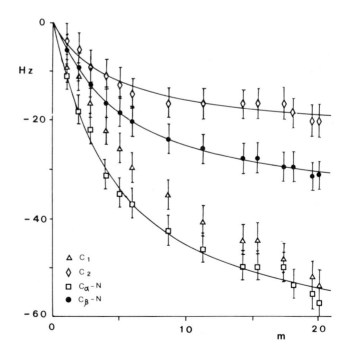

Figure 6. Plots of ^{13}C shifts of DMA induced by increasing molalities of C_6D_6 in $CDCl_3$. The shifts of the carbonyl carbon are not interpolated by any curve and were excluded from the minimization procedure that yielded the equilibrium parameters.

without any interpolating curve. The reason for the discrepancy can be probably ascribed to the strong interaction that the (acidic) hydrogen of chloroform is known to exibit toward carbonyl groups in many compounds. The unique behaviour of the chloroform solutions is reflected also by some of the equilibrium parameters. Table 1 summarizes the formation quotients and the limiting chemical shifts of all complexes between DMA and benzene in the three solvents used. The overall picture given by table 1 shows that the behaviour of the system DMA-benzene in the three solvents is markedly different. However the spread of the formation quo-

Table 1.   Equilibrium parameters of the complexes between DMA and benzene in three different solvents. All Δ's are in Hz. Q_1's are in kg mol^{-1}, Q_2's are in kg^{2}mol^{-2}.

D M A

| | CDCl$_3$ | | CCl$_4$ | | C$_6$D$_{12}$ | |
	Δ_1°	Δ_2°	Δ_1°	Δ_2°	Δ_1°	Δ_2°
H (C$_2$)	−8.7 ±2.6	−134.3 ±0.8	−77.7 ± 4.8	−75.1 ±3.0	−77.2 ±0.9	—
H (C$_\alpha$)	−35.1 ± 3.2	−128.5 ±0.9	−166.6 ±12.1	−238.4 ±5.1	−197.8 ±2.2	—
H (C$_\beta$)	−16.8 ± 3.2	−239.8 ±1.2	− 62.5 ± 3.8	−54.0 ±3.1	− 63.3 ± 0.8	—
C$_1$	—	—	37.4 ± 3.9	72.3 ±6.7	44.5 ±5.5	—
C$_2$	− 2.9 ±1.4	− 23.2 ±1.0	0 ±6.0	37.1 ±7.5	0 ±6.0	—
C$_\alpha$	− 5.5 ±2.0	− 67.54±1.1	−29.5 ±2.9	−29.3 ±5.3	− 54.1 ±5.6	—
C$_\beta$	− 1.7 ±1.0	− 38.30±0.9	−11.6 ±2.2	0 ±6.0	0 ±6.0	—
Q$_1$	1.701 ±0.380		0.350±0.025		0.194 ±0.003	
Q$_2$	0.190 ±0.004		0.065 ±0.016		—	

tients of the 1:1 complexes is so large (from 0.2 to 1.7) to cast some doubts on the fact that the differences may be attributed solely to solvent effects. Accordingly we rechecked the minimization procedure that leads to the equilibrium parameters in a very rigorous way. In fact it is possible that the choice of the starting set of parameters (Δ's and Q's) may influence the entire minimization process, thus leading to a

relative minimum (19 b). In the case of cyclohexane, that exibits a linear behaviour, it was an easy matter to vary the starting set $(Q_1, \Delta_{i,1})$
in a very large interval. Even by starting with a Q_1 ten times larger
and Δ's ten times smaller than those of table 1 the same minimum was
reached. The much larger number of equilibrium parameters involved when
we have the coexistence of 1:1 and 1:2 complexes renders checks of this
sort rather impractical. On the other hand, the possibility of reaching
false minima is much higher in these cases, mainly because of the difficulty of discriminating between the contributions of the two complexes
to experimental chemical shifts at high benzene concentrations. That is,
it is possible that the limiting Δ_1's be underestimated as indicated by
the fact that in some literature cases they are much lower than the largest observed shifts. Such an artifact may also be favoured by the nature of the function used in the minimization procedure, since it depends
mainly from the products $Q\Delta$ rather than from individual Q's and Δ's,
so that as long as these products are reasonably correct, the minimum
may have very good values of the standard deviations. It is possible to
minimize the possibility of artifacts with the aid of experimental procedures.

In the case of DMA the equilibrium parameters found for the cyclohexane solutions, that reflect a genuine linear behaviour and correspond
to the absolute minimum, were used as starting parameters for the 1:1
complexes of carbon tetrachloride and chloroform solutions, hoping that
they could drive the minimization procedure toward the absolute minimum.
In the case of the system caffeine-benzene it was not easy to find a solvent in which it behaves as single 1:1 complex; besides we wanted to
compare the parameters with those of previous studies (11,18) in carbon
tetrachloride. Thus we relied simply on the extension of the measurements to a larger number of nuclei (the carbons) and on the better quality
of the proton measurements at higher field as possible means of avoiding
relative minima.

The parameters of carbon tetrachloride and chloroform solutions
shown in table 1 are indeed the results of many calculations, with different starting sets containing the final parameters of the cyclohexane
solutions and show that the parameters characterizing the 1:1 complexes
in the three solvents are really different from one another, as a result
of solvent effects, although the Δ_1's have similar trends in all three
cases and those of carbon tetrachloride solutions are very similar to
the corresponding Δ's in cyclohexane.

Accordingly we inferred that only the Δ_1°'s and Q_1's of the 1:1
complexes in carbon tetrachloride and in cyclohexane can be considered
reasonably free both from specific solvent interactions and from contri-

butions of 1:2 complexes. Hence the use of the Δ_1°'s in both solvents for structural determinations.

The chemical shift changes induced by benzene on the proton and carbon resonances of caffeine are reported in figures 7 and 8 respectively. As for the system DMA-benzene, the solid curves that interpolate the experimental points represent the result of a best fitting procedure that employs all chemical shift data (i.e. ^1H and ^{13}C) simultaneously to calculate the Q's and the Δ's. A good fitting of the experimental data could be obtained only on the assumption that the solutions contain a mixture of 1:1 and 1:2 complexes, as already found for the ^1H data at 2.3 T. The absolute values of the equilibrium parameters however are rather different from those found at 2.3 T, in particular for the 1:1 complex.

Table 2 shows the equilibrium parameters for the system caffeine-benzene, calculated from the proton and carbon data at 6.3 T. The value

Table 2. Equilibrium parameters of the complexes between caffeine and benzene in carbon tetrachloride. The Δ's are in Hz. Q_1 is in kg mol^{-1}, Q_2 is in kg^{2}mol^{-2}

CAFFEINE		
	Δ_1°	Δ_2°
CH_3-N1	-11.2 ± 0.6	0 ± 1.0
CH_3-N3	-27.9 ± 2.3	-40.1 ± 1.0
CH_3-N7	-137.4 ± 11.8	-251.2 ± 3.0
$H-C8$	-142.3 ± 11.9	-245.1 ± 2.1
$C-N1$	-16.8 ± 1.3	0 ± 6.0
$C-N3$	0 ± 6.0	0 ± 6.0
$C-N7$	-24.1 ± 3.4	-69.1 ± 10.5
$C8$	4.6 ± 13	0 ± 6.0

$$Q_1 \quad 0.729 \pm 0.063$$

$$Q_2 \quad 0.169 \pm 0.021$$

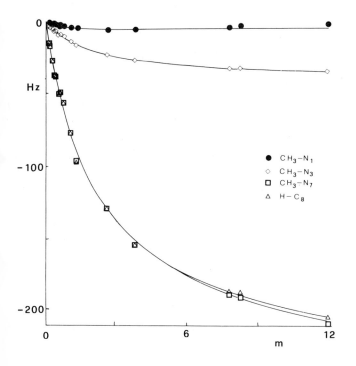

Figure 7. Plots of ^1H shifts of caffeine induced by increasing molalities of C_6D_6 in CCl_4.

● CH_3-N_1
◇ CH_3-N_3
□ CH_3-N_7
△ $H-C_8$

Figure 8. Plots of ^{13}C shifts of caffeine induced by increasing molalities of C_6D_6 in CCl_4.

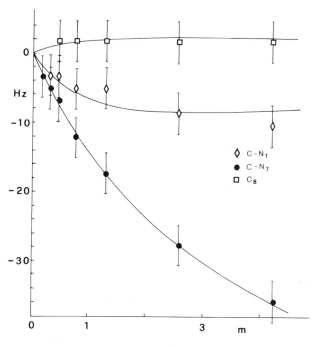

◇ $C-N_1$
● $C-N_7$
□ C_8

of Q_1 is much smaller than that previously calculated from the data at 2.3 T (11), i.e. 0.73 kg mol^{-1} instead of 3.5 kg mol^{-1}, a discrepancy that can not be reasonably attributed to experimental errors. In fact, the high value of Q_1 obtained from the 100 MHz data had already led us to suspect that these equilibrium parameters belong to a relative minimum in spite of the very good values of the standard deviations of the che- mical shifts (11). This circumstance is confirmed by the fact that if we use the values of table 2 as starting parameters for the best fitting procedure it is possible to reach a similar minimum (with Q_1 = 0.60 kg mol^{-1}) also with the 100 MHz chemical shifts. The possibility of reaching false minima has been discussed at length also by other authors (19b); we can only emphasize here that the only conceivable way of avoiding such pitfalls may reside in experimental approaches. It is reassuring, in the case of the system caffeine-benzene, that the value of Q_1 found with more (and better) experimental data is closer to that of the system TMU-benzene that consists of a simple 1:1 complex (7).

It is also very important to note that even the Δ_1°'s corresponding to the false minimum have approximately the same relative values (11) as those reported in table 2 and might thus be used in a structural cha- racterization to yield a similar geometrical arrangement of caffeine and benzene.

GEOMETRY

DMA

The procedure we used for the determination of the geometry of a complex is based on a least-squares fitting of the observed changes in chemical shifts with the values calculated from ring-current effects. Three simple assumptions were held valid throughout all calculations: 1) the changes in chemical shifts of the "acceptor" molecule are attri- buted solely to ring-current effects (17); 2) all molecules taking part in complexation behave as rigid bodies and finally 3) all rotating group (such as $-CH_3$'s) have complete rotational freedom so that the chemical shifts can be averaged over all accessible positions with equal weights.

Assumption 2) implies that we must determine at least six limiting chemical shifts of the complex since the relative orientation of two ri- gid bodies is defined by six independent parameters; hence the need of using ^{13}C as well as ^1H resonances as already pointed out in the intro- duction. The number of parameters may be reduced to five if we assume that the field connected with the ring-current of benzene has an effec- tive cylindrical symmetry as is the case when we calculate the shielding

by means of the equations of Waugh-Fessenden and Bovey (17). Accordingly we chose the five parameters as the ϱ and z coordinates of the origin of a mobile coordinate system, plus three Euler's angles (16) to define the orientation of the system centered on the acceptor molecule
with respect to that (fixed) on the benzene ring. The shifts of the (rotating) methyl groups were obtained by means of an integration based on
the method of Gauss and making use of Legendre polynomials of 32nd degree (35).

Before proceeding to the fitting of the shifts it is essential to
choose a model, i.e. it is necessary to define the maximum number of
1:1 complexes that we are going to use in the fitting procedure. In fact,
it has been shown by Mulliken and Orgel (36) that the experimental measurements may reflect any number of 1:1 complexes (i.e. Q_1 may be a weighted average of numerous Q_1's and the Δ_1's the mean of the Δ_1's). Accordingly we might choose to fit the experimental chemical shifts with
a weighted average of calculated shifts corresponding to many different
1:1 complexes. Such a model, while very flexible for the fitting procedure, may be meaningless from a physical point of view and thus for the
interpretation of the forces responsible of the complexation. Besides,
although there is no reason, in principle, against the existence of many complexes, in most actual cases there are geometrical and/or energetic restrictions that limit the number of possible complexes. We chose
to try the fitting of the experimental shifts using a single 1:1 complex
that is not rigid from any point of view but represents the average arrangement of the two molecules.

In turn, the possibility of fitting the equilibrium parameters using
a model based on a single 1:1 complex may be taken as an indication of
the predominance of a single type of interaction forces.

Exceptionally, it may be reasonable to use a very limited number
of 1:1 complexes, provided we had some simple guideline for this choice
and, above all, enough experimental data to calculate all the necessary
degrees of freedom. In fact this last condition is very difficult to fulfill in most actual cases since it is usually already difficult to have
enough experimental data for the rigorous definition of a single 1:1 complex. A useful compromise is possible in the case of DMA, if we take advantage of the fact that this molecule has a pseudo C_2 symmetry around
the axis of the amide bond when only the volume of the groups is considered, disregarding their chemical identity. That is, the bulkiness of
the side of DMA with two methyl groups is similar to that of the side
with one methyl and one carbonyl oxygen. Thus it is possible to use a
model with two 1:1 complexes in which the DMA molecule has the same o-
rientation with respect to benzene but for the fact that the side facing
the benzene ring is interchanged, as shown in the following scheme

$$1-w \qquad\qquad\qquad\qquad w$$

In such a model we do not need to double the degrees of freedom as would
be for two completely independent complexes; the only further variable
we must introduce is the fractional population w. The model based on a
single 1:1 complex will henceforth referred to as model a whereas that
based on the mixture of two isosteric complexes will be called model b.

In all cases the final molecular models were subjected to empirical
internal energy calculations to test their consistency with respect   to
the geometrical arrangement corresponding to the minimum of internal e-
nergy (see experimental section).

Table 3 shows the final geometrical parameters and the corresponding calculated chemical shifts obtained for the complex DMA-benzene in
cyclohexane using model a. It is seen that while the proton chemical
shifts are reproduced with great accuracy the carbon data are characterized by large differences between experimental and calculated shifts.

Figure 9 shows the molecular model of the 1:1 complex in cyclohexane of table 3; the corresponding arrangement for the complex in carbon tetrachloride is nearly identical, reflecting the similarity of the
limiting chemical shifts of the two 1:1 complexes. The main feature of
the molecular model of figure 9 is that the planes of DMA and benzene
form an angle of ca. 90°, whereas the geometrical arrangement corresponding to the absolute minimum of internal energy is characterized by a
nearly parallel disposition of the two planes, as shown pictorially in
figure 10. On the other hand the complex corresponding to the absolute
energy minimum, or any single complex with approximately parallel mole-

Table 3. Geometrical parameters and calculated chemical shifts for the best simulated structures of the complexes DMA-benzene based on a single 1:1 complex (a) and on two isosteric DMA-benzene complexes (b). The coordinates ϱ and z are in Å.

D M A $-$ C_6D_{12}

	Δ_i° (exp.)	Δ_i° (calc.) a	Δ_i° (calc.) b
H (C_a)	$- 77.2 \pm 0.9$	$- 77.1$	$- 77.1$
H (C_a)	-197.8 ± 2.2	-198.1	$- 198.1$
H (C_β)	$- 63.3 \pm 0.8$	$- 63.3$	$- 63.3$
C_1	44.5 ± 5.5	$- 3.2$	0.5
C_a	0 ± 6.0	$- 3.2$	$- 2.4$
C_α	$- 54.1 \pm 5.6$	$- 7.5$	$- 6.6$
C_β	0 ± 6.0	$- 3.1$	$- 2.2$

	ϱ	z	φ	ϑ	η	w
a	4.5	5.3	125.4°	106.3°	129.0°	—
b	4.9	3.8	1.0°	-5.3°	75.4°	0.72

Figure 9. Molecular model of the complex between DMA and benzene obtained through the simulation of the chemical shifts of a single 1:1 complex by means of ring-current effects.

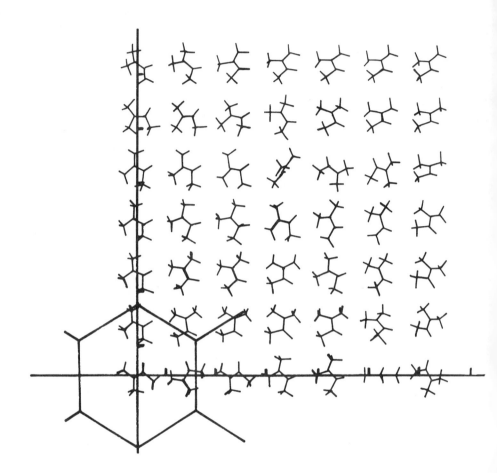

Figure 10.   Typical geometrical arrangements of DMA and
benzene in the neighbourhood of the minimum of internal
energy obtained from empirical energy calculations.   The
absolute minimum corresponds to the origin of the graph.
Rows and columns of DMA molecules are   1 Å apart and are
made up with DMA molecular models five times as small as
the benzene model.

cules shows a very poor agreement not only for carbon but also for proton shifts.

Use of model b on the contrary leads to a mixture of two 1:1 complexes in which the molecular planes are nearly parallel and the agreement between experimental and calculated shifts is at least as satisfactory as that of the best complex of model a (table 3).

The test of internal energy is of course much more favourable for
the geometrical arrangement given by model b (figure 11) than for that
of model a (figure 9) since all points around the minimum of figure 10
feature nearly parallel DMA and benzene molecules.

The poor fit obtained for carbon shifts with both models requires
however further discussion and a critical reexamination of the whole
strategy of structure determination, since it was based on the very use
of carbon shifts. The errors of the experimental carbon shifts data are

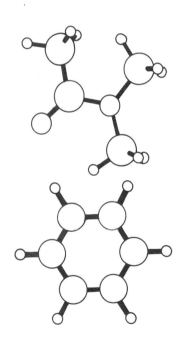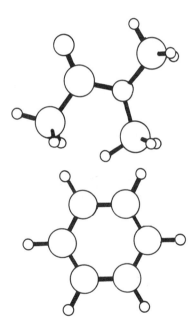

Figure 11.   Molecular models of the two isosteric
complexes between DMA and benzene corresponding to
the chemical shifts of table 3. The relative populations are 0.72 and 0.28.

certainly much larger than those of the protons and it is certainly fair
to say that the standard deviations of tables 1 and 2 represent an opti-
mistic view. It is clear however that the disagreement between the experimental shifts and those calculated with both models is too large to be
entirely attributed to errors. There are essentially two possibilities
to account for the observed discrepancies: a trivial one is to invoke
the existence of several 1:1 complexes of comparable stability, alternatively it is possible that the changes in chemical shifts (37) are not
entirely due to ring current effects. Although the coexistence of many
1:1 complexes is a very likely event (and in fact it was always implied
every time we spoke of an average 1:1 complex) the assumption that they
give rise to completely different sets of chemical shifts implies that
they correspond to drastically different geometrical arrangements of
comparable stability. This is tantamount to saying that different intermolecular forces give rise to various 1:1 complexes simultaneously, a
rather unlikely situation indeed. It seems more likely to postulate that
complexation may induce small changes in the electronic density of the
acceptor molecule that can exert direct influences on the chemical shift.
Such effects would undoubtedly influence the carbon shifts to a much lar
ger extent than the proton shifts.

If this view is accepted it is possible to have some confidences in
the structures found by means of the simulation of the ring-current
effects. In fact, the carbon shifts induced solely by ring current effects are expected to be small at distances of the order of 6÷7 Å, as
in the case with three of the carbon shifts (both experimental and calculated). The only two (experimental) shifts that deviate substantially
from zero can be considered as a direct consequence of the mentioned
changes in electronic density induced by complexation.

Caffeine

The complex caffeine-benzene shows a behaviour very similar to that
of DMA when we try to simulate the geometry by means of the calculation
of the shift changes induced by the ring-current of benzene.

Table 4 shows that the best geometrical arrangement corresponds to
a very good reproduction of the four proton shifts but only one of the
carbon shifts has a calculated value close to the experimental shift. It
is likely that before using carbon shifts for structural determination
it is necessary to check ring-current shifts with model compounds in
which a clear distinction between ring-current effects and electronic
effects can be made.

At present, if we take properly into account the large errors of
all carbon shifts and the fact that five out of eight shifts are correc-

Table 4.   Geometrical parameters and calculated chemical shifts for the best simulated structure of the complex caffeine-benzene. The coordinate ϱ and z are in Å.

CAFFEINE

	Δ_i° exp.	Δ_i° calc.
CH$_3$ N 1	- 11.2 ± 0.6	- 10.1
CH$_3$ N 3	- 27.9 ± 2.3	- 28.6
CH$_3$ N 7	-137.4 ±11.8	- 137.7
H C 8	-142.3 ±11.9	- 142.1
C N 1	- 16.8 ± 1.3	- 0.5
C N 3	0 ± 6.0	- 1.2
C N 7	- 24.1 ± 3.4	- 6.3
C 8	4.6 ± 1.3	- 13.2

ϱ	z	φ	ϑ	η
4.6	4.9	103.7°	-38.4°	66.9°

tly reproduced (see table 4), it is possible to have some confidence in the geometry of the complex caffeine-benzene depicted in figure 12. The internal energy test showed that this arrangement, although not corresponding to the absolute minimum, has a value of the internal energy very close to it.

The most notable feature of the structure of figure 12 is that the main planes of the two molecules of the complex form an angle far from 0°, a circumstance that is difficult to reconcile with many types of interaction forces. A more detailed discussion is not warranted by the quality of the data and ought to be postponed to after we get a better understanding of the factors influencing carbon shifts. It is in order however to comment on previous structural interpretations and corresponding interpretations of interaction forces.

Most of these interpretations were based on qualitative estimates of very few limiting proton shifts that in turn were derived from che-

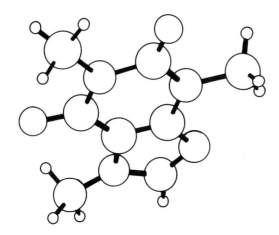

Figure 12.   Molecular
model of the complex be-
tween caffeine and ben-
zene corresponding   to
the best simulation of
the shifts by means of
ring-current contribu-
tions.

mical shift data collected in small concentration ranges, and at very
low fields.

In view of the problems encountered in interpreting solution equilibria based on higher quality data and on the even greater difficulties
found for the interpretation of the shifts in terms of ring-current effects it is only fair to consider all previous structural conclusions
on collision complexes at least doubtful.

REFERENCES

1) Foster, R., editor: 1973, "Molecular Complexes", Vol 1 and Vol 2,
   Elek P., London, 1973
2) Mulliken, R. S.: 1952, J. Amer. Chem. Soc. 74, pp. 811-824
3) Bhacca, N. S., Williams, D. H.: 1964, "Applications of N.M.R. Spectroscopy in Organic Chemistry", Holden-Day, San Francisco, Ca.
4) Sandoval, A. A., Hanna, M. W.: 1966, J. Phys. Chem. 70, pp. 1203-
   1206
5) Yonezawa, T., Morishima, I., Takeuchi, K.: 1967, Bull. Chem. Soc.
   Japan, 40, pp. 1807-1813

6) Ronayne, J., Williams, D. H.: 1967, J. Chem. Soc. (B), pp. 541-552

7) Andriessen, H. J. M., Laarhoven, W. H., Nivard, R. J. F.: 1972, J. Chem. Soc. (Perkin II), pp. 861-868

8) Donesi, A., Paolillo, L., Temussi, P. A.: 1976, J. Phys. Chem., 80, pp. 279-282

9) Dodson, B., Foster, R., Bright, A. A. S., Foreman, M. I., Gorton, J: 1971, J. Chem. Soc. (B), pp. 1283-1293

10) Chudek, J., Foster, R.:1976, Tetrahedron, 34, pp. 2209-2211

11) Andini, S., Ferrara, L., Temussi, P. A., Lelj, F., Tancredi, T.: 1979, J. Phys. Chem., 83, pp. 1766-1770

12) Tancredi, T., Lelj, F., Andini, S., Ferrara, L., Temussi, P. A.: 1979, J. Phys. Chem., 83, pp. 2902-2906

13) Damiani, A., De Sanctis, P., Giglio, E., Liquori, A. M., Puliti, R., Ripamonti, A.: 1965, Acta Cristallogr., 19, pp. 340-344

14) Damiani, A., Giglio, E., Liquori, A. M., Puliti, R., Ripamonti, A.: 1966, J. Mol. Biol., 20, pp. 211-217

15) Prout, C. K., Kamenar, B.: 1973, "Crystal Structure of Electron Donor-Acceptor Complexes", Cap. 4 in "Molecular Complexes" Vol 1, Foster, R., editor, Elek, London

16) Goldstein, H.: 1955, "Classical Mechanics". Addison Wesley, New York, N. Y.

17) a) Waugh, J. S., Fessenden, R. W.: 1957, J. Amer. Chem. Soc., 79, pp. 846-849
    b) Jhonson, C. E., Bovey, F. A.: 1958, J. Chem. Phys., 29, p. 1012

18) Hanna, M. W., Sandoval, A.: 1968, Biochem. Biophys. Acta, 155, pp. 433-436

19) a) Sillen, L. G.: 1956, Acta Chem. Scand., 10, pp. 803-805.
    b) Rosseinsky, D. R., Kellawi, H.: 1969, J. Chem. Soc. (A), p. 1207

20) Kan, L., Borer, P. N., Cheng, D. M., Ts'o, P.O.P.: 1980, Biopolymers, pp. 1641-1654

21) a) Drago, R. S., Carlson, R. L., Rose, N. J., Wenz, D. A.: 1961, J. Amer. Chem. Soc. 83, pp. 3572-3576
    b) Takkar, A. L., Tensmeyer, L. G., Hermann, R. B., Wilham, W. L.: 1970, Chem. Commun., pp. 524-525

22) a) Lochmann, R.: 1977, Int. J. Quantum Chem. 12, pp. 795-801 and references therein
    b) Lochmann, R., Hofmann, H. J.: 1977, Int. J. Quantum Chem., 11, pp. 427-431

23) Cristinziano, P. L., Lelj, F.: to be published

24) Kitaigorodskij, A. I.: 1978, Chem. Soc. Rev., 7, pp. 133-175

25) Pitzer, L.: 1959, Adv. Chem. Phys., 2, pp. 50-54

26) Hopfinger, A. J.:1973, in "Properties of Macromolecules" Academic Press, New York, N. Y.

27) Momany, F. A., Carruthers, L. M., McGuire, R. F., Sheraga, H. A.: 1974, J. Phys. Chem., 78, pp. 1595-1601

28) Lifson, S., Hagler, A. T., Dauber, P.: 1979, J. Amer. Chem. Soc., 101, pp. 5111-5118

29) Mantione, M. J.: 1968, Theoret. Chim. Acta, 11, pp. 119

30) Mantione, M. J.: 1969, Theoret. Chim. Acta, 15, p. 141

31) Mantione, M. J.: 1969, Int. J. Quantum Chem., 3, p. 185

32)Cristinziano, P. L.: 1981, Thesis, Università di Napoli
33)Anet, F. A. L., Bourn, A. J. R.: 1965, J. Amer. Chem. Soc., 87,
   pp. 5250-5255
34)Benesi, H. A., Hildebrand, J. H.: 1949, J. Amer. Chem. Soc., 71,
   p. 2703
35)Krylov, V. I.: 1962, "Approximate Calculations of Integrals",
   Mac Millan, N. Y., London, pp. 100-111, pp. 337-340
36)Orgel, L. E., Mulliken, R. S.: 1957, J. Amer. Chem. Soc., 79, pp.
   4839
37)Andriessen, H. J. M., Laarhoven, W. H., Nivard, R. J. F.: 1972,
   J. Chem. Soc. (Perkin II), pp. 861-868

ELECTROSTATIC AND TOPOLOGICAL INTERACTIONS IN DNA

Marc Le Bret^ and Bruno H. Zimm^^
^Laboratoire de Physicochimie Macromoléculaire
LA CNRS 147; INSERM U 140 Institut Gustave Roussy
94800 Villejuif France
^^Department of Chemistry B017; University of Ca-
lifornia, San Diego; La Jolla CA 92093 U.S.A.

The distribution of ions free to move about DNA is
determined using a Monte-Carlo method. It is then compared
to the distribution obtained through the Poisson-Boltzmann
equation. The importance of the ions correlation and of the
short range interactions is evaluated. The influence of the
interactions between the different DNA molecules on the activity coefficients of the co-ions and of the counter-ions
is studied. The Poisson-Boltzmann equation is integrated
numerically for a torus to give an estimate of the electrostatic contribution to the persistence length. Finally, we
study how the torsional and the bending flexibilities and
the excluded volume of double-standed DNA contribute to
the supercoiling energy of knots of various types.

INTRODUCTION

In this communication, we are mostly interested in the
long or very long range interactions in double stranded DNA.
namely 1) the electrostatic interactions which decrease only
as the inverse of the distance and 2) the interactions which
are mediated by the macromolecule itself. In the latter
group we find the torsional and the bending flexibilities,
the excluded volume effects and the topological constraints
caused either by the linking of the two strands or by the
knot type of the DNA axis. Of course, all these interactions
are modulated not only by one another but also by the small
range interactions.

DISTRIBUTION OF THE IONS ABOUT DNA.

It is, in principle, easy to be determined from the
study of the canonical ensemble. At a given temperature T,

B. Pullman (ed.), Intermolecular Forces, 257–271.
Copyright © 1981 by D. Reidel Publishing Company.

in a given volume V, we consider one DNA molecule (or just a piece of it) a given number of water molecules, ν_+ counter-ions and ν_- co-ions. The volume V is bounded by the outer surface Σ. We consider all the configurations of the system that is, all the possible distribution of ions and water molecules around the fixed DNA molecule and determine for each of them its total energy E. This total energy, of course, is a multivariable function. It depends on the position of all the particles studied. Each configuration contributes to the equilibrium configuration by the weight $\exp(-E/kT)$.

The many terms contributing to the total energy may be divided into two broad categories: a) the long range terms contain the coulombic interactions which are inversely pro-portional to the distance. These terms must be fully taken into account in the problems we study here. b) the short range terms are the polarizabilities, the interactions between the induced dipoles and multipoles and the Lennard Jones potentials.

It is a pity that all the terms contributing to the total energy cannot be taken into account because of the size and the slowliness of available "rapid" computers. That is why some simplifications must be done. In this communi-cation the solvent is treated as a continuum of given dielec-tric constant D_s. All ions-solvent and solvent-solvent inter-actions are neglected. The mobile ions and the fixed charges on the poly-ion are hard spheres. Their short range inter-actions are thus partially taken into account. The counter-ions have the radius a_+ and the valency z_+, the co-ions have the radius a_- and the valency z_-. The fixed charges on the poly-ion have a radius of 1 Å and bear one elementary elec-tronic charge. Two models of the DNA molecule have been studied. In the first model, the fixed charges are set along the DNA axis and their axial distance is

$$b = 1.7 \text{ Å} \tag{1}$$

In the second model, the fixed charges are set more realis-tically along two helices drawn on a cylinder of radius 10 Å Their other cylindrical coordinates are:

$$z_i = i \times 3.4 \text{ Å} \tag{2}$$

$$\theta_i = i \times 36 \pm 90 \quad \text{degrees} \tag{3}$$

In both models, the poly-ion has an impenetrable radius of a The centers of the mobile ions are free to move between two coaxial cylinders of radii $a + a_i$ and R. The energy of a configuration is

$$E = \ell_B \sum_{i<j} z_i z_j / r_{ij} \tag{4}$$

where ℓ_B is the Bjerrum length, that is the distance at
which the coulombic interaction of two protons is equal
to the thermal factor:

$$\ell_B = e^2/4\pi\varepsilon_o D_s kT \qquad (5)$$

Here, e is the charge of a proton, ε_o is the permittivity
of the vacuum and D_s is the dielectric constant of the solvent. In an aqueous medium and at room temperature it is
equal to 7.135 Å. On the other hand, the probability of a
configuration which is not compatible with the impenetrable
radii is set to zero. For instance, if one of the distances
r_{ij} between two mobile ions or one mobile ion and one of
the fixed charge is less than the sum of the radii of the
hard spheres, the energy E is set to infinity. If the radial distance of the center of the i-th mobile ion to the
revolution axis of the cylindrical poly-ion is not comprised between $a + a_i$ and R, the energy is also set to
infinity.

The scheme we have just outlined would certainly give
the distribution of the ions, however it would require a
fantastic number of configurations. A very efficient
short-cut is the Metropolis algorithm [1]. Before presenting some results obtained with that Monte-Carlo method
we would like to give a rapid sketch of the more traditional
Poisson-Boltzmann approach.

The Poisson-Boltzmann equation.

Instead of considering the energy of each configuration
which is a function of the coordinates of the fixed charges
and of all the mobile ions studied, we consider the average
on all configurations of the electrostatic potential at r:
$\Psi(r)$. In the most simplified version of this mean field
theory, the correlation between the ions is completely
ignored and so are the short range interactions. This
approach leads to a second order differential equation, that
has been derived by Gouy [2,3] and that is called the
Poisson-Boltzmann equation. As it is not linear, it is inconsistent as far as the free energy is concerned [4]. However, it may easily be integrated when the shape of the
poly-ion and its charge distribution have a high degree of
symmetry.

Some authors have tried to generalized the simplified
Poisson-Boltzmann equation in order to take into account
the finite size of the mobile ions [5,6,7,8], the correlation between the ions [8], the polarizabilities [8]
and the dependence of the dielectric constant on the local
electric field strength [7,8]. Most of these studies concern
spherical poly-ions. The generalized Poisson-Boltzmann equations will no more be considered here.

The volume V, which is bounded by the outer surface Σ is now separated into two regions by the surface S of the poly-ion: inside, we have no mobile charges and the dielectric constant is D_i; outside, there are mobile charges. At the surface S, there is at each point a local number of protonic charge per unit surface σ . The distribution of fixed charges may be either a set of point charges (Dirac functions) or a smooth distribution. The system is then completely determined by D_i, D_s, z_+, z_-, ν_+, ν_-, σ and the geometry of S and Σ . The reduced potential y, defined as

$$y = e\psi/kT \tag{6}$$

is a continuous function made of two functions y_i and y_s respectively defined in the poly-ion and in the solvent. The continuity of y implies:

$$y_i(S) = y_s(S) \tag{7}$$

Because of the charges its normal derivative jumps:

$$\partial y_s/\partial n - \varepsilon \, \partial y_i/\partial n = -4\pi \ell_B \sigma \tag{8}$$

where n is the external normal to S, and ε is the ratio of the internal to the solvent dielectric constants. Along Σ

$$y_s(\Sigma) = 0 \tag{9}$$

We call $n_+(r)$ and $n_-(r)$ the numbers of counter-ions and of co-ions per unit volume at r. On the outer surface these numbers are constant since   is an equipotential and are equal to N_+ and to N_-. Of course, N_+ and N_- are unknown. Because of the Boltzmann law, we have the two formulae:

$$n_\pm(r) = N_\pm \exp \mp z_\pm y_s(r) \tag{10}$$

The values of $N_\pm$ and of y_s are related to the known total number of counter-ions and co-ions by the two formulae:

$$\nu_\pm = N_\pm \iiint \exp(\mp z_\pm y_s(r)) \, dV \tag{11}$$

where the integrals are extended to the volume between S and Σ. In the solvent, the local number of charges per unit volume is:

$$\rho = z_+ n_+(r) - z_- n_-(r) \tag{12}$$

When the dielectric constant is a ... constant, the Poisson formula reads:

$$\Delta\psi = -\rho e/\varepsilon_o D \tag{13}$$

The Boltzmann formulae (10) and the Poisson formula (13) yield the so-called Poisson-Boltzmann equation which is valid in the solvent:

$$\Delta y_s = 4\pi\ell_B \ (\ z_- N_- \exp z_- y_s - z_+ N_+ \exp -z_+ y_s)\ (14)$$

As there are no charges inside the poly-ion:

$$\Delta y_i = 0 \tag{15}$$

The set of equations (14,15) with the boundary conditions (7,8,9) and the two integrals (11) define a unique function y, as may be rigorously shown. Integrating (14,15) and using Eqs. (7,8,9,10,11) we get:

$$4\pi\ell_B \ \iint \ \partial y_s / \partial n \ d\Sigma = z_- \nu_- - z_+ \nu_+ - \iint \sigma d S \tag{16}$$

This equation (16) could have been directly obtained using the Gauss' theorem. It does not add anything to the determination of y, but may be used to replace one of the two integrals (11).

From now on, the volume V, as a whole, is supposed to be neutral. In this very special case, the left hand-side of Eq. (16) is equal to zero. Then, y can rigorously be shown to have the same sign throughout the volume V. This sign is the sign of the total fixed charge of the poly-ion, which is the surface integral written at the right end of Eq. (16). Therefore, for a negatively charged poly-ion such as DNA:

$$z_- \nu_- < z_- N_- V < z_+ N_+ V < z_+ \nu_+ \tag{17}$$

We call X, the absolute number of net protonic charge at the surface of the poly-ion per added co-ionic charge. Moreover, the activity coefficients are written Y_+ and Y_-:

$$X = |\ \iint \sigma \ d S\ |\ /\ \nu_- z_- \tag{18}$$

$$Y_+ = N_+ V\ /\ \nu_+ \tag{19}$$

$$Y_- = N_- V\ /\ \nu_-$$

Because of the electric neutrality, the Eqs. (16-19) give:

$$1 < Y_- < (1+X)\, Y_+ < 1+X \tag{20}$$

The activity coefficients give a rough idea of the distribution of the ions. The activity coefficient of the co-ions is always greater than unity, because the co-ions are repelled from the poly-ion. The activity coefficient of the counter-ions is greater than unity because they are attrac-

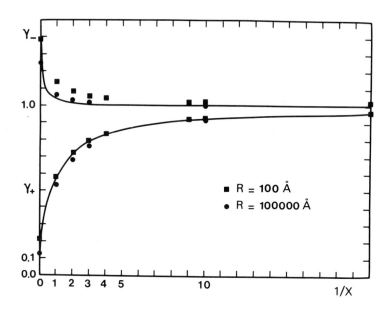

Figure 1. Activity coefficients for a 1-1 salt as a
         function of the number of added co-ion per
         fixed charge. The poly-ion is a uniformly
         charged cylinder simulating DNA: a = 10 Å
         b = 1.7 Å two different values for R. The
         solid lines represent the Guéron & Weisbuch
         [13] formulae.

ted by the poly-ion. If an infinite number of co-ions is added, X tends to zero and both activity coefficients tend to
unity. We set:

$$\kappa^2 = 4\pi \ell_B (N_+ + N_-) \tag{21}$$

When X tends to infinity, y_s tends to Y_s, the solution of:

$$\Delta Y_s = \kappa^2 \sinh Y_s \tag{22}$$

All that has been written up to here is valid for any poly-
ions provided that V is finite, and S does not intersect Σ .

Uniformly charged cylinders.

    The radius a and the axial distance b between the fixed
negative charges of the poly-ion are related by:

$$\sigma_{cyl} = -1/2\pi ab \tag{23}$$

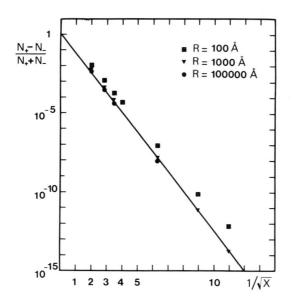

Figure 2. Variation of $(N_+ - N_-)/(N_+ + N_-)$ for a model DNA (see fig. 1) with the number of co-ions per fixed charge. The least square linear regression is shown by the solid line. R=100Å corresponds to 0.03 M "phosphates".

Moreover, the true charge parameter is defined as

$$\xi = \ell_B / b \qquad\qquad (24)$$

In this case, the outer surface is a co-axial cylinder of radius R. When R is made infinite, the potential Y_s may be easily integrated numerically from (22) [9,10,11]. In the absence of co-ions, y_s is known in closed-form [12] for any value of R. Some formulae have been proposed [13] for the activity coefficients from approximations of the limit of Y_s when R is made infinite, that is, when the phosphate concentration is exactly zero, or when the interactions between different DNAs are neglected. Figure 1 shows that the activity coefficients we have computed from y_s vary not only with X, but also with the concentration of the fixed charges. The formulae for the activity coefficient of the counter-ions [13] are astonishingly correct when R is small. Those corresponding to the co-ions are correct as expected when R is very large, but are rough under-evaluation when the interactions between the different DNAs become important. The figure 2 shows that y_s tends exponenyially to Y_s, as the square root of X decreases.

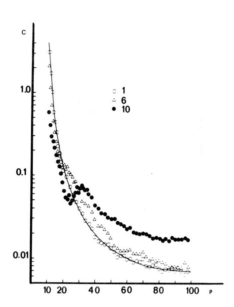

Figure 3. Monte-Carlo distribution of 10 counter-ions
of various radii about the model 1 of DNA
(see text). The radii are such that $a_+ + a = 10$ Å
and there are no co-ions so that the Poisson
Boltzmann solution (solid line) is the same
in all cases and known analytically [12].

Now, what is the validity of the Poisson-Boltzmann approach ? Figure 3 shows a Monte-Carlo computation and the
corresponding Poisson-Boltzmann solution. It is clear that
small ions do have the predicted distribution. Bulkier ions
form layers. The potential at the surface of the 'DNA'
increases as the ions are bulkier, and so does the excess
free energy. Neglecting the ion correlations is thus less
important than neglecting the short range interactions.
Similar results have been obtained with the more realistic
model 2 of DNA, or if co-ions have been introduced, or if
bivalent counter-ions were present.

FLEXIBILITY

The total free energy which is necessary to bend a
short rod-like piece of a poly-ion is:

$$\Delta G_t = P_t \, \alpha^2 \, b \, / \, 2a^2 \tag{25}$$

where α is the ratio of the radius a of the poly-ion to the

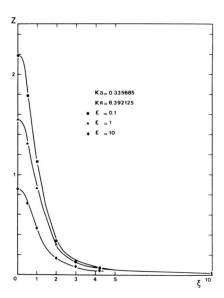

Figure 4. Values of Z versus the true charge parameter
for various values of the ratio of the
internal to solvent dielectric constants.
The molar concentration of the 1-1 salt is
approximately 0.01M if a=10 Å. The radius
of curvature of the torus is 250 Å.

radius of curvature of the bent rod. P_t has two contributions: a purely elastic one P_o, and an electrostatic one
P_e. Intuitively, P_e increases when the ionic strength decreases because the fixed charges are less screened so
that their mutual repulsions increase during the bending.
We have integrated numerically the Poisson-Boltzmann equation (22) for a torus [14]. The charge distribution on
the cross section depends on the angle ϕ :

$$\sigma_{torus} = \sigma_{cyl} / (1 + \alpha \cos \phi) \qquad (26)$$

Once the average on the angle ϕ in the cross-section of the
potential of the torus is obtained, we compute Z

$$Z(\xi, \varepsilon, \kappa a, \alpha) = (y_{torus}^{aver} - y_{cyl}) / 4\xi \kappa^2 a^2 \qquad (27)$$

Of course, Z depends on ξ, ε, κa, α. Some results are shown
on figure 4. Then

$$P_e = \xi^2 \int_o^1 Z(v\xi, \varepsilon, \kappa a, \alpha) \, v \, dv / 2 \kappa^2 \ell_B \qquad (28)$$

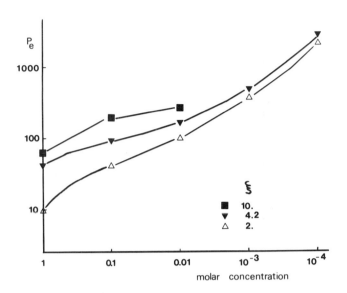

Figure 5. Electrostatic contribution to the persis-
          tence length (in Angstroms) as a function
          of the molar concentration of a 1-1 salt.
          The poly-ion has a radius of 10 Å, and has
          various values for its true charge parame-
          ter (Eq. 24). The ratio of the dielectric
          constants is 0.1.

Some results concerning the electrostatic contribution to
the persistence length are shown on figure 5. Now, we
should have:

$$P_t = P_o + P_e \tag{29}$$

If the experimentally total persistence length is plotted
versus our electrostatic contribution we should get a
staight line of slope unity. The problem here is to separate the stiffening of the chain caused by an increase of
the persistence length, from the expansion of the random
coil caused by an increase of the excluded volume. That
is why most experimentalists present their data with
respect to some reference. Since we have here absolute
numbers for the electrostatic contribution, we may look
for the value of the persistence length at 0.2M NaCl such
that the values measured at other salt concentrations fit
the equation (29). For instance, Harrington's data [15]
fit our 'Poisson-Boltzmann' values for the electrostatic
contribution and the equation (29) when his total persis-

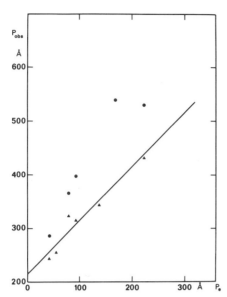

Figure 6. The total persistence lengths (Angstroms)
for DNA at various NaCl concentrations are plotted versus
our electrostatic contribution to the persistence length.
(Charge parameter = 4.2; dielectric ratio = 0.1). Triangles
Harrington's data with the reference fitting the best the
Eq.(29) (solid line). Squares: the data of Borochov et al.
[36] are slightly above the solid line. Hagerman's value
[20] (1800Å for our 2300Å) would be below the solid line.

tence length at 0.2M NaCl is equal to 323 Angstroms. The
'elastic' or non-electrostatic contribution is then equal
to 214.7 Angstroms [16]. Figure 6 shows the plot for a
slightly different value of the reference. If both the salt
concentration and the true charge parameter are exactly
equal to zero, then Z is equal to unity, whatever is the
ratio of the dielectric constants. At this limit, the Odijk
[17] and Skolnick and Fixman [18] formula is obtained. However, if their formula is assumed to be valid for non zero
ion concentrations and for non zero charge parameters,
Harrington's data and the equation (29) remain incompatible
whatever the chosen reference. Note that our results agree
with the experimental data without invoking any specific
binding of sodium to the phosphates,as it has recently been
shown experimentally [19]. Values of the total persistence
length at low ionic strength are very important to check
the validity of our computations. The value obtained by
Hagerman [20] with restriction fragments is 1800 A at
0.0001M NaCl, which is slightly less than our expectation:
2500 Å. Several reasons may be invoked. 1) Our expected
value is longer than the total length of Hagerman's frag-

ments and cannot be obtained with so short DNAs. 2) the
charge distribution is assumed to be given by the Eq. (26).
It is known that the average angle between two base pairs
varies with the ionic strength [21]. If the angles between
the base pairs remain the same in the bent rod as in the
straight one, our charge distribution remains valid but for
the smearing. However, we have often found large differences in the potential at the surface of the torus along a
cross-section. The absolute value of the potential on the
cross-section is all the greater as the point is closer to
the revolution axis of the torus. Therefore a distortion
of the DNA such that the charges move far from the revolution axis of the torus is energetically favorable and make
the persistence length decrease.

TOPOLOGICAL INTERACTIONS

        Let us first recall that catenanes and knots do exist
[22,23]. In order to characterize completely the topology
of a set of closed strands of DNA, we must determine how
the strands are catenated (figure 7) then, the knot class
of isotopy of each strand. We restrict our analysis to double helical DNAs. However, the word 'helix' has too precise a connotation and should be avoided. We restrict our
analysis to those DNAs that have a 'duplex axis'. Here,
'duplex axis' is defined after differential geometry considerations [24], as a differentiable continuous line such
that each strand of the double stranded DNA intersects once
and only once the interior of any small circle say of radius
20 A centered on the 'duplex axis' and the plane of which
is perpendicular to the 'duplex axis'. The following may be
rigorously shown: 1) cruciform [25], lasso-like [26] and
so-called 'denatured' DNAs have no 'duplex axis' and are
excluded from our analysis. 2) the two strands and the duplex axis are isotopic and belong to the same knot class if
there is a 'duplex axis'. 3) If the 'duplex axis' exists,
then the two strands of a double stranded DNA cannot be
catenated as in figure 7c. Moreover catenation and linking
(that is their Gauss' integral) are equivalent: If the
strands are catenated they are linked and if they are linked
they are catenated. 4) Calugareanu's formula [22,35] holds
if there is a 'duplex axis'. Written in its generalized
version [27] it reads:

$$\alpha = w + \beta \tag{30}$$

where w is the writhing number of the duplex axis [28], β
is the sum (in turns) of all the angles between adjacent
base pairs and α is the linking number. The number of
constraint turns is c defined by [29,30] :

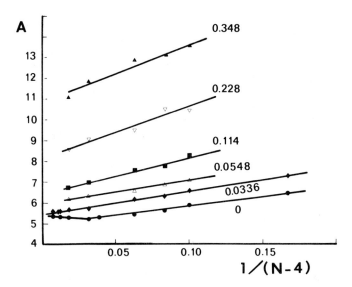

Figure 8. Values of A in Eq. 34 for closed unknotted chains of N bonds which have been generated on a computer. The ratio of the distance of closest approach between two different bonds to the length of the bond is written on the lines. From Le Bret [30].

labels [32,33] and from its theoretical interpretation [34] Then K/C is equal to 0.55. As b is 1.7 Å, the parameter A may be computed from (35) once the total persistence length is known. Assuming it is comprised between 400 and 600 Å, A lies between 9.5 and 6.3. Figure 8 shows then that the distance of closest approach is comprised between 190 and 66 Angstroms. Obviously, the 'distance of closest approach' may be re-interpreted as the distance at which the repulsions between two DNA parts cannot any more be ignored.

This work has been supported by the Fondation pour la Recherche Médicale Française, the Délégation Générale pour la Recherche Scientifique et Technique and the Université Pierre et Marie Curie (Paris 6). A NATO fellowship has been appreciated.

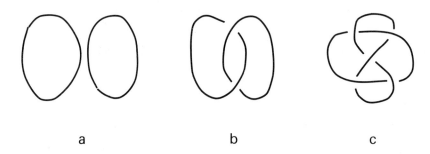

a   b   c

Figure 7. The closed lines are catenated in b and c but not in a. They are linked in b but hot in a and c.

$$c = w - w_o + \beta - \beta_o \qquad (31)$$

where w_o and β_o are assumed to be independent from c and w. Experimentally [31] the free energy of supercoiling is:

$$\Delta G_{supercoiling} = KRTc^2/L \qquad (32)$$

where K is equal to 1000 in 'low' salt (0.01 M NaCl) for a 10000 base pair DNA, and L is the number of base pairs of the double stranded DNA. We assume that:

$$\Delta G_{twist} = CRT(\beta-\beta_o)^2/L \qquad (33)$$

For a chain of $N=Lb/P_t$ bonds of length $2P_t$ we showed [30]

$$\Delta G_{writhing} = AP_t RT(w-w_o)^2/Lb \qquad (34)$$

The values of the parameter A are shown on figure 8. More-over the parameters A,C,K are related by:

$$K/C = 1 - Kb/AP_t = \beta - \beta_o/c = 1 - (w-w_o)/c \qquad (35)$$

C is known from the torsional brownian motion of DNA as mo-nitored by the fluorescence anisotropy decay of intercalated

REFERENCES

[1] Metropolis,N.,Rosenbluth,A.W.,Rosenbluth,M.N.,Teller,A.
     H. & Teller,E.:1953,J.Chem.Phys.21,pp1087-1092.
[2] Gouy,G.:1908,Ann.Chim.Phys.8,p.294.
[3] Gouy,G.:1917,Ann.Phys. série 9,7,pp.129-184.
[4] Fowler,R. & Guggenheim,E.A.:1949 in 'Statistical Thermo-
     dynamics' Cambridge University Press chapter 9.
[5] Sparnay,M.J.:1972,J.Electroanal.Chem.37,pp.65-70.
[6] Gregor,H.P. & Gregor,J.M.:1977,J.Chem.Phys.66,pp.296-306
[7] Frahm,J. & Dieckmann S.:1979,J.Colloid Interf.Sci.70,pp.
     440-447.
[8] Fixman,M.:1979,J.Chem.Phys.70,pp.4995-5003.
[9] Sugai,S. & Nitta,K.:1973,Biopolymers 12,pp.963-1376.
[10] Stigter,D.:1975,J.Colloid Interf.Sci.53,pp.296-306.
[11] Guéron,M. & Weisbuch,G.:1980,Biopolymers 19,pp.353-382.
[12] Fuoss,R.M.,Katchalsky,A. & Lifson,S.:1951,Proc.Nat.Acad
     Sci.USA 37,pp.579-589.
[13] Guéron,M. & Weisbuch,G.:1979,J.Phys.Chem.83,pp.1991-
     1998.
[14] Le Bret,M.:1981,C.R.Acad.Sci.Paris 292 II,pp.291-294.
[15] Harrington,R.E.:1978,Biopolymers 17,pp.919-936.
[16] Harrington,R.E.:Personnal communication.
[17] Odijk,T.:1977,J.Polym.Sc.,Polym.Phys.Ed.15,pp.477-483.
[18] Skolnick,J. & Fixman,M.:1977,Macromolecules 10,pp.944-
     948.
[19] Bleam,M.L.,Anderson,C.F. & Record,M.T.:1980,Proc.Nat.
     Acad.Sci.USA 77,pp.3085-3089.
[20] Hagerman,P.:1981,Biopolymers in press.
[21] Wang,J.C.:1969,J.Mol.Biol.43,pp.25-39.
[22] Riou,G. & Delain,E.:1969,Proc.Nat.Acad.Sci.USA 62,pp.
     210-217.
[23] Brown,P.O.,Cozzarelli,N.R.:1981,Proc.Nat.Acad.Sci.USA
     78,pp.843-847.
[24] Calugareanu,G.:1961,Czech.Math.J.11,pp.588-625.
[25] Panayotatos,N. & Wells,R.D.:1981,Nature,289,pp.466-470.
[26] Bratosin,S.,Laub,O.,Tal,J.,Aloni,Y.:1979,Proc.Nat.Acad.
     Sci.USA 76,pp.4289-4293.
[27] White,J.:1969,Am.J.Math.91,pp.962-728.
[28] Fuller,F.B.:1971,Proc.Nat.Acad.Sci.USA 68,pp.815-819.
[29] Le Bret,M.:1979,Biopolymers 18,pp.1709-1725.
[30] Le Bret,M.:1980,Biopolymers 19,pp.619-637.
[31] Depew,R.E. & Wang,J.C.:1975,Proc.Nat.Acad.Sci.USA 72,
     pp.4275-4279.
[32] Millar,D.P.,Robbins,R.J. & Zewail,A.H.:1980,Proc.Nat.
     Acad.Sci.USA 77,pp.5593-5597.
[33] Wahl,P.,Paoletti,J. & Le Pecq,J.B.:1970,Proc.Nat.Acad.
     Sci.USA 65,pp.417-421.
[34] Barkley,M.D. & Zimm,B.H.:1979,J.Chem.Phys.70,pp.2991-
     3007.
[35] Pohl,W.F.:1968,J.Math.Mech.17,pp.975-985.
[36] Borochov,N.,Eisenberg,H.& Kam,Z.:1981,Biopolymers 20,
     pp.231-235. and Biopolymers 20 in press.

^{1}H NMR STUDY OF THE NATURE OF BONDING INTERACTIONS INVOLVED IN COMPLEXES BETWEEN NUCLEIC ACIDS AND INTERCALATING COMPOUNDS.

Delbarre A.[★], Gaugain B.[★], Markovits J.[☆], Vilar A.[★], Le Pecq J.B.[☆] and Roques B.P.[★]

[★]Département de Chimie Organique (ERA 613 CNRS et SCN 21 INSERM) UER des Sciences Pharmaceutiques et Biologiques 4 avenue de l'Observatoire, 75006 Paris, France.
[☆]Laboratoire de Physicochimie Macromoléculaire (LA 147 CNRS et U 140 INSERM) Institut Gustave Roussy, 94800 Villejuif, France.

Abstract

Simple monointercalating compounds (acridine, ethidium, ellipticine) self-associate in aqueous solution with association constant in the range of 100 M^{-1} to 5000 M^{-1} depending on structure (nature of the ring and substituent) and environment (pH and salt concentration). Bis-intercalators folding and aggregation were studied and the importance of the rigidity of the linking chain to prevent the autostacking was demonstrated. The sequence specificity of intercalating agents was studied at the level of the minihelix formed by autocomplementary ribo-dinucleotides. Chloro-2 methoxy-6 amino-9 acridine exhibits a strong pyrimidine 3'-5' purine preference and the geometry of its intercalated complex with CpG was determined using isoshielding curves. The addition to this acridine ring of a 9 substituted pentylcarboxamido chain confers a specificity of this compound for guanine. The involvement of hydro-gen-bonds between the carboxamide group born by the acridine lateral chain and the guanine was clearly demonstrated from temperature depen-cies of NH protons and from the variation of the linewidths of the pentylcarboxamido chain during the fusion of the intercalated complex.

The chemical characterization of DNA-sequences involved in reco-gnition of regulating proteins allows to envision the development of synthetic molecules able to bind specifically to these sequences and therefore to inhibit selectively the genetic expression. The high DNA-affinity required for such a purpose could be theoretically obtained with bis-intercalating compounds (1) and the sequence-selectivity using molecules able to form hydrogen-bonds with specific base-pairs (2,3). The rational design of such synthetic models needs a precise knowledge of the different kinds of bonds involved in nucleic acids recognition, i.e., intermolecular stacking interaction, intramolecular self-associa-tion of dimers, formation of intercalated minihelices, hydrogen bonding

B. Pullman (ed.), Intermolecular Forces, 273–283.

schemes etc... This can be performed at the oligonucleotide level using
NMR (4) or crystallographic techniques (5). However, ^{1}H NMR allows to
investigate this problem in solution giving information on the dynamics
and the mode of DNA-binding of mono- and bis-intercalating derivatives.
As an illustration of the possibilities of ^{1}H NMR technique in this
field, we will report in this paper studies concerning self-association
of monomeric intercalating agents, conformation of bis-intercalators,
sequence preference of intercalating rings and base recognition through
hydrogen bond formation.

I. Self-association of intercalating compounds.

Owing to their planar and hydrophobic properties, all DNA-inter-
calating compounds form stacked aggregates in aqueous solution, as
shown by the concentration dependence of the aromatic protons. By extrapolation of the chemical shifts at zero concentration the self-association constants (Kas) are computed assuming that the successive stacking
constants are identical and that the effects of magnetic anisotropy are
additive (6). Moreover, stacking geometries can be proposed using
reported isoshielding curves for ethidium and acridine (7) or molecular frameworks for ellipticine.
From these results it appears that :
i) All the aromatic moieties are stacked one on each other with
one ring inverted as illustrated on Figure 1. This geometry corresponds
to a favourable orientation of the quadrupole axis of these compounds
in two parallel planes and to a minimization of the repulsion between
the positive charges (8 , 9 ,10).

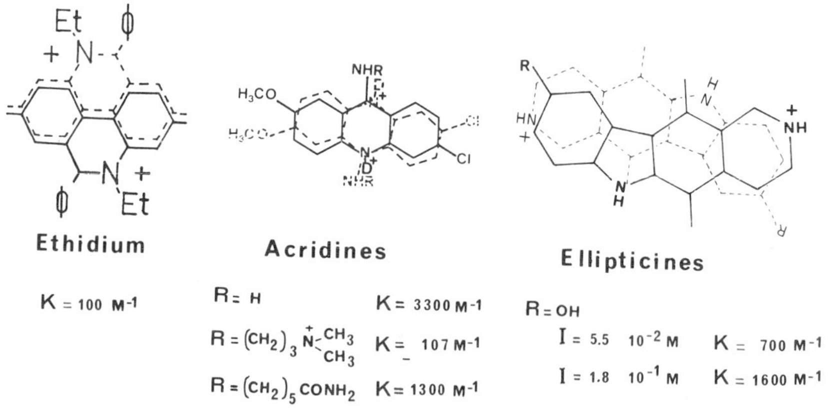

Ethidium

$K = 100$ M^{-1}

Acridines

R = H $K = 3300$ M^{-1}

R = (CH$_2$)$_3$ $\overset{+}{N}$-CH$_3$ $K = 107$ M^{-1}

R = (CH$_2$)$_5$CONH$_2$ $K = 1300$ M^{-1}

Ellipticines

R = OH

I = 5.5 10^{-2} M $K = 700$ M^{-1}

I = 1.8 10^{-1} M $K = 1600$ M^{-1}

Figure 1. Proposed stacked geometries for different intercalating
agents and association constants (Kas in M^{-1}).

ii) The association constants are strongly modulated by the charge
born by the lateral chain as shown by the differences in Kas values for
acridines with R = H and R = -(CH$_2$)$_3$-N$^+$(CH$_3$)$_2$.

iii) The increase of Kas at high ionic strength could be related to the reduction of the electrostatic repulsions between the positively charged rings by the solvated surrounding anions. This feature might be of biological relevance when positively charged intercalators bind to the polyanionic double helix of DNA.

iv) The stronger self-association of acridine with R = -(CH_2)_5CONH_2 could be related to an hydrogen bond formation between the protonated pyridinic nitrogen of one molecule and the carbonyl of the carboxamide group of another acridine.

Interactions of these intercalating compounds with nucleic acids requires a disruption of the aggregates prior to the intercalation process. Consequently it was of interest to determine the statistical repartition of molecules within the aggregates as a function of the concentration for different values of Kas. As shown on Figure 2 for Kas > 10^3 M^{-1}, the proportion of n-mers remains relatively high at the concentrations (5 x 10^{-4} to 10^{-3} M) currently used for ^{1}H NMR study of intercalation in oligonucleotides. Therefore, the occurence of these remaining aggregates must be taken into account for the determination of the geometry of intercalated minihelices (vide infra).

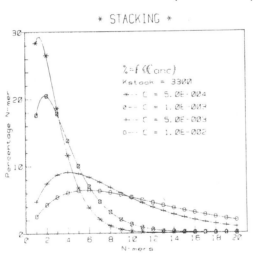

Figure 2. Statistical repartition of n-mers as a function of the concentration for different values of Kass (computed using equations reported in 10).

II. Conformation and association of bis-intercalators.

The aromatic rings of bis-intercalators linked by aminoalkyl chains can lead to intermolecular aggregates as their monomeric parents or to intramolecular stacked forms. The presence of such folded conformations can be easily evidenced by ^{1}H NMR study at various temperature and concentrations. Indeed, to unfolding processes which occur with temperature increase correspond classical melting curves (8). Such studies performed on acridine (8), ethidium (10), ellipticine (11) and pyrido-

carbazole dimers lead to the following remarks :
    i) At low concentration (10^{-4}M) and pH $\geqslant$ 7 where the linking chain
of the dimers are deprotonated, the protons linewidths (at 270 MHz)
remain sharp. Therefore, assuming that the chemical shift difference
between folded and unfolded forms can reach 1 to 1.5 ppm (computed
from isoshielding curves) we found that the rate of this equilibrium
is > 2 x 10^3 s^{-1}.
    ii) The proximity between the charged intercalating rings induces
a decrease of their pKa's which are 1 to 3 units lower than that of
the related monomers. The same is true for the nitrogen of the linking
chain. As expected the folding process, strongly depends on the pH and
the ionic strenght of the solution. Thus at physiological pH and in
the phosphate buffer conditions used for the study of DNA-interaction,
all these flexible dimers are almost completely folded. All these
results have lead us to synthesize new bis-intercalators made up of
more rigid linking chains (12). At the same low concentration (10^{-4}M),
the spectra of acridine monomer and rigid dimer are similar whereas
the protons of the flexible dimer are upfield shifted due to the occurence of intramolecular self-stacking (Figure 3).

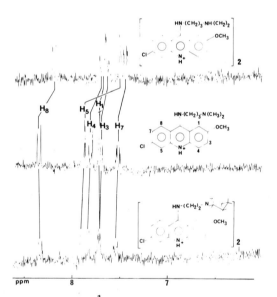

Figure 3. Aromatic part of ^{1}H spectra of acridine monomer, flexi-
          ble and rigid dimers performed at 270 MHz in D$_2$O
          (10^{-4}M, pH = 5.6, temp 21°C, HMDS as int. ref.).

    Nevertheless, it must be noticed that, whatever the nature of
their linking chains, all the dimers are able to form intermolecular
aggregates at high concentration (> 10^{-3}M, pH $\geqslant$ 7). However, as evidenced from temperature studies, the kinetics of these associations
are very different (Figure 4). Indeed the melting curves of the

aggregates is much more cooperative for the dimers with rigid chains than for the dimers with flexible linkers. This phenomenon is clearly evidenced by the difference in the linewidths of the aromatic protons in the two kinds of dimers at the same concentration (Figure 4). These features could be related with their different DNA-binding processes (1).

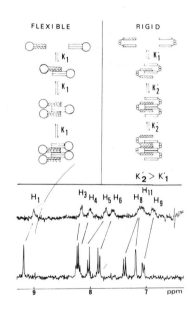
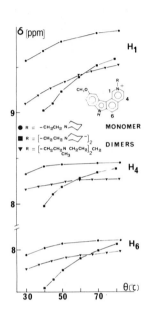

Figure 4. *Right* : Spectra of rigid (top) and flexible (bottom) dimers of pyridocarbazoles in H_2O ($10^{-3}M$, pH = 7.5, temp = 40°C). Schemes of intermol. associations *Left* : Temperature dependencies of chemical shifts for monomer and dimers of pyridocarbazoles in H_2O ($10^{-3}M$, pH = 7.5).

III. Sequence preference of intercalating rings and base recognition through hydrogen bond formation.

As already discussed, selective base-recognition could occur through hydrogen bonding. This specificity could be reinforced by a sequence preference of the intercalating ring. To check this hypothesis we synthesized the 2-methoxy 6-chloro 9 (5-carbamoylpentylamino) acridine $\underline{1}$. In this compound, the alkyl chain holding the carboxamide group is long enough to permit the formation of specific hydrogen bonds with either guanine in the small groove or adenine in the large groove depending on the orientation of the acridine ring intercalated into the DNA helix. The titration curves of $\underline{1}$ (2 mM) with CpG in D_2O clearly demonstrate the strong pyrimidine 3'-5' purine preference of this compound (Figure 5).

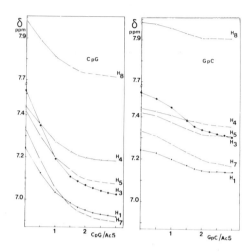

Figure 5. Titration of 1 with CpG and GpC in D$_2$O, 0.04 M sodium
deuteroacetate buffer pH 5.6, temp = 25°C.

The upfield shifts following the addition of CpG and the break in
the titration curves for the 2/1, CpG to acridine ratio evidence the
formation of an intercalated minihelix complex induced by the intercalating agent. The sigmoidal shapes of the titration curves with GpC
could indicate a preliminary formation of a 1/1, GpC/1 complex followed
by a cooperative occurence of an intercalated minihelix. The geometry
of the 2/1, CpG/1 intercalated complex was estimated using both the
experimental induced shieldings Δδ$_i$ and those, Δδ$_c$ computed from the
ring currents effects of the different bases and acridine rings proposed by Geissner-Pretre and Pullman (7 ,13). As shown in Table 1, the
best fits are obtained for a complex characterized by a quasi-symetric
stacking of the chloromethoxy aminoacridine ring on the base-pairs
plane (Figure 6).

Table 1. Comparison between computed Δδ$_c$ and experimental Δδ$_i$
upfield shifts for the intercalation of 1 in CpG
according to the model of Figure 6.

	H$_1$	H$_3$	H$_4$	H$_5$	H$_7$	H$_8$
Δδ$_c$	0.89	0.71	0.57	0.84	0.71	0.67
Δδ$_i$	0.79	0.64	0.60	0.77	0.60	0.66

The strong upfield shifts of H$_1$, H$_8$ and the 9-NH-CH$_2$ group located
in the plane of the acridine ring can only be explained assuming that
the 9-amino group lies in the minor groove of the minihelix.

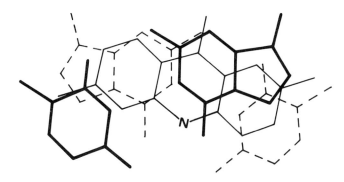

Figure 6. Proposed geometry for the intercalation of $\underline{1}$ with CpG (14).

The introduction of chloro and methoxy substituents on the acridine ring leads to a strong pyrimidine 3'-5' purine preference evidenced by the selective interaction of 2-methoxy 6-chloro 9-aminoacridine, $\underline{2}$, with CpG (14) contrasting with the reported purine-pyrimidine preference of unsubstituted 9-aminoacridine, $\underline{3}$, (15). This latter result seems to differ from theoretical calculations which predict that intercalation should occur more easily in pyrimidine 3'-5' purine sequence than in the reverse one (16). Such differences in the sequence preference of $\underline{2}$ and $\underline{3}$ at the dinucleotide level could explain the differences in the DNA binding mode of dimers made up with these acridine moieties linked through their 9-amino group by a -$(CH_2)_6$- alkyl chain. Indeed, the dimer derived from $\underline{2}$ behaves as a monomeric intercalating agent at the DNA level whereas the corresponding dimer obtained from $\underline{3}$ intercalates its two rings between adjacent base-pairs (17). These features could be related to a sequence specificity effect but most probably to the smaller size of the unsubstituted 9-aminoacridine ring which could interact with pyrimidine-purine as well as purine-pyrimidine sequences. This implicates a relative mobility of the 9-aminoacridine ring in the intercalating sites according to the presence of two kinds of arrangment of the heterocycle intercalated in minihelices of ICpG (18). Likewise the large degree of freedom of this unsubstituted 9-aminoacridine in autocomplementary minihelices is supported by the lack of clear steochiometries in the titration curves of a 9-aminopentylamidoacridine with both CpG and GpC and therefore to the absence of a fixed geometry for such complexes (14).

To evidence the role of the steric hindrance of chloro and methoxy groups of acridine in the inhibition of bis-intercalation, the heterodimer, $\underline{4}$, was synthesized.

Preliminary results of hydrodynamic measurements seem to indicate the DNA bis-intercalating ability of 4. Therefore, one can assume that in 4, the larger substituted acridine ring which has a higher affinity, first intercalates into DNA allowing the unsubstituted acridine moiety to freely search for the energy minimized configuration in the adjacent intercalating site. Such a required conformational freedom is limited when the two acridine rings are substituted.

On the other hand, the crucial role of hydrogen bonding in base-recognition is illustrated by ^{1}H NMR studies of the 2.5/1 solution of CpG/1 in H$_2$O. As shown in Figure 7, the occurence of an intercalated minihelical complex is clearly shown by the upfield shifts of all the aromatic protons occuring in the mixture.

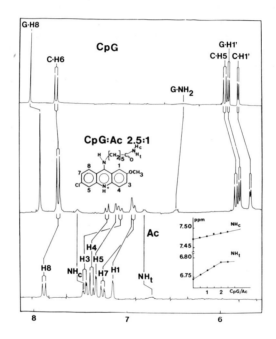

Figure 7. ^{1}H NMR spectra (270 MHz) of : CpG (top), CpG/1, 2.5/1 complex (middle), 1 (bottom). Acridine conc. = 10^{-2}M. Spectra were run in H$_2$O, 0.4 M NaCD$_3$CO$_2$ pH5.6. Inset = titration curves of the *trans* and *cis* NH amide protons with CpG in the above conditions (redrawn from Gaugain et al., 14).

Furthermore, the complexation process leads to a narrowing in the linewidths of the GNH$_2$ and amide protons. This effect corresponds to a decrease of their exchange rates which probably results from hydrogen bonding in the intercalated complex.

Further evidence of the formation of a 2/1, CpG/1 microhelical complex stabilized by additional hydrogen bonds was obtained following

the restricted motion of the acridine side chain in the complex. Indeed,
the linewidth of the CH$_2$ groups is dependent upon both the overall
molecular reorientation of the complex and the internal segmental
motion of the chain which reflects the degree of freedom of each methylene. Therefore, the involvement of the alkyl chain in hydrogen bonding
scheme can be followed by examination of the linewidth of the methylene
groups during the breakdown of the intercalated complex induced by a
temperature increase. In the free acridine 1, linewidths of these methylene vary regularly within a very narrow range in function of the
temperature (results not shown). In contrast, in the 2.1/1, CpG/1
complex, the linewidths of these CH$_2$ are affected differently in function of their position when the temperature rises from 20 to 70°C
(Figure 8). The regular decrease in the linewidth of CH$_2$ (b) reflects
the dissociation of the complex whereas the faster narrowing of the
CH$_2$ (d) and CH$_2$ (e) in the 20°C to 40°C temperature range could be due
to the breakdown of the hydrogen bonds between the amide group and the
guanine which precedes the fraying of the minihelix. This interpretation is reinforced by the behaviour of the CH$_2$ (c), located in the
middle of the chain, whose linewidth remains practically temperature
independent. This melting process can be followed from the classical
sigmoidal curves corresponding to the deshielding of the aromatic
protons in function of the temperature. In the 2.5/1, CpG / 1 complex
this feature is shown in Figure 8 for the H$_3$ acridine proton and leads
to a fraying curve with a mid-point around 45°C.

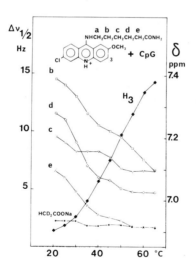

Figure 8. Temperature variations of the linewidth at half-height
of the CH$_2$ groups and of the chemical shift of H$_3$ in
the 2.5/1, CpG/1 complex (conc. 1 : 10^{-2}M, deuteroacetate buffer 0.4M, pH = 5.6).

Finally, the involvement of the $CONH_2$ group in hydrogen bonds with guanine is confirmed by the position of the methylene groups of the side chain which are differently affected by the complexation. Thus, the CH_2 (b) is clearly shielded whereas the CH_2 (e) is slightly deshielded in the complex (not shown here). These opposites effects are related to the perpendicular orientation of the side chain born by the acridine ring with the terminal amide group directed towards the guanine ring. In such a geometry, the CH_2 (a) and CH_2 (b) experience shielding current shifts from the superimposed base-pairs while the CH_2 (e) group undergoes a deshielding due to its position in the plane of the base-pairs.

All these results strongly support an immobilization of the carbo-xamide group at the level of the guanine ring where it forms hydrogen bonds with this base.

Conclusion

^{1}H NMR allows to experimentally determine the types of interactions involved in nucleic acid complexes provided that simple models such as autocomplementary oligonucleotides and mono- or bis-intercalating agents are used. The geometries of intercalated minihelical complexes can be accuratly determined from comparison between experimental indu-ced shieldings and theoretical upfield shifts computed from isoshielding curves.

The possibility of base sequence recognition through the formation of a pair of hydrogen bond between a carboxamide group and guanine was demonstrated. Such kind of studies would be very usefull in the ratio-nal design of pharmacological agents for specific DNA sequences.

Acknowledgements

The authors thank Mrs A. Bouju for typing.

This work was supported by grants from the Délégation Générale à la Recherche Scientifique et Technique, the Fondation pour la Recher-che Médicale Française, the Association pour le Développement de la Recherche sur le Cancer, the Université René Descartes (PARIS V) and the Université Pierre et Marie Curie (PARIS VI).

References

1. Markovits, J., Gaugain, B., Roques, B.P., and Le Pecq, J.B.:
    1981, this Symposium.
2. Seeman, N.C., Rosenberg, J.M., and Rich, A.: 1976, Proc. Natl.
    Acad. Sci. USA, 73, pp. 804-808.
3. Hélène, C.: 1977, FEBS Letters, 74, pp. 10-13.
4. Krugh, T.R., and Nuss, M.E.: 1979, in Biological Applications of
    Magnetic Resonance (Shulman, R.G., ed) Academic Press, New-York,
    pp. 113-175.
5. Sobell, H.M., Tsai, C-C., Jain, S.C., and Gilbert, S.G.: 1977,
    J. Mol. Biol., 114, pp. 333-365.
6. Dimicoli, J.L., and Hélène, C.: 1973, J. Amer. Chem. Soc., 95,
    pp. 1036-1044.
7. Giessner-Prettre, C., and Pullman, B.: 1976, C.R. Acad. Sci. Paris,
    283, pp. 675-677.
8. Barbet, J., Roques, B.P., Combrisson, S., and Le Pecq, J.B.: 1976,
    Biochemistry, 15, pp. 2642-2650.
9. Roques, B.P., Barbet, J., Oberlin, R., and Le Pecq, J.B.: 1976,
    C.R. Acad. Sci. Paris, 283, Serie D, pp. 1365-1367.
10. Delbarre, A., Roques, B.P., Le Pecq, J.B., Lallemand, J.Y., and
    Nguyen-Dat-Xuong: 1976, Biophys. Chem., 4, pp. 275-279.
11. Gaugain, B., Barbet, J., Oberlin, R., Roques, B.P., and Le Pecq,
    J.B.: 1978, Biochemistry, 17, pp. 5071-5078.
12. Pelaprat, D., Delbarre, A., Le Guen, I., Le Pecq, J.B., and Roques,
    B.P.: 1980, 23, pp. 1336-1342.
13. Giessner-Prettre, C., and Pullman, B.: 1976, Biochem. Biophys. Res.
    Commun., 70, pp. 578-581.
14. Gaugain, B., Markovits, J., Le Pecq, J.B., and Roques, B.P.: 1981,
    Biochemistry, in press.
15. Reuben, J., Baker, B.M., and Kallenbach, N.R.: 1978, Biochemistry,
    17, pp. 2916-1919.
16. Nuss, M.E., Marsh, F.J., and Kollman, P.A.: 1979, J. Amer. Chem.
    Soc., 101, pp. 825-833 and references cited herein.
17. Wright, R.G.M., Wakelin, L.P.G., Fieldes, A., Acheson, R.M., and
    Waring, M.J.: 1980, Biochemistry, 19, pp. 5825-2836.
18. Sakore, T.D., Reddy, B.S., and Sobell, H.M.: 1979, J. Mol. Biol.,
    135, pp. 763-785.

DNA MONO AND BISINTERCALATORS AS MODELS FOR THE STUDY OF PROTEIN NUCLEIC ACID INTERACTIONS: ORIGIN OF THE HIGH AFFINITY AND SELECTIVITY.

Markovits J.*,Gaugain B.+,Roques B.P.+,and Le Pecq J.B.*
*Laboratoire de Physicochimie Macromoleculaire
(LA 147 CNRS et U 140 INSERM),Institut Gustave Roussy
94800 Villejuif France.
+ Laboratoire de Chimie Organique (ERA 613 CNRS) UER
des Sciences Pharmaceutiques et Biologiques,4 avenue
de l'Observatoire,75270 Paris Cedex-06 France.

Abstract

In order to analyse the role of symmetry and dimerization on protein DNA interaction, DNA mono and bisintercalating molecules have been prepared and studied as model compounds. Bisintercalating molecules were obtained which have DNA binding affinity one hundred thousand times higher than the corresponding monomer. This increase corresponds almost to the maximum theoretical one. These data show that the mere symmetrical dimerization of a ligand can lead to a very large increase of its binding affinity for a receptor. On the other hand,monointercalating molecules bearing a side chain terminated by a carboxamide group have been synthetized . These compounds can bind to DNA by intercalating their aromatic ring,while the side chain lies in the small groove. The terminal carboxamide group then forms a pair of hydrogen bonds with guanine. This interaction promotes a specificity for GC rich DNA and is responsible for an additional interaction of -1 Kcal/mole.The contribution of similar hydrogen bonds to specificity in protein DNA complexes is discussed.

Various ligands are able to reversibly bind to proteins or to nucleic acids with a very high affinity. These strong interactions are involved in many important biological processes. That is the case, for instance, of the binding of antigens to antibodies,of hormones and drugs to their receptors, of proteins to nucleic acids etc...In all these cases, the interaction is characterized by a high specificity and a high binding constant (from 10^9 to 10^{13} M^{-1}).

The interaction between proteins and nucleic acids has been particularly well studied (1). The proteins which regu-

B. Pullman (ed.), Intermolecular Forces, 285–298.

late the genetic expression are of course of special importance. They bind to DNA to specific sites. The concentration of these DNA-sites in cells is small (< 10^{-9} M).
Therefore the binding constant of the regulatory proteins
for these DNA-sites must be higher than 1010 - 1012 M .
On the other hand the quantity of DNA contained in a given
site is very small compared to the total quantity of DNA
in a cell (about one part per million). Therefore the interaction must be highly specific. In these systems, two
important factors relevant to the strength and the specificity of these interactions have been underlined (1).

   a) The dimeric and symmetrical structure of the interacting species is an important feature (2) . The DNA
piece which is recognized has often a palindronic sequence
with a center of symmetry. The binding protein has also a
corresponding two fold axis of symmetry.

   b) The involvement of pairs of hydrogen bonds between,
carboxamide, guanidinium groups born by glutamine asparagine and arginine side chains in proteins, and the base
pairs in DNA, insures the specificity of the interaction
(3,4).

   During the last few years (5-16), we have been pre-
paring simple DNA binding compounds which can be used as
models for the study of the role of these two parameters.
In this paper, it will be shown that the dimerization of a
ligand can indeed lead to a large increase of affinity for
its receptor. In addition the study of the interaction of
small DNA binding molecules, able to form a single pair of
hydrogen bonds with base pairs, permits the determination
of the degree of specificity conferred by such type of
interactions (16).

   I) The importance of being dimeric.

   Let us compare the interaction of a dimeric ligand
made up of two identical subunits with a complementary
dimeric structure and the interaction of the monomeric
ligand with its receptor as symbolysed on Figure 1.

   If we know the binding constant associated with the
monomeric system K_{mon} , it is possible to very simply compute a maximum value for the binding constant of the
dimeric system : $(K_{dim})_{max}$ .If $(\Delta G_{dim})_{max}$ and ΔG_{mon}
are the maximum free energy of interaction of the dimeric
complex and the free energy of interaction of the monomeric complex respectively :

$$(\Delta G_{dim})_{max} = 2 \Delta G_{mon} - R T Ln X \qquad [1]$$

   The second term RTlnX (where X is the molar concentration of water in solution) arises because the entropies of
mixing do not cancel out in the addition of the subunits

free energies (17). Therefore:

$$R \ T \ Ln \ (K_{dim})_{max} = 2 \ R \ T \ Ln \ K + R \ T \ Ln \ X \qquad [2]$$

$$(K_{dim})_{max} = 55.6 \ K^2_{mon} \qquad [3]$$

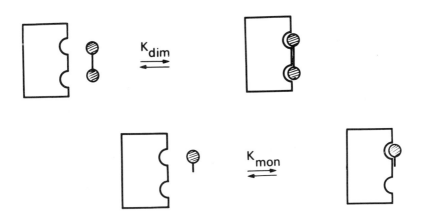

Figure 1. Schematic representation of a monomeric
and dimeric interacting system.

The relationship between the binding affinity of mono-
saccharides and disaccharides for lyzosyme has been studied by Chipman and Sharon (17). The same relationship
has been shown to apply.

The dimerization of a ligand when accompanied with
the symmetrical dimerization of its sites on the receptor
is therefore expected to lead to a large increase of the
binding affinity of the system. However the actual measured binding constant of the dimeric system will always be
smaller than the maximum one, because some free energy
associated with the change of conformation and mobility
of the dimer will be lost on dimerization. It is therefore of interest to compare on a simple system the relative binding constants of a monomeric and a corresponding
dimeric system and to study the effect of the dimer conformation on its binding constant.

In order to compare the behavior of dimeric and
monomeric systems, we have prepared molecules made up of

two DNA intercalating moieties linked by chains of various
structures. Under certain conditions, these molecules can
bind to DNA and intercalate their two subunits between base
pairs as schematized on Figure 2. In that case the comparison of the DNA binding affinity of the monomer and dimer
can be done. Furthermore the importance of the nature of
the linking chain which controls the conformational freedom of such molecules can be studied.

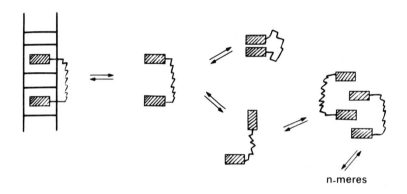

n-meres

        Figure 2. Schematic representation of the binding
                 of a bifunctional intercalator to DNA and
                 of the species involved in the equilibrium.

To illustrate this problem, the DNA binding affinity of
a series of dimers behaving as bifunctional intercalators
will be compared to that of the corresponding monomer.

        II) Comparison of the affinity of bifunctional and
monofunctional intercalators for DNA.

        At pH 5.0 the acridine dimers shown in Figure 3 are
fully ionized (pK $\sim$ 6.5). They are not aggregated and the
equilibrium between folded and unfolded conformation is
faster than the rate of DNA binding. In these conditions,
it was shown (11) that the binding process occurs in a
single step.

            Dimer + DNA $\rightleftarrows$ Dimer-DNA   [4]

The ratio of the onrate constant k_1 and off rate

constant k $_{-1}$ has been shown to be equal to the binding constant K (11).

$$K = k_1 / k_{-1} \qquad\qquad [5]$$

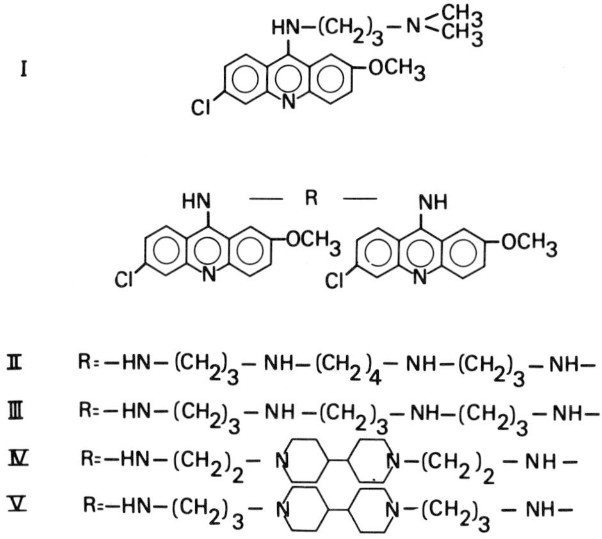

I

HN—(CH$_2$)$_3$—N$<$CH$_3$/CH$_3$
—OCH$_3$
Cl

HN —— R —— NH

—OCH$_3$ —OCH$_3$
Cl   Cl

II   R=—HN—(CH$_2$)$_3$—NH—(CH$_2$)$_4$—NH—(CH$_2$)$_3$—NH—

III   R=—HN—(CH$_2$)$_3$—NH—(CH$_2$)$_3$—NH—(CH$_2$)$_3$—NH—

IV   R=—HN—(CH$_2$)$_2$—N⟨⟩N—(CH$_2$)$_2$—NH—

V   R=—HN—(CH$_2$)$_3$—N⟨⟩N—(CH$_2$)$_3$—NH—

Figure 3. Structure of the molecules studied in this
          paper.(I) is the reference acridine monomer.
          The acridine dimers with a spermine chain
          (II),a thermine chain (III),a diethyl-bispi-
          peridine chain (IV), and a dipropyl-bispi-
          peridine chain (V),are abbreviated :
          spermine AcDi,thermine AcDi,diethylbispiper
          AcDi and dipropylbispiper AcDi respectively.

The chloromethoxyacridine fluorescence increases by
a large factor on binding to poly[d(A-T)] (5). The onrate
constant k$_1$ can then be simply measured by the stopped
flow technique using fluorescence detection (11). The
offrate constant k$_{-1}$ can be measured by displacing the
dye from its complex with poly[d(A-T)], by an excess of a
G-C rich DNA. Because the dye is not fluorescent when
bound to a G-C rich DNA, the displacememt reaction is
accompanied by a large decrease of fluorescence. The rate
constant of this displacememt reaction is equal to k$_{-1}$
(11). In Figure 4, we compare the rate of fluorescence
change associated with the displacement of the dye from

its poly[d(A-T)] complex by an excess of Microccocus luteus
DNA ,for the reference acridine monomer (I) and the thermine
dimer (III).

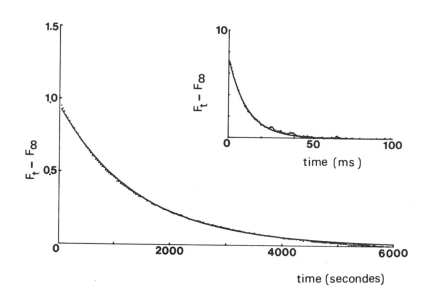

Figure 4. Comparison of the rate of displacement of
the poly[d(A-T)] thermine acridine dimer(III)
complex (lower left) and of the poly[d(A-T)]
acridine monomer (I) complex (upper right) by
an excess of Microccocus luteus DNA (72 % GC)
Dyes are first equilibrated for 24 hours
with poly[d(A-T)] (acridine dyes $2.\times10^{-7}$ M
poly[d(A-T)] 10^{-5} M base pair, in
sodium acetate buffer 0.2 M pH 5.0). After
addition of a 50 fold excess of sonicated
Micrococcus luteus DNA, the fluorescence of
the acridine monomer is recorded in a Durrum
stopped flow apparatus linked to a Digital
Minc computer insuring an analogic to digital conversion, the fluorescence of the dimer is measured in a SLM 8000 photoncounting
spectrofluorometer, the registered number
of counts being transmitted at regular intervals to a Digital Minc computer. The
difference between the fluorescence at time
t,(F_t) and at infinite time ($F \infty$),is
plotted as a function of time. (Compare
the time scale in the two cases). The experimental data (points) are fitted by linear
regression with one exponential (continuous
line).Rate constants are reported in Table 1.

Table 1. DNA (poly [d(A-T)]) binding parameters.
Onrate constants were measured by the stopped
flow technique using pseudo-first order analysis as already described (11). In all cases
k_1 values were found equal to 2×10^7 M^{-1} s^{-1} .
Offrate constants were measured as described
in the legend of Figure 3. At pH 5.0 buffer
was Na Acetate 0.2 M ; at pH 7.4 buffer was
Tris HCl 0.1 M, NaCl 0.1 M .

Compound	pH 5.0		pH 7.4	
	k_{-1} (s^{-1})	k_1/k_{-1} (M^{-1})	k_{-1} (s^{-1})	k_1/k_{-1} (M^{-1})
Ac Mono- mere	93	2.5×10^5	160	1.2×10^5
spermine AcDi	1.4×10^{-3}	1.4×10^{10}	1.4×10^{-2}	1.4×10^9
Thermine ACDi	7.1×10^{-4}	2.8×10^{10}	7.7×10^{-2}	2.6×10^8
diethylbis- piper AcDi	1.5×10^{-3}	1.3×10^{10}	1.7×10^{-2}	1.2×10^9
dipropylbis- piper AcDi	2.6×10^{-3}	7.6×10^9	3.6×10^{-3}	5.5×10^9

Because k_1 values are equal for monomers and dimers
(11) the ratio of the displacement rate constants is equal
to the inverse of the ratio of the corresponding binding
constants. It appears that in this particular case the
binding constant of the dimer is about 10^5 fold higher
than that of the monomer and is only 100 fold smaller
than the maximum computed value.

$K_{mon} = 2.5 \times 10^5$ $K_{dim} = 2.8 \ 10^{10}$ $(K_{dim})_{max} = 3.5 \times 10^{12}$ (M^{-1}).

Other dimers differing by their linking chains have
been prepared (18) and studied in the same way. The results obtained for these dimers are reported in Table 1.
At pH 7.4 (in 0.1M Tris HCl, 0.1M NaCl) the dimers
loose at least one proton. The dimers with a flexible chain
(spermine or thermine) are therefore in their folded con-

formation and aggregate as soon as the concentration is increased. Their solubility is dramatically lowered then the
salt concentration is increased. In contrast the dimers
with a rigid chain (diethyl bispiperidine for instance)
stay in their unfolded conformation as shown by NMR study
(19) and their tendency to aggregate is much less. When
a solution of a flexible dimer is mixed in the stopped
flow instrument with DNA ,the rate of DNA binding is limited by the rate of disruption of the aggregates which
is quite slow (a few secondes can then be needed for
equilibrium to reach)(11). Therefore the onrate DNA binding constant (k_1) cannot be measured. In identical conditions such a phenomenon is not observed with the rigid
chain dimers.

    To circumvent this aggregation problem in measuring
DNA binding parameters, the following experiments were
done. The dyes were first dissolved in a low concentration acidic buffer (Na acetate 0.005 M, pH 5.0). In these
conditions the dyes are protonated and do not aggregate.
The dye solutions were later mixed in the stopped flow
instrument with an equal volume of DNA solution at pH 7.4.
The buffer of this solution is concentrated two fold. After mixing the solution is at pH 7.4, in 0.1M Tris-HCl
0.1M NaCl as in the preceeding experiment. The true DNA
binding rates at pH 7.4 could then be measured, because
the flexible chain dimer has no time to aggregate before
DNA binding. The results obtained for the different dimers are reported in Table 1.

    Several conclusions can be drawn from these results:
    i) The dimerization process increases in all cases
the binding constant by a very large factor. The binding
constants of the dimers are relatively close to their maximum theoretical values. Such a result was also obtained
in other series of bisintercalators (9,10,13,14).

    ii) At a given pH, the flexibility of the linking
chain does not appear to have a large influence on the
binding constants.

    iii) The flexibility of the linking chain seems to
alter mainly the ability of the dimers to aggregate at
higher pH. When the salt concentration is increased, there
is an increasing competition between the aggregating and
the binding process (Figure 2).

    Some dimers made up of pyridocarbazole rings were
found to have very high antitumor activities (12-14).These
activities were restricted to the dimers which have a rigid
chain. The relationship between the structure of the linking chain and the pharmacological properties of these molecules is still unclear, since DNA binding properties are
not greatly influenced by these parameters.

III) Recognition of a DNA base by a pair of hydrogen bonds.

Seeman et al. (3) and Hélène (4) have proposed that carboxamide groups born by glutamine or asparagine could be involved in protein DNA sequence recognition. Such a carboxamide group could form two hydrogen bonds with guanine from the small groove of the DNA double helix or with adenine from the large groove of DNA. To estimate the possible contribution of such an interaction in DNA protein complexes, we have prepared chloro-methoxy-acridine derivatives bearing a polymethylene side chain of varying length terminated by a carboxamide group (16). Such molecules are able to bind to DNA by intercalating their chloro-methoxy-acridine rings between two adjacent base pairs. In this situation, the dye side chain, which bears the carboxamide group, lies in either one of the two DNA grooves. If the chain lies in the small groove, an interaction of the carboxamide group of the acridine side chain with the guanine base will promote a guanine specificity. In the opposite case, an adenine specificity will be observed. The quantitative comparison of the binding of such molecules, to DNA of various G-C content, could permit an evaluation of the contribution of such hydrogen bonding pattern to the total free energy of interaction.

NMR studies of the interaction of such molecules with autocomplementary dinucleotides CpG and GpC have been performed (15). These studies have shown that at the level of the mini double helix formed by these autocomplementary dinucleotides, the carboxamide group of these acridine derivatives formed a pair of hydrogen bonds with the guanine base from the small groove, as schematized on Figure 5 .

At the level of DNA, it was observed that some of these molecules displayed a preferential affinity for G-C rich DNA. The binding behavior of these compounds to DNA could be interpreted in the following way:

The acridine rings intercalate mainly between adjacent base pairs of which sequence is pyrimidine 3′-5′purine. Therefore three kinds of sites are mainly occupied by these dyes on DNA. They correspond to the sequences TpA, TpG and CpG which have binding affinities K_1 , K_2 and K_3 respectively. The guanine specificity of these dyes can be estimated from the measurements of the ratios of the binding constants K_2/K_1 and K_3/K_1 which can be deduced from DNA binding studies (16). It was observed that :

i) The two ratios K_2/K_1 and K_3/K_1 were strongly dependent on the length of the side chain bearing the carboxamide group.These results are reproduced in Figure 6.

ii) In all cases,the guanine specificity of these dyes

was completely abolished when carboxamide groups were
replaced by methyl groups.

iii) When the chain was longer than five methylene
groups, guanine specificity was abolished when the carboxamide group was blocked by two methyl groups or when it
was replaced by an ester group, indicating the necessity
of two hydrogen bonds for obtaining guanine specificity in
this case.

iiii) When the chain was only two methylene groups
long, the guanine specificity was observed even with chains
terminated by ester groups or dimethyl carboxamide groups,
indicating the requirememt of only one hydrogen bond for
obtaining guanine specificity in this case.

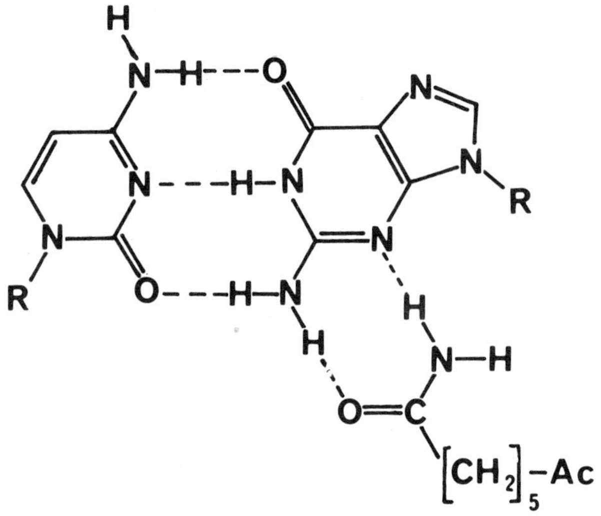

Figure 5. Schematic representation of the interaction
          of the carboxamide group born by the side
          chain of the acridine molecule with the
          guanine base. Redrawn from Markovits et
          al. (16).

Model building studies (16) showed that, when such
an acridine molecule is intercalated adjacent to a guanine,
the carboxamide group of its side chain can form two hydrogen bonds with guanine, only if its chain is longer or
equal to five methylene groups. When the side chain is
two methylene group long, the carbonyl group of the side
chain is rigidly held in an arrangememt similar to that
observed for the L-threonine carbonyl in the peptide side
chain of Actinomycin D as seen on Figure 7. In that case

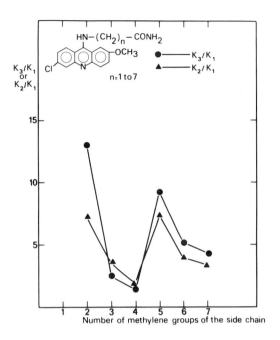

Figure 6. Variation of the guanine specificity of the
acridine derivatives as function of their
side chain length. (K_1 , K_2 and K_3
refer to the binding affinity of sites corresponding to sequences TpA, TpG, CpG respectively.
Redrawn from Markovits et al. (16).

the carbonyl group of the side chain is at the contact of
the guanine amino group and can easily form one strong
hydrogen bond.Likewise in Actinomycin D, the L-threonine
carbonyl has been shown to interact with the amino group
of guanine (2).

The conclusion of this study are:

i) When hydrogen bonding groups are born by flexible
chains, two hydrogen bonds are required to confer a significant guanine specificity.

ii) In the opposite case, when the mobility of the
hydrogen bonding group is restricted, a single hydrogen
bond seems sufficient to promote a significant guanine
specificity. This phenomenon accounts for the variations
of guanine binding preference of these compounds as a
function of the length of their side chain, as seen in
Figure 6.

Our results indicate that a single hydrogen bond between a carbonyl group of which mobility is restricted, or
a pair of hydrogen bonds between a carboxamide group held

on a flexible chain with guanine contribute -1 Kcal/mole
to the free energy of interaction. This value can be compared to that computed by Gresh and Pullman (20). As expected, the experimental value is smaller than the computed one. Nevertheless it is interesting to note that the
calculations predict that guanine will be a preferential
site of recognition.

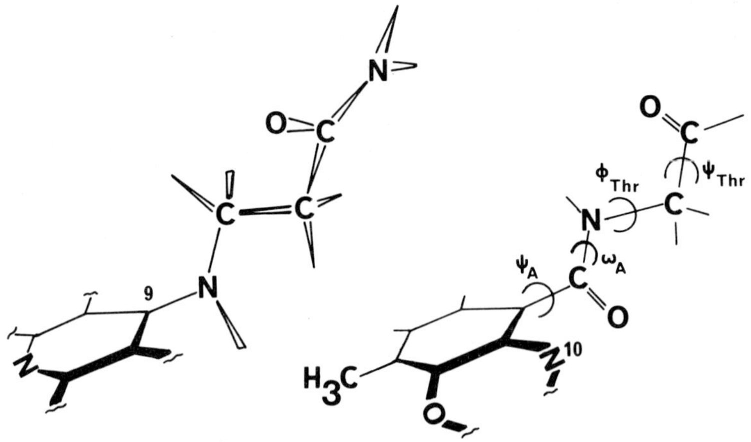

Figure 7. Comparison of the respective position of
          the carbonyl group in the acridine-ethyl-
          amide (left) and the L-threonyl carbonyl
          in Actinomycin D (right). Redrawn from
          Markovits et al.(16).

Conclusion

    The studies which have been reviewed in this paper un-
derline two basic physical-chemical processes which could
play an important role in the formation of biological complexes of high stability and specificity.
    i) The dimerization of the interacting species.
This simple process can lead directly and simply to a very
large increase of binding affinity as shown by the study of
our simple model compounds.Such process can derive directly
from duplication of genes or DNA segments in the case of
protein DNA interactions.
    ii) The involvement of hydrogen bond pairs with bases
in the DNA double helix. This could account for the specificity of the protein nucleic acid interactions as suggested

already by several authors (3,4,20).

The contribution of such hydrogen bonds to the free energy of interaction (-1 Kcal/mole) is less than the computed enthalpy change (-10.5 Kcal/mole) (20). This is expected because it has been seen that this contribution was very dependent on the conformation and rigidity of the chain holding the hydrogen bonding groups. The study of the lac-repressor or RNA polymerase binding to DNA shows that these proteins bind to nonspecific sites with a binding constant 10^6 fold less than that corresponding to specific sites on DNA (1,21-22).

If hydrogen bonds between guanine or adenine and asparagine or glutamine are involved in protein nucleic acid recognitions,and if these hydrogen bonds are similar to those studied here,5 to 8 specific contacts would be sufficient to confer to these interactions such a high specificity.Finally,we would like to underline that this kind of study was determinative for the rational design of a new series of antitumor compounds (12-14).

Acknowledgments

This work has been supported by grants from D.G.R.S.T., I.N.S.E.R.M. (SCN 21),Fondation pour la Recherche Médicale Francaise,Association pour le Developpement de la Recherche sur le Cancer,Université Pierre et Marie Curie (Paris VI).
The authors thank Mrs J. Couprie for its excellent technical assistance.

References

1- Jovin, T.M.: 1976, Ann. Rev. Biochem. 45,pp. 889-920.
2- Sobell, H.M., and Jain, S.C.: 1972, J.Mol.Biol. 68, pp.21-34.
3- Seeman, N.C., Rosenberg, J.M., and Rich, A.: 1976, Proc. Natl. Acad. Sci. USA, 73, pp. 804-808.
4- Hélène, C.: 1977, FEBS Letters, 74, pp. 10-13.
5- Le Pecq, J.B., Le Bret, M., Barbet, J., and Roques, B.P.: 1975, Proc. Natl. Acad. Sci. USA, 72, pp. 2915-2919.
6- Barbet, J., Roques, B.P., Combrisson, S., and Le Pecq J.B.: 1976, Biochemistry, 15, pp. 2642-2650.
7- Roques, B.P., Barbet, J., Oberlin, R., and

Le Pecq, J.B.: 1976, C.R.Acad. Sci. Paris 283, Serie D, pp. 1365-1367.

8- Delbarre, A., Roques, B.P., and Le Pecq J.B.: 1977, C.R.Acad. Sci. Paris, 284, Serie D, pp. 81-84.

9- Gaugain, B., Barbet, J., Oberlin, R., Roques, B.P., and Le Pecq, J.B.: 1978, Biochemistry, 17, pp. 5071-5078.

10- Gaugain, B., Barbet, J., Capelle, N., and Roques, B.P., and Le Pecq, J.B.: 1978, Biochemistry, 17, pp. 5078-5088.

11- Capelle, N., Barbet, J., Dessen, Ph., Blanquet, S., Roques, B.P., and Le Pecq, J.B.: 1979, Biochemistry, 15, pp. 3354-3362.

12- Roques, B.P., Pelaprat, D., Le Guen, I., Porcher, G., Gosse, Ch., and Le Pecq J.B.: 1979, Biochem. Pharmacol. 28, pp. 1811-1815.

13- Pelaprat, D., Oberlin, R., Le Guen, I., Roques, B.P., and Le Pecq, J.B.: 1980, J. Med. Chem. 23, pp. 1330-1335.

14- Pelaprat, D., Delbarre, A., Le Guen, I., Roques, B.P., and Le Pecq, J.B.: 1980, J. Med. Chem. 23, pp. 1336-1343.

15- Gaugain, B., Markovits, J., Le Pecq, J.B., and Roques, B.P.: 1981, Biochemistry, in press.

16- Markovits, J., Gaugain, B., Barbet, J., Roques, B.P., and Le Pecq, J.B.: 1981, Biochemistry, in press.

17- Chipman, D.M., and Sharon, N.: 1969, Science, 165, pp. 454-465.

18- Markovits, J.: 1980, These de Doctorat d'Etat es Sciences. Universite P.M. Curie (Paris VI).

19- Delbarre, A.,Gaugain, B., Markovits, J., Vilar, A., Le Pecq, J.B., and Roques, B.P.: 1981,This Symposium

20- Gresh, N., and Pullman,B.: 1980 Biochim.Biophys.Acta, 608, pp47-53

21- deHaseth,P.L., Gross, C.A., Burgess R.R. and Record, M.T. Jr.: 1977, Biochemistry,16 ,pp 4777-4783

22- deHaseth, P.L., Lohman, T.M., Burgess,R.R. and Record, M.T. Jr.: 1978, Biochemistry,17 ,pp 1612-1622

EMPIRICAL MODELS OF HYDRATION OF SMALL PEPTIDES

Françoise VOVELLE, Monique GENEST and Marius PTAK
Centre de Biophysique Moléculaire (C.N.R.S.) et Université
d'Orléans, 1A avenue de la Recherche Scientifique,
45045 ORLEANS Cedex, France.
Bernard MAIGRET
Laboratoire de Biophysique Moléculaire (E.R.A. 828),
Université de Nancy I, CO 140, 54037 NANCY Cedex, France.

Abstract.- Simple empirical potentials have been used to calculate hydration sites of C_5, C_{7eq}. and α_R conformers of Ac-L-Ala-NHMe. The formation of specific water bridges connecting two peptide groups are then
displayed. The coexistence of individual sites and of water bridges
have been then analyzed for a tetrapeptide : Ac-L-Thr-L-Thr-L-Ser-LAsp-
NHMe folded into a β-turn I and for a cyclic dipeptide c(L-Thr-L-His).
Side chain - side chain, side chain - backbone and backbone - backbone
single or double water bridges, are proposed to be specific modes of
hydration of peptides, which are not necessarily maintened in optimized
configurations of the first hydration shells. The present structural
aspects of hydration will serve for further static and dynamic studies
of peptides conformations in crystals and in solution.

INTRODUCTION.

In many biological processes in which they are involved, peptides are
in an aqueous environment. *In vitro*, their conformations often depend
on interactions with surrounding water in crystals as well as in solution and it is now one of the major aims of conformational analysis to
evaluate environmental effects on conformations, basically determined
by intramolecular interactions. The considerable difficulties arising
from dynamic descriptions of solvent-solute systems are amplified by
the diversity of chemical composition and by the size and the flexibility of peptides. Nevertheless, internal structural fluctuations of a
dipeptide in solution and dynamic properties of the surrounding water
have been recently described with many details by using a molecular dynamics computer simulation method [1,2]. More global insights into the
organization of peptide-water systems in crystalline state [3] and in
solution [4] have been given by using Monte Carlo simulation methods.
In such dynamic descriptions, structural aspects of peptides hydration
can be somewhat difficult to reach, because of the necessity to do a
probabilist analysis and a very adequate selection of the more

B. Pullman (ed.), Intermolecular Forces, 299–315.

significant configurations. A completely different approach can be proposed to describe the particular modes of interaction of water with peptides. Preferential positions of one or of a few water molecules around a solute and then, optimized organization of the first hydration shells can be determined by minimizing the intermolecular peptide-water energy. Among different methods [5] used, the supermolecule method especially well developed by Pullman and Pullman [6] has to be mentioned as well as approximate empirical methods [7]. In spite of obvious limits, such *static* methods deserve to address two fundamental questions refering to the specificity of solvation : how specific sites are maintened in the organization of the first hydration shells and how optimized configurations of these shells are maintened in the fluctuating organization of the global system ? In hydrated peptide crystals, there is generally a limited number of water molecules interacting with the peptide and contributing to the packing. A comparison with hydration sites calculated for isolated molecules would allow to evaluate the effects of the tridimensional lattice constraints on these sites and on H-bonds (hydrogen bonds) geometry. Such an analysis can possibly be transposed to specific hydration cases of some parts of proteins. A last justification of the use of empirical methods is the necessity to dispose of the most simple and most accurate potentials before to undertake simulations in which a very high number of configurations of the peptide-water system must be generated.
This paper[1] reports on theoretical empirical determinations of hydration sites of small peptides held in given conformations. Linear and cyclic dipeptides and a linear tetrapeptide have been chosen in order to analyze specifically the hydration of peptide bonds on one hand and the hydration of some polar or charged side chains on the other. Such calculations represent only a step in the construction of a model including both structural and dynamic aspects of peptide hydration.

METHODS.

Most of the calculations of water interactions with carboxyl and amino groups involved in CONH bonds of the peptide backbone have been carried out for simplified models such as formamide and N-methylacetamide [8-12]. Starting from such models, we have extended our calculations to a protected dipeptide (dipeptide means two peptide bonds, here) of alanine : Ac-L-Ala-NHMe locked in different standard conformations : C_5, $C_{7eq.}$, α_R. As this peptide has been the subject of dynamic studies [2, 4], two concepts of hydration sites have been compared. A cyclic dipeptide : c(L-Thr-L-His) carrying two polar side chains has been chosen to investigate more specifically, simultaneous hydration of side chains and of peptide bonds. Crystal structure [13], conformations in solution [14], calculated conformations [15] and water accessibilities [16] have been previously established for this special peptide. A more general case has been analyzed, in which water interacts with a β-turn : the protected tetrapeptide : Ac-L-Thr-L-Thr-L-Ser-L-Asp-NHMe is a model of the 61-64 sequence of α-chymotrypsin in the crystal state [17]. Among the simplest empirical potentials used in calculations of water-water

and water-solute interactions, the Rowlinson potential [18] and the
Stillinger and Rahman ST2 6-12 potential [19] are more frequently used,
possibly with slight modifications for dynamics studies of peptide [2].
We have recently proposed a simplified *ad hoc* potential [12,20], noted
EMPWI , that enables us to simulate rather satisfactorily static peptide-water interactions by improving the directivity of H-bonds as compared with other potentials. This effective potential was first adjusted
by taking the water dimer as model. Its transferability was then tested
by comparison with quantum mechanical [6] or some refined empirical
[20] calculations of hydration sites of simple models or small peptides.
To date, the potential we propose is the following for peptide-water
(pw) dimer (A', C' coefficients are slightly different from those initially proposed in ref.12) and for water-water dimer :

$$U_{pw} = \frac{1}{4\pi\varepsilon_0} \sum_i^p \sum_j^w \frac{q_i q_j}{R_{ij}} + \sum_i^p \sum_j^w \left(\frac{A}{r_{ij}^{12}} - \frac{C}{r_{12}^6} \right) + \sum_i^p \sum_j^w \left(\frac{A'}{r_{ij}'^{12}} - \frac{C'}{r_{ij}'^6} \right)_{HB}$$

R_{ij} = distances between q_i (peptide) and q_j (water) point charges
     (atoms and lone pairs).

r_{ij} = distances between i (peptide) and j (water) atoms non engaged in
     H-bonds.

r_{ij}' = distances between i (peptide) and j (water) atoms engaged in
     H-bonds.

	A'	C'
O......O	4.10^5	367
O......N	6.10^5	365
O......H	10^3	0
N......H	10^3	0

A' : $kcal.mol^{-1}Å^{-12}$
C' : $kcal.mol^{-1}Å^{-6}$

The obvious failure of such potential is to do not reproduce correctly
the well known non-pair-additivity of H-bonds involving water molecules.
An improvement has been proposed by Caillet and Claverie [21] by adding
a polarisation term to electrostatic and dispersion-repulsion terms.
A simple version of this method has then been used to improve calculations involving several water molecules :

$$U_{pw} = \sum U_{elec.} + \sum U_{pol.} + \sum U_{disp.-rep.}$$

$U_{elec.}$ = electrostatic interaction energy between atomic point charges.
$U_{pol.}$ = $- \frac{1}{2} \sum \alpha_i (\mathcal{E}_i)^2$ = atomic polarisation energy, α_i being the polarisability of the atom i located in the electric field $\mathcal{E}_i$ created
by all other atoms.
$U_{disp.-rep.}$ = dispersion-repulsion energy (Buckingham 6-exp. potential).
Special α_i and A,C coefficients (6-exp. potential) being used for
atoms engaged in H-bonds.

RESULTS and DISCUSSION.

Hydration sites of the peptide backbone.

EMPWI potentials enable us to determine rather satisfactorily the preferential interaction modes of one water molecule with CO and NH groups

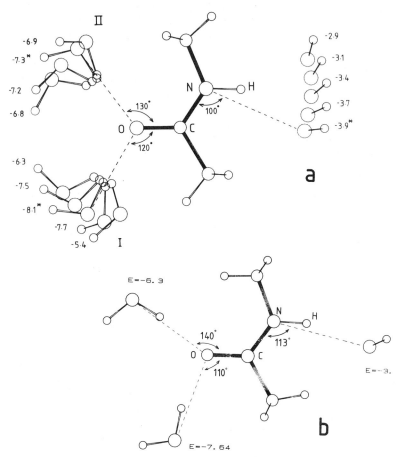

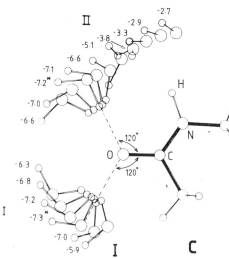

Figure 1. Hydration sites of *trans* (a,b) and *cis* (c) N-methylacetamide calculated with EMPWI potentials. Energies are given in kcal.mol^{-1}. The minima are in the CONH plane. b) shows effects of methyl rotations upon geometries and energies of H-bonds.

of *trans* N-methylacetamide, usually taken as a model of peptide bonds. According to well recognized theoretical data [8-10], two sites are found for the CO group in the vicinity of the directions of carbonyl oxygen sp^2 lone pairs (fig.1). The dimer I, in which the H-bond is *trans* to nitrogen is the stablest one, with $\widehat{COO}_W = \theta \simeq 120°$ when $\theta \simeq 130°$ for the dimer II. For the NH group, a broad minimum is found in a region including the N-H axis. The optimal H-bond is not quite linear and is very less stable than those found for the CO group. The geometries and energies of both kinds of H-bonds significantly depend on the rotations of terminal methyl groups. Such effects of short solute-water interactions on H-bonds will appear subsequently as a fundamental feature in peptide hydration.

Experimentally, a preferential hydration of the CO group of amides has been recently demonstrated by Wolfenden [22] in a vapor phase analysis, which remarkably parallels the higher frequency of CO...water H-bonds in peptide crystals [23]. Most of experimental data to which one could refer for the directionality of H-bonds are open to discussion. In some cases, constraints of crystal packing can severely distort H-bonds [24]. This observation was often used in the past to question theoretical predictions. A more recent analysis of peptide-water associations [23] clearly shows that the mean geometry of CO...water and NH...water H-bonds in crystals is very close to the predicted one [10]. As a matter of fact, a crucial point is to evaluate distortions depending on three main factors : a) the shape of the potential hypersurface near the minima ; b) the short interactions other than H-bond interactions ; c) the constraints due to intermolecular interactions, i.e., to water-water interactions (solution, crystal) and to peptide-water interactions (in which water interacts with several peptide molecules in crystal). Such effects will be revealed in the course of this study.

The hydration sites of *cis* N-methylacetamide have been determined in order to see the role of peptide bond geometry. *cis* CONH bonds are only found in synthetic and natural diketopiperazines and in some larger peptides in which such bonds are N-methylated [25]. As seen in fig.1, the site I does not mainly depend on the configuration of the peptide bond whereas the site II becomes common to the neighbouring CO and NH groups. Such proximity effects are typical and will be often displayed for peptides.

In sequences containing several CONH bonds, the individual accessibilities of CO and NH groups primarily depend on the mode of folding of the peptide backbone. *A priori*, there will be a competition between CO...NH *intra*molecular H-bonds and peptide-water *inter*molecular H-bonds. Three standard stable conformations [26] of Ac-L-Ala-NHMe have been selected: a) the C_5 conformer ($\phi = -150°$, $\psi = 150°$) belongs to the broad family of extended forms ; b) the $C_{7eq.}$ conformer ($\phi = -80°$, $\psi = 80°$, intermediate between β and α_R regions of the conformational map) in which a distorted $1 \leftarrow 2$ H-bond stabilizes the backbone folding ; c) the α_R conformer ($\phi = -80°$, $\psi = -50°$) which is the conformational unit of right-handed helices (note that in a dipeptide there is no internal H-bond since four consecutive CONH bonds are required to have a C_{13} helix turn).

For the C_5 conformer (fig.2), two sites have been found for the CO and NH groups involved in the five membered cycle, on one side of the

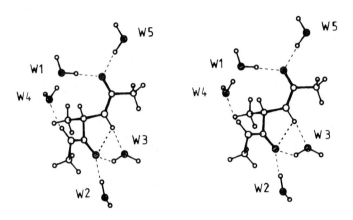

Figure 2. Hydration sites for the C_5 conformer ($\phi = -150°$, $\psi = 150°$) of Ac-L-Ala-NHMe (energies are given in the text).

extended backbone. The W_2 (-8.0 kcal.mol^{-1}) water molecule is very close to the site II previously found for an isolated CO group in *trans* N-methylacetamide. The W_3 (-10.1 kcal.mol^{-1}) water molecule bridges the CO proton acceptor and NH proton donor groups through two distorted H-bonds. On the other side of the backbone, the CO and NH groups are more independent. Therefore, hydration sites are nearly individual : sites I : W_5 (-8.9 kcal.mol^{-1}) and II : W_1 (-11.3 kcal.mol^{-1}) for CO and site W_4 (-9 kcal.mol^{-1}) for NH groups.

Important stabilization due to proximity effects has to be noted for W_3 on one hand and W_1 and W_4 on the other. Such effects which basically depend on conformations have been found for the two other conformers. As expected, the site I : W_1 (-8.4 kcal.mol^{-1}) is present for the internal CO group involved in the seven membered ring of $C_{7eq.}$ (fig.3) as well as the sites II : W_5 (-8.4 kcal.mol^{-1}) and I : W_2 (-9.1 kcal.mol^{-1}) are present for the external CO group. The existence of an intramolecular H-bond does not suppress the possibility of hydration since a simple water bridge : W_4 (-7.5 kcal.mol^{-1}) is found on one side of the C_7 cycle. A second simple water bridge W_3 (-9.6 kcal.mol^{-1}) connects the external CO and NH groups. Such a bridge has been previously described by Pullman and Pullman [6] and proposed as stabilizing C_7 conformations in water solution.

The coexistence of individual sites and of simple water bridges appears also as a rule for the α_R conformer (fig.4). Individual sites : W_4 (-9.8 kcal.mol^{-1}) and W_3 (-9.4 kcal.mol^{-1}) exist for CO groups whereas bridges : W_1 (-11.2 kcal.mol^{-1}) and W_2 (-13.0 kcal.mol^{-1}) connect both NH groups and CO groups respectively. The exceptionnal stability of the W_2 site has been noted.

This study clearly shows how peptide conformation determines the formation of simple water bridges which will be then considered as specific hydration sites. Among the three conformers described here, the α_R and then C_5 ones have the stablest specific sites. For isolated peptide-

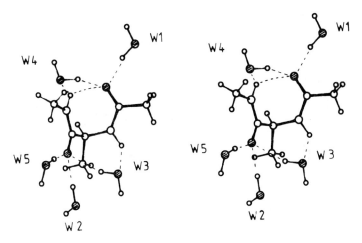

Figure 3. Hydration sites for the C_{7eq}. ($\phi = -80°$, $\psi = 80°$) of Ac-L-Ala-NHMe (energies are given in the text).

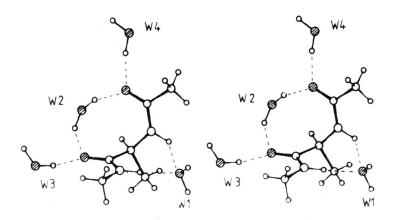

Figure 4. Hydration sites for the α_R conformer ($\phi = -80°$, $\psi = -50°$) of Ac-L-Ala-NHMe (energies are given in the text).

water dimers, water bridges appear as stabilizing structures. This does not justify to conclude about the relative stability of peptides in solution. *A priori*, it could be a coincidence to find experimentally the best stability for the α_R conformer [27]. Many other factors such as accessibilities [28,29] and solvent dynamics certainly contribute to determine the minimum of the solvent-solute free energy. Let us also recall that α_R conformer of a dipeptide cannot be taken as a model of a complete α_R helix in which a regular network of H-bonds is formed. This does not suppress completely interactions with water [6] but excludes

formation of bridges. Ac-L-Ala-NHMe reproduces better the behaviour of
chain termini in which peptide bonds remain accessible [29].

Hydration sites of polar and charged side chains.

Among the twenty residues constituting proteins, nine have side chains
carrying polar or charged groups capable of interacting with water : OH
(Ser,Thr), COO^-(Asp,Glu), NH_3^+(Lys), $CONH_2$(Asn,Gln), $NHC^+H(NH_2)_2$(Arg)
$N\diagdown\diagup NH$(His). Hydration sites of most of these groups have been calculated by using quantum mechanical methods [6]. Empirical poten-
tials enable us to reproduce the essential features of these side chain
water interactions. As expected, the stablest sites are found for carboxylate and ammonium charged groups. The amide groups of Asn and Gln
behave nearly as model amides. One site exists for neutral imidazole as
seen in fig.5 depicting some selected examples.

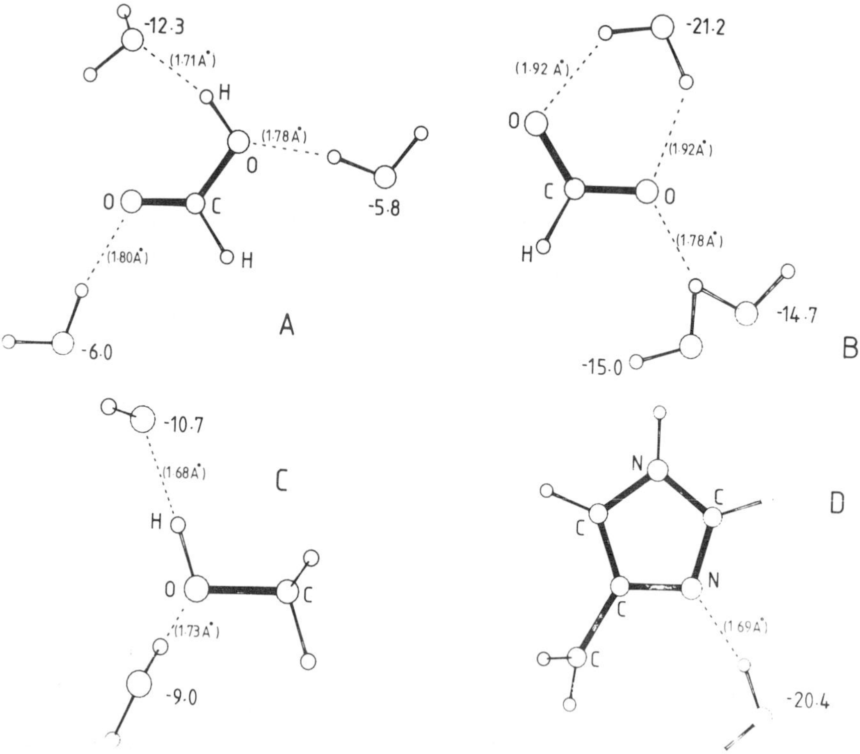

Figure 5. Hydration sites of : A) H COOH ; B) H COO^-, another site
exists symmetrically to the H-C bond ; C) CH_3OH ; D) $5-CH_3$-imidazole.
Energies are given in $kcal.mol^{-1}$.

Some complementary informations have been recently given in an analysis

of H-bonds in crystals structures [30] of amino acids, peptides and related molecules. Agreement between theoretical and experimental data is
generally satisfactory, in the limit of constraints packing effects on
H-bonds geometry. A systematic study of hydration of side chains as a
function of their conformations would be very tedious and cannot be separated from a study of backbone-water interactions, as seen below.

Hydration sites of peptides.

In a peptide sequence containing polar or charged side chains, the possibilities of specific hydration are considerably increased : proton
donor and proton acceptor peptide sites can be combined in order to
form backbone-backbone, backbone – side chain and side chain – side chain
bridges. Several backbone-backbone simple water bridges have been previously described for Ac-L-Ala-NHMe conformers. Another typical example
[23] has been found in the crystal structure of antanamide [31] where
one water molecule closes a cycle formed by a folded backbone looking
like a β-turn (fig.6). Such particular mode of interaction has been recently found in the crystal structure of a linear protected L-Pro-L-
Ala peptide [32] having a central *cis* N-methylated bond. A typical side

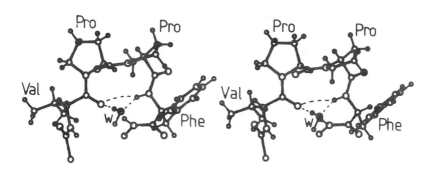

Figure 6. Calculated backbone-backbone simple water bridge in a fragment of antanamide folded as in the crystal [31].

chain – side chain simple water bridge has been found for c(L-Thr-L-His)
locked in a conformation depicted in fig.7. This case, which has been
previously analyzed [12,20], clearly illustrates the typical mode of
formation of a specific hydration site. For a stable conformation corresponding to an intramolecular energy minimum, two peptide groups are
in positions allowing the formation of a water bridge. Minimization of
the total energy induces a small rotation of the His side chain and
leads to a very stable structure which is similar to the crystal one
[13]. The coexistence of different kinds of simple water bridges has
been analyzed for a protected tetrapeptide : Ac-L-Thr-L-Thr-L-Ser-L-
Asp-NHMe folded into a β-turn [17]. The role of such foldings in the
three dimensional organization of proteins has been well recognized[33].

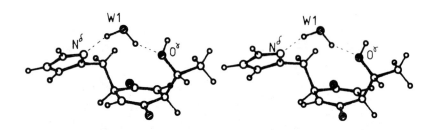

Figure 7. Calculated side chain – side chain simple water bridge in c(L-Thr-L-His) (see also ref.12,20). Side chain conformations His : $X_1 = 305°$, $X_2 = 90°$; Thr : $X_1 = 70°$, $X_2 = 240°$. Energy $= -15.4$ kcal.mol^{-1}.

These chain reversing regions are generally located at the surface of proteins and then should participate to their hydration. Among the different calculated sites (fig.8) of this tetrapeptide, two call attention. The W_1 water molecule bridges both Asp and Thr side chains in a very specific way (-28.9 kcal.mol^{-1}, three H-bonds). The W_2 water molecule forms a backbone-backbone bridge (-13.7 kcal.mol^{-1}) connecting one
CO group involved in the β-turn and the C-terminal CO group. Such a position was proposed to exist in the crystal of α-chymotrypsin [17]. Such kind of simple water bridges seems to be one of the preferential modes of hydration of β-turns. NH...water...CO bridges have been found in crystals of cyclic peptides[34,35] that look like very much the previous one. Many other interesting possibilities have been discovered for this very hydrophilic tetrapeptide which will be analyzed elsewhere [36]. Among the main features, let us emphasize on the low energies of water molecules interacting with Asp carboxylate group and a remaining possibility of hydration of the CO group engaged in the intramolecular H-bond closing the β-turn.
A somewhat different mode of hydration has been displayed by considering several water molecules interacting with neighbouring polar peptide groups.
A typical case is depicted in fig.9, in which two water molecules form a double bridge between the side chains of c(L-Thr-L-His). The individual (His)N^δ and (Thr)O^γ sites interact with water molecules which are themselves linked by a H-bond. Energetically, this situation is only slightly less favourable than those described in fig.7, in which the W_1 molecule bridging the peptide side chains would interact with an external water molecule. This is a striking illustration of two competitive modes of hydration involving simple and double water bridges. An analogy could be proposed with chelates where water bridges form cyclic structures bonded stereo specifically to peptides. A number of such structures are found in peptide crystals in which they are compatible with an optimal crystal packing. In solution, multiple complementary water-water interactions have to be taken into account, which certainly interfere with organization of water in and around peptide groups.

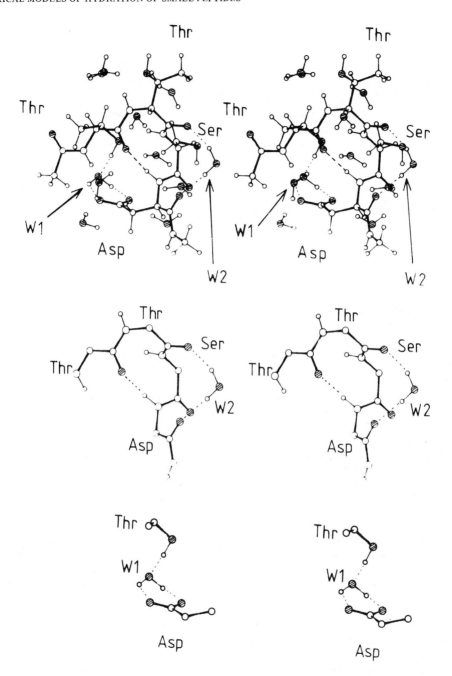

Figure 8. Calculated hydration sites of Ac-L-Thr-L-Thr-L-Ser-L-Asp-NHMe folded into a β-turn I. Asp side chain is ionized. W_1 interacts with (Asp)COO⁻ and (Thr)OH. W_2 with (Thr)CO (β-turn) and with (Asp) CO.

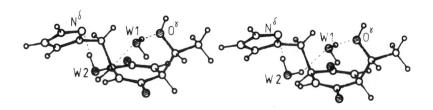

Figure 9. A calculated double water bridge in c(L-Thr-L-His). Side chains conformations are the same as in fig.7. Caillet-Claverie's potentials have been used. W_2 (-19.1 kcal.mol^{-1}), W_1 (-8.6 kcal.mol^{-1}).

Hydration shells.

For molecules completely surrounded by bulk water, two classes of peptide-water interactions will be distinguished at the molecular level : a) *polar interactions* in which a proton donor or proton acceptor group of the peptide is an anchoring point for one or two water molecules, from which the solvent organizes itself in the potential gradient of these groups ; b) *apolar interactions* in which water molecules have to organize themselves around hydrophobic groups, according to the classical concept of "structured" solvent near non polar solutes [37]. For peptides containing several non polar groups, the so-called hydrophobic clustering can be very effective for determining their conformations. The organization of hydration shells can be approached by different ways. One can first determine the hydration sites previously defined and then one can add the number of water molecules required for covering completely the surface of the peptide, according to steric hindrance conditions. A more systematic way [7] that has been used here is to put water molecules in the centers of Van der Waals spheres covering the peptide surface in order to have an optimal packing and then to construct the successive shells by starting from such Van der Waals layers. The minimization of all peptide-water and water-water interactions leads theoretically to the optimal organization of water shells. A third method, especially well adapted to dynamic simulations [1-4] is to place one solute molecule in a cubic box containing some hundred of water molecules. As a matter of fact, these simulations clearly show that the influence of the solute is nearly limited to the first hydration shell. Nevertheless, the presence of a second and even of a third layer is necessary to determine correctly the configuration of the first one. This effect has been clearly proved by calculating the optimized configurations of the first hydration layer of c(L-Thr-L-His). As seen on fig.10, such configurations can maintain the principal hydration site which is the W_1 simple water bridge previously described (fig.7). The tendency of water to form a layer stabilized by a H-bond lattice is clearly seen under the diketopiperazine ring (fig.10). In

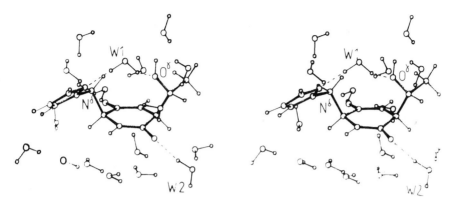

Figure 10. Optimized configuration of the first hydration shell of
c(L-Thr-L-His) (Caillet-Claverie's potentials). The side chain - side
chain one water bridge W_1 is present. W_2 is an example of individual site.

other configurations, the W_1 site is absent, whereas a double water
bridge is formed (fig.11). As a matter of fact, a variety of double water bridges (and even triple bridges) can be formed (compare fig.9 and

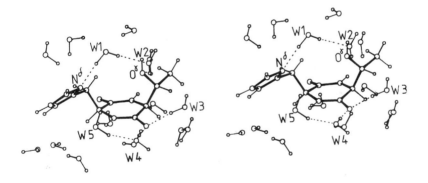

Figure 11. Optimized configuration of the first hydration shell of
c(L-Thr-L-His) (Caillet-Claverie's potentials). A double water bridge
W_1,W_2 connects the peptide side chains.

fig.11) which are energetically rather similar (1-2 kcal.mol^{-1}).
A main difficulty in this procedure is the convergence of the minimization process. As water molecules are allowed to rotate and/or to trans-
late, they can form a number of H-bonds lattices having nearly the same
minimal energy. By starting from a configuration in which a main hydration site exists, one maintains generally this site. On the other hand,
when a water dimer is first formed, the previous hydration site is not
obligatorily then found. This is due to the model itself [7] in which

water-water interactions are limited to the first hydration shell more
than to important entropic effects. This creates a kind of "surface
pressure" which partly disappears when a second and a third hydration
shells are introduced. Unfortunately, the minimization process becomes
then rather hard working because of the number of water molecules that
have to be taken into account. Nevertheless, some qualitative rules
seem to emerge from our preliminary investigations : a) the main hydration sites (water bridges) are generally maintained in the organization
of hydration shells ; b) near the polar groups, water tends to have a
radial organization along directions corresponding to standard H-bonding ; c) near the non polar groups, water tends to organize in layers
stabilized by H-bonds which can be somewhat different from "ice-like"
structures. It is quite clear that all these rules will take their full
significance in a comparison with dynamics simulations which are now
undertaken.

General discussion.

Schematically, three kinds of hydration sites have been found by opti-
mizing peptide-water interactions : a) *individual sites* for isolated
groups in which H-bonds have nearly standard geometries and energies ;
b) *simple water bridges* in which the solvent molecule connects two or
three backbone and/or side chains groups ; c) *double* (or even triple)
water bridges which connect two peptide groups. Clearly the second
class of sites is the more specific one,considering that the solute
must have a very special conformation for inducing their formation. It
is worthwhile to emphasize that the geometries and energies of all H-
bonds involved in these three kinds of sites depend on : a) local peptide-water interactions other than H-bond interactions which can significantly contribute (1-2 kcal.mol^{-1}) to stabilize solvent molecules ;
b) water-water interactions. The corresponding distortions are generally more marked for NH...water than for CO...water H-bonds. In extreme
cases, the hydration of amino groups can be removed.
In the present state of dynamics descriptions, a critical discussion of
our present data is rather limited, if properties of aqueous solutions
are considered. For small peptides, the most stable hydration sites
should correspond to maxima of the probability density maps established
in simulations. Mont-Carlo simulations have been recently carried out
by Romano and Clementi [39] for glycine and serine. It is effectively
observed that several maxima of probability density maps are very near
from the optimal positions of water predicted by static discrete methods. For larger peptides, the possible existence of several quasi-
isoenergetic configurations of water in the first hydration shell obscures the relations between static optimal configurations and dynamic
more probable configurations. Hydration sites found in the present work
for the α_R conformer of Ac-L-Ala-NHMe do not appear systematically as
maxima of probability density in the Monte-Carlo simulation of Hagler
et al [4]. A typical example is the simple water bridge described in
fig.4 which can be replaced by a less specific double water bridge. In
the same way, some differences seem to exist between hydration sites
found for the $C_{7eq.}$ conformer and the typical positions of water

molecules determined by molecular dynamics simulation [2]. There is not any simple picture of peptides hydration that emerges from these different approaches. A polar group of peptide can take the place of one water molecule in the interactions with the rest of the solvent, whereas a non polar tends to induce a structuration of the solvent. Maximal hydration of polar groups can destabilize a peptide conformation whereas water bridges can stabilize it. It is quite clear that in solution, multiple equilibria occur and contribute to determine the enthalpy and the entropy terms of free energy.
More relevant are the comparisons with hydration in peptide crystals. A number of peptide-water optimized configurations determined theoretically are found experimentally [23] without drastic changes. That illustrates the remarkable capacity of water to fill all available space around a peptide in order to have an optimal packing and without dramatic constraints. Except cases where special strong intermolecular interactions (such as ionic interactions) are present, a tridimensional crystal lattice can be quite compatible with stable molecular conformations including a few water molecules. One could compare these conclusions to those recently reached by Finney [38] in his analysis of hydration of protein crystals.

CONCLUSION.

In spite of serious limits in the use of empirical potentials describing peptide-water interactions, a comprehensive view of the basic modes of interaction of water with small peptides has been given. The general coherence of our data and frequent concordances with experimental situations in peptide crystals show that these potentials have an acceptable transferability and that they allow to reproduce, at least qualitatively, structural properties. Such potentials will be used subsequently in Monte-Carlo simulations in which we will focuss more especially our attention on structural aspects.
Our static description does not give direct informations about the conformational stability of peptides in water. On the other hand, it shows clearly the elementary energetic processes which contribute to determine this stability. A very different approach (and very less expansive than simulations) could be proposed which combines an analysis of accessible surface areas [16,28,29], a statistical analysis of conformations [40] and an analysis of hydration sites. Indeed, one has to consider that simulations of flexible natural peptides such as hormones, antibiotics, toxins, etc... which contain one, two or three dozens of residues is completely out of scope at present. Furthermore, several of these peptides carrying charged groups should be immersed in a ionic solution and not in pure water.
That the conformations existing in aqueous solution are the biologically active forms is now dogmatic (obviously one eliminates all the peptides interacting with membranes and interfaces). Some corrections could be proposed by considering that the really active forms are those involved in the specific interactions of a peptide with its receptor. It is quite clear that such interactions strongly modify hydration of

this peptide which is then in a conformation only partly allowed to interact with water. The concept of stable hydration sites is certainly important from this point of view, inasmuch as hydration water can be considered as belonging to the peptide structure and as partly determining its specificity of interaction (as for hydrated ions). Therefore, it seems necessary to develop simultaneous different approaches of structural and dynamic aspects of hydration of peptides in relation with their interaction properties.

---

1. This paper is the number III of a series entitled : Hydration of peptides (see ref.16,20).
2. ϕ,ψ angles are the same as in ref.27 in which stabilities of Ac-L-Ala-NHMe conformers have been experimentally investigated. The α_R conformer does not correspond exactly to a unit of α_R helix ($\phi \simeq -60°$, $\psi \simeq -50°$). Slight changes of ϕ,ψ values do not modify drastically hydration sites.

References.

1. Rossky, P.J., Karplus, M., and Rahman, A. : 1979, Biopolymers 18, pp.825-854.
2. Rossky, P.J., and Karplus, M. : 1979, J. Am. Chem. Soc. 101, pp. 1913-1937.
3. Hagler, A.T., Moult, J., and Osguthorpe, D.J. : 1980, Biopolymers 19, pp.395-418.
4. Hagler, A.T., Osguthorpe, D.J., and Robson, B. : 1980, Science 208, pp.599-601.
5. For a review see : Environmental effects on molecular structure and properties, Pullman, B., ed. : 1976, Reidel, D., Dordrecht.
6. Pullman, A., and Pullman, B. : 1975, Quart. Rev. Biophysics 7, pp. 505-566.
7. Langlet, J., Claverie, P., Pullman, B., and Piazzola, D. : 1979, Int. J. Quantum Chem. : Quantum Biol. Symposium 6, pp.409-437 and references therein.
8. Pullman, A., Alagona, G., and Tomasi, J. : 1974, Theor. Chim. Acta 33, pp.87-80.
9. Scheiner, S., and Kern, C.W. : 1977, J. Am. Chem. Soc. 99, pp. 7042-7050.
10. Del Bene, J.E. : 1978, J. Am. Chem. Soc. 100, pp.1387-1394.
11. Hinton, J.F., and Harpool, R.D. : 1977, J. Am. Chem. Soc. 99, pp. 349-353.
12. Vovelle, F., and Ptak, M. : 1979, Int. J. Peptide Protein Res. 13, pp.435-446 and references therein.
13. Cotrait, M., Ptak, M., Busetta, B., and Heitz, A. : 1976, J. Am. Chem. Soc. 98, pp.1073-1076.
14. Ptak, M., Dreux, M., and Heitz, A. : 1978, Biopolymers 17, pp. 1129-1148.
15. Genest, M., and Ptak, M. : 1978, Int. J. Peptide Protein Res. 11, pp.194-208.

16. Genest, M., Vovelle, F., Ptak, M., Maigret, B., and Premilat, S. : 1980, J. Theor. Biol. 87, pp.71-84.
17. Birktoft, J.J., and Blow, D.M. : 1972, J. Mol. Biol. 68, pp.187-240.
18. Rowlinson, J.S. : 1951, Trans. Faraday Soc. 47, pp.120- .
19. Stillinger, F.H., and Rahman, A. : 1974, J. Chem. Phys. 60, pp. 1545-1557.
20. Vovelle, F., Genest, M., Ptak, M., Maigret, B., and Premilat, S. : 1980, J. Theor. Biol. 87, pp.85-95.
21. Caillet, J., and Claverie, P. : 1975, Acta Cryst. A31, pp.448-461.
22. Wolfenden, R. : 1978, Biochemistry 17, pp.201-204.
23. Yang, C.-H., Brown, J.N., and Kopple, K.D. : 1979, Int. J. Peptide Protein Res. 14, pp.12-20.
24. Donohue, J. : 1968, in Structural Chemistry and Molecular Biology, Rich, A., and Davidson, N., ed., W.H. Freeman and Co, San Francisco, pp.443-465.
25. Iitaka, Y., Nakamura, H., Takada, K., and Takita, T. : 1974, Acta Cryst. B30, pp.2817-2825.
26. Two selected reviews :
Ramachandran, G.N., and Sasisekharan, V. : 1968, Adv. Protein Chem. 23, pp.283-437.
Pullman, B., and Pullman, A. : 1974, Adv. Protein Chem. 28, pp. 347-526.
27. Madison, V., and Kopple, K.D. : 1980, J. Am. Chem. Soc. 102, pp. 4855-4863.
28. Ponnuswamy, P.K., and Manavalan, P. : 1976, J. Theor. Biol. 60, pp.481-486.
29. Manavalan, P., Ponnuswamy, P.K., and Srinivasan, A.R. : 1977, Biochem. J. 167, pp.171-182.
30. Vinogradov, S.N. : 1979, Int. J. Peptide Protein Res. 14, pp.281-289.
31. Karle, J.L., and Duesler, E. : 1977, Proc. Natl. Acad. Sci. N.Y. 74, pp.2602-2606.
32. Aubry, A., Vitoux, B., Boussard, G., and Marraud, M. : personal communication.
33. Chou, P.Y., and Fasman, G.D. : 1977, J. Mol. Biol. 115, pp.135-175.
34. Karle, I.L., Gibson, J.W., and Karle, J. : 1970, J. Am. Chem. Soc. 92, pp.3755-3760.
35. Hossain, M.B., and Van der Helm, D. : 1978, J. Am. Chem. Soc. 100, pp.5191-5198.
36. Vovelle, F., Genest, M., and Ptak, M. : to be published. A detailed analysis of H-bonds energies and geometries and of water coordinates will be presented.
37. For a review see : Franks, F. : 1975, in Water, a comprehensive treatise. Franks, F., ed., Plenum Press, New York, vol.4, chap.I.
38. Finney, J.L. : 1979, in Water, a comprehensive treatise. Franks, F., ed., Plenum Press, New York, vol.6, pp.47-122.
39. Romano, S., and Clementi, E. : 1978, Int. J. Quantum Chem. 14, pp. 839-850 and 1980, idem 17, pp.1007-1021.
40. Englert, A., and Leclerc, M. : this symposium.

STACKING INTERACTIONS IN OLIGOPEPTIDE-NUCLEIC ACID COMPLEXES

Tula Behmoaras[+], Judith Fidy[+], Claude Hélène[+o], Gérard Lancelot[o], Trung Le Doan[+], Roger Mayer[o], Thérèse Montenay-Garestier[+] and Jean-Jacques Toulmé[+]

+ Laboratoire de Biophysique, INSERM U201, Muséum National d'Histoire Naturelle, 61, rue Buffon, 75005 Paris
o Centre de Biophysique Moléculaire, CNRS, 45045 Orléans

ABSTRACT

Oligopeptides containing basic and aromatic residues bind to single-stranded and double-stranded nucleic acids and oligodeoxynucleotides. Two types of complexes are formed which both involve electrostatic interactions. Stacking of the aromatic residue with nucleic acid bases is shown to take place in one of these complexes. Nuclear magnetic resonance, fluorescence lifetime and polarization measurements, phosphorescence analysis have been used to characterize oligopeptide-nucleic acid (oligonucleotide) complexes. A review of these data is presented.

1. INTRODUCTION

Protein-nucleic acid associations play a crucial role in all living cells. They are involved in DNA replication, repair and transcription, in the processing and translation of messenger RNA, in the three-dimensional organization of ribosomes and viruses ... The regulation of gene expression is under the control of specific proteins whose interplay is very delicately balanced to reach optimum functioning of the biological system.

The recognition of nucleic acid structures (single strands, loops, A, B or Z forms of DNA ...) and that of nucleic acid base sequences requires well-defined interactions between functional groups of nucleic acids and proteins. Electrostatic interactions are involved in many protein-nucleic acid complexes. As a consequence these complexes dissociate when ionic strength increases (1). Hydrogen bonding between nucleic acid donor or acceptor groups (base, sugar, phosphate)

317

and amino acid side chains (or peptidic bond) provides the most important contribution to the recognition of nucleic acid base sequences
(2). Very little is known about hydrophobic interactions in nucleic
acid-protein associations even though the large entropy increase which
accompanies complex formation suggests that their contribution might
be important (2). Stacking interactions which are strongly favored in
aqueous solutions may be classified as "hydrophobic" even though they
do not follow the "classical" definition of this type of interaction. As
stacking between nucleic acid bases themselves (3), they are characterized by a negative reaction enthalpy and a negative entropy change
(4).

Evidence for stacking interactions between the aromatic residues of
proteins and nucleic acid bases has been provided by a proton magnetic resonance study of oligopeptide-polynucleotide (nucleic acid) complexes (5-10). On the basis of fluorescence studies, a model has been
proposed for the binding of oligopeptides containing basic and aromatic residues to nucleic acids (11). This model postulates the formation
of two different complexes which both involve electrostatic interactions
of the basic amino acids (α and   ε-amino groups) with phosphates.
The aromatic residue does not participate in the binding process in
complex (I) whereas it is involved in stacking interactions in complex
(II) :

$$(1) \quad \text{Nucleic Acid + Peptide} \underset{}{\overset{K_1}{\rightleftharpoons}} \text{Complex (I)} \underset{}{\overset{K_2}{\rightleftharpoons}} \text{Complex (II)}$$

            ("outside complex")   ("stacked complex")

A review of the experimental data which provide evidence for stacking
interactions is presented below together with dynamic studies which
strongly support the existence of two types of complexes.

2. STACKING INTERACTIONS IN OLIGOPEPTIDE-NUCLEIC ACID
   (OLIGONUCLEOTIDE) COMPLEXES

2.1. Nuclear magnetic resonance studies

Stacking interactions are expected to give rise to upfield shifts of the
aromatic proton resonances due to ring-current shielding effects.
Large upfield shifts of tryptophyl, tyrosyl and phenylalanyl proton resonances are observed when oligopeptides such as Lys-Trp-Lys,
Lys-Tyr-Lys and Lys-Phe-Lys bind to single-stranded poly(A) at pH 7
(5-10). Double-stranded DNA has no effect on the chemical shifts of
tyrosine protons (8, 10) whereas tryptophan resonances are shifted upfield although less than in the case of single-stranded DNA (8). The
conclusion reached from these studies - which have been recently reviewed (12) - is that stacking interactions are strongly favored in
single-stranded nucleic acid structures.

In order to obtain a more detailed picture of stacked complexes we
have undertaken a study of oligopeptide binding to self-complementary
oligodeoxyribonucleotides d(ATGCAT) and d(AATTGCAATT). These
oligonucleotides form a duplex structure at low temperature (0°C) and
dissociate at higher temperature (melting temperatures 41°C and
54°C for the hexamer and decamer, respectively, at 10 mM concentration). The resonances of all aromatic protons can be followed separately (13, 14). These oligonucleotides therefore provide a good system to determine the behavior of each individual base (or base pair)
in their complexes with an oligopeptide. Figure 1 shows the PMR

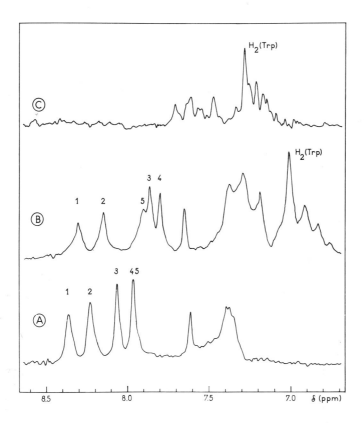

Figure 1 : Proton magnetic resonance spectra of 2.6×10^{-3} M
oligodeoxynucleotide d(ATGCAT) (spectrum A), of 2.6×10^{-3} M
Lys-Trp-Gly-Lys OtBu alone (spectrum C), and of their mixture (spectrum B, 2.6×10^{-3} M each) at 280 K in D_2O (pD=7)
in the presence of 0.5 mM EDTA. The resonances have been
attributed to 1:H8(A5), 2:H8(A1), 3:H2(A5), 4:H2(A1), 5:H8(63)
(see ref. 13). Resonances are measured with respect to an internal DSS reference.

spectra at 280 K of the hexanucleotide d(ATGCAT), the tetrapeptide
Lys-Trp-Gly-Lys OtBu and their equimolar mixture. Complex formation is accompanied by upfield shifts of the aromatic protons of the
tryptophyl residue together with upfield shifts of some of the oligonucleotide proton resonances. Table 1 gives the upfield shift observed
for the H(2) proton resonance of tryptophan at different peptide/oligo-
nucleotide ratios, at 275 K when the oligonucleotide has a predominant
duplex structure, and at 320 K where this duplex structure is melted.
The upfield shifts of the tryptophyl proton resonances are much more
important at low than at high temperature. At 320 K the chemical shifts
of the peptide do not depend on the concentration ratio indicating that
all peptide molecules are bound up to a 1:1 ratio. At 275 K the upfield
shift slightly decreases when the concentration ratio increases suggesting either that some of the peptide remains free in solution or
that the structure of the complex changes with peptide concentration.
(Table 1). However the differences are small (less than 8 %) when the
concentration ratio increases from 0.1 to 1 indicating that nearly all
peptide is bound to the oligonucleotide and that the structure of the
stacked complex does not change greatly with peptide concentration.

As shown in figure 2 the upfield shifts of the oligonucleotide resonances also change with temperature, decreasing when temperature
increases. The upfield shifts of oligonucleotide proton resonances at
280 K are reported in Table 2 for 1:1 complexes with the oligopeptides
Lys-Trp-Gly-Lys OtBu and Lys-Ala-Ala-Lys NHEt. The second peptide gives only electrostatic interactions whereas the first gives rise

Table 1 : Upfield shifts of the H_2 proton of tryptophan
in Lys-Trp-Gly-Lys OtBu at different temperatures
in presence of d(ATGCAT) (3×10^{-3} M in hexamer) for
different [peptide] / [oligonucleotide] ratios in D_2O.

[Peptide]/[Oligonucleotide]	275 K	320 K
0.1	0.358	0.078
0.2	0.348	0.078
0.33	0.343	0.078
0.5	0.335	0.078
1	0.330	0.078

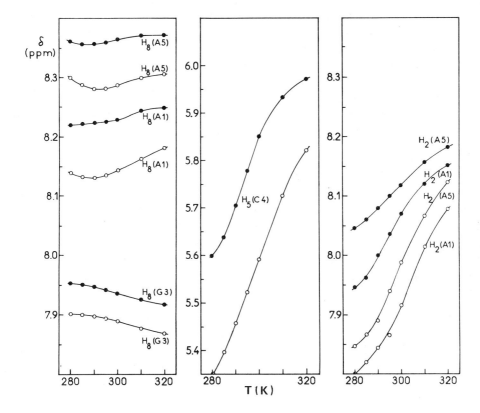

Figure 2 : Temperature dependence of chemical shifts
for 2.6×10^{-3} M hexamer $d\binom{1\ 23\ 456}{ATGCAT}$ in the absence (●)
and in the presence of 2.6×10^{-3} M Lys-Trp-Gly-Lys OtBu
(o) in D_2O at pH 7 in the presence of 0.5 mM EDTA.

to both electrostatic <u>and</u> stacking interactions. Three protons are
strongly affected by the binding of the tryptophan-containing peptide :
The H2 protons of adenines in positions 1 and 5, and the H5 proton of
cytosine in position 4. However it can be seen that this last proton is
very sensitive to electrostatic binding of Lys-Ala-Ala-Lys NHEt which
induces a conformational change in the oligonucleotide and affects the
stacking of the base pairs in the middle of the duplex structure. The
large upfield shifts observed for the H2 resonances of adenines 1 and
5 suggest that the tryptophyl residue of the peptide stacks with the
two AT base pairs at the end of the oligonucleotide. Since the H2 protons are located in the small groove of the duplex structure, the NMR
results also suggest that stacking takes place in the small groove
rather than in the large groove of the double helix. The resonances of

Table 2 : Change in chemical shifts of protons of d(A T G C A T) $\overset{1\,2\,3\,4\,5\,6}{}$
(3x10^{-3}M in hexamer) in the presence of Lys-Trp-Gly-Lys-
OtBu (3x10^{-3}M) at 280 K and 320 K and Lys-Ala-Ala-Lys-
NHEt (3x10^{-3}M) at 280 K in D_2O. A (-) sign indicates an
upfield shift.

	Lys-Trp-Gly-Lys OtBu		Lys-Ala-Ala-Lys NHEt
	280 K	320 K	280 K
H_8 A(5)	- 0.062	- 0.065	+ 0.009
H_8 A(1)	- 0.082	- 0.072	+ 0.010
H_2 (A 5)	- 0.201	- 0.073	- 0.024
H_2 (A 1)	- 0.172	- 0.105	- 0.020
H_8 (G3)	- 0.057	- 0.053	0.0
H_5 (C 4)	- 0.28	- 0.16	- 0.15*
CH_3 T(2)	- 0.05*	-	- 0.025
CH_3 T(6)	- 0.05*	-	- 0.025

    * The values were obtained by extrapolation of data
       obtained at other temperatures or concentrations.

the H8 protons are much less affected, and those of thymine methyl
groups even less. Purine H8 resonances are not very sensitive to the
structure of the oligonucleotide as shown in figure 2 from the temperature dependence of the chemical shifts in the absence of peptide.

The oligodeoxynucleotide dATGCAT remains in a double-stranded
conformation upon peptide binding at 280 K. This is demonstrated by
the observation of the base-pair exchangeable proton resonances in
H_2O solution. Fast exchange with H_2O protons in single-stranded
structures eliminates these resonances. The N(1)H resonance of G-C
base pair is observed at 12.78 ppm in the free oligodeoxynucleotide
and is shifted upfield by - 0.08 ppm in the presence of Lys-Trp-Gly-
Lys OtBu (1 peptide per hexanucleotide). Corresponding values are
13.78 ppm and - 0.14 ppm, respectively, for the N(3)H resonance
of the internal AT base pair (thymine T(2)). The N(3)H resonance of
the external AT base pair (thymine T(6)) is observed at 13.05 ppm
but line broadening makes it difficult to obtain accurate values in the
presence of the peptide. The magnitude of the upfield shift of the

N(3)H proton of T(2) reinforces the conclusion reached above that tryptophan stacks primarily with the terminal AT base pairs.

In the single-stranded oligonucleotide (at 320 K) all oligonucleotide resonances are shifted upfield by similar amounts (Figure 2, Table 2). This tends to suggest that stacking of the tryptophyl ring occurs at all possible sites along the oligonucleotide. The smaller upfield shifts of the tryptophan protons as compared to the duplex structure (Table 1) might be explained by the large differences in ring-current effects of the different bases (15). Adenine has the largest effect and this might be responsible for the large upfield shifts observed in the case of the duplex structure if the tryptophyl ring is located near the two terminal AT base pairs (see above). In the single-stranded structure, only an average over all stacked complexes is measured if fast exchange occurs - as expected - between the different complexes.

In the case of long nucleic acids it was concluded that stacking was strongly favored in single-stranded structures. Based on the upfield shifts observed for tryptophan in the complexes of oligopeptides with oligonucleotides it might be concluded that stacking is favored in the duplex structure. However, as recalled above, two effects must be taken into account : i) end effects are, of course, predominant with small oligonucleotides ; ii) only the average between the different complex structures is measured. A study of complexes involving oligonucleotides of different lengths and base sequences is required to clarify these different points.

2.2 Luminescence measurements

Evidence for stacking interactions was also obtained from two types of experiments carried out at low temperature (77 K) :
i) energy transfer at the triplet level is very efficient in single-stranded poly(A) (16). Since the triplet state of tryptophan is lower in energy than the triplet state of adenine, tryptophan should act as a trap for the delocalized excitation energy. This has been experimentally observed from phosphorescence (17) and ODMR (optically detected magnetic resonance) (18) measurements at 77 K. The tryptophyl ring of Lys-Trp-Lys acts as a trap for the excitation energy migrating at the triplet level in poly(A). Such a triplet transfer requires a good orbital overlap and is therefore very efficient when donor and acceptor are stacked. Similar results have been recently obtained with oligopeptides such as Lys-Trp-Gly-Lys-OtBu and Lys-Gly-Trp-Lys-OtBu (unpublished results) indicating that stacking interactions also occur with these peptides.
ii) if a nucleic acid base is substituted by a heavy atom (such as mercury or platinum) the spectroscopic properties of a tryptophyl ring

located in its vicinity should be strongly perturbed. This heavy atom
effect is characterized by a quenching of the tryptophan fluorescence,
an increase of the phosphorescence intensity and a shortening of the
phosphorescence lifetime (19). This has been observed for complexes
of Lys-Trp-Lys with poly 5-mercuriuridylic acid (20). The phosphorescence lifetime of tryptophan is reduced from ≈ 6 s   in the free
peptide to ≈ 5 ms in the complexed peptide. Measurement of the phosphorescence lifetime is thus a very sensitive method to probe the presence of a tryptophyl residue near a heavy atom-substituted base. A
Van-der-Waals contact between the heavy atom and the perturbed aromatic ring is required. This effect is still observed when the mercury
atom in the fifth position of the pyrimidine ring is linked to a mercap-
tan (cysteine for example). It is unlikely that the heavy atom might
come close enough to the tryptophyl ring in another structure than the
stacked complex formed by the indole and pyrimidine rings.

3. EVIDENCE FOR TWO TYPES OF COMPLEXES IN OLIGOPEPTIDE-NUCLEIC ACID ASSOCIATION

Based on a thermodynamic analysis of fluorescence and circular di-
chroism data, it was previously concluded that a two-step model was
necessary to describe the binding of oligopeptides containing basic and
aromatic residues to polynucleotides and nucleic acids (see equation 1).
This model was originally proposed on the basis of fluorescence quantum yield and lifetime measurements (11). It was known that the fluorescence of tryptophan or tyrosine stacked with nucleic acid bases was
completely quenched (21). Therefore complex II of equation (1)
("stacked complex") was expected to have a fluorescence quantum
yield reduced to zero. On the other hand complex I ("outside complex")
was supposed to be characterized by a fluorescence quantum yield and
a fluorescence lifetime identical to that of the free peptide since the
fluorescence decay of the complex could not be distinguished from that
of the free peptide (11). In order to obtain more information on complex I and complex II we have undertaken a series of dynamic measurements which are summarized below.

3.1. Fluorescence decay measurements

Using synchrotron radiation available at LURE (Orsay, France), the
fluorescence decay of Lys-Trp-Lys was measured in the absence and
in the presence of different nucleic acids (22). The decay of the free
peptide could be represented as a superposition of two exponentials
whose lifetimes at 10 °C were 1.17 and 3.33 ns, respectively. In the
presence of nucleic acids these lifetimes were slightly altered as well
as their respective contributions but the average lifetime

$$\langle\tau\rangle = \frac{C_1 \tau_1 + C_2 \tau_2}{C_1 + C_2}$$ was not greatly affected ($< 10\,\%$) upon nucleic

acid binding even when fluorescence quenching was very efficient as shown in Table 3. These results confirm that complex II does not emit any detectable fluorescence and that complex I has fluorescence characteristics (average lifetime and quantum yield) very similar to those of the free peptide.

Table 3 : Fluorescence decay analysis at 10°C of Lys-Trp-Lys according to a two-exponential decay law $I = C_1 \exp - t/\tau_1 + C_2 \exp - t/\tau_2$. The average lifetime $\langle\tau\rangle$ is $(C_1\tau_1 + C_2\tau_2) / (C_1 + C_2)$ (data from reference 22). I_F and I_F° are the fluorescence intensities of the peptide in the presence and in the absence of the nucleic acid, respectively

	I_F/I_F°	τ_1(ns)	C_1	τ_2(ns)	C_2	$\langle\tau\rangle$
Lys –Trp-Lys alone	1	1.17	0.49	3.33	0.51	2.27
Lys-Trp-Lys + native DNA	0.70	1.02	0.43	3.35	0.57	2.35
Lys-Trp-Lys + denatured DNA	0.16	0.95	0.44	2.87	0.56	2.03
Lys-Trp-Lys + apurinic DNA	0.44	1.44	0.53	3.45	0.47	2.38
(2 % apurinic sites)						

3.2. Accessibility of tryptophan in "outside complex" I

Acrylamide is a good neutral quencher of tryptophan fluorescence. Quenching involves both a static and a dynamic component which can be dissociated as previously reported (23). Quenching data for Lys-Gly-Trp-Lys-OtBu either free or complexed to native DNA are summarized in Table 4. This system was chosen because binding of the tetrapeptide to native DNA was nearly complete at low ionic strength (1 mM Na cacodylate, 1 mM NaCl, 0.2 mM EDTA, pH 6) and fluorescence was not strongly inhibited upon binding to native DNA therefore allowing acrylamide quenching experiments to be carried out. The conclusion of this study is that tryptophan in complex I is readily accessible to an external fluorescence quencher.

Table 4 : Quenching of Lys-Gly-Trp-Lys-OtBu fluorescence at 20°C by acrylamide. K_{SV} and V are the Stern-Volmer constant (dynamic quenching) and the ground-state association constant (static quenching). K_{SV} and V were obtained by fitting the fluorescence quenching data according to the following equation $I_0/I = (1+K_{SV}[Q])\exp(V Q)$ (see reference 23)

	$K_{SV}(M^{-1})$	$V(M^{-1})$
Peptide	12	0.58
Peptide + DNA	11	0.25

3.3 Dynamics of bound peptide

If tryptophan does not interact with nucleic acid bases in complex I, its rotational correlation time should not be much longer than in the free peptide. This correlation time was determined from fluorescence anisotropy measurements according to Perrin's law (eq. 2) assuming a spherical symmetry :

$$(2) \qquad \frac{1}{r} = \frac{1}{r_0} (1+\tau \frac{kT}{\eta V}) = \frac{1}{r_0} (1 + \frac{\tau}{\theta})$$

where r_0 is the limit anisotropy reached when the peptide is completely immobilized, τ is the fluorescence lifetime, $\theta = \eta V/kT$ is the rotational correlation time, η is the viscosity and V the volume of the equivalent sphere.

Two methods were used to determine θ : i) the fluorescence lifetime τ of the peptide was decreased at constant viscosity by addition of acrylamide which - as shown above - reduces the quantum yield of the peptide. Part of this quenching is due to a dynamic process which reduces the fluorescence lifetime (see table 4) ; ii) the viscosity η was increased by adding glycerol to the complex. Glycerol does not induce dissociation of the complex but changes the fluorescence quantum yield and lifetime of the peptide. This was taken into account by rewriting equation (2) as

$$(3) \qquad \frac{1}{r} = \frac{1}{r_0} (1 + \frac{kT}{V} \frac{\tau}{\tau_0} \frac{\tau_0}{\eta})$$

where τ_0 is the fluorescence lifetime of the complex in the absence of glycerol ($\eta = 10^{-3}$ Pa. s) and τ is the lifetime in the presence of glycerol at viscosity η ($> 10^{-3}$ Pa.s). A plot of $1/r$ <u>versus</u> $\tau / \tau_0 \eta$ should give a straight line whose slope is $(1/r_o)(\tau_o kT/V)$ and y-axis intercept is $1/r_o$. This has been experimentally observed (figure 3) for both the free peptide (Lys-Gly-Trp-Lys-OtBu) and the peptide bound to native DNA. From these data rotational correlation times were estimated to be 147 ps and 303 picoseconds for free and bound peptide, respectively. These results clearly show that tryptophan in complex I has a mobility which is not compatible with its being involved in strong interactions with nucleic acid bases. A similar conclusion was reached using the first method (reduction of τ through acrylamide fluorescence quenching).

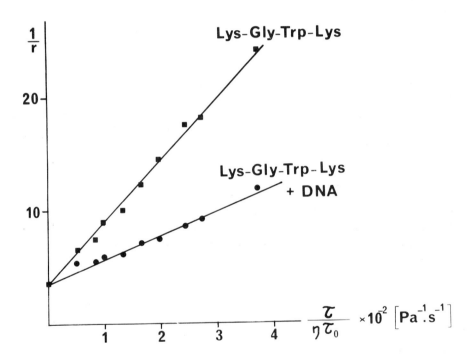

Figure 3 : Plot of $1/r$ versus $\tau/\tau_o \eta$ as defined in the text (see equation 3) for 9×10^{-5}M Lys-Gly-Trp-Lys OtBu in the absence (■) and in the presence (●) of 8.3×10^{-4}M DNA at 20°C in a 1 mM Na cacodylate buffer (pH 6), 1 mM NaCl and 0.2 mM EDTA. Viscosity (η) was changed by adding glycerol. τ_o is the average fluorescence lifetime of the peptide (1.6 ns at 20°C) and τ its value in the presence of glycerol.

3.4. Binding of oligopeptides to apurinic DNA

If stacking interactions are involved in complex II then it is expected that single-stranded structures will lead to a higher K_2 value (eq. 1) because less energy is required to unstack bases in a single-stranded than in a double-stranded structure. As a matter of fact K_2 is about 20 times higher in denatured DNA $(K_2 \approx 6)$ than in native DNA $(K_2 \approx 0.3)$. A highly favorable situation for stacking could be created in a double helix if a purine was removed because the room left by this purine could accommodate the indole ring of tryptophan which has a similar size. No energy will be required to unstack bases since the two bases which are on either side of an apurinic site are separated by ≈ 0.68nm, if it is assumed that the DNA structure is not markedly altered. A study of K_2 as a function of the number of apurinic sites in DNA has led to the conclusion that each removed purine creates a strong binding site for Lys-Trp-Lys with an average value of $K_2 \approx 200$ (24). This means that the overall association constant K_1 $(1 + K_2)$ is about two orders of magnitude higher for an apurinic site as compared to a native site.

It was then observed that oligopeptides bound to an apurinic site through replacement of the purine by the aromatic residue of the peptide could cleave the phosphodiester bond at the apurinic site. The lysine α- or ε-amino group is involved in the cleavage reaction. Therefore an oligopeptide such as Lys-Trp-Lys mimics an apurinic endonuclease : it exhibits both the specificity of binding and the catalytic activity which are characteristic of an enzyme. Although it should not be concluded that the cleavage mechanisms are identical, it is interesting to notice that a simple tripeptide can use one residue for its binding specificity (tryptophan) and one residue for the chemical reaction (lysine α- or ε-amino group). Analysis of the reaction mechanism is presently under study.

4. CONCLUSION

In conclusion, the above results demonstrate that complex I (eq. 1) is an "outside complex" where the aromatic residue does not engage interactions with nucleic acid bases and which owes its stability to electrostatic interactions only. These data are in agreement with the two-step model presented above in which the "outside complex" I behaves as the free peptide from the fluorescence point of view whereas complex II has a completely quenched fluorescence due to stacking of the tryptophyl ring with nucleic acid bases. The equilibrium is shifted toward complex II in single-stranded DNA $(K_2 \approx 6)$ and even more at apurinic sites in native DNA $(K_2 \approx 200)$.

Equilibrium data cannot distinguish between a two-step mechanism and two separate equilibria. Only kinetic experiments should be able to make the distinction. Data have already been published (25) which support the two-step model originally proposed (11). However there is still a discrepancy as to the values of K_2 which have been determined from kinetic and equilibrium data on Lys-Trp-Lys binding to poly(A). Further kinetic experiments are required to explain this discrepancy. All fluorescence data which we have presented above have been obtained at much lower peptide/DNA ratios than those used in kinetic measurements. It is also possible that field-jump relaxation experiments detect complexes which are different from complexes I and II as defined in equation (1).

Table 5 gives the free energy difference between stacked complex II and "outside complex" I together with K_1 and K_2 values (it should be noted that the equilibrium binding constant for complexes I and II are K_1 and K_1K_2, respectively, while the overall binding constant is equal to $K_1(1+K_2)$). It can be seen that stacking is energetically unfavorable in double-stranded DNA whereas its contribution is important in binding to single-stranded DNA and even more to apurinic sites in DNA.

Table 5 : Values of association constants K_1 and K_2 and change in free energy $\Delta (\Delta G)$ at 278 K due to stacking in complex II as compared to complex I (eq. 1). Association constants have been measured in a pH 6 Na cacodylate buffer (1 mM) in the presence of 1 mM NaCl and 0.2 mM EDTA.

	$K_1 (M^{-1})$	K_2	$\Delta (\Delta G)$ (kJ. mol^{-1})
Native DNA	1.2×10^4	0.33	+ 2.6
Denatured DNA	0.8×10^4	6	- 4.15
Apurinic site in double stranded DNA	1.0×10^4	200	-12.25

ACKNOWLEDGMENTS

This work was supported in part by DGRST (MRM 79.7.1039), INSERM (CRL 79.403.32 and ATP 77.79.109), Ligue Nationale Française contre le Cancer and Fondation pour la Recherche Médicale.

REFERENCES

(1) Record, M. T., Anderson C. F. and Lohman, T. M. : 1978, Quart.
       Rev. Biophys. 11, pp. 103-178
(2) Hélène, C. and Lancelot, G. : 1981, Progr. Biophys. Mol. Biol.
       (in press)
(3) Porschke, D. and Eggers, F. : 1972, Europ. J. Biochem. 26,
       pp. 490-498
(4) Dimicoli, J. L. and Hélène, C. : 1971, Biochimie 53, pp. 331-345
(5) Hélène, C. and Dimicoli, J. L. : 1972, FEBS Letters 26, pp. 6-10
(6) Gabbay, E. J., Sanford, K. and Baxter, C. S. : 1972, Biochemis-
       try 11, pp. 3429-3435
(7) Gabbay, E. J., Sanford, K., Baxter, C. S. and Kapicak, L. : 1973
       Biochemistry 12, pp. 4021-4029
(8) Dimicoli, J. L. and Hélène, C. : 1974, Biochemistry 13, pp. 714-73(
(9) Gabbay, E. J., Adawadkar, P. D., Kapicak, L., Pearce, S. and
       Wilson, W. D. : 1976, Biochemistry 15, pp. 152-157
(10) Mayer, R., Toulmé, F., Montenay-Garestier, T. and Hélène, C.
       1979 J. Biol. Chem. 254, pp. 75-82
(11) Brun, F., Toulmé, J. J. and Hélène, C. : 1975, Biochemistry
       14, pp. 558-563
(12) Hélène, C. and Maurizot, J. C. : 1981 CRC Crit. Rev. Biochem.
       10, pp. 213-258
(13) Patel, D. J. : 1975, Biochemistry 14, pp 3984-3989
(14) Lancelot, G., Mayer, R., Thuong, N. T., Chassignol, M. and
       Hélène, C. : 1981, submitted for publication
(15) Giessner-Prettre, C., Pullman, B., Borer, P. N., Kan L. S.
       and Ts'o, P. O. P. : 1976, Biopolymers 15, pp. 2277-2286
(16) Hélène, C. and Longworth, J. W. : 1972, J. Chem. Phys. 57,
       pp. 399-408
(17) Hélène, C. : 1973, Photochem. Photobiol. 18, pp. 255-262
(18) Co, T. T. and Maki, A. H. : 1978, Biochemistry 17, pp. 182-186
(19) Hélène, C. : 1979, C. R. Acad. Sci. 288, pp. 433-436
(20) Hélène, C., Le Doan, T. and Toulmé, J. J. : 1979, Nucl. Ac.
       Res. 7, pp. 1945-1954
(21) Montenay-Garestier, T. and Hélène, C. : 1971, Biochemistry
       10, pp. 300-306
(22) Montenay-Garestier, T., Brochon, J. C. and Hélène, C. : 1981
       Int. J. Quant. Chem. (in press)
(23) Eftink M.R. and Ghiron, C. : 1976, J. Phys. Chem. 80, 486-493
(24) Behmoaras, T., Toulmé, J. J. and Hélène, C. : 1981, Proc. Nat.
       Acad. Sci. USA, 78, pp. 926-930
(25) Porschke, D. : 1980, Nucl. Ac. Res. 8, pp. 1591-1612

INTERACTION MODELS FOR WATER IN RELATION TO PROTEIN HYDRATION

H.J.C.Berendsen, J.P.M.Postma, W.F.van Gunsteren and
J.Hermans*
Laboratory of Physical Chemistry, the University of
Groningen, Nijenborgh 16, 9747 AG Groningen, the Netherlands
*on leave from Department of Biochemistry, University of
North Carolina, Chapel Hill, N.C., USA.

ABSTRACT
For molecular dynamics simulations of hydrated proteins a simple yet reliable model for the intermolecular potential for water is required. Such a model must be an effective pair potential valid for liquid densities that takes average many-body interactions into account. We have developed a three-point charge model (on hydrogen and oxygen positions) with a Lennard-Jones 6-12 potential on the oxygen positions only. Parameters for the model were determined from 12 molecular dynamics runs covering the two-dimensional parameter space of charge and oxygen repulsion. Both potential energy and pressure were required to coincide with experimental values. The model has very satisfactory properties, is easily incorporated into protein-water potentials, and requires only 0.25 sec computertime per dynamics step (for 216 molecules) on a CRAY-1 computer.

1. INTRODUCTION

The study of hydrated proteins by computer simulation methods such as Monte Carlo (MC) or Molecular Dynamics (MD) requires the use of a simple interaction model for water. The model must be suitable for a sufficiently accurate simulation of the liquid state of water, but must also be easily extendable to interaction of water with various molecular groups on the protein molecule.

The requirement of simplicity is related to the complexity of a simulation of a hydrated protein. The straightforward simulation of a small protein such as bovine pancreatic trypsin inhibitor (BPTI) with 58 amino acids in aqueous solution, using three-dimensional periodic boundary conditions, requires the inclusion of about 2000 water molecules if the protein is to be kept out of the interaction range of its own images. Representing all methyl and methylene groups as united atoms, the simulation comprises 458 non-hydrogen atoms of the protein, but about 6600 atoms in total including the water molecules. The MD simulation of such a system is a gigantic computing task, not

B. Pullman (ed.), Intermolecular Forces, 331–342.

only because of its size but also because water molecules tend to become trapped in favourable sites and long simulations are required to attain sufficient statistical accuracy.

Several approximations are possible that aim at considerable reduction of computer effort for the simulation of a hydrated macro-molecule. It is obvious that one should try to avoid the detailed simulation of water molecules far removed from the protein surface. The crudest approximation (next to simulations of a protein in vacuum [1-4]) is to replace the interaction with water molecules by potentials of mean force, representing the mean free energy of interaction between surface groups and water. The next approximation is the use of stochastic dynamics which, in addition to a mean force, mimics the dynamic character of the interaction with water by stochastic forces. Such methods are extendable to inclusion of a limited number of water molecules as required to simulate those water molecules that belong to the specific hydration[5] of the protein.

Although such approximate methods are under present development, they are not yet available in a reliable form and they will require extensive testing against more fully detailed models. Therefore for some time to come reliable simulations of fully hydrated proteins incorporating several thousand water molecules will be required. In order to be feasible, even on a modern supercomputer, such simulations require a simple water model.

In the following sections we will consider various available models for water against the background of the philosophies on which these models are based. We then arrive at a new model for liquid water that is both simple and reliable for the purpose for which it is intended.

2. EFFECTIVE PAIR POTENTIALS

An ideal interaction potential should be derived from ab-initio quantum calculations and predict reliably all known experimental data for all phases of water. Such potentials do not exist because of two reasons: a) ab-initio methods are not accurate enough, b) the interaction potentials are not pairwise additive. The first reason is not serious because we can adjust parameters of the model according to empirical data (the model then becomes semi-empirical). If only ice and vapour data were used and liquid properties would then be well-predicted, we could speak of a satisfactory model. Unfortunately, such models cannot be devised. The deviation from pairwise additivity is quite strong [6,7] and can easily reach values of -5 to -8 kJ/mol. Thus any model that is expected to predict properties of liquid water correctly must be either a non-pairadditive model or it must use an *effective pair potential* that includes the average non-additive character of the interaction.

Non-pairadditive potentials can best be realised by introducing polarisability into the model[7,8]. This, unfortunately, makes the use of such a model in MC or MD simulations quite expensive. Hence, and in view of our present aim, we will not consider polarisable models further. Thus restricting ourselves to effective pair potentials, we necessarily restrict the applicability of a given model to a certain range of density, temperature and external force fields. We choose the liquid density near zero pressure, around room temperature and without external fields. The model we propose in Section 4 has been tested for liquid water at a density of $1g/cm^3$ and a temperature of 300K. In a future publication we will also evaluate the model at conditions differing from those mentioned above.

3. SOME PROPERTIES OF AVAILABLE MODELS

Early empirical models made use of both solid and gas phase data. Parameters were adjusted to ice energy, ice lattice constants, gas phase dipole moment and gas phase second virial coefficients. The Rowlinson model[9] uses four point charges plus Lennard-Jones 6-12 interactions centered on the charges. The positive charge of 0.3278 e is situated on the proton positions while the negative charges are situated 0.025 nm above and below the oxygen positions, perpendicular to the molecular plane. Parameters were adjusted to ice data and to the dipole moment and second virial coefficient of the vapour. The BNS model[10] also uses four point charges (0.1956e), each in tetrahedral positions at 0.1nm from the oxygen. There is a single Lennard-Jones 6-12 potential centered on the oxygen nuclei, and the electrical potential is smoothly switched off at short range, depending on the oxygen-oxygen distance. Parameters were adjusted to ice data and the second virial coefficient.

When the BNS model was used for the first MD simulation on liquid water[11], it turned out not to be entirely satisfactory for the liquid state. An adjustment was made in the parameters equivalent to a scaling of temperature. This increases the binding energy of a pair and causes a deviation of the second virial coefficient to larger (absolute) values[12]. Later it was found necessary to make further adjustments to the model on the basis of MD simulations. The well-known ST2 model was proposed[13] which has been quite successful for the simulation of liquid water. The negative charges were pushed towards the oxygen over 0.02 nm and the parameters were readjusted. The second virial coefficient now deviates quite considerably from the experimental value[7,12], being larger by almost a factor of two. Also the dipole moment of the model is considerably larger than the gas phase dipole moment (see Table 1).

TABLE 1   SOME PROPERTIES OF WATER MODELS

Model	Ref	type	μ debye gas = 1.85	pair — minimum	
				E kJ.mol^{-1}	r nm
1. Rowlinson	9	gas + solid	1.85	22.6	0.269
2. BNS	10	eff.pair	2.11	27.2	0.276
3. BNS scaled	11	eff. pair	2.17	28.8	0.276
4. ST2	13	eff. pair	2.35	28.6	0.285
5. MCY	14	pure pair	2.19	24.6	0.287
6. MCY scaled	15	eff. pair	2.34	28.0	0.287
7. PE	7	polarisable	1.85	20.9	0.300
8. SPC(this work)	–	eff. pair	2.27	27.6	0.276

The configuration-interaction potential of Matsuoka, Clementi and Yoshimine[14] (MCY), although it has a higher dipole moment than the isolated water molecule, does not yield correct results if used with MC[16] or MD[15]. A reasonable first peak in the radial distribution function is obtained, but the second neighbour peak is at too short a distance, while the overall appearance resembles a higher temperature behaviour. The energy of the liquid is too low and the pressure is too high. Better results are obtained when the energy is scaled[15] by a factor of 1.14.

The adjustments necessary to obtain reasonable results for the liquid state clearly point to the inadequacy of a pure pair potential. The effective pair potential has as salient features:
a.   the potential well between hydrogen-bonded pairs is about 25% deeper than for an isolated pair,
b.   the dipole moment is about 25% larger than the dipole moment of the isolated molecule.
Introducing polarisability into a water model has revealed that similar average changes in energy and dipole moment indeed occur[7,8].

4. A SIMPLE POINT CHARGE MODEL

Based on the requirement of simplicity we have devised a new effective pair potential for liquid water. Molecular dynamics simulations are most economic if carried out on a rigid model (without internal degrees of freedom) using cartesian coordinates and employing a model consisting of the smallest possible number of point charges. The point charges should preferably coincide with the positions of the atomic masses, thus avoiding the reconstruction of charge centers and the redistribution of forces and torques. Efficient methods are available to carry out MD in cartesian coordinates on molecules with internal constraints[17,18].

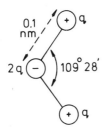

Fig. 1. Geometry and charges
of the SPC model

We tried to parameterise a three-point charge model according to Fig.1, with a Lennard Jones interaction between oxygen centers

$$E_{LJ} = -(\frac{A}{r})^6 + (\frac{B}{r})^{12}; \ r = r_{OO}$$

If we fix the attractive r^6 term of the oxygen-oxygen interaction to the "experimental" value derived from the London expression:[19]

$$A = 0.37122 \ nm \ (kJ.mol^{-1})^{1/6}$$

only two parameters remain to be adjusted: q and B. These should be derived from MC or MD simulations of liquid water. We chose to fit both the interaction energy and the pressure of the liquid to coincide with the experimental values at 300K, while the form of the radial distribution function g(r) was monitored as well.

The two-dimensional parameter space was sampled by 12 M.D. runs at 300K on 216 water molecules in a periodic box with density 1 g/cm^3. Each run was carried out, after equilibration, over a time of 2 ps, using a cut-off radius for the potential of .6 nm. The time step in the simulation was   .002 ps. Both the potential energy E and the pressure turned out to be sufficiently linear in the parameters q and B. A plot of E and pV/NkT is given in Fig.2, as obtained from a least squares fit to the 12 sample data. It turned out that only in a certain region of this parameter space radial distribution functions with water-like structures can be obtained. Those points are indicated by + in Fig. 2, while the points marked - did not yield resolved second neighbour peaks.

The target point in parameter space is E = 41.8 kJ.mol^{-1} and p = o. This value of the potential energy has been derived from the heat of vaporization at 25°C, making proper corrections for the work upon volume change and for intra- and intermolecular quantum vibrations.[20] For the final choice of parameters allowance was made for small corrections in E and p due to the finite cut-off radius used in the

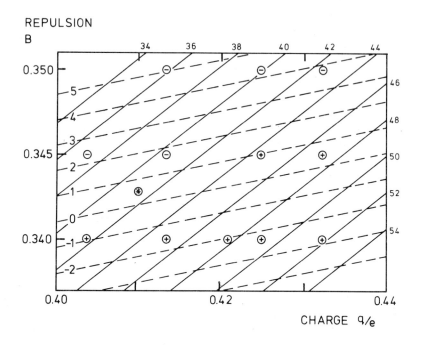

REPULSION
B

Fig.2. Potential energy and pressure of liquid water at 300 K in parameter space of charge on hydrogen q and oxygen-oxygen repulsion parameter B. B is expressed in units nm. (kJ/mol)$^{-1/12}$. Drawn lines: potential energy in kJ/mol; broken lines: pressure expressed as pV/NkT (one unit corresponds to 1385 bar). Circles represent MD simulations. + has a resolved second neighbour peak in the radial distribution curve; - has not. * : final parameters: q = 0.41 e and B = 0.3428

MD simulations. The final parameters are:
        q = 0.41 e
        B = 0.3428 nm (kJ.mol^{-1})$^{-12}$
With these values the radial distribution curve still shows structure. We note, however, that particularly the value of the repulsion is critically close to the structureless region.

    Subsequently a full MD run was carried out on 216 molecules, with a small time step and a large cut-off radius. The results are summarised in Table 2.

TABLE 2.   PROPERTIES OF SPC WATER

MD run characteristics:
density $\rho = 1 \text{ g cm}^{-3}$
temperature $T = 300 \text{ K}$
nr of mols $N = 216$
time step $\Delta t = 0.0005 \text{ ps}$
time span $t = 12.5 \text{ ps}$
cut-off radius $R_c = 0.85 \text{ nm}$

MD run results:
potential energy $-E = 42.2 \text{ kJ.mol}^{-1}$ (41.8*)
pressure $pV/NkT = -0.36$ (0.0)
diffusion constant $D = 3.6 \times 10^{-9} \text{ ms}^{-2}$ (2.7)

Numbers in parenthesis indicate experimental values

* including quantum corrections for intermolecular vibrations.

The radial distribution function is given in Fig. 3, together with the experimental g(r) from X-ray diffraction[21] and the g(r) obtained from a MD run on ST2 water. The static and structural properties are quite well reproduced; the radial distribution is better than that of the ST2 model which overemphasises the tetrahedral structure of the liquid.

Dynamic properties studied thus far include the diffusion constant and the velocity correlation function of the centre of mass. The diffusion constant, obtained from the slope of the molecular squared displacement plotted against time, is somewhat larger than the experimental value[22] (see Table 2). The velocity correlation function (fig. 4) has an oscillatory character with negative tail indicative of rebounce against the molecular cage formed by neighbours. Its fourier transform (fig. 4) reveals the experimental peaks at 60 and 175-200 cm^{-1} as observed by neutron scattering[23]. The oscillatory character is some- what less pronounced than for the ST2 model[11].

Compared with the ST2 potential, the SPC potential has a very similar hydrogen-bonded pair energy curve.(fig.5). There is substantial difference when the second molecule is rotated about an axis perpendicular to the plane of the first molecule. No double minimum is observed, as it is for the rather tetrahedral ST2 model. It is likely that this feature of the ST2 model is unrealistic.

We conclude that the SPC model yields quite satisfactory results for liquid water, which are an improvement on the ST2 model. It is highly efficient for computer simulations. One dynamic step for 216 molecules requires 2.7 seconds CPU time on the CYBER 170/760 of the Computer Centre of the University of Groningen, about 0.25 seconds on a

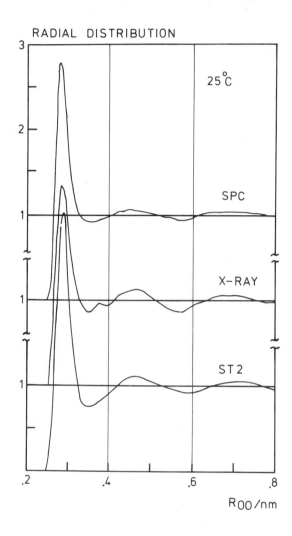

Fig. 3. Radial distribution curve g(r) for the SPC model, compared with X-ray data of Narten (ref 21). For comparison a g(r) of ST2 water at 300K is given.

CRAY 1 computer at Daresbury, U.K. and about 0.2 seconds on a CYBER 205 (tested at CDC, Minneapolis).

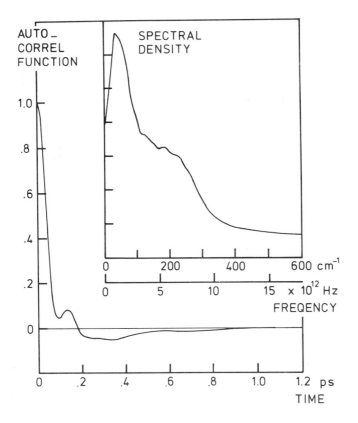

AUTO –
CORREL
FUNCTION

SPECTRAL
DENSITY

Fig. 4. Autocorrelation function of the center of mass velocity and its fourier transform.

5. WATER-PROTEIN POTENTIALS

The SPC potential is quite suitable for interaction with molecular groups on proteins when potentials of the 6-12-1 type are used. Since there are no virtual charges on the water molecule, it is also not necessary to introduce virtual charges on oxygen molecules of the hydrogen bond acceptors. Simple combination rules can be applied to obtain Lennard-Jones parameters for interactions between water and atoms of the protein. It is necessary to include those hydrogens on the protein that are involved in hydrogen bond donors; it is not necessary to include special hydrogen bonding potentials. Parameters can be chosen such that both the hydrogen-bonding energy and separation between donor and acceptor correspond to experimental data or to values obtained from calculations[24]. The sparseness of reliable data limits the reliability of the Lennard-Jones parameters on hydrogen-bond donors and acceptors.

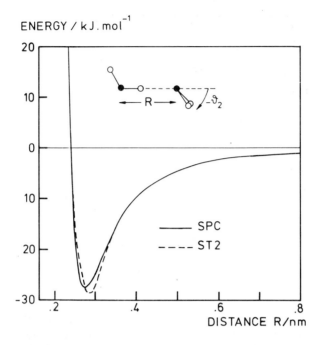

Fig. 5. Pair energy for the most favourable hydrogen-bonded pair for the SPC and ST2 models, as a function of molecular separation. The rotation angle θ_2 of the second molecule is always such that the energy is minimal. For ST2 θ_2 is -54°; for SPC θ_2 is -27°.

We are presently engaged in molecular dynamics simulations of small hydrated proteins, using the SPC water model.

Acknowledgements: This research has been supported by the Foundation of Chemical Research (SON) under the auspices of the Netherlands Organisation for Pure Research (Z.W.O.). J.H. acknowledges a partial fellowship obtained from ZWO during 1980. The Computer Centre of the University of Groningen has provided the means to carry out this research on the CDC Cyber 170/760. P. van der Ploeg, CRAY Research at Daresbury and Control Data Corporation at Minneapolis have kindly provided help for testing our program on two supercomputers.

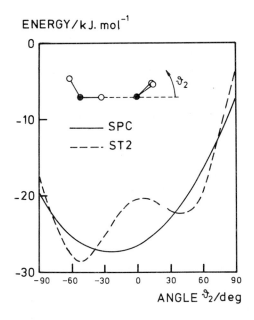

Fig. 6. Pair energy for the most favourable hydrogen-bonded pair for the SPC and ST2 models, as function of the rotation of the second molecule about an axis perpendicular to the plane of the first molecule. The distance between the molecules always corresponds to an energy minimum.

References:

1. McCammon, J.A., Gelin,B.R., and Karplus,M: 1977, Nature 267, 585; Karplus, M and McCammon, J.A.: 1979, Nature 277, 578.
2. McCammon,J.A., Wolynes,P.G., and Karplus,M: 1979, Biochem. 18, 927
3. Levitt, M.: 1980, in *Protein Folding*, Ed. R. Jaenicke, Elsevier/ North-Holland, Amsterdam; Levitt, M.: 1981, Nature, to be published.
4. van Gunsterén, W.F. and Karplus, M.: 1981, J.Am.Chem.Soc., to be published.
5. Berendsen, H.J.C.: 1975, in *Water, A Comprehensive Treatise*, Vol.5, 293, Ed. F.Franks, Plenum Press, New York.
6. Stillinger, F.H.: 1975, Adv.Chem.Phys. 31, 1
7. Barnes, P., Finney,J.L., Nicholas, J.D., and Quinn,J.E.: 1979, Nature 282, 459.
8. Berendsen, H.J.C.: 1972, in *Molecular Dynamics and Monte Carlo Calculations on Water*, Report CECAM Workshop, Orsay, France, p.63
9. Rowlinson, J.S.: 1951, Trans.Faraday Soc. 47, 120.
10. Ben-Naim, A. and Stillinger, F.H.: 1972, in *Structure and Transport Processes in Water and Aqueous Solutions*, Ch.8, Ed.R.A. Horne, Wiley-Interscience, New York.

11. Rahman, A. and Stillinger, F.H.: 1971, J.Chem.Phys.55, 3336.
12. Edam, A., Grigera, J.R. and Berendsen,H.J.C.: 1976, unpublished results.
13. Stillinger,F.H. and Rahman,A.: 1974, J.Chem.Phys.60, 1545.
14. Matsuoka, O., Clementi,E. and Yoshimine,M.: 1976, J.Chem.Phys. 64, 1351.
15. Impey, R.W., Klein,M.L., and McDonald,I.R.: 1981, J.Chem.Phys.74,647
16. Lie,G.C., Clementi,E., and Yoshimine,M.: 1976, J.Chem.Phys.64, 2314; Mezei,M., Swaminathan,S. and Beveridge,D.L.: 1979, J.Chem.Phys. 71, 3366.
17. Ryckaert,J.-P., Ciccotti,G., and Berendsen,H.J.C.: 1977, J.Comp. Phys.23, 327.
18. van Gunsteren,W.F. and Berendsen, H.J.C.: 1977, Mol.Phys.34, 1311.
19. Zeiss,G.D. and Meath,W.J.: 1975, Mol.Phys.30, 161.
20. A full description of the model and its properties will be given elsewhere.
21. Narten, A.H. and Levy,H.A.: 1971, J.Chem.Phys. 55, 2263; Narten,A.H. 1972, personal communication
22. Mills,R.: 1973, J.Phys.Chem. 77, 685
23. See, e.g., work quoted in Eisenberg, D. and Kauzmann, W.: 1969 in *The Structure and Properties of Water*, Clarendon Press, Oxford.
24. Clementi, E., Cavellone, F., and Scordamaglia, R.: 1977, J.Am.Chem.Soc. 99, 5531.

INVESTIGATIONS ON THE ROLE OF ELECTROSTATIC
INTERMOLECULAR FORCES IN LIQUIDS. GROUND STATE
PROPERTIES OF AMIDES IN SOLUTION.

Jean-Louis RIVAIL and Daniel RINALDI
Laboratoire de Chimie théorique (E.R.A. C.N.R.S. n° 22)
Université de NANCY I, 54037 NANCY Cedex - FRANCE

ABSTRACT : The electrostatic interaction of a solute with a great number
non polar solvent molecules is investigated with the help of a model in
which the solvent is replaced by a continuum of low dielectric constant,
which permits detailed quantum chemical calculations. Ground state properties of formamide, acetamide and their N mono and dimethylated derivatives are computed using the MINDO/3 method, especially equilibrium
geometries, rotational isomerism, C=O and NH stretching force constants
and dipole moment derivatives, electronic population, ^{14}N and ^{17}O nuclear
quadrupole couplings.

1. INTRODUCTION : SCOPE OF THE STUDY.

The fact that matter is built up with electrically charged particles of
both signs makes it quite sensitive to an electric field at the macroscopic scale and at the atomic or molecular one as well. At the molecular
level the distribution of charges in space is generally quite intricated
and dependent on the geometry and the electronic structure of the molecule.

When two molecules approach each other, the interaction is classicaly
divided into three terms (1). The first one consists of the electrostatic
energy of each molecule in the electric field created by the other, which
we call the electrostatic term (2). Due to the quantum behaviour of the
electrons, this term is a first approximation which represents rather
well the whole interaction energy at rather long intermolecular distances.
At shorter distances, the intermolecular potential desserves a more
complete treatment developed in the study of Van der Waals forces and
one distinguishes two other contributions to the interaction energy :
dispersion and exchange energies.

In condensed phases, these three parts of the intermolecular potential
are all very important and are responsible for the magnitude and the
variety of the modifications one usually observes on the structure and
the properties of molecules when the data obtained in a low pressure gas

B. Pullman (ed.), Intermolecular Forces, 343–360.

phase and in the liquid or solid state are compared. A complete and accurate theory of these effects is very far from being achieved and one
is still forced to explore the doorstep of this huge and fascinating
domain.

Up to now various approaches have been tried and their results indicate
that Van der Waals interaction energy can be reasonably well split into
atom-atom additive contributions, regardless of the accurate electronic
structure of the molecule. Therefore when no "chemical" interactions
(such as hydrogen bonding or charge transfer) occur between a solute and
the molecules of the solvent, the average value of the sum of the dispersion and the exchange energies may be considered as nearly constant when
the geometry of the solute molecules changes, as far as the overall molecular shape is not too modified. This assumption is probably acceptable
in the case of two rotational isomers of small molecules which never
depart too much from a roughly spherical shape. The comparison of the
intramolecular potentials studied on the isolated molecule and on a molecule in solution is then able to give information on the influence of
the only energy term thought to vary : the electrostatic one.

Besides, the fact that this decomposition in electrostatic plus
Van der Waals terms is theoretically founded incites us to study separately these terms, especially among a chemical series of molecules, and
the electrostatic one appeared to us as the most interesting because of
the possibility of finding some relationships between the behaviour of
molecules in solution and some of their physical (electrical) properties.

But the detailed study of electrostatic interactions in the liquid state
is still a very ambitious aim for many reasons. The solvent is a collection of discrete entities which can create in their surroundings a very
strong and anisotropic electric field, especially when they are highly
dipolar, and the evaluation of their effect on the solute needs a statistical treatment. The solute itself may exhibit a very odd distribution
of electric charges, and this distribution is not fixed : the polarizability of the solute under the influence of the solvent molecule may
modify greatly its properties. Since the electric field acting on this
molecule is expected not to be homogeneous at all one cannot use the
usual dipolar polarizability tensor to reproduce correctly the modification of the electronic distribution. The best way for treating this phenomenon seems to be the computation of the molecular electronic structure
after having introduced into the molecular hamiltonian the external electric potential, provided that this potential is known, at least in some
approximate way. Focusing our attention on the solute molecules, we chose
to appeal to a model in which the solvent is represented by a continuum.
These models have now been extensively studied (3-9). They are liable to
a mathematically rigorous treatment and the solute-solvent interaction
energy has been shown to represent the electrostatic contribution to the
Gibbs free energy of solvation (10). In addition it seems important that
any local modification occurring in the electronic structure of the solute
should be taken into account in the study of the interaction. One of the

easiest way for doing this is to use a multipolar expansion of this interaction energy, which can be developed as far as useful (when it is conver-
gent) due to the existence of easy recurrence formulae well adapted to
quantum chemical calculations. This choice led us to the very irreatistic
spherical cavity model. We shall try to limit the inconveniences of this
model by chosing a series of small molecules in which the chemically
important pattern is placed in the middle. In addition, in order to make
the continuum representation of the solvent more acceptable we chose a
medium which corresponds to an ensemble of non polar, isotropically polarizable molecules such as carbon tetrachloride or cyclohexane. In such
a medium the electric moments induced by the solute are small but not
randomly oriented. Therefore their interactions are additive to a larger
extent and the electric field created by these induced moments at the
level of the solute molecule (called the reaction field) is far from
being negligible.

Finally, the series of small amides formamide (FA), acetamide (AA) and
their N methylated derivatives : N methyl formamide (NMFA), NN dimethyl
formamide (DMFA), N methyl acetamide (NMAA) and NN dimethyl acetamide
(DMAA), appeared to us as a set of good candidates for this study. Their
characteristic pattern is very polar, very polarizable and quite asymetric
Besides, the fact that this pattern is found in polypeptides and proteins
has incited many scientists, theoreticians and experimentalists to study
some of these molecules considered as models of the peptidic bond. In
particular, several very accurate quantum chemical calculations have been
performed on small amides and their interaction with water molecules have
been thoroughly studied (11-15).

This work does not pretend by any mean to give an accurate account of the
modification introduced on the molecular properties by the electrostatic
interactions with the solvent. Its aim is only to try to bring out the
great trends of the phenomena. Our greatest ambition is to provide all
those who have to deal with liquid phase phenomena with some guideline
which may help them in their interpretations. Lacking this guideline,
the great number of experimental facts related to molecular properties
in solution is rather confusing.

Solvent effects on electronic spectra may be very important and are often
referred to as typical examples of the modifications of the molecular
properties induced by the solvent. This problem is still very complex
because it involves two different electronic states of the molecule. We
chose to focus our attention on the electronic ground state only and the
results presented here are intended to illustrate two kinds of consequences of these interactions.

- the modification of the intramolecular potential (equilibrium geometries,
rotational isomerism, barriers to internal rotation, vibrational spectra)

- the perturbation of the electronic distribution (population analysis,
^{14}N and ^{17}O nuclear quadrupole couplings).

2. COMPUTATIONAL DETAILS.

The quantum chemical application of the spherical cavity model has been described previously (6). The cavity is used to evaluate the electric field due to the polarization of the solvent molecules by the distribution of charges of the solute. The resulting electric potential is added to the electronic and nuclear potential in the hamiltonian. The approximate eigenfunctions of this operator, in the form of L.C.A.O. molecular orbitals are obtained through modified S.C.F. equations corresponding to a free molecule experiencing a non homogeneous electric field. This field depends on the molecular electric moments, on the volume of the cavity and on the electric permittivity of the continuum.

This permittivity has been chosen equal to 3 times the permittivity of a vacuum. This low dielectric constant corresponds to a very dilute solution of a polar substance in a non polar solute such as carbon tetrachloride or any alkane.

The volume of the cavity has been chosen equal to the molecular volume of the amide at room temperature in the liquid state or in a solution. The values used are given in table 1.

Table 1. Volume of the cavities ($\overset{\circ}{A}{}^3$)

FA	NMFA	DMFA	AA	NMAA	DMAA
95.94	97.00	128.50	92.19	122.00	154.44

The quantum chemical method used in the computations has to meet with several prerequisites. In particular, a correct evaluation of the electrostatic interactions demands the charge distribution in the molecule to be realistically described. Therefore the method must at least reproduce correctly the dipole moment, which is the first non zero multipole moment of the charge distribution, and the only one easily obtainable experimentally. In addition the study of the solvent effect on the molecular geometry requires a method allowing an easy optimization of the molecular geometry with respect to all internal coordinates. Finally because of the oversimplified model used to account for solute-solvent interactions, one is led to study a series of molecules and then a large number of computations. For all these reasons, we chose a semi-empirical method : MINDO/3 which reproduces pretty well the experimental as well as the fully optimized *ab initio* geometries (15) on the isolated molecules and predicts dipole moments much closer to the experimental ones than those obtained by standard *ab initio* methods. The results relative to the isolated molecules are given in appendix.

The geometry optimization procedure has been described earlier (16) and is the same as that used in the GEOMO programme (17).

Force constants and dipole moment derivatives have been calculated analy tically by means of an original method (18) based upon a theoretical treatment due to Bratos (19). These calculations need the computation of the first derivative of the density matrix with respect to the coordinates of interest, which is performed by a self-consistent iterative procedure.

Nuclear quadrupole coupling constant have been computed by using the procedure proposed by Dewar *et al.* (20).

3. RESULTS.

3.1. Modifications of the intramolecular potential by electrostatic solute-solvent interactions.

The hypersurface representing the variations of the molecular energy in terms of the internal degrees of freedom may be greatly modified by the interaction of the solute with the surrounding solvent molecules. We shall examine this effect through some aspects.

3.1.1. Modification of the equilibrium geometry.

Although the atoms directly bonded to the amidic nitrogen and carbon atom do not get out noticeably from the plane defined by these atoms and the oxygen one, bond lengths and bond angles are significantly modified under the influence of the solvent. In the whole series one notices that :

- the amidic N-C bond is shortened of a few thousandths of $\mathring{A}$.
- the C=O, N-H and/or N-CH$_3$ bond are lengthened of few thousandths of $\mathring{A}$.
- the XNY angles (X,Y = H or CH$_3$) are reduced by about 1°
- the NCO angle is increased by about 1°.

3.1.2. E-Z conformational equilibrium.

N-methyl formamide and N-methyl acetamide undergo E-Z rotational isomerism. As found previously (12) our results indicate that for the isolated molecules the Z isomer (in which N-H and C=O occupy trans positions) are the most stable by 0.35 kcal/mole and 1.62 kcal/mole respectively. These isomers have the greatest dipole moment also. Therefore from a short reasoning based upon the dipole moment alone, one would conclude that the stability of this preferred conformation would be improved in a solution. The opposite is observed in the case of N-methyl formamide. The E isomer appears definitively more stable in the liquid state (1.5 kcal lower than the Z isomer) and this for two typical reasons :

1) in both isomers the charge distribution is far from being only dipolar but the importance of the quadrupole and octupole moments is greater in the E one as it appears in table 2.

Table 2. Contribution of the first multipole
moments to the electrostatic solvation energy
of N-methyl formamide (kcal/mole)

	E	Z
dipole	3.99	3.91
quadrupole	2.40	1.16
octupole	1.82	0.83
hexadecapole	0.65	0.36

2) although the dipole moment is smaller in the free molecule , the dipolar contribution to the solvation energy is greater for the E isomer
because of a greater polarizability. A variational calculation of this
quantity (21) shows that in both isomers the principal axis of polarizability is almost colinear with the dipole moment, but that the magnitude of the corresponding component of the tensor is about 7 % greater
for the E isomer, the average polarizability being almost independent
of the isomerization. The fact that such a small variation of one component of the polarizability tensor is able to bring a large modification
in the solvation energy gives an idea of the magnitude of the electric
field acting on the solute, even in a low dielectric constant solvent.

The same effects occur in the case of N-methyl acetamide but the fact
that the molecule is larger reduces the magnitude of the solvation energies (especially for higher moments) and the Z isomer remains slightly
preferred (0.3 kcal/mole lower than E instead of 1.62 in the isolated
state).

3.1.3. Modification of the barrier to inertal rotation.

Denoting by Θ the angle between the plane of the molecule and the plane
bissecting the two CNH or CNMe planes the molecular energy has two minim
for Θ = 90 and 270° and two maxima for Θ = 0 and 180°. The interaction
with the solvent modifies the energy vs Θ curve in a way which is difficult to predict without detailed calculations because of the many factor
which may compete. The comparison of formamide and N-methyl acetamide
(figure 1) illustrates this point.

One first notices that the solvation energies are greater in formamide.
This is due to the molecular size, smaller in this case. But the effect
is more pronounced when this molecule is in the twisted configuration
(i.e. when the molecular energy is maximum). Then the barriers are lower
to a noticeable extent (6.2 and 9.7 kcal/mole instead of 11.8 and 12.4
kcal/mole). From the analysis of this fact, it appears that the permaner
moments play a great role in this differential solvation. When Θ = 0, th
molecule has a large dipole moment because of the additive contributions
of the lone pair on the nitrogen atom and of the carbonyl group, and the
solvation energy (12 kcal/mole) is mainly due to this term. When Θ = 180

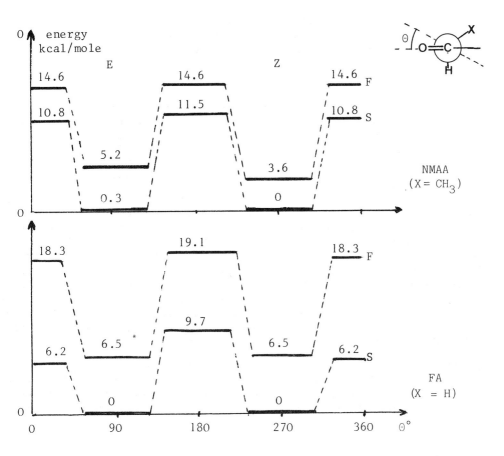

Figure 1. Energies (relative to the most stable rotational
isomer) of the free (F) and the solvated (S) molecules.

the dipolar contribution to the solvation energy is less important but
the higher moments, especially the quadrupole, take a large part to the
phenomenon.

In the case of N methyl acetamide, the largest solvation affects the E
isomer, because of the multipole moments and the large anisotropic polarizability mentionned above. In the other conformations, the solvent effect is of the same order of magnitude : 3 to 4 kcal/mole.

3.1.4. Vibrational spectra.

The Infra-red and Raman spectra of amides in the gas phase, pure liquid
state or in solution have been thoroughly studied (22-25) and have proved

being very rich in information concerning the structure of the molecule and its interactions.

In this work, the force constants and the derivative of the dipole moment have been computed for the C=O and N-H stretching vibrations. The results are given in table 3.

The force constant for the C=O stretching vibration of the isolated molecules is of the same order of magnitude for all the compounds. In the presence of the solvent, our results predict a noticeable decrease of this quantity, as observed experimentally (24,25). Meanwhile, the modulus of the dipole moment derivative is increased to a rather large extent.

The N-H stretching force constants are not very affected by the presence of the solvent although the experimental foundings indicate a slight decrease. One notices that the electrostatic solvent effect seems to have a general tendency to reduce the cis N-H stretching frequencies and to increase the trans N-H ones, but the positive electric charge of the hydrogen atom in primary and secondary amides makes it quite sensitive to small chemical interactions like H bonding. It is therefore difficult to pay greater attention to this remark. The modulus of the dipole moment derivative is in most cases decreased under the influence of the solvent, except in the E conformers of N methylated molecules in which a net increase is observed. Again it is difficult to built a law from these isolated examples in the absence of reliable experimental data.

3.2. Perturbations of the molecular electronic distribution.

The modifications brought by the solvent to the molecular structure can be separated mentally into two steps :

1) a polarization of the electrons by the reaction field
2) a displacement of the nuclei. In order to distinguish the effect of these two steps we have performed SCF calculations in three typical state

1. the fully optimized geometry of the free molecule (F)
2. the solvated molecule constrained to keep the previous geometry (Sc)
3. the fully optimized (or relaxed) geometry in the solvent (Sr).

3.2.1. Population analysis.

The results of a population analysis in the three states mentionned above are reported in table 4 for the O,C,N atoms of the amide pattern. The data illustrate the intramolecular charge transfers induced by the solvent In all cases the most important variation is observed during the first step of solvation imagined in this work (Sc). In the great majority of the cases, the relaxation of the nuclei increases slightly the charge transfers, but the electronic population of the central carbon atom decreases in the first step and increases slightly in the second one. These results illustrate clearly the increasing importance of a charge distribution of the type :

Table 4 : Electronic population and π bond orders

		Q_O	Q_C	Q_N	P^{π}_{N-C}	$P^{\pi}_{C=O}$
FA	F	6.5358	3.3314	5.2226	0.5541	0.7407
	Sc	6.6104	3.3100	5.2183	0.5964	0.7001
	Sr	6.6210	3.3172	5.2051	0.6092	0.6891
NMFA (E)	F	6.5282	3.3453	5.1944	0.5297	0.7529
	Sc	6.6189	3.3143	5.1786	0.5723	0.7083
	Sr	6.6333	3.3211	5.1663	0.5905	0.6930
NMFA (Z)	F	6.5277	3.3450	5.1796	0.5328	0.7522
	Sc	6.5929	3.3259	5.1696	0.5639	0.7197
	Sr	6.6039	3.3311	5.1759	0.5766	0.7093
DMFA	F	6.5228	3.3510	5.1759	0.5087	0.7638
	Sc	6.5853	3.3315	5.1660	0.5387	0.7334
	Sr	6.5958	3.3361	5.1584	0.5528	0.7227
AA	F	6.5692	3.3601	5.2373	0.5386	0.7342
	Sc	6.6326	3.3388	5.2355	0.5713	0.7000
	Sr	6.6404	3.3438	5.2249	0.5813	0.6904
NMAA (E)	F	6.5623	3.3734	5.2114	0.5128	0.7455
	Sc	6.6264	3.3489	5.2021	0.5414	0.7133
	Sr	6.6345	3.3539	5.1941	0.5530	0.7036
NMAA (Z)	F	6.5623	3.3745	5.1955	0.5185	0.7439
	Sc	6.6088	3.3566	5.1944	0.5369	0.7205
	Sr	6.6143	3.3596	5.1886	0.5453	0.7139
DMAA	F	6.5578	3.3825	5.1876	0.4910	0.7569
	Sc	6.6011	3.3656	5.1832	0.5092	0.7357
	Sr	6.063	3.3680	5.1788	0.5164	0.7273

Table 3 : Electrostatic solvent effect on C = O and N - H stretching vibrations.

(1) square root of the ratio of the computed force constants $(\frac{ks}{kf})^{1/2}$

(2) ratio of the experimental frequencies solution in CCl_4/gaz phase

(3) ratio of the computed derivatives of the dipole moment

		$\nu_{C=O}$			ν_{N-H} [a]				
		1	2	3	1 c	1 t	2	3 c	3 t
FA		0.967	(e) 0.989	1.131	1.000	1.000	(c)(d) 0.996 0.993	0.947	0.616
NMFA	E	0.959	0.988	1.171	1.000		(c) 0.995	1.026	
	Z	0.969		1.146		1.000			0.895
DMFA		0.971	(e) 0.992	1.554					
AA		0.972	0.976	1.116	0.998	1.003	(c)(d) 0.999 0.994	0.975	0.864
NMAA	E	0.972	(c) 0.972	1.138	1.000		(c) 0.994	1.127	
	Z	0.980	0.987	1.124		1.004			0.996
DMAA		0.982	(e) 0.982	1.116					

(a)   indices c and t represent NH bonds occupying cis and trans positions respectively to the carbonyl group.

(b)   solution in chloroform ref (24)

(c)   ref (24)

(d)   symmetric and antisymmetric vibrations

(e)   ref (25)

when passing from the free molecule to the solvated species. The interpretation is confirmed by the analysis of the π bond order of the N-C and
C=O bonds. They also fit qualitatively the observed solvent effect on ^{15}N
NMR (26) which would deserve a more detailed experimental and theoretical
study.

3.2.2. Nuclear quadrupole coupling constant.

The nuclear quadrupole coupling constant is a physical quantity which is
extremely sensitive to small perturbations in the electronic distribution
around the nucleus of interest. This property makes it very difficult to
compute accurately. Therefore we do not pretend to predict exactly the
value which would be measured and we shall look at the relative variation of the nuclear quadrupole coupling constants of ^{14}N and ^{17}O isotopes
which are obtained by multiplying the computed electric field gradient eq
computed at the nucleus by the value of the appropriate nuclear quadrupole moment eQ.

The quadrupole coupling is defined by a traceless tensor whose principal
values are $\chi_{xx} = (e^2qQ)_{xx}$; $\chi_{yy} = (e^2qQ)_{yy}$ and $\chi_{zz} = (e^2qQ)_{zz}$. Convention-
naly $|\chi_{zz}| > |\chi_{yy}| > |\chi_{xx}|$. The coupling constant χ is defined by χ_{zz} and
the asymetry parameter by : $\eta = (|\chi_{yy}| - |\chi_{xx}|)/|\chi_{zz}|$

These results are given in table 5.

They cannot be compared with many experimental data. ^{14}N coupling in formamide has been measured in the gas phase by microwave spectroscopy (27).
In the principal axes of inertia the quadrupole coupling tensor has three
diagonal terms with $\chi_c = -3.848 \pm 0.004$ MHz and $\eta \simeq 0.019$. Going to the
principal axes of the nuclear quadrupole coupling tensor, one expects
$\chi_{zz} < \chi_c$. Therefore our results ($\chi = -5.201$ MHz $\eta = 0.091$) are compatible
with this measurement owing to the great uncertainty which affects the
theoretical results, especially when they are obtained with a semi empiri
cal method. No safe liquid state value of this constant can be proposed
but a N.Q.R. study of crystalline formamide gave $|\chi| = 2.274$ MHz and
$\eta = 0.378$ (28). This spectacular modification of the coupling is usually
explained by the existence of H bonds in the crystal. In order to check
this assumption we have performed a computation on a trimer (see figure 2
chosen by analogy with the cristalline pattern of acetamide (29), which
gave $\chi = -4.906$ MHz and $\eta = 0.158$. The effect is of the same order of
magnitude as the electrostatic solvent effect and both phenomena probably play a role in the solid. But the only other available experimental
result concerns ^{14}N coupling in NN dimethyl formamide. In this case hydro
gen bonding on nitrogen cannot be invoked to the same extent as in formam:

Table 5 : ^{14}N and ^{17}O nuclear quadrupole coupling

			FA	NMFA E	NMFA Z	DMFA	AA	NMAA E	NMAA Z	DMAA
^{14}N	χ	F	− 5.201	− 5.049	− 5.074	− 4.940	− 5.230	− 5.065	− 5.090	− 4.975
		Sc	− 4.896	− 4.829	− 4.883	− 4.802	− 4.995	− 4.920	− 4.977	− 4.894
		Sr	− 4.821	− 4.741	− 4.823	− 4.743	− 4.938	− 4.868	− 4.945	− 4.866
	η	F	0.091	0.100	0.085	0.062	0.086	0.136	0.084	0.084
		Sc	0.143	0.123	0.116	0.085	0.119	0.142	0.103	0.088
		Sr	0.156	0.136	0.131	0.098	0.344	0.145	0.110	0.095
^{17}O	χ	F	−10.010	− 9.749	− 9.735	− 9.498	− 9.960	− 9.649	− 9.727	− 9.375
		Sc	−10.143	− 9.809	− 9.845	− 9.584	−10.073	− 9.732	− 9.813	− 9.454
		Sr	−10.478	−10.219	−10.137	− 9.881	−10.344	−10.011	−10.002	− 9.638
	η	F	0.666	0.716	0.717	0.765	0.733	0.783	0.780	0.842
		Sc	0.573	0.612	0.638	0.687	0.645	0.694	0.713	0.776
		Sr	0.545	0.568	0.608	0.651	0.616	0.660	0.690	0.753

and the values $|\chi|$ = 2.81 MHz and η = 0.48 (30) are quite comparable
with the previous ones. Since the computed coupling constants are not
very different for the free molecules - 5.2 and - 4.8 MHz, one expects
for the gas phase coupling constant of NN dimethyl formamide a value of
the order of - 3.2 MHz. Thus the variation of this quantity when passing from the gas to the solid state is still greater than our prediction. To our opinion the strongly anisotropic electric field found in
molecular crystals is responsible for this difference for two reasons :
the electronic structure is more perturbed in a solid than in a liquid
and the contribution of the crystal field to the electric field gradient
at the nucleus is far from being negligible. In our calculations the
contribution of the reaction field to this gradient is neglected.

To conclude with ^{14}N quadrupole coupling, it can be stated, with some
confidence, that electrostatic interactions in the liquid state modify
noticeably the coupling constant and the anisotropy parameter, to an
extent comparable with that due to hydrogen bonding. The actual values
in the liquid state are probably intermediate between the couplings
measured in the gas phase and in the solid state.

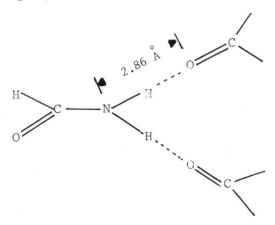

Figure 2. Model of H bonded formamide in the crystal

To our knowledge, there is no experimental ^{17}O nuclear quadrupole coupling
constant available in the case of amides. Experimentalists sometimes use
the value measured in the gas phase on formaldehyde (31) which indicates
a very anisotropic coupling (η = 0.694) and a positive coupling constant
(χ = 12.37 MHz). Our results are in agreement with this anisotropy but
the coupling constants are all negative. In order to check our method
on this last point, we have performed the same computation on formaldehyde and the results are : X = 8.89 MHz and η = 0.636. The comparison
gives an idea of the poor accuracy of the computation but the change of
sign when passing from formaldehyde to formamide becomes more credible.

A comparative examination of the whole set of results shows several characteristic features. The magnitude of the solvent effect is larger in
small molecules (formamide) than in the bigger ones. In our model, this

effect is a direct consequence of the radius of the cavity which governs the reaction field. But this result seems physically meaningful since the nitrogen atom, and to a smaller extent the oxygen one are more protected by the methyl groups fixed on the molecule. Great care has been taken in order not to overestimate the effect of the solvent. The spherical cavity prevents the continuum from being too close to the atoms in the middle of the molecule, and the radius of this cavity has been chosen at a reasonable upper limit. Therefore the magnitude of all the effects is probably a lower limit and the electrostatic solvent effect on ^{14}N and ^{17}O nuclear quadrupole coupling in amides is probably rather important. Regarding the direction of the effects, one notices that in all cases the electrostatic interactions decrease the modulus of the ^{14}N quadrupole coupling constant and increase the anisotropy parameter, and that ^{17}O behaves exactly at the opposite.

These observations, together with the relatively small effect of the nuclei relaxation compared with the electronic polarization, are consistent with the modifications of the electronic distribution observed above

4.   GENERAL DISCUSSION.

Whatever the degree of confidence in the model be, a qualitative conclusion emerges from this investigation concerning the importance of electrostatic interactions between a polar solute and the non polar molecules of a low dielectric constant solvent. These interactions modify greatly most of the molecular properties. In the case of amides studied here, they appear to be responsible, at least in part, for the noticeable decrease of the C=O stretching force constant and for important changes in ^{14}N and ^{17}O nuclear quadrupole coupling constant and asymetry parameter. As expected these effects increase when increasing the dielectric constant of the solvent. Table 6 summarizes the evolution of some properties of formamide when passing from the isolated molecule to a solvent of dielectric constant 3 and then 78.

Table 6. Variation of some properties of formamide when increasing the solvent polarity.

Relative dielectric constant	$\nu_{C=O}$ force constant a.u.	dipole moment derivative $D.\text{Å}^{-1}$	^{14}N coupling X/MHz	η	^{17}O coupling X/MHz	η
1	0.950	4.236	− 5.201	0.091	− 10.010	0.666
2	0.884	4.789	− 4.821	0.156	− 10.478	0.545
3	0.815	5.362	− 4.389	0.235	− 10.877	0.434

All these quantities undergo large variations under the influence of the solvent, but the magnitude of these variations is of the same order when passing from the free molecule to a non polar solvent than when the dielectric constant of the solvent is raised from 3 to 78. This well known behaviour illustrates the importance of having good gas phase data and experimental values obtained in a non polar inert solvent better than in a highly polar liquid to get information on electrostatic interactions. Indeed, solvents of high dielectric constant may be misleading for at least two reasons : it is often difficult to consider them as inert and the importance of the electric field in the neighbourhood of large dipole moments makes the use of a continuum model very questionable.

From a purely qualitative point of view, the main conclusion which can be drawn from our results regards the importance of the electronic polarizability of the solute, especially the intramolecular charge transfers. These electronic displacements cannot be described completely by using the usual dipolar polarizability tensor due to the anisotropy and the strength of the local electric field in liquids which necessitates the use of multipole polarizabilities and hyperpolarizabilities (2). These quantities can now be computed with some accuracy on an isolated molecule (32) and be added to the set of molecular characteristics which is usually used to predict the behaviour of a molecule interacting with the solvent's ones. But like the multipole moments, these quantities are defined at a reference point and they do not describe very accurately the electronic behaviour far apart from this point. Therefore such a description of the solute is expected to be poor in the case of large prolate molecules. It is then possible to replace it by a set of local multipole expansions of the permanent moments and polarizabilities (3) but the partitionning of an electronic structure is still a very questionable operation.

This discussion shows that it seems very difficult to describe correctly the effect of intermolecular forces in liquids on many molecular properties of great interest for a chemist or a biochemist by considering a molecule as a classical body and using an approximate intermolecular potential in a simulation of the liquid by a Monte-Carlo or molecular dynamics technique. The assumption of transferability of the intermolecular potential computed with great details on a pair of molecules and which has proved being very useful in predicting structural of thermodynamical properties of liquids and solutions (33) may be very bad for the prediction of the effect of the surroundings on some local molecular properties because the polarization effects may change greatly the local electronic density.

Quantum chemistry is the only approach which avoids this oversimplification of the molecular reality. But it is still difficult to include properly the effect of the surroundings in the computation except, probably, in some cases when strong and well defined interactions (such as hydrogen bonds) occur. And even in this case, the electrostatic long range forces which concern a very large number of molecules are present and they are not well described by a small aggregate. We chose to emphasize this point and to simplify the description of the liquid in order to be able to pay great attention to the solute molecule. We therefore considered a "true" molecule in a model liquid. The opposite approach, represented by the present state of Monte-Carlo or Molecular dynamics computations, can be sketched as a study of model molecules in a "true" liquids. We all long for the theoretical tools wich enable us to deal with true molecules in true liquids.

REFERENCES.

(1) Claverie, P. : in Intermolecular Interactions : from Diatomics to Biopolymers, ed.Pullman, B.(Wiley, New-York 1978), pp. 69-305

(2) Buckingham, A.D. : 1967, Adv.Chem.Phys., 12, pp. 107-142
     Buckingham, A.D. : in Intermolecular Interactions : from Diatomics to Biopolymers, ed.Pullman, B.(Wiley, New-York 1978), pp.1-67

(3) Huron, M.J., and Claverie, P. : 1972, J.Phys.Chem., 76, pp.2123 ; 1974, J.Phys.Chem., 78, pp. 1853-1862

(4) Rinaldi, D., and Rivail, J.L. : 1973, Theor.Chim.Acta, 32, pp.57-70

(5) Tapia, O., and Goscinski, O. : 1975, Mol.Phys., 29, pp.1653

(6) Rivail, J.L., and Rinaldi, D. : 1976, Chem.Phys., 18, pp. 233-242

(7a) Hylton-Mc Creery, J., Christoffersen, R.E., and Hall, G.C. : 1976, J.Am.Chem.Soc., 98, pp.7191

(7b) Burch, J.L., Raghuveer, K.S., and Christoffersen, R.E. : in Environmental effects on Molecular structure and properties, ed.Pullman, B.(Reidel, Dordrecht 1976), pp.17

(8) Tapia, O., Poulain, E., and Sussman, F. : 1975, Chem.Phys.Lett., 33, pp.65
     Tapia, O., Poulain, E. : 1977, Int.J.Quant.Chem., 11, pp.473
     Lamborelle, C., and Tapia, O. : 1979, Chem.Phys., 42, pp.26

(9) Miertus, S., Scrocco, E., and Tomasi, J. : 1981, Chem.Phys., 55, pp.117

(10a) Claverie, P., Daudey, J.P., Langlet, J., Pullman, B., Piazzola, D., and Huron, M.J. : 1978, J.Phys.Chem., 82, pp. 405-418
(10b) Claverie, P. : in Quantum Theory of Chemical Reactions (Vol. III) ed. Daudel, R., Pullman, A., Salem, L., and Veillard, A. (Reidel, Dordrecht) to be published

(11a) Dreyfus, M., Maigret, B., and Pullman, A., : 1970, Theor.Chim.Acta, 17, pp.109
(11b) Dreyfus, M., and Pullman, A. : 1970, Theor.Chim.Acta, 19, pp.20

(12) Perricaudet, M., and Pullman, A. : 1973, Int.J.Peptides Prot.Res., 5, pp.99

(13) Pullman, A., Alagona, G., and Tomasi, J. : 1974, Theor.Chim.Acta, 33, pp.87

(14) Ottersen, T. : 1975, J.Mol.Struct., 26, pp.365
(15) Fogarasi, G., Pulay, P., Török, F., and Boggs, J.E. : 1979, J.Mol. Struct., 57, pp.259

(16)   Rinaldi, D. : 1976, Comput.and Chem.,1,pp. 109

(17)   Quantum Chemistry Program Exchange n° 290

(18)   Rinaldi, D., and Rivail, J.L. : J.Chim.Phys. (to be published)

(19)   Bratoz, S. : in Calcul des Fonctions d'onde moléculaires - Colloques internationaux au CNRS - LXXXII (CNRS, Paris 1958), pp. 287

(20)   Dewar, M.J.S., Kollmar, H.W., and Sung Ho Suck : 1975, J.Am.Chem. Soc., 97, pp. 5590

(21)   Rinaldi, D., and Rivail, J.L. : 1974, Theor.Chim.Acta, 32, pp. 243

(22)   Rey-Lafon, M., Lascombe J., and Josien, M.L. : 1963, Ann.Chim., 8, pp. 493

(23a) Fillaux, F., and de Lozé, C. : 1972, Biopolymers, 11, p.2063 ; 1972, J.Chim.Phys., 69, pp. 36
(23b) de Lozé, C., Baron, M.H., and Fillaux, F. : 1978, J.Chim.Phys., 75, pp. 631

(24)   Cutmore, E.A., and Hallam, H.E. : 1969, Spectrochim.Acta, 25.A, pp. 1767

(25)   Guiheneuf, G. : 1974, Thesis (Université de Nantes)

(26)   Martin, G.J., Bertrand, T., Le Botlan, D., and Letourneux, J.M. : 1979, J.Chem.Res., 5, pp. 408

(27)   Kukolich, S.G., and Nelson, A.C. : 1971, Chem.Phys.Lett., 11, pp.383

(28)   Guibé, L., and Lucken, E.A.C. : 1978, C.R.Acad.Sci., 263.B, pp.815

(29)   Senti, F., and Harker, H. : 1940, J.Am.Chem.Soc., 62, pp. 2008

(30)   Guibé, L. : Personal Communication

(31)   Flygare, W.H., and Lowe, J.L. : 1965, J.Chem.Phys., 43, pp.3645

(32)   Rivail, J.L., and Cartier, A. : 1978, Mol.Phys., 36, pp. 1085
       Rivail, J.L., and Cartier, A. : 1979, Chem.Phys.Lett., 61, 469
       Cartier, A. : Theor.Chim.Acta (in the print)

(33)   See for instance Computer Modeling of Matter, ed. Lykos,P.,A.C.S. Symposium series n° 86 (American Chemical Society, Washington 1978)

APPENDIX

Fully optimized geometries of amides obtained by the MINDO/3 method.
Comparison with *ab initio* results.

	FA [a]	NMFA [a]	DMFA	AA [a]	NMAA [a]	DMAA
C=O	1.207 (1.215)	1.206	1.205 (1.219)	1.204	1.220 (1.218)	1.219	1.218 (1.222)	1.217
C-N	1.334 (1.356)	1.352	1.351 (1.351)	1.370	1.348 (1.364)	1.369	1.365 (1.358)	1.389
$C-X_1$	1.016 (0.995)	1.028	1.412 (1.465)	1.432	1.017 (0.994)	1.029	1.411 (1.464)	1.437
$C-X_2$	1.015 (0.992)	1.410	1.027 (0.993)	1.434	1.017 (0.992)	1.411	1.031 (0.992)	1.435
C-Y	1.140 (1.083)	1.139	1.141 (1.084)	1.141	1.499 (1.519)	1.498	1.501 (1.519)	1.502
<O=C-N	126.6 (125.1)	125.8	127.2 (124.2)	127.4	120.2 (122.6)	118.6	121.1 (121.9)	119.9
$<C-N-X_1$	125.7 (119.3)	117.2	132.5 (120.2)	124.1	125.4 (118.6)	115.4	134.1 (120.4)	123.1
$<C-N-X_2$	123.9 (121.8)	131.0	115.6 (120.1)	109.9	126.3 (122.0)	134.8	116.5 (120.2)	114.8
<N-C-Y	110.0 (112.5)	111.4	109.9 (113.3)	114.8	116.0 (113.8	118.9	115.9 (114.8)	119.9
μ (b)	3.867 (3.71)	3.442	3.792 (3.8)	3.307 (3.8)	3.991 (3.7)	3.408	3.830 (3.7)	3.361 (3.8

(a) data between brackets ( ) refer to fully optimized
   ab initio geometries from ref (15)

(b) ( ) experimental data are gas phase values from
   L. Mc Clellan, Tables of experimental dipole moments
   Vol. 2 (Rahara Entreprises, El Cerrito 1974)

ON THE ROLE OF THE SIGNAL PEPTIDE IN THE INITIATION OF
PROTEIN EXPORTATION

BEDOUELLE Hugues and HOFNUNG Maurice
Unité de Programmation Moléculaire et Toxicologie Génétique,
Institut Pasteur, 28 rue du Dr. Roux, 75015 PARIS (FRANCE)

SUMMARY

Protein exportation is a process whereby at least a portion of a
polypeptide chain is transfered across a lipid bilayer. Most exported
proteins are made as precursors with a N-terminal extension -the signal
peptide- which is removed upon exportation.

We review briefly current ideas about the rôle of the signal peptide.
We discuss in particular the case of bacterial signal peptides and
define a parameter -the Hydrophobic Axis Length (H.A.L.)- which allow
to account for the properties of the mutations presently known in
bacterial signal peptides.

PROTEIN LOCALIZATION AND PROTEIN EXPORTATION

In the living cell, a newly made protein emerging from a ribosome has
to find its way to its own particular destination (1,2,3,4). Some
protein species stay in the cytoplasm, some are secreted outside the
cell, some end up in a membrane, others find their way to structures
such as the mitochondria or the lysosomes. It is now generally believed
that the polypeptide chain, which is the primary product of a structural
gene, determines in the cellular context the final localization of the
mature protein. Depending on the protein, the steps which lead to the
maturation and the localization may take place during or after synthesis.
The nature and the order of these steps may also differ. However, the
processes by which some proteins are secreted through membranes
(secreted proteins) and the processes by which other proteins are
integrated into membranes (integral membrane proteins) share several
properties. In particular, the localization of both types of proteins
(exported proteins) involves the transfer of at least a portion of the
polypeptide chain across a lipid bilayer.

Exported proteins are frequently -but not always- synthesized as precursors with an N-terminal extension called the signal peptide (3).
This extension is cleaved in the course of the localization and
maturation processes. The signal peptide is usually defined as the part

361

of the precursor that is removed upon exportation. One should be aware
that this may be only part of the N-terminal portion of the precursor
needed for exportation (5,6). In fact, despite the increasing attention
devoted to the process of exportation and in particular to the rôle
of signal peptides, our understanding of the molecular mechanisms
involved is very limited. Although the amino-acid sequences of many
signal peptides have been determined, ideas about the relation between
their sequences and functions are still very speculative. We present
here some of these ideas by focusing on the case of bacterial exported
proteins. The reason for this choice is that the rôle of the signal
peptide of bacterial exported proteins has been brought recently
within the reach of genetic analysis by the isolation and characterization of mutations which affect signal peptides (7,8). It should
however be noted that procaryotic and eucaryotic signal peptides are
very similar in their general structure (8,9,10,11). They may, at least
in certain cases, be substituted one to the other and stay functional
(12).

BACTERIAL SIGNAL PEPTIDES

The amino-acid sequences of bacterial signal peptides differ notably
but present striking similar features (fig.1).

i) The extreme N-terminal end constitutes a charged "tail" of two to
eight amino-acids which includes usually one or two basic amino-acids.

ii) The tail is followed by a stretch of uncharged mostly hydrophobic
amino-acids. This hydrophobic "core" has a length of 15 to 18
amino-acids.

iii) The last C-terminal amino-acid has usually a short side chain. It
is mostly ala (12 out of 14 cases) or gly, or ser (cys is also found
in eukaryotic signal peptides).

Several views have been expressed for the mode of action of signal
peptides. One, mostly based on studies with eucaryotic systems,
assumes that the signal peptide allows the binding of the nascent
polypeptide -and subsequently eventually of the ribosomes-to a
proteinaceous receptor in the membrane (3). This view has been
substantiated by the isolation of components of this receptor (13,17).
Another one, based upon studies with the precursor of the coat protein
of phage M13, assumes that the signal peptide is essential for keeping
the nascent protein in a water soluble conformation and that dissolution
of the hydrophobic core of the signal peptide into the lipid bilayer,
together with or followed by its cleavage, triggers conformation
changes which allow the protein to cross the membrane (4). A third
view -the so-called loop model- based on the sequences of signal
peptides proposes that their action involves three successive steps:
i) binding of the positively charged tail to the negatively charged
inner face of the membrane, ii) looping of the hydrophobic core through

```
PROTEIN      CHARGED TAIL                           HYDROPHOBIC CORE

lamB     met met ile thr leu arg lys |leu pro leu ala val ala val ala ala gly val met ser ala gln ala met ala| val asp phe

ompF             met met lys arg asn ile leu ala val ile val pro ala leu leu val ala gly thr ala asn ala| ala glu ile

ompA                 met lys lys |thr ala ile ala ile ala val ala leu ala gly phe ala thr val| ala gln ala ala pro lys

lpp          met lys ala thr lys |leu val leu gly ala val ile leu gly ser thr leu leu ala gly| cys ser ser


malE     met lys ile lys thr gly ala arg |ile leu ala leu ser ala leu thr thr met met phe ser ala ser ala leu ala| lys ile glu

phoA                     met lys |gln ser thr ile ala leu leu pro leu leu phe thr pro val thr lys ala arg| thr pro

araBP        met lys  X  thr lys |leu val leu gly ala val ile leu thr ala gly leu ser  X  gly ala  X| ala glu asn leu

LIVBP    met asn ile lys gly lys |ala leu leu ala gly cys ile ala leu ala phe ser asn met ala leu ala|

LSBP     met lys ala asn ala lys |thr ile ile ala gly met ile ala leu ala ile ser his thr ala met ala| asp asp ile

HisBP                met lys lys |leu ala leu ser leu ser leu val leu ala phe ser ser ala thr ala ala phe ala|

LAOBP               met lys lys |thr val leu ala leu ser leu leu ile gly leu gly ala thr ala ala| ser thr ala

β-lactamase  met ser ile gln his phe arg |val ala leu ile pro phe phe ala ala phe cys leu pro val phe ala| his pro glu


fd major met lys lys ser leu val leu lys |ala ser val ala val ala thr leu val pro met leu ser phe ala| ala glu gly

fd minor            met lys lys |leu leu phe ala ile pro leu val val pro phe tyr ser his ser ala|
```

Figure 1 : Amino-acid sequences of bacterial signal peptides

The signal peptides have been aligned on the last charged amino-acid of the "tail". Cleavage points are indicated by an arrow. The designation of the corresponding protein is indicated in the left column.

Outer membrane proteins : lamB : λ receptor (λrec) (22) ; ompF : protein 1a (Hall, M., Berman, M. and Silhavy T. personal communication); ompA : pompA (11) ; lpp : lipoprotein (11).

Periplasmic proteins : malE : maltose binding protein (MBP) (7) ; phoA : alkaline phosphatase (Inouye, H., Barnes, W. and Beckwith, J. personal communication) ; LIV : leucine isoleucine valine binding protein (21) ; LS : leucine specific binding protein (21) ; His : Histidine binding protein from Salmonella Typhimurium (21) ; LAO : lysine arginine ornithine binding protein from Salmonella Typhimurium (21) ; β-lactamase : penicillinase,product of the amp gene of plasmid pBR322(11).

Inner membrane proteins : fd major and fd minor are respectively the major and minor coat proteins of phage fd (8).

Unless otherwise indicated the proteins are from E.coli. The reference given here for each signal peptide corresponds generally to review papers where the reference of the original determination of the sequence may be found. X means that the residue has not been identified.

the lipid bilayer,iii) cleavage of the signal peptide. A fourth
view includes thermodynamic considerations. It assumes that the free
energy of dissolution of the hydrophobic core of the signal peptide
into the lipid bilayer is critical for its action. This energy of
dissolution includes the transfer from a hydrophilic to a hydrophobic
medium of each amino-acid residue plus the establishement of the
secondary structure of the peptide in the hydrophobic environment. The
secondary structure is usually assumed to be alpha-helical in the
membrane. With such evaluations authors have attempted to predict the
behaviour of a nascent protein of known sequence in presence of a
membrane (16,17). Although involvement of a protein receptor is not
considered in this view (16,17),it is not excluded and could for
example occur at a later step.

Indeed the different views are not always mutually exclusive. In
particular if one envisions the rôle of the signal peptide as including
several steps, it is possible that one view or another prevails
depending on the step. At any rate it is possible to examine what can
be learnt on the relation between sequences and functions from
mutations affecting the sequences of signal peptides.

MUTATIONS IN SIGNAL PEPTIDES : THE NOTIONS OF H.A.L. AND t.H.A.L.

Series of mutations which affect the export process have been isolated
in the bacterial signal peptides of the λ receptor (λrec), the product
of gene lamB (plamB) and of the maltose binding protein (MBP), the
product of gene malE (pmalE). The λ receptor is located in the outer
membrane and the maltose binding protein in the periplasmic region :
both proteins are thus normally exported across the cytoplasmic
membrane of E.coli K12. All the mutations studied affect the hydrophobic
core of the signal peptide. According to the amino-acid changes,the
mutations which are totally included in the signal peptide fall into
two classes (fig.2).
Class I : changes of an uncharged to a charged amino-acid in the
hydrophobic core of the signal peptide. Class I comprises 8 different
mutations corresponding to 16 independant mutational events all
located between residues 14 and 19.
Class II : comprises 3 mutations. Change of a leu to pro at position
10 in the MBP signal peptide and two deletions S78 (residues 10 to 13)
and S60 and S84 (residues 10 to 21) in the λrec signal peptide.

The continuous stretch of uncharged amino-acids which includes the
hydrophobic core of the signal peptide and generally extends slightly
beyond the cleavage site is most probably a critical factor for the
function of the signal peptide. Class I mutations divide this stretch
into two parts. The location of the mutations in the 14-19 region
of the signal peptide is such that they never leave a continuous
stretch of more than 12 uncharged amino-acids. We assume that if class
I mutations have not been found outside of the 14-19 region it is
because they would then not efficiently block export. This would mean

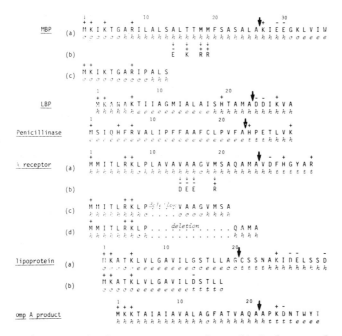

Figure 2 : Amino terminal sequences and predicted secondary structures
for six bacterial preproteins and signal peptide mutants. For each
preprotein we have represented the amino-acid sequence using here the
one letter code. Numbering starts at the N-terminal residue. + and –
indicate amino-acid residues with respectively positively and negatively
charged side chains. The sequences have been lined up on the most distal
charged residue of the N-terminal charged tail. The peptide bond cleaved
upon release of the signal sequence is indicated by a thick arrow (↓).
Secondary structure predictions were made according to the Chou and
Fasman method (19). They are written in italics: h,e,t,c, stand for
α-helix, extended or β-sheet, β-turn and aperiodic or random coil
structure, respectively. We have used the most recent available values
for the conformational parameters. This results in predictions which are
sometimes different from those published in earlier works (9). Our
predictions for the lipoprotein is very similar to that published by
Chan et al (23) and Garnier et al (10), who used another method. MBP =
a) wild type sequence b) class I point mutations c) leu → pro mutation
which introduces a random coil region. LBP = (see LSBP fig. 1).
Penicillinase = (see β-lactamase fig. 1). λ receptor = a) wild type
b) class I point mutations c) deletion S78 which induces formation of
a random coil region d) deletions S60 and S84 which do not induce formation of a random coil region but lower the H.A.L. under the threshold
value. Other deletions which extend outside the signal peptide are not
represented here (8). Lipoprotein, product of gene lpp (see fig. 1)
a) wild type b) mutant. ompA (see fig. 1).

that the minimal functional size of the largest continuous uncharged
stretch should be 12 amino-acids.

Such an hypothesis does not account for two of the class II mutations :
S78 leaves a continuous stretch of 15 uncharged residues, the leu → pro
transition leaves a continuous stretch of 18 uncharged residues. One
could say that these two mutations block another step than that affected
by class I mutations. However, certainly the number of amino-acids in
the continuous uncharged stretch is unsufficient to account for its
functions. We thus tried to incorporate other properties of the
hydrophobic core of signal peptides so as to have a single explanation
for all the mutations. The use of the hydrophobic index (18) or of the
free energy of dissolution (16) of the largest uncharged continuous
stretch of amino-acids could not account for the leu → pro mutation
and depending on the procedure used for calculation could or could not
account for the S78 deletion (data not shown).

We defined a parameter which allows to account for the properties of
all the known mutations in signal peptides (11). This parameter -which
we call the Hydrophobic Axis Length (H.A.L.)-makes use of the predicted
secondary structures of the hydrophobic core of signal peptides. It
rests on the following views.

1) The hydrophobic cores of signal peptides are predicted as having
essentially a periodic structure (fig.2). It is for example α-helical
for the malE and lamB products and extended for the lpp product (11).
In all the cases we have examined, the hydrophobic cores of the signal
peptides are in fact devoid of β turns and of extensive random coil
regions, both of which could result in a reversal of the polypeptide
chain on itself (fig.3, e). The hydrophobic core extends thus along
the axis of this periodic structure (fig.3, a).

We call Hydrophobic Axis Length (H.A.L.) of a signal peptide the
physical length of the largest stretch of uncharged amino-acids
measured along the axis of the periodic structure.

Our hypothesis is : the efficiency of exportation is related to the
H.A.L. value ; under a certain threshold value which we call the
t.H.A.L. essentially no export occurs (fig.3).

2) Class I mutations do not change the predicted secondary structure.
We take as H.A.L. the length of the longest of the two hydrophobic
subsegments defined by each mutation. This allows to define a threshold
value (t.H.A.L.) of 18Å -i.e. 12 amino-acids in α-helical structure.

In the case of class II mutations, one (S60) does not change the
predicted secondary structure. It brings the H.A.l. down to 10.5Å
(7 amino-acids in α-helix), well under the t.H.A.L.. The leu → pro
mutation at position 10 in malE, because it induces formation of a
random coil region is susceptible to bring the N-terminal charged tail
in the vicinity of the 14-19 region. This could result in an H.A.L.

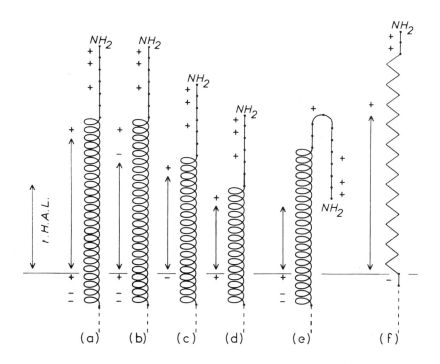

Figure 3 : A schematic illustration of the notion of H.A.L. Residues
are schematically represented in helical (ℓ), extended ($\wedge$) and coil
($\longleftrightarrow$) conformation. + and - indicate amino-acid residues with positively
and negatively charged side chains respectively. The polypeptide chains
have been aligned on the first charged amino-acid at the C-terminal
end of the longest uncharged segment in the wild type or mutant signal
sequences. ($\longleftrightarrow$) indicates the H.A.L. of this uncharged segment. Export
initiation is assumed to be efficient when the H.A.L. is greater than
the t.H.A.L. value. a, b, c, f = efficient export initiation. d, e, =
no export initiation.   (a) to (e) wild type and mutant N-terminal parts
of the MBP : (a) wild type ; (b) hypothetical mutation Ala → Glu at
position 11 ; (c) hypothetical mutation Ala → Asp at position 22 ; (d)
mutation Met → Arg at position 18 ; (e) mutation Leu → Pro at position
10. (f) mutant N-terminal part of the lipoprotein ; the mutation
Gly → Asp is at position 14. This schema does not prejudge on 1) the
orientation of the signal sequence in the membrane ; 3) the rôle and the
fate of the N-terminal charged tail ; 4) the rôle and the site of the
signal sequence cleavage ; 5) the timing of translocation ; 6) the
rôle of ribosomes.

value below the t.H.A.L. The same argument holds for deletions
S78 and S84 (fig.3, e).

3) One mutation is known in the signal peptide of the lipoprotein. This
mutation, a gly → asp change located at position 14, does not prevent
efficiently export. This is readily explained by the t.H.A.L. hypothesis
since the H.A.L. of the mutant lipoprotein correspond to 8 amino-acid
in an extended structure (28.8Å), and is well over the t.H.A.L.
(fig.3, f).

REMARKS ON THE NOTIONS OF H.A.L. AND t.H.A.L.

1) Interpretation of the H.A.L. and of the t.H.A.L.

The notions of H.A.L. and t.H.A.L. can be incorporated in many different views of the mode of action of the signal peptide. They fit very
well with the idea that the step which is affected by the mutations
involves a direct interaction between the hydrophobic core of the signal
peptide and the apolar part of the lipid bilayer. However this does
not exclude the intervention of a receptor protein at other steps (10)
or even at this step (3,23). They fit also with a loop model or a
direct transfer model as illustrated on fig.4.

The t.H.A.L. we have calculated, 12 amino-acids in α-helix structure,
corresponds to 18Å which can be compared to 25Å to 35Å for the size
of the apolar part of the lipid bilayer. If the periodic structure was
3_{10} instead of α-helical as suggested by others (17), we would calculate
24 Å for the t.H.A.L., very near the minimal size of the apolar part of
the lipid bilayer. It is interesting to mention that all the H.A.L. we
have evaluated, including those for eukaryotic signal peptides with
short hydrophobic cores {insulin, VSV glycoprotein (10)} (data not
shown), are always above the t.H.A.L. we have calculated using the
lamB and malE mutations.

One should observe that the notion of H.A.L. implicates two factors :
the absence of charged amino-acids and the predicted secondary
structure. The predicted secondary structure of a polypeptide chain
reflects in part the hydrophobicity of the component amino-acids. This
can be seen in the correlation which exists between the hydrophobic
indices of the individual amino-acids (18) and the parameters Pα and
Pβ used in secondary structure predictions (19). The models used to
calculate the free energy of transfer of the hydrophobic core from
a hydropholic to a hydrophobic medium either do not take into account
the contribution of the secondary structure (18) or assume that it is
the same for all residues, -i.e. an α-helical structure (16,17).
This may be why they do not allow to account for the class II mutations
which change the predicted secondary structure.

A direct interaction between the hydrophobic core and apolar parts of
the membrane could influence the secondary structure of the signal

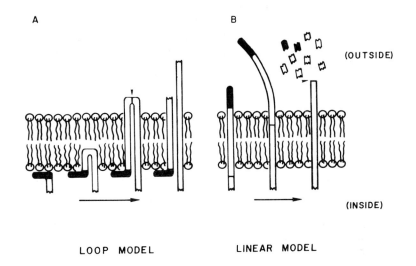

Figure 4 : Two views of the mode of insertion of the signal peptide into a membrane (from ref. 15).

(A) Loop model (B) Linear model. The charged tail of the signal peptide is drawn in black. The hydrophobic core is the following blank segment. The cleavage sites are shown as small arrows.
These views do not refer to the secondary structure of the polypeptide nor to the rôle of other factors such as receptor proteins or ribosomes (see text).

sequence : apolar media are known to promote helix formation at the
expense of both random coil and extended conformations. The hydrophobic
cores that we predict in an extended structure might then adopt a
partially helical conformation which would lower their H.A.L.'s,while
those that we predict   in an α-helical structure should stay α-helical.
However the structure of the hydrophobic cores that we predicted in an
extended conformation could be stabilized by hydrogen-bonded interaction
with integral membrane proteins or homologous signal sequences. The
region defined by the residues 21 to 39 in the fd major coat protein
provides an example of a peptide segment which spans the membrane in
an extended conformation (24). At any rate there is at present very
little experimental evidence on the conformation of the hydrophobic core
of signal peptides in the membrane (20) and theoretical considerations
do not exclude β extended structures (16,17,23).

Finally we should mention that the notion of H.A.L. does not account
for the observation that some residual export exists in class I mutants
(7, 26). We observed that there was a good correlation between the
efficiency of the block and the hydrophobicity of the amino-acid (18,24)
which has been replaced in the 14-19 region. If such a correlation
proves still valid for other mutations,it could indicate the need for
a second order parameter to interpret the effect of the mutations.

2) Predictions issued from the notions of H.A.L. and t.H.A.L.

Among the predictions that can be made for malE and lamB some are
listed below.

a) Other single base pair mutations could affect export by introducing
charged amino-acids in sites of the 14-19 region which have not yet
been affected. From the DNA sequence (7,22) we predict for example
in lamB : gly_{17} → asp or arg ; val_{18} → glu ; and in malE : thr_{16} → lys
or arg.

b) Mutations changing an hydrophobic for a charged amino-acid outside
the 14-19 region should not block export efficiently.

c) Mutations changing an hydrophobic for another hydrophobic amino-acid
should not block export if the secondary structure is unaltered.

d) Deletions in the hydrophobic core of more than 7 amino-acids for
lamB and of more than 6 amino-acids for malE should prevent export.

e) Secondary sites revertants should be found which restore an H.A.L.
over the t.H.A.L. For example insertion of uncharged amino-acids near
the mutational site. In the case of deletion S78,reversion could also
occur by restoration of a periodic predicted structure for the hydrophobic core. The prediction of the random coil region is mainly due
to the fact that deletion S78 brings closer residues pro_9 and gly_{17}.
Mutations substituting   uncharged amino-acids usually found in
periodical regions· shoul restore the exportation of the S78 mutant.

For example : $\underline{gly}_{13} \rightarrow \underline{cys}$ or $\underline{ser}$; $\underline{pro}_9 \rightarrow \underline{leu}$ or $\underline{thr}$ or $\underline{ala}$ or $\underline{ser}$.

CONCLUSION

The initiation of protein exportation is a complex process which involves certainly several steps. At least one of these steps depends on the hydrophobic core of the signal peptide of the exported protein. We introduced the notion of Hydrophobic Axis Length of this hydrophobic core. This parameter seems to be critical for the action of the signal peptide.

The notion of H.A.L. allows to account for the properties of the known mutations in signal peptides and to predict other mutations or secondary site revertants. A minimal value (t.H.A.L.) can be defined for a functional signal peptide.

The step depending on the hydrophobic core and blocked by the mutations can be simply accounted for by a direct interaction between the hydro-phobic core of the signal peptide and the apolar part of the membrane. As already mentionned this view does not exclude the possibility that a protein receptor or other factors are involved. For example supressor mutations which restore the export of the λ receptor with a deletion of 12 amino-acids in the signal peptide (S60) have been found. They have been allocated to a ribosomal protein (25). Such a protein -and possibly   the ribosome- could then play a direct or indirect rôle in the determination of the t.H.A.L. which would be lower in the supressor strain than in the wild type strain.

ACKNOWLEDGEMENTS

We thank Inouye H., Barnes W. and Beckwith J. - Hall M., Bernam M. and Silhavy T. - Landick R. and Oxender D.L. for the communication of results prior to publication and Garnier J. for useful discussions.

H.B. is the recipient of a special stipend from the Institut Français du Pétrole. This work was supported by grants from the C.N.R.S. (ATP N°4248 and 5021) and from the D.G.R.S.T. (ATP N°0664).

REFERENCES

1 - Marx, J.L.:1980, Science,pp. 164-167. 207.
2 - Davis, B.D., Tai, P.C.:1980, Nature,pp. 433-438. 283
3 - Blobel, G.:1980, Proc. Nat. Acad. Sci. USA,pp.1496-1500.
4 - Wickner, W.:1980, Science,pp. 861-868. 210.
5 - Silhavy, T., Bassford, P.J., Beckwith, J.:1979, in "Bacterial outer membrane: biogenesis and function" J.Wiley and Sons,pp.203-254.
6 - Moreno, F., Fowler, A.V., Hall, M., Silhavy, T.J., Zabin, I., Schwartz, M.:1980, Nature,pp. 356-359. 266.

7 - Bedouelle, H., Bassford, P.J., Fowler, A.V., Zabin, I., Beckwith, J.
    Hofnung, M.:1980, Nature,pp. 78-81. 285.
8 - Emr, S.D., Hedgpeth, J., Clément, J.M., Silhavy, T.J., Hofnung, M.:
    1980, Nature,pp. 82-85. 285.
9 - Austen, B.M.:1979, F.E.B.S. Lett.,pp. 308-313. 103.
10- Garnier, J., Gaye, P., Mercier, J.C., Robson, B.:1980, Biochimie,
    pp. 231-239. 62.
11- Bedouelle, H., Hofnung, M.:1981, in "Membrane transport an
    Neuroreceptors", A.R. Liss Inc New York, in press.
12- Talmadge, K., Kaufman, J. and Gilbert, W.:1980, Proc. Nat. Acad.
    Sci. USA, pp. 3988-3992.
13- Warren, G. and Dobberstein, B.:1978, Nature,pp. 569-571.273.
14- Walter, P. and Blobel, G.:1980, Proc. Nat. Acad. Sci. USA,
    pp. 7112-7116.
15- Inouye, M., Halegoua, S.:1980, CRC Critical Reviews in Biochem.,
    pp. 339-371.
16- von Heijne, G.:1980, Eur. J. Biochem.,pp. 431-438.103.
17- Engelman, D.M. and Steitz, T.A.:1981, Cell,pp. 411-422. 23.
18- Segrest, J.P., Feldmann, R.J.:1974, J. Mol. Biol.,pp. 853-858. 87.
19- Chou, P.Y., Fasman, G.D.:1978, Adv. Enzymol.,pp. 45-148. 47.
20- Rosenblatt, M., Beaudette, N.V., Fasman, G.D.:1980, Proc. Acad.
    Sci. USA,pp. 3983-3987.
21- Landick, R., Oxender, D.L.:1981, Membranes and Transport,
    (A. Martonosi ed.) Plenum Press, in press.
22- Hedgpeth, J., Clément, J.M., Marchal, C., Perrin, D., Hofnung, M.:
    1980, Proc. Acad. Sci. USA,pp. 2621-2625.
23- Chan, S.J, Patzeld, C., Duguid, J.R., Quinn, P., Labrecque, A.,
    Noyes, B., Keim, P., Heinrikson, R.L., Steiner, D.F.:1979, in
    "From Gene to Protein : Information transfer in normal and abnormal
    cells" eds Russel, T.R., Brew, K., Gaver, H. and Schultz, J.,
    New York A.C.,pp. 361-378.
24- Nozaki, Y., Chamberlain, B.K., Webster, R.E., Tanford, C.:1976,
    Nature,pp. 335-337. 259.
25- Emr, S.D., Hanley-Way, S., Silhavy, T.J.:1981, Cell,pp. 79-81. 23.
26- Emr, S.D., Silhavy, T.J.:1980, J. Mol. Biol., pp. 63-90. 141.

MONTE CARLO CALCULATIONS OF THE DIMENSIONS OF MODEL PEPTIDES AND PEPTIDE HORMONES RELATED TO ENERGY TRANSFER.

Anne Englert,   Marc Leclerc and Jean Paul Demonte
Laboratoire de Chimie Generale I, Faculté des Sciences,
Université Libre de Bruxelles, B-1050 Brussels,Belgium.

Properties related to non-radiative energy transfer derived from statistical models and semi-empirical potential functions for model peptides, for adrenocorticotropic hormone and for enkephalin analogues in aqueous solution are discussed. Intramolecular conformational energy includes interactions between all atoms ; solvent-peptide interactions are taken into account by rough modifications of potential functions in one of the models and in another model, by including a semi-empirical conformation-dependent hydration energy, as estimated in the literature. The agreement between properties computed with the former model and the experimental values shows that the state of adrenocorticotropin and enkephalin analogues in aqueous solution can be represented by a population of conformers.

1. INTRODUCTION.

The dimensions of molecules of relatively low molecular weight such as peptide hormones can be derived from the efficiency of electronic energy transfer by the Förster mechanism (Förster,1948) if suitably chosen luminophores are attached to distant sites of the molecule. A number of experimental determinations of the efficiencies in various hormones have been reported in recent years (Eisinger,1969 ; Schiller, 1972 ; Schiller et al,1977 ; Schiller,1977 ; Schiller et al,1978 ; Schiller and St-Hilaire,1980). The experiments, performed in aqueous solution, provide valuable information on the state of biological molecules in water, as determined by intramolecular forces and by interactions with the solvent. On the other hand, the conformation of small molecules is amenable to computational analysis by fast computers in terms of semi-empirical potential functions yielding values of various conformation-dependent properties, such as the efficiency of non-radiative energy transfer or NMR coupling constants, which can then be compared with experimental values. Calculations of this type have been performed in our laboratory for adrenocorticotropic hormone (Leclerc et al,1977) and for a number of enkephalin analogues (Demonte et al,1981). These calculations are being discussed in the present paper.

B. Pullman (ed.), Intermolecular Forces, 373–382.

The main question regarding the conformation of many peptide hormones in aqueous solution is whether they adopt a unique low-energy structure or if they exist as a population of conformers. Arguments in favour of both assumptions have been presented in the literature (see below), while the results reported here are derived by Monte Carlo methods (Metropolis et al,1953) for peptides present as a population of conformers. A general discussion of properties related to energy transfer in flexible model peptides with various degrees of polymerisation is first given. In order to take solvent-peptide interactions into account, three different models are considered the unperturbed chain model (Flory,1969) (model(a)); a model, in which interactions between all atoms are included in the calculation of the conformational energy, with rough modification of the van der Waals energy of a pair of atoms when at least one of the atoms in the pair belongs to a hydrophilic residue (Premilat and Maigret,1977)(model(b)); a model, in which the conformational energy is composed of all intra-molecular interactions and of a semi-empirical conformation-dependent hydration energy as estimated by Hodes et al,1979 (Demonte,1981) (model(c)).

The properties of peptide hormones, computed with model (b), are in agreement with the experimental values . However, considering that interactions with solvent are not taken into account explicitely, calculations reported can only serve as a guide for more elaborate models. The aim of the present paper is mostly to draw the attention both of experimentalists and of theoreticians to the wealth of information on molecular dimensions provided by studies of non-radia-tive energy transfer.

2. METHODS.

The conformational intramolecular energy is derived from the semi-empirical potential functions of Scheraga,1968, composed of pairwise van der Waals and electrostatic interaction energies and of torsional potentials around valence bonds. Hydrogen bonds are not included. The dielectric constant is 3.5, except in model (c), where it is 7. In model (b) a distinction is introduced between hydrophobic and hydrophilic residues: the van der Waals energy of a pair of atoms including at least one atom belonging to a hydrophilic residue is set to zero in the range of distances where otherwise it would be attracti-ve (Premilat and Maigret,1977).

A representative sample of N chains is obtained as described previously (Premilat and Hermans,1973; Premilat and Maigret,1977) Chain configurations are constructed from conformational states of individual amino-acid residues, selected from a list of torsion angles (taken in increments of 20° for ψ and ϕ and in increments of 30° for χ) defining local low-energy conformations (3.75 kcal above the minimum) of corresponding peptide units. In model (c) the local con-formational energy includes contributions due to hydration. The total conformational energy of a chain i is equal to the seem of local energies in model (a), to the sum (E_i) of interactions between all atoms and of torsional potentials in model (b), to the sum of E_i (using the unmodified van der Waals potential functions) and of the hydration energy $E_{H,i}$ in model (c).

The average value of a property $<X>$ is obtained from the values of that property associated with each chain in the sample weighted by a probability factor, equal to the Boltzmann factor in model (a) and to N^{-1} in models (b) and (c), where the algorithm of Metropolis is used. In the calculations reported here N is equal to 10.000.

3. DIMENSIONS OF MODEL OLIGOPEPTIDES AND ENERGY TRANSFER.

The efficiency E of non-radiative energy transfer between a donor D and an acceptor A luminophores, separated by a constant distance r_1 is (Förster, 1948)

$$E = (R_o^6/r_1^6) / (1 + R_o^6/r_1^6) \qquad (3.1)$$

where R_o is the Förster's critical distance ; R_0^6 is equal to the product of the orientation factor κ^2 and of a spectroscopic constant. When the luminophores are characterized by single transition dipole moments, the orientation factor κ^2 is

$$\kappa^2 = (\bar{D}.\bar{A} - 3(\bar{D}.\bar{R})(\bar{A}.\bar{R}))^2 \qquad (3.2)$$

where $\bar{D}$, $\bar{A}$ and $\bar{R}$ are unit vectors along the transition dipole direction of D and A and along the D-A separation, respectively ; the values of κ^2 range from 0 to 4. The average value of the orientation factor for luminophores rotating isotropically, $<\kappa^2>$ran is 2/3.

If the D-A separation is not unique, as is the case for a population of conformers, the experimentally determined efficiency is an average value $<E>_s$ or $<E>_d$ corresponding to a static or a dynamic averaging, respectively. When the lifetime of a chain conformation is much longer (shorter) than the lifetime of the excited state of the donor, the averaging regime is static (dynamic).

The efficiencies $<E>_s$ and $<E>_d$ are (Dale and Eisinger,1975)

$$<E>_s = \int_o^\infty \int_o^4 f(\kappa^2, r_1) \; E(\kappa^2, r_1) \; d\kappa^2 dr_1 \qquad (3.3)$$

with

$$E(\kappa^2, r_1) = (R_o^6/r_1^6)/(1 + R_o^6/r_1^6)$$

and

$$<E>_d = <R_o^6/r_1^6> / (1 + <R_o^6/r_1^6>) \qquad (3.4)$$

with

$$<R_o^6/r_1^6> = \int_o^\infty \int_o^4 f(\kappa^2, r_1) \; (R_o^6/r_1^6) d\kappa^2 dr_1$$

In flexible molecules the luminophores are separated by a large number of valence bonds, rotating more or less freely and it is reasonable therefore to assume that the distances r_1 and the orientations κ^2 are not correlated. When this is the case, equation (3.3) becomes

$$<E>_s^{ran} = \int_0^\infty \int_0^{ran} f(r_1) f(\kappa^2) \; E \; (\kappa^2, r_1) \; d\kappa^2 dr_1 \qquad (3.5)$$

where $f^{ran}(\kappa^2)$ is a random distribution of orientations. For luminophores characterized by single transition dipole moments, an analytical expression for $f^{ran}(\kappa^2)$ has been given (Tompa and Englert,1979). For molecules when at least one of the luminophores has a low polarisation $f^{ran}(\kappa^2)$ has been derived in numerical form (Haas et al,1978b). The comparison of $<E>_s$ (3.3) with $<E>_s^{ran}$ (3.5) obtained from models (a) and (c) for Tyr-(Ala)$_4$-Tyr (the luminophore Tyr being characterized by a single transition dipole moment (Ten Bosch and Knopp,1969))as a function of the ratio of the average distance $<r_1>$ to the Förster's critical distance computed with $<\kappa^2>^{ran}$, R_0^{ran}, shows that (3.5) is a satisfactory approximation for $<E>_s$.

$<r_1>/R_0^{ran}$	0.5	0.75	1	1.25	1.5
		Model(a)			
$<E>_s$	.88	.66	.43	.26	.15
$<E>_s^{ran}$	.88	.65	.40	.23	.12
		Model(c)			
$<E>_s$	.88	.69	.50	.36	.25
$<E>_s^{ran}$	.86	.64	.44	.30	.20

On the other hand, when luminophores undergo fast rotational motion, the factor R_0^6 is constant and can be computed with $<\kappa^2>^{ran}$. The averaging regime corresponding to such a dynamic rotational and a static translational averaging is referred to as $<E>_s'$. In flexible chains $<E>_s$ and $<E>_s'$ do not differ very much. The efficiencies $<E>_s'$ computed with models (a) and (c) for chains Tyr-(Ala)$_n$-Tyr equal to 4, 9 and 14 are shown in Figure 1 as functions of $<r_1>/R_0^{ran}$. The efficiency E for a rigid molecule is also shown. The corresponding intraluminophore separations $<r_1>$ and standard deviations σ_1^2 computed with models (a), (b) and (c) are given below. These values in conjunction with computed efficiencies are useful for the interpretation of experimentally determined efficiencies in terms of various models. The efficiencies $<E>_s'$ derived from the unperturbed chain model for chains with n equal to 4, 9 and 14 fall on the same curve (see Figure 1), which can be represented by the equation $ax^m(1 + ax^m)^{-1}$, in which a and m are constants equal to 1.137 and 4.891, respectively and x is $R_0^{ran}/<r_1>$.

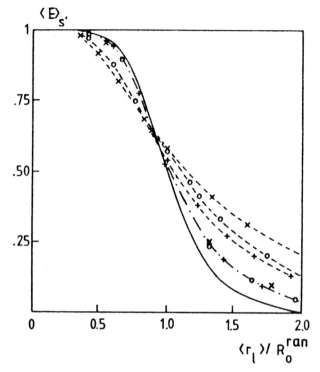

Figure 1. Efficiencies $<E>_{s'}$ as functions of $<r_1>/R_0^{ran}$ computed for Tyr-$(Ala)_n$-Tyr for n=4(+) ; n=9(o) ; n=14(x) with model (a) ($- \cdot - \cdot -$) and model (c) ($- - -$). E ($————$) is given for reference.

n	4	9	14
	Model (a)		
$<r_1>$,(Å)	16.0	26.6	35.7
σ_1^2 , (Å^2)	11.4	29.6	70.9
	Model (b) (hydrophobic chains)		
$<r_1>$,(Å)	8.3	9.4	11.0
σ_1^2 , (Å^2)	12.3	21.6	33.9
	Model (c)		
$<r_1>$,(Å)	11.6	14.0	16.0
σ_1^2 , (Å^2)	18.3	39.7	71.5

Because of its dependance on $\langle r_1^{-6} \rangle$, $\langle E \rangle_d$, the efficiency in the
dynamic regime, is drastically dependent on short D-A separations,
(Leclerc et al,1978) and its value exceeds that of $\langle E \rangle_s$.

It is of interest to underline that the existence of a population
of conformers could be demonstrated by changing the averaging regime
from static to dynamic or vice versa.

The type of averaging is assessed by computing Δr_1, the change in
the D-A separation r_1 during the time interval equal to the lifetime
of the excited state of the donor luminophore, τ_D. This change Δr_1 is
equal to $(D_{tr}\sigma_D)^{1/2}$ (Eisinger et al,1969) where D_{tr} is the translational diffusion coefficient of chain ends, determined recently in peptides
Dns-$(G)_n$-Naph of various degrees of polymerization n. The
repeating unit G is the N^5-(2-hydroxyethyl)-L-glutamine residue and
D_{ns} and N_{aph} are the dansyl and naphtyl groups (Haas et al,1978a).
The diffusion coefficient D_{tr}, which increases with n, is equal to
$5\times10^{-7}cm^2sec^{-1}$ for n equal to 8 in a solvent of low viscosity. Therefore
Δr_1 is 3Å and 13Å for a donor luminophore such as Tyr ($\tau_D \sim$ 2 nsec
(Cowgill,1966)) and N_{aph} ($\tau_D \sim$ 60 nsec (Haas et al,1975)) respectively.
For such a peptide spanned by a donor and acceptor, separated roughly
by 20Å (Haas et al,1975) the averaging regime is static if the donor
is Tyr and dynamic if it is N_{aph}. Decreasing the lifetime of the
excited state of the N_{aph} luminophore by quenchers(1) would also change
the regime from dynamic to static.

4. DIMENSIONS OF PEPTIDE HORMONES AND ENERGY TRANSFER.

4.1. Adrenocorticotropic hormone.

The efficiencies of energy transfer from tyrosine to tryptophan
in aqueous solution have been determined for adrenocorticotropin (1-16)
(ACTH(1-16)) and for ACTH ($Gly_{1,2,3}$ 4-24) (Eisinger,1969). No large
conformational changes have been observed between fragments when
isolated or when incorporated in the fully active ACTH(1-24) (see
Figure 2)

(2)   (9)
H-Ser-Tyr-Ser-Met-Glu-His-Phe-Arg-Trp-Gly-Lys-Pro-
Val-Gly-Lys-Arg-Arg-Pro-Val-Lys-Val-Tyr-Pro(OH)
                        (21)   (23)

Fig.2 Primary sequence of human ACTH(1-24)

The efficiencies of transfer from tryptophan to dansyllysine[21] in
dansylated ACTH(1-24) have also been determined in water and in other
solvents (Schiller,1972). These studies, as well as a number of other
investigations by various experimental methods such as dialysis
(Craig et al,1965), CD (Holladay and Puett,1976) and hydrogen exchange
(Li,1956) indicate that the behaviour of ACTH in solution is best
explained by a mobile equilibrium involving different shapes, although
arguments in favour of the existence of a helical-type structure in

the N-terminal part of the hormone have been put forward as a result of measurements by NMR (Toma et al,1976).

The properties related to energy transfer have been computed using model (b) for fragments as defined by the acceptor and the donor (Leclerc et al,1977). In this model the amino-acid side chains (except for Gly, Lys (Dns), Phe, Pro, Trp and Tyr) have been replaced by a composite atom representing a methyl group.Hydrophilic amino-acids (Zimmerman et al, 1968) do not contribute to the attractive van der Waals energy (see Methods). The full charges of side-chains which are ionized at neutral PH have been decreased by half in order to represent the screening effect of water and counter-ions. The charge distribution on the dansyl side-chain has been computed by a quantum-mechanical program. From the values shown below it is evident that the agreement between experiment and properties computed with model (b) is very satisfactory, considering that the uncertainty on experimental efficiencies is $\pm$ 0.1 and that the statistical error on computed values of $<E>_s$ is $\pm$ 0.03.

Luminophores	Calculated		Experimental			Ref
	$<E>_s$	$<r_1>$ (A)	R_0 (A)	E	r_1 (A)	
Tyr^2-Trp^9	0.47	12.	9.8	0.5	10.	(1)
Tyr^{23}-Trp^9	0.20	24.3	11.0	0.15	>19.	(1)
Trp^9-$Lys(Dns)^{21}$	0.44	21.7	19.4	0.45	20.1	(2)
Trp^9-$Lys(Dns)^{21(3)}$	0.57	19.9	21.9	0.64	20.0	(2)

(1) Eisinger,1969
(2) Schiller,1972
(3) All amino and carboxyl groups protected by Boc and Bu^t groups respectively

On the other hand, transfer efficiencies, derived from the unperturbed chain model (e.g. 0.03 and 0.05 for transfer from Tyr^2 to Trp^9 and from Tyr^{23} to Trp^9, respectively) are far below the experimental values.

4.2. Enkephalin analogues.

The state of enkephalin in water has been interpreted in terms of an ensemble of conformations on the basis of NMR (Higashijima et al, 1979), CD (Spirtes et al,1978) and on the basis of recent investigations by Raman Laser spectroscopy (Han et al,1978), while some of the earlier NMR studies concluded to the existence of a single preferred structure of β-bend type, similar to the one found in non-polar solvents (Anteunis et al,1977). In the theoretical analysis reported here (Demonte et al,1981) all side-chains are represented in full detail and glycine is considered as a hydrophobic amino-acid.

From the table below, showing experimental values determined in water and properties derived from model (b) for three dipolar $[X^2, Trp^4, Met^5]$ enkephalin analogues, it is apparent that the distances r_1 in these molecules are not very different. However, in the inactive $[L-Ala^2, Trp^4, Met^5]$-enkephalin the luminophores are closer than in the active Gly^2 or $D-Ala^2$ derivatives. The efficiencies computed with model (b) are in excellent agreement with experimental values and reproduce perfectly the above trend of luminophore separations.

| X^2 | Computed | | Experimental[1][2] | | | |
	$<T>_s$	$<r_1>$ (Å)	Act[3]	R_o	T	r_1 (Å)
Gly	0.64	9.3	129	10.5	0.71	9.1
D-Ala	0.57	9.6	34	10.2	0.62	9.4
L-Ala	0.78	7.8	2	10.4	0.81	8.2

(1) Schiller et al, 1978
(2) Experimental error less than 0.05
(3) Activity relative to Met^5-enkephalin

While the average distance between the luminophores in Gly^2 and $D-Ala^2$ derivatives are larger than in the analogue with L-Ala at position 2, the average end-to-end distances in either of the former molecules is smaller than in the latter (Demonte et al, 1981). Replacement of Gly or D-Ala by L-Ala brings the luminophores to proximity as a consequence of increased end-to-end distance. These differences in dimensions are related to local interactions of the constituent amino-acid residues, while the stabilization of local conformational states is critically dependent on interactions between distant atoms.

The effieicncies computed with model (a) do not vary significantly with the composition of analogues considered here.

5. CONCLUSION

Transfer efficiencies computed for oligopeptides of various degrees of polymerization can be related to chain dimensions derived from different theoretical models. Excellent agreement between experimental values of transfer efficiencies and the values derived from a statistical model, based on the evaluation of conformational energy from all intramolecular interactions, using rough modifications of potential functions in order to take water-peptide interactions into account is obtained for adrenocorticotropic hormone and for a number of enkephalin analogues. The comparison of experimental and computed values of the efficiency of transfer permits a discrimination between the unperturbed chain model and a model with all interactions included. Despite the fact that the efficiency is determined by the overall conformation of the molecule, its value is sensitive to conformational states of individual amino-acid residues.

The experimental determination of the efficiency appears to be a useful guide for conformational studies.

[1] This suggestion is due to Professor I.Z. Steinberg.

References

Anteunis,M.,Lala,A.R.,Garbay-Jaurequiberry,C. and Roques,B.P.:1977, Biochemistry 16,pp.1462-1466.

Cowgill,R.W.:1966,Biochem.Biophys.Acta 133,pp.6-18.

Coy,D.H.:1978,Biochem.Biophys.Res.Comm. 81,pp.602-609.

Craig,L.C.,Fisher,J.D. and King,T.P.:1965,Biochemistry 4,pp.311-313.

Dale,R.E. and Eisinger,J.:1974,Biopolymers 13,pp.1573-1605.

Demonte,J.P.,Guillard,R. and Englert,A.:1981,submitted.

Demonte,J.P.:1981,in preparation.

Eisinger,J.:1969,Biochemistry 8,pp.311-318.

Eisinger,J.,Feuer,B. and Lamola,A.A.:1969,Biochemistry 8,pp.3908.

Flory, P.J.:1969,Statistical Mechanics of Chain Molecules Interscience,New York,pp.274-286.

Förster,T.:1948,Ann.Phys.2,pp.55-75.

Haas,E.,Wilchek,M.,Katchalski-Katzir,E. and Steinberg,I.Z.:1975,Proc. Natl.Acad.Sci.USA 72,pp.1807-1811.

Haas,E.,Katchalski-Katzir,E. and Steinberg,I.Z.:1978a,Biopolymers 17, pp.11-31.

Haas,E.,Katchalski-Katzir,E. and Steinberg,I.Z.:1978b,Biochemistry 17, pp.5064-5070.

Han,S.L.,Stimson,E.R.,Maxfield,F.R. and Scheraga,H.A.:1980,Int.J. Peptide Protein Res.16,pp.173-182.

Higashijima,T.,Kobayashi,J. and Miyazama,T.:1979,Eur.J.Biochem.97, pp.43-57.

Hodes,Z.I.,Nemethy,G. and Scheraga,H.A.:1979,Biopolymers 18,pp.1565-1610.

Holladay,L.A. and Puett,D.:1976,Biopolymers 15,pp.43-59.

Leclerc,M.,Premilat,S. and Englert,A.:1977,Proc.Vth.Amer.Peptide Symposium (Goodman,M. & Meienhofer,J.,Eds.) Wiley,New York,pp.364-367.

Leclerc,M.,Premilat,S.,Guillard,R.,Renneboog-Squilbin and Englert,A.:
1977,Biopolymers 16,pp.531-544.

Leclerc,M.,Premilat,S. and Englert,A.:1978:Biopolymers 17,pp.2459-
2473.

Li,C.H.:1956,Adv.Prot.Chem. 11,pp.101-190.

Metropolis,N.,Rosenbluth,A.W.,Rosenbluth,M.N.,Teller,A.H. and Teller,
E.J.:1953,J.Chem.Phys. 21,pp.1087-1092.

Premilat,S. and Hermans,J.,Jr.:1973,J.Chem.Phys. 59,pp.2601-2612.

Premilat,S. and Maigret,B.:1977,J.Chem.Phys. 66,pp.3418-3425.

Scheraga,H.A.:1968,Advan.Phys.Org.Chem. 6,pp.103-104.

Schiller,P.W.:1972,Proc.Natl.Acad.Sci.USA 69,pp.975-979

Schiller,P.W.:1977,Can.J.Biochem. 55,pp.75-82.

Schiller,P.W.,Yam,C.F. and Lis,M.:1977,Biochemistry 16,pp.1831-1838.

Schiller,P.W.,Yam,C.F. and Prosmanne,J.:1978,J.Med.Chem. 21,pp.1110-
1116.

Schiller,P.W. and St.Hilaire,J.:1980,J.Med.Chem.23,pp.290-294.

Spirtes,M.A.,Schwartz,R.W.,Mattice,W.L. and Coy,D.H.:1978,Biochem.
Biophys.Res.Comm. 81,pp.602-609.

Ten Bosch,J.J. and Knopp,J.A.:1969,Biochem.Biophys.Acta 188,pp.173-
184.

Tompa,H. and Englert,A.:1979,Biophysical Chemistry 9,pp.211-214.

Zimmerman,J.M.,Eliezer,N. and Simha,R.:1968,J.Theoret.Biol.21,
pp.170-201.

ON THE RELATION BETWEEN CHARGE REDISTRIBUTION AND INTERMOLE-CULAR FORCES IN MODELS FOR MOLECULAR INTERACTIONS IN BIOLOGY.

Harel Weinstein, Sid Topiol and Roman Osman

Department of Pharmacology, Mount Sinai School of Medicine Of the City University of New York
New York, New York   10029

Investigations of enzyme-substrate interactions are of special interest in the application of quantum chemistry to the study of biological mechanisms because the structures and the properties of enzymes have been studied extensively and in detail by many experimental techniques. The basic information is therefore available to elucidate the molecular details of the mechanisms and to understand the specific functional roles of the structural components of the enzyme and of the substrates or inhibitors. Because the intermediate stages of the mechanisms proposed for the function of many enzymes are often not well delineated, theoretical studies can contribute directly to the elucidation of these fundamental processes by combining the information obtained experiment-ally from a variety of sources and by analyzing it in in a unified for-malism. This can be achieved by modeling the structural components and by simulating mechanisms of interaction between the enzyme, and substrates or inhibitors. These calculations can be expected to reveal the nature of the intermolecular forces involved in these interactions and to provide a useful basis both for the comparison of proposed mechanisms and for the description of the roles played by the functional groups. In studies of the enzyme carboxypeptidase we have modeled the interaction of the zinc-containing active site with different ligands in order to elucidate the functional role of the transition metal [1], and have simulated the contribution of other functional groups in the active site to the catalytic process [2]. The analysis of the inter-molecular forces involved in the interaction of the active site model with the ligands indicated the nature of the effect that the various components may have in the process of hydrolysis. The elements of this analysis consisted of the decomposition of the interaction energy, the the analysis of the electron charge redistribution induced by the inter-action, and the evaluation of molecular reactivity characteristics such as the electrostatic potentials.

The interaction of the active site of carboxypeptidase with a model substrate, formamide, was modeled by the system shown in Figure 2. The zinc containing complex $[Zn(NH_3)_2(OH)]^+$ was chosen to represent the portion of the active site in which Zn^{2+} is coordinated to His 69,

B. Pullman (ed.), Intermolecular Forces, 383–396.

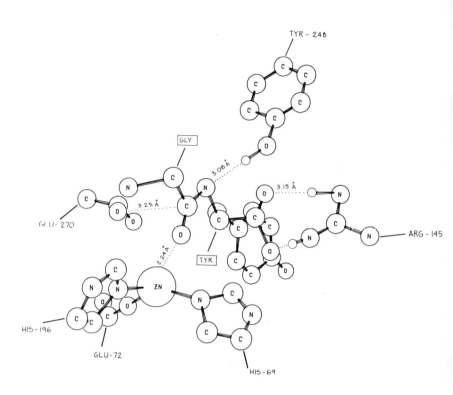

Fig. 1. The structure of Gly-Tyr at the active site of carboxypeptidase A. Coordinates are from the crystal structure of
the enzyme inhibitor complex [6].

His 196, and Glu 72. Other functional groups in the active site of
carboxypeptidase, i.e., the nucleophile (either Glu 270 or an activated
H_2O), and Tyr 248, were modeled by water molecules (Wn and We, respectively). The geometries of these models and the considerations leading
to their choice and their relation to the crystal structure of the enzyme complex with the inhibitor Gly-Tyr (Figure 2) were discussed in
detail elsewhere [1,2]. The results of the simulations with these models defined a sequence of interrelated molecular interactions that delineated specific functional roles for the various constituents of the
active site [2]: We found that the attack on the carbonyl carbon of
the model substrate was indeed facilitated by the polarization of the
substrate bound to the zinc-containing complex. It appeared, however,
from these calculations that the scissile C-N bond is not weakened but,
rather, is strengthened by the binding to the zinc complex. The formation of a hydrogen bond between the model of Tyr 248 (i.e., We in Fig.
2) and the pyramidalized nitrogen in the peptide bond helped dissipate

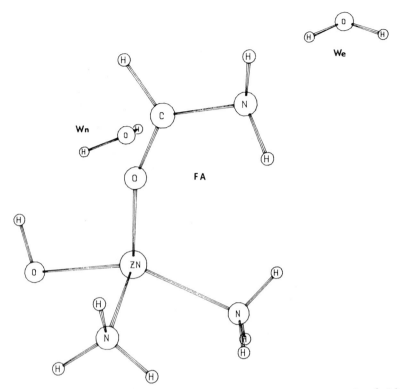

Fig. 2. A model for the interaction of formamide (FA) with models for some components of the active sites of carboxypeptidase shown in Fig. 1: The Zn-containing active site is modeled by $[Zn(OH)(NH_3)_2]^+$, Tyr 248 is modeled by a water molecule--We, and a nucleophile (either Glu 270 or an activated water molecule) is modeled by Wn. The position of the model substrate (FA) relative to the zinc complex is taken from the structure in Fig. 1. The positions of We and Wn relative to FA were optimized as described in [2].

the charge accumulated in the C-N bond and thus helped in its weakening. The results also indicated that the nucleophilic attack on the carbonyl carbon is synergistic with the formation of this hydrogen bond, i.e., that each interaction facilitates the other. The key role that the zinc-containing cationic site has in this sequence of interactions became evident from the simulation. The question of the involvement of the d-electrons in the molecular mechanism arose therefore because the cationic center of the active site contains a transition metal. This question was especially relevant because previous previous theoretical studies of this enzyme used models in which the zinc ion was not specifically considered [3,4,5]. We showed previously that the Zn ion confers on the model active site properties that are not well mimicked by other cationic centers [1]. The question of the specific role of the d-orbitals in the effects of zinc remained, however, unanswered. We describe here an anaylsis of this question and discuss its general implications to the study of intermolecular forces and electron charge redistributions in molecular interactions.

METHODS

The geometries of the molecules and complexes were described in detail [1,2]. The bond lengths and angles were taken from the crystal struc- ture of the carboxypeptidase complex with Gly-Tyr [6], or were opti- mized [2]. (See legends to Figs. 1 and 2). The calculations were performed with the Coreless Hartree Fock Effective Potential (CHFEP) method described in detail elsewhere (see Reference 7 and bibliography therein). Like other effective potential methods this is a treatment of valence electrons only, in which an effective core potential is con- structed and considered to remain unchanged by the interaction of the molecules. This approach greatly simplifies the calculations by reduc- ing the number of orbitals that are explicitly considered in the calcu- lation of electronic structure. The success of this method in faith- fully reproducing ab-initio results from all electron calculations is well documented (see [7] and references therein). The basis set used in all the calculations was the energy optimized, split valence LP-31G basis for first row atoms, and an energy optimized split valence basis for the zinc atom, as described before [2]. The quality of the LP-31G basis and its import to the calculation of the transition metal were discussed [7,8].

For the CHFEP calculations on first row atoms we used the published effective potentials [9,10]. For Zn, we used two types of effective potentials (EP). The EP constructed for the twelve valence electron calculation (VAL-12) replaces the Ar core of Zn and leaves a twelve valence electron configuration composed of the ten 3d orbitals, and the 4s, and 4p orbitals [9,10]. The two valence electron CHFEP calcu- lation (VAL-2) is constructed in the same way, but incorporates the 3d orbitals in the EP leaving only the 4s and 4p orbitals as a valence shell.

The involvement of the d-electrons of zinc in the functional role of the cationic site of the enzyme can be elucidated by direct compari- sons of results obtained with the two models, VAL-12 and VAL-2, in the CHFEP scheme. Thus, the electron density distributions calculated with VAL-12 and VAL-2 are directly comparable if the ground state 3d-elec- trons of the zinc atom are added to the VAL-2 wavefunction. This is possible because VAL-2 and VAL-12 differ in that the VAL-2 has kept a frozen representation of the 3d orbitals of the ground state Zn atom by including them in the core as part of the effective potential. The same valence orbitals are used in the rest of the valence field of VAL-2 and VAL-12. By adding back the ground state 3d orbitals to the VAL-2 wavefunction, the frozen 3d orbital approximation is retained and the d-electrons are explicitly included in the valence region, as in VAL-12. It must be noted that this procedure does not compare VAL-2 results to a simple frozen 3d orbital approximation because the CHFEP method will create small differences due to the local representation of the exchange interaction between 3d and other valence orbitals and due to orthogonalization effects.

The functional counterpoise correction scheme of Boys and Bernardi [11] was used to evaluate the effects of the Basis Set Superposition Error (BSSE). This method uses the basis set of two interacting molecules to calculate each of them. The orbitals centered on molecule B when molecule A is being calculated are termed "ghost functions."

RESULTS AND DISCUSSION

1. The energies of interaction.

The energies of the active site components ([Zn], We, Wn) and of the model substrate (FA) represented in Figure 2 are given in Table 1.

TABLE 1.

SYMBOL	COMPLEX	ENERGY[a]
[Zn]	$[Zn^{2+}(NH_3)_2(OH^-)]^+$	-103.053245 (-39.966124)
FA	$HCONH_2$	-32.775347
[Zn]·FA	$[Zn^{2+}(NH_3)_2(OH^-)(HCONH_2)]^+$	-135.875562 (-72.764491)
We	H_2O; H-bond to N of FA	-16.937029
Wn	H_2O; attack on C of FA	-16.937066

[a] Energies are in hartrees (a.u.). The calculations are with the CHFEP method and the LP-31G basis set. Numbers in parentheses are energy values calculated with the same method but in the VAL-2 approximation with the d-electrons of Zn included in the core EP.

The energies of interaction, calculated with the two schemes of the CHFEP method, are listed in Table 2. The synergism in the interaction of We and Wn with FA bound to the [Zn] complex is evident from these results (see also Reference 2), both in the VAL-12 and in the VAL-2 schemes. The same energies of interaction are predicted by the two schemes for the complexes of Wn and We with the [Zn]·FA complex. However, the stabilization energy of the [Zn]·FA complex is calculated with the VAL-12 scheme (-29.5 Kcal/mole) to be more than twice as large than that calculated with the VAL-2 scheme (-14.4 Kcal/mole).

TABLE 2. Comparison of interaction energies calculated with the Coreless Hartree-Fock Effective Potential (CHFEP) method in the 12 valence electron scheme (VAL-12) and the 2 valence electron scheme (VAL-2)[a]

INTERACTION	VAL-12	VAL-2
[Zn] ... FA	-29.5	-14.4
[Zn]·FA···We	2.1	2.1
[Zn]·FA···Wn	-14.2	-14.1
We···[Zn]·FA···Wn	-13.4	-13.3

[a]Energies are in Kcal/mole

It appears, therefore, that the explicit inclusion of the d-electrons of zinc has a great effect on the calculated stabilization of the complex between the zinc-containing cationic site and formamide. Nevertheless, the energy of interaction of bound formamide with the nucleophile and with the hydrogen bonded water molecule appears to be unaffected by the mode of treatment of the d-electrons on zinc. This lack of sensitivity would indicate that the electronic structure of FA, near the sites at which We and Wn interact, is not affected much by the differences in the schemes used here for the calculation of the [Zn]·FA complex. It is therefore interesting to establish the origin of the VAL-12 and VAL-2, by studying the electronic charge redistribution induced by the formation of the complexes.

2. The electronic charge distributions.

The difference between the electron density distributions calculated for the [Zn]·FA complex in the VAL-12 and the VAL-2 scheme is shown in Figure 3. Due to the functional importance of the interaction between the carbonyl oxygen and the zinc center of the cationic site, this difference map (Fig. 3) was calculated in a plane that contains the atoms in the Zn....O=C bond and is positioned in the perspective shown in Figure 4. Clear differences in the density distributions obtained from VAL-12 and VAL-2 wavefunctions are observable both near the FA component and centered around the Zn. The two main sources for these differences in description by the two CHFEP shemes are: 1) the difference in the description of the component systems, FA and [Zn]; and 2) the difference in the description of the interaction of the component systems. In both cases the origin of the differences is related to the d-electrons of the zinc.

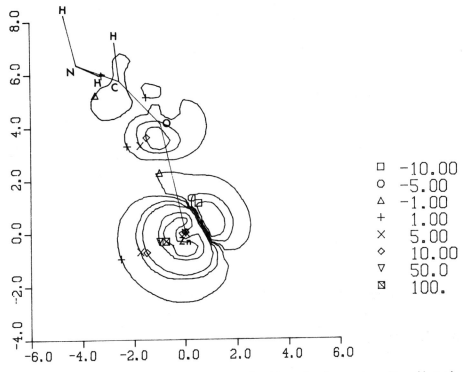

Fig. 3. Map of the difference in the electron density distribution of the [Zn]·FA complex calculated with the VAL-12 and
the VAL-2 schemes [i.e., (VAL-12)-(VAL-2)]. The plane contains the atoms of the Zn...O=C bonds as described in Fig. 4.
Contour values are $n \times 10^4$ electrons/bohr^3.

TABLE 3. Net charges in the constituent groups of the [Zn]
complex calculated with the VAL-12 and VAL-2 schemes of the
CHFEP method.

GROUP[a)	VAL-12	VAL-2
Zn	1.0937	1.0519
$(NH_3)_{69}$	0.1652	0.1763
$(OH)_{72}$	-0.4550	-0.4344
$(NH_3)_{196}$	0.1961	0.2060

a) Group indices refer to the residues in the active site of
   carboxypeptidase.

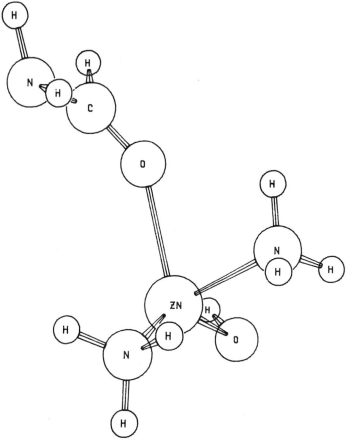

Fig. 4. Spatial perspective on the plane, containing the atoms of the Zn...C=0 bonds, in which the electron density distribution maps are calculated.

The FA component is described identically in the two calculations, VAL-12 and VAL-2. The Basis Set Superposition Error (BSSE) [11] could, however, be different in the two schemes. It is not, as discussed below.

The [Zn] component is described differently by the two schemes and the addition of the ground state 3d-electrons to the density calculated from VAL-2 is necessary, as described in METHODS, in order to compare directly the results from the two calculations. The difference between the electron density distributions calculated for [Zn] with the VAL-12 and with the VAL-2 scheme is shown in Figure 5; it is strikingly similar to the difference between the densities calculated by the two schemes in the [Zn] component of the [Zn]·FA complex (Fig. 3). The major difference is on the zinc atom, as indiated also by the comparison of net atomic charges calculated from VAL-12 and VAL-2 wavefunctions (Table 3) by a Mulliken population analysis [12]. From

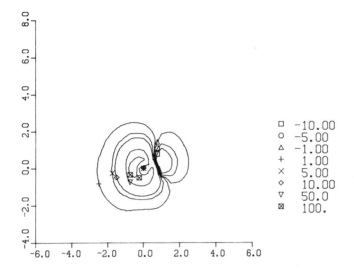

Fig. 5. Map of the difference in the electron density distribution of the [Zn] complex calculated with the VAL-12 and the the VAL-2 schemes [i.e., (VAL-12)-(VAL-2)]. The plane is described in Fig. 4. The contours are $n \times 10^4$ electrons/bohr^3.

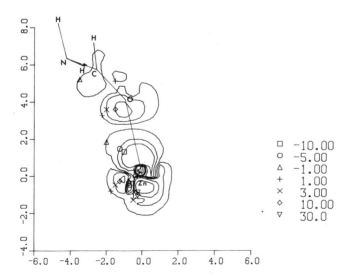

Fig. 6. Map of the difference in the redistribution of the electronic charge due to the formation of the [Zn]·FA complex calculated with the VAL-12 and the VAL-2 schemes (i.e., redistribution with VAL-12 minus redistribution with VAL-2). Plane and contours are as in Fig. 3.

Fig. 5 it is evident, however, that the greatest difference in the description of [Zn] in the VAL-12 and VAL-2 schemes cannot be inferred
from the charges in Table 3 because it consists mainly of the polarization of the valence electrons of the zinc by the ligands in [Zn].
     To learn the effects of the d-electrons on the charge redistribu-
tion induced by the formation of the [Zn]·FA complex we analyzed the
difference between the redistribution maps calculated by VAL-12 and
VAL-2. The nature of the redistribution calculated with VAL-12 was
discussed elsewhere [2]. Figure 6 shows the calculated difference between the redistributions obtained with the two CHFEP schemes. In the
FA part this map is nearly identical to the one in Fig. 3. This indicates that the charge distributions calculated for the [Zn]·FA complex
with VAL-12 and VAL-2 are different (Fig. 3) because the electron charge
redistribution induced in FA by the interaction with [Zn] is not identical in the VAL-12 and VAL-2 calculations. The difference between these
results from the two schemes is overall quite small (it represents only
5 percent of the total redistribution in FA due to interaction with
[Zn]); both from Fig. 6 and from Table 4 it appears to be largest on the
oxygen of formamide.

TABLE 4. Charge redistribution (ΔQ) in formamide (FA) upon
formation of the [Zn]·FA complex. Comparison of results from
the two CHFEP approximations.

ATOM	FA	VAL-12		VAL-2	
		[Zn]·FA	ΔQ	[Zn]·FA	ΔQ
O_{FA}	-0.5045	-0.6544	-0.1499	-0.6184	-0.1139
$(CH)_{FA}$	0.5886	0.7242	0.1356	0.7179	0.1293
$(NH_2)_{FA}$	-0.0841	0.0208	0.1050	-0.0052	0.0789

     Inspection of the [Zn] part of Fig. 6 indicates that in the VAL-2
scheme there is more polarization towards FA than in the VAL-12 calculation. Comparison with Fig. 5 indicates that except in the immediate
vicinity of the Zn....O bond with FA, this difference in the way VAL-12
and VAL-2 describe the polarization by FA is smaller than the difference
in the representation of [Zn] itself by the two schemes. This is also
evident from the following considerations based on a comparison of
Tables 3 and 5: Both schemes give very similar ΔQ values in Table 5.
The difference between these ΔQ values is smaller than the difference
between the net charges predicted by the two schemes for the groups of
the isolated [Zn] (Table 3). As a result, the difference between the
net charges on the [Zn] groups in [Zn]·FA predicted by VAL-12 and VAL-2
(Table 5) is nearly identical to the difference predicted by the two
schemes in the net charges of isolated [Zn] (Table 3).

TABLE 5. Charge redistribution (ΔQ) in the $[Zn(NH_3)_2(OH)]^+$ ion ($[Zn]$) upon complexation with formamide (FA), calculated with the CHFEP method in the VAL-12 and VAL-2 schemes.

| GROUP | VAL-12 | | VAL-2 | |
	$[Zn] \cdot FA$	ΔQ	$[Zn] \cdot FA$	ΔQ
Zn	1.1187	0.0250	1.0704	0.0185
$(NH_3)_{69}$	0.1232	-0.0420	0.1350	-0.0413
$(OH)_{72}$	-0.4968	-0.0418	-0.4779	-0.0434
$(NH_3)_{196}$	0.1642	-0.0318	0.1785	-0.0275

3. The effects of basis set superpositions

The nature of the charge redistributions upon formation of molecular complexes may sometimes be obscured by the spurious effects of charge rearrangement that is solely due to the truncated basis sets used in the calculations (for a recent discussion see [13]). This Basis Set Superposition Error (BSSE) has been characterized and methods for its correction have been proposed [11]. Some examples were recently described in which rather large BSSE effects on the charge distribution in molecular complexes resulted in negligible contributions to the interaction energy due to a fortuitous cancellation of effects [14]. Because in the present discussion we compare molecular complexes calculated with basis sets that are not larger that split valence, and because VAL-12 and VAL-2 differ in the size of the basis sets used, i.e., with and without 3d orbitals, we had to determine whether the BSSE affects the conclusions obtained above. The counterpoise correction was used for this analysis, as described in the METHODS section. A comparison of results in Tables 2 and 6 shows that the contribution of BSSE to the energy of interaction between FA and [Zn] is small and that the difference between the stabilization energies of the [Zn]·FA complex calculated with the VAL-12 and the VAL-2 schemes is practically unchanged by the BSSE. The contribution of the 3d orbitals to the BSSE can be inferred from the fact that the improvement in the energy of FA is small, and independent of whether the "ghost functions" do or do not include the 3d orbitals. Nearly the same can be said for the counterpoise correction to [Zn].

Unlike other cases in which the energy of interaction was insensitive to the BSSE but major changes in the electron density distribution were observed with the counterpoise correction [14], results in Table 7 indicate that the rearrangement of the charge of [Zn] into the FA ghost

TABLE 6. Counterpoise correction of the Basis Set Superposition Error (BSSE) in the stabilization of the [Zn]·FA complex calculated by the CHFEP method in the VAL-12 and VAL-2 schemes.

SYSTEM	BSSE	
	VAL-12	VAL-2
[Zn] + FA ghost functions	-1.77	-0.99
FA + [Zn] ghost functions	-3.01	-2.93
Corrected [Zn]·FA stabilization	-24.69	-10.53

[a] Energies in Kcal/mole

TABLE 7. Effect of Basis Set Superposition Error (BSSE) on the charge distribution in the [Zn]·FA complex calculated with the CHFEP method in the VAL-12 and VAL-2 approximations.

SYSTEM	CHARGE IN GHOST FUNCTION	
	VAL-12	VAL-2
[Zn] + FA ghost function	0.009	0.001
FA + [Zn] ghost function	0.059	0.060
Total charge transfer from FA to [Zn]	0.091	0.093

is very small. The rearrangement of FA into the "ghost functions" of [Zn] is slightly larger, in agreement with the larger effect on the energy (Table 6). Also in good agreement with the BSSE effect on the energy, it is clear from the results in Table 7 that this rearrangement of charge is independent of whether the "ghost functions" contain the 3d orbitals of zinc. It can be concluded, therefore, that the BSSE has very little effect on the differences between the results obtained from the VAL-12 and VAL-2 calculations.

CONCLUSIONS

The results show that the main contribution of the d-electrons of zinc to the nature of the intermolecular interaction between $[Zn(OH)(NH_3)_2]^+$ and $HCONH_2$ comes from their role in the description of the electronic structure of the [Zn] complex. The description of the electronic structure of [Zn] obtained with the d-electrons explicitly included in the valence (VAL- 2), differs from that obtained from the scheme in which the d-electrons are kept in the core (VAL-12). This difference in the description of the electronic structure of [Zn] has a marked effect on the interaction with FA. Although the explicit inclusion of 3d-orbitals seems to affect mainly the electrostatic and polarization components of the interaction between [Zn] and FA, the present findings are consonant with our earlier conclusions [2] that an electrostatic cationic site is not sufficient to reproduce the properties and reactivity of the zinc-containing complex at the active site of carboxypeptidase. We also find here, in agreement with these earlier conclusions, that the effects of the fourth ligand on the active site would not be well represented if the active site mechanism were to be simulated by models in which the contribution of the d-electrons is neglected.

It is important to note, however, that the susceptibility of FA in the [Zn]·FA complex to a nucleophilic interaction at the carbonyl carbon (i.e., by Wn) and to an electrophilic interaction at the pyramidalized nitrogen (i.e., by We) was found not to be sensitive to the representation of the [Zn], i.e., with or without d-orbitals in its valence shell (Table 2). It is thus evident that it is possible to obtain a large effect on the interaction of two components of a complex without a major change in some of their reactivity characteristics. Other reactivity characteristics may, however, be more affected by the representation of the interaction. It is therefore not necessarily the magnitude of the intermolecular forces that must be of primary concern for the accurate simulation of biological mechanisms, but rather the nature of these interaction forces as evidenced by changes induced in the electron density distributions of the reactants.

ACKNOWLEDGEMENTS

This work was supported by the National Institute on Drug Abuse (NIDA) under grants DA-01875 and DA-02534. H. Weinstein is recipient of an Irma T. Hirschl Career Scientist Award and a Research Scientist Development Award (K02 DA-00060) from NIDA. S. Topiol was supported by a departmental training grant T32 DA-07135 from NIDA.

A generous grant of computer time from the University Computing Center of the City University of New York is gratefully acknowledged.

REFERENCES

1.   Osman, R. and Weinstein, H., Isr. J. Chem. $\underline{19}$:140 (1980).
2.   Osman, R., Weinstein, H. and Topiol, S., Ann. N.Y. Acad. Sci. (1981) in press.
3.   Scheiner, S. and Lipscomb, W.N., J. Amer. Chem. Soc. $\underline{99}$: 3466 (1977).
4.   Hayes, D.M. and Kollman, P.A., J. Amer. Chem. Soc. $\underline{98}$: 3335 (1976).
5.   Hayes, D.M. and Kollman, P.A., J. Amer. Chem. Soc. $\underline{98}$: 7811 (1976).
6.   Lipscomb, W.N., Hartsuck, J.A., Reeke, G.N., Quiocho, F.A., Bethge, P.A., Ludwig, M.L., Steitz, T.A., Muirhead, H. and Coppola, J.C., Brookhaven Symp. Biol. $\underline{21}$:24 (1968).
7.   Topiol, S., Osman, R., and Weinstein. H., Ann. N.Y. Acad. Sci. (1981) in press.
8.   Topiol, S. and Pople, J.A., Int. J. Quantum Chem. $\underline{S\ 12}$: 493 (1978).
9.   Topiol, S., Moskowitz, J.W. and Melius, C.F., J., Chem. Phys. $\underline{70}$:3008 (1979).
10. Topiol, S., Moskowitz, J.W. and Melius, C.F., J. Chem. Phys. 68:2364 (1978).
11. Boys, S.F., and Bernardi, F., Mol. Phys. $\underline{19}$:553 (1970).
12. Mulliken, R.S., J. Chem. Phys. $\underline{23}$: 1833 ($\underline{1955}$).
13. Kolos, W., Theor. Chim. Acta Berlin 54:187 (1980).
14. Osman, R., Topiol, S., and Weinstein, H., J. Comput. Chem. $\underline{2}$:73 (1981).

INTERMOLECULAR INTERACTIONS IN AN EXTERNAL ELECTRIC FIELD : APPLICATION
TO THE ANALYSIS OF THE EVALUATION OF INTERACTION ENERGIES FROM FIELD
MASS SPECTROMETRY EXPERIMENTS

by Jacqueline LANGLET, Pierre CLAVERIE and
Françoise CARON
Laboratoire de Biochimie Théorique associé au CNRS
Institut de Biologie Physico-Chimique
13, rue Pierre et Marie Curie
75005 PARIS (FRANCE).

ABSTRACT

After a brief survey of experimental methods available for getting information about molecular complexes in vacuo, we concentrate our attention
upon the use of field mass spectrometry for measuring the dimerization
equilibrium constants. The study of a simple model, namely two interacting dipoles embedded in an external electric field, shows that a strong
enough field may seriously disturb the configuration of the complex, and
therefore its stabilization energy. We then perform a detailed quantitative study for the case of the interactions between nucleic acid bases
(A, T, U, G, C), for which such field mass spectrometric experiments
have been performed. Our theoretical results allow us to propose more
refined interpretations of the experimental data. We conclude that, for
a safe application of field mass spectrometry to the evaluation of the
binding energy of molecular complexes, low field limit values should be
systematically evaluated.

I. INTRODUCTION

Obtaining reliable and detailed information about the interaction potential between two atoms or molecules through experimental methods is a
long-standing problem in the study of intermolecular interactions. The
trouble is that, when we want more and more detailed information about
the two body potential, it becomes more and more difficult to obtain.
Thus, bulk matter properties are easily measured, but the information
that they provide is very indirect, since these properties involve either
a sum of a large number of elementary interactions (case of a crystalline solid), or (still worse !) a plain statistical averaging over the
interaction potential (case of a solid or a gas). Moreover, except for
dilute gases, the molecules are sufficiently close together for making
non-additivity effects possibly non-negligible ; at least, some appropriate analysis should be made for assessing in each case the degree of
validity of the neglect of three-body, four-body ... effects. Therefore,
experimental techniques dealing directly with molecular complexes (which

397

B. Pullman (ed.), Intermolecular Forces, 397–429.

may involve two, or a few, molecules or atoms) are of special interest. Thus, for interactions between atoms (rare gases), scattering experiments have been widely used in order to determine the interaction energy as a function of the interatomic distance (see e.g. the reveiw by Leonas [1]). The method may in principle be applied to interactions involving small molecules instead of atoms, but the intermolecular potential then depends on several variables, and a full derivation of it from the scattering data (which would require a complete solution of the "inverse scattering problem) becomes extremely difficult. Actually, until recently, it is only the orientation-averaged potential (function of the distance only) whose determination was attempted in such cases, and even so, very serious discrepancies may appear between the orientation-averaged potentials deduced from different experiments (see for example the discussion by Kołos et al. [2] concerning the methane-methane interaction). This situation clearly indicates the difficulty of the problem.

Since the experimental determination of the interaction energy hypersurface for two molecules appears very difficult, it is rather natural to focus one's attention to a more restricted information, namely that concerning the minimum (or minima) of the interaction energy, and this corresponds to the study of the so-called van der Waals molecules (see e.g. the reviews [3, 4, 5]), a study which essentially relies upon spectroscopic, electron diffraction, and mass spectrometric methods. One of the main procedures consist in determining the amount of dimers with respect to monomers in the gas phase (the dimers may be identified through spectroscopic, or mass spectrometric, methods [6-8]). The equilibrium constant K_d corresponding to dimerization thus obtained indeed gives some information about the dimer, but it must be emphasized that this information is not necessarily derived in a straightforward way. To be more precise : if we apply the usual equations of chemical thermodynamics we obtain the dimerization enthalpy ΔH_d as :

$$\Delta H_d = - R \frac{d}{d(1/T)} (\text{Log } K_d) \qquad\qquad (I - 1)$$

but the validity of this chemical scheme is limited. The correct general derivation of K_d through a statistical-mechanical treatment has been investigated noticeably by Hill [9, 10], Stogryn and Hirschfelder [11] . For the case of interactions between rare gas atoms or very small molecules, the depth of the potential well is not large with respect to RT, and the simple relationship (I - 1) does not hold, as it may easily checked from the values of the mole fraction of dimers calculated by Stogryn and Hirschfelder [11, table IV] . It must be emphasized that these theoretical values have been found in satisfactory agreement with the experimental values obtained for rare gas dimers [6b]. Despite the above quoted works, this limited validity of the relation (I - 1) does not yet seem very well-known. Fortunately, this relation should actually be vali for strong complexes, such as ion-molecule associations [7] or, to a lesser extent, hydrogen-bonded complexes [8], since the depth of the potential minimum is of at least 10 kcal/mole is all these cases, i.e. much larger than RT. But another problem must be considered, namely the possible perturbing effect of the measuring process upon the molecular complexes involved, since even strong complexes have binding energies significantly weaker than those associated with usual chemical reactions. Thus, when the mole fraction of dimers with respect to monomers is measured by using usual mass spectrometry (i.e. the ionization is provoked by collisions with a suitable electron beam), the two following causes of systematic bias are listed, (among others, less fundamental) by Leckenby and Robbins [6b, p. 392, points 2 and 3] :

(a) the ionization cross section of the dimer is greater than that of the monomer (possibly by a factor of order 2). This factor would modify the value of K_d, but, if it is independent of the temperature T, this will not affect the value of ΔH_d.

(b) the ionization process may cause dissociation of too weak complexes, due to the energy transfered from the colliding electron to the complex (molecular recoil energy $\Delta E = (m/M)E$, where m is the electron mass, M the mass of the collided molecule, and E the kinetic energy of the electron). For E = 50 eV and M = 40, ΔE = 0.016 kcal/mole [6b], which is, fortunately, rather small.

This example shows that a careful analysis of the experimental procedure is really necessary. Let us now consider the field ionization mass spectrometric method, used by Yanson et al. [8] for studying the complexes between nucleic acid bases. Here, the ionization is due to a strong electric field, which provokes the tunneling of an electron away from the molecule. The electric field is created by a thin electrode (radius of about 1000 Å) supplied with a high potential (1.3 to 3 kV) (further details may be found in the monographs by Beckey [12, 13], and references therein). The possible causes for systematic bias are rather different : the ionization probability per unit time (inverse of the mean life-time τ of the neutral molecule), denoted by P, is roughly given by [12, chap. 1, p. 3, eq. (I. 4)] :

$$P = 1/\tau \simeq 10^{16} \exp (- 0.68 \ I^{3/2}/F) \qquad (I - 2)$$

where P is expressed in s^{-1}, τ in s, I (ionization potential of the neutral atom or molecule) in eV, and F (electric field) in V/Å (a more accurate treatment may be found in [14]). Thus this ionization probability will not vary from molecules to complexes in the same way as it did in the previous case (ionisation by collision). Moreover, the problem of the possible dissociation of the complex due to the recoil energy now disappears. But it is replaced by another problem, possibly even more serious from the quantitative point of view, namely the disturbance of the molecular complex (as concerns both its energy and configuration) due to the strong external electric field, a disturbance which cannot be excluded as soon as the molecules involved have a non-zero dipole moment. Indeed, every dipolar molecule will tend to orient in such a way as to have its dipole $\vec{\mu}$ parallel to the field $\vec{F}$, and consequently two such molecules would tend to have their dipoles $\vec{\mu}_1$ and $\vec{\mu}_2$ parallel to each other. Therefore, if the optimum configuration of the complex in the absence of the electric field corresponds to non parallel dipoles, we must expect a disturbing influence of the external electric field $\vec{F}$, as

soon as it becomes non negligible with respect to the field associated
with the intermolecular interaction, i.e. the field created by one molecule at the position of the other. Now, what are the orders of magnitude
involved ? The external field F, in the neighbourhood of the electrode,
is of the order of 0.1 to 1 V/Å [8, 12, 13], i.e. very nearly 0.002 to
0.02 a.u., and the field created by a dipole μ = 1 a.u. (i.e. 2.54 Debye) at a distance R = 6.4 a.u. (nearly 3.4 Å, which is a typical distance between stacked molecules) in a direction perpendicular to the
dipole, is μ/R^3 = 0.0038 a.u., near the lower end of the previous interval. Thus, the values of the external electric field used in field mass
spectrometry definitely are of the same order of magnitude as the intermolecular electric fields in complexes, about 0.01 a.u.. This situation
is quite natural : if we look at the formula (I - 2) for the life-time
τ, we see that, in order to have τ = 1 a.u. = 0.243 X 10^{-16} s, we must
have F of the order of 1 a.u. (in the case of the hydrogen atom, an
exact treatment [15] gives τ = (F/4) exp (2F/3), with τ and F expressed
in a.u.) ; if we want to have non negligible ionic current J = NP, where
N is the number of neutral molecules in the ionization zone [8], we cannot allow too large values of τ, i.e. too weak values of F (in atomic
units).

Thus the conclusion is clear : under the conditions usual in field mass
spectrometry, the disturbing effect of the electric field upon complexes
of dipolar molecules cannot be neglected. A striking confirmation of
this statement may actually be found in the paper by Yanson et al. [8,
p. 1161-1162] : for the Cytosine-Cytosine dimer, they found markedly
different values of ΔH_{CC} corresponding to different values of the electrode voltage V_e (and therefore different values of the field F), namely
for V_e = 3.0 kV, ΔH_{CC} = 12 kcal/mole ; for V_e = 2.0 kV, ΔH_{CC} = 16 kcal/
mole ; for V_e = 1.3 kV, ΔH_{CC} = 17 kcal/mole.

Obtaining such differences is in complete agreement with our qualitative
analysis presented above. In actual fact, Yanson et al. [8, p. 1162]
rightly state that the correct value of the binding energy should be obtained by plotting ΔH_d as a function of F (or in practice V_e), and taking
the limit for zero field. But it is not clear to which extent they carried out such a procedure : they state that no dependence of ΔH on V_e
was observed for the complexes A-U, U-U and U-T, but nothing is stated
for G-C. On another hand, Verkin et al. [16] made a similar study concerning methylated nucleic acid bases (which should give stacked complexes
rather than hydrogen bonded ones), and gave no information about an analysis in terms of a varying V_e (and F). Now, they obtain rather low values of ΔH_d, with a non-negligible dependence on the nature of the metal
of which the electrode is made (Platinum (Pt) or Tungsten (W)) :

$m^{1,3}$ U - $m^{1,3}$ U : - 2.72 (Pt) and - 3.56 (W)
$m^{1,3}$ U - $m^{1,3}$ T : - 2.9 (Pt) and - 3.92 (W)
$m^{1,3}$ T - $m^{1,3}$ T :   - 3.9 (W)

Such a dependence on the nature of the electrode should obviously disap-
pear in the zero field limit, and the marked dependence observed here

suggest that we are far from this limit. But the example of the C-C dimer mentioned above then suggests that the above quoted values of ΔH_d for the methylated bases could be too low simply because they would not correspond to the zero-field limit.

We recently achieved a theoretical study of the interaction between nucleic acid bases (A, T, U, G, C) <u>in vacuo</u> through several kinds of formulae [17], and we found (table VI of [17]), for the associations U-U and T-T, values already larger than the previously quoted values of ref. [16] (- 6.3 and - 7.3 kcal/mole respectively) ; methylation is expected to result in a further slight increase of the theoretical values due to the increase of the dispersion term of the interaction energy), and the discrepancy would be still larger. On the contrary, for hydrogen-bonded associations, our theoretical results (corresponding to the best methods A, A', A") were in satisfactory agreement with the results of Yanson et al. [8]. It therefore appeared interesting to perform a theoretical study of the interactions between the molecules in the presence of an external electric field, and this study is initiated in the present work. At this point, it seems worth emphasizing that the problem considered here is part of a very general topic, namely the disturbing effect of the environment upon a two-body complex, or upon the conformation of a single molecule. Previous works in our laboratory were devoted to such studies for a crystalline environment [18 – 24] and for a liquid environment (solvent effect) [25 - 28], and in many cases, it was found that the environment could indeed have a non negligible effect. In the present investigation, the disturbing environment is just the external electric field, and, as we shall see, here too, the disturbance due to the environment may be important.

II. A SIMPLE MODEL : TWO INTERACTING DIPOLES EMBEDDED IN AN EXTERNAL ELECTRIC FIELD

We consider for simplicity two dipoles of equal magnitude μ , with zero polarizability. In the external electric field F, each of the isolated dipoles will orient itself parallel to F, with a corresponding energy

$$E_d^F = - \mu F \qquad (II.1)$$

Similarly, the complex will orient itself in such a way as to make its <u>total</u> dipole moment $(\vec{\mu}_1 + \vec{\mu}_2)$ parallel to $\vec{F}$, thus giving an energy

$$E(dd,F) = -(\vec{\mu}_1 + \vec{\mu}_2).F = -\left|\vec{\mu}_1 + \vec{\mu}_2\right| F = -(2\mu\cos\frac{\theta}{2})\, F \qquad (II.2)$$

where θ denotes the angle $(\vec{\mu}_1,\vec{\mu}_2)$ of the two dipoles (F bissects the angle $(\vec{\mu}_1,\vec{\mu}_2)$). On the other hand, the interaction energy between the two dipoles themselves is :

$$E_{dd} = \frac{1}{R^3} [\vec{\mu}_1.\vec{\mu}_2 -3 (\vec{\mu}_1 \cdot \frac{\vec{R}}{R}) (\vec{\mu}_2 \cdot \frac{\vec{R}}{R})] \qquad (II.3)$$

where R denotes the vector going from μ_1 to μ_2; the total energy of the complex in the field will be :

$$E_{dd}^F = E_{dd} + E\ (dd,F)$$ (II.4)

and we shall search for (local) minima of E_{dd}^F. Now, the perturbation of the complex due to the field will be quite different according to the relative orientation of the two dipoles in the unperturbed complex. We shall consider the following two illustrative cases (fig. 1) :

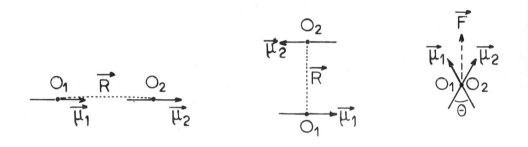

(A) the two dipoles   (B) $\vec{R}$ is perpendicular   (C) case (B)
     are colinear.   to each of the dipoles   viewed along
                              and these are antipara- O_1O_2 for $F \neq 0$
                              lel (for F = 0).

Figure 1. Two configurations for interacting dipoles.

A) The two dipoles are colinear (for F = 0).
Here the situation is very simple, because the unperturbed complex has a configuration which is optimal as concerns the interaction with the field : introducing a non-zero angle θ between the two dipoles would reduce (in absolute value) both E_{dd} and E (dd,F), hence the zero-field configuration will remain unchanged. Moreover, since $E(dd,F) = 2\mu F = 2 E_d^F$, the stabilization energy ΔE_{dd}^F of the complex with respect to the isolated molecules in the presence of the field, namely

$$\Delta E_{dd}^F = E_{dd}^F - 2\ E_d^F = E_{dd} + E(dd,F) - 2E_d^F$$ (II.5)

will reduce to $E_{dd} = -2\ \vec{\mu_1}\cdot\vec{\mu_2}\ /\ \vec{R}^3$, namely its value for F = 0. Thus in the present case, the external electric field brings no perturbation, either to the configuration or to the stabilization energy of the complex.

B) The two dipoles are antiparallel (for F = 0).
The situation is now quite different : since $\vec{\mu_1} + \vec{\mu_2} = 0$, $E(dd,F) = 0$, hence the stabilization energy of the complex for this configuration :

$$\Delta E_{dd}^{F} = E_{dd} - 2 E_{d}^{F} = E_{dd} + 2\mu F \qquad (II.6)$$

decreases linearly as a function of F, and we must study the possibilities of configurational change. We shall assume that each dipole may
rotate around its fixed center while remaining in the plane perpendicular to R. This assumption would correspond, for example, to the case of
two planar molecules in a stacked configuration, with the molecular planes kept at a fixed distance R and the "centers" O_1 and O_2 of the molecular dipoles remaining fixed, with the molecules free to rotate around
their respective "centers" (a possibility for defining such a center
consists in considering the barycenters of the positive and negative net
atomic charges, respectively, and then taking the middle of the segment
joining these two points).
With such an assumption, the configuration of the complex depends on the
angle θ only, as indicated in Fig. 1C (which represents the projection
of the complex into a plane perpendicular to O_1O_2). We then have :

$$E_{dd} (\theta) = (\mu^2/R^3) \cos \theta \qquad E(dd,F) = - 2\mu F \cos \frac{\theta}{2} \qquad (II.7)$$

hence

$$E_{dd}^{F} = \frac{\mu^2}{R^3} \cos \theta - 2\mu F \cos \frac{\theta}{2} = \frac{\mu^2}{R^3} (2 u^2 -1) - (2\mu F)u \qquad (II.8)$$

where we introduced the variable $u = \cos \frac{\theta}{2}$, which may vary from 0 ($\theta =$
π, antiparallel dipoles) to 1 ($\theta = 0$, parallel dipoles). The quadratic
function E_{dd}^{F} (u) would reach its minimum $E_{min}^{F} = - (F^2R^3/2 + \mu^2/R^3)$ for
$u_{min} = FR^3/2$, but these values are physically acceptable only if u_{min}
< 1, i.e. $F< 2\mu/R^3$. For larger values of F, the function E_{dd}^{F} (u) monotonously decreases from $E_{dd}^{F}(0) = - \mu^2/R^3$ to $E_{dd}^{F}(1) = \mu^2/R^3 - 2\mu F$, and
we must therefore take $u_{min} = 1$ and $E_{min}^{F} = E_{dd}^{F}(1)$. Summing up these results we therefore have :

for $F < 2 \mu / R^3$
$$\begin{cases} \cos \dfrac{\theta_{min}}{2} = u_{min} = \dfrac{R^3}{2\mu} F \\[2mm] E_{min}^{F} = - (F^2R^3/2 + \mu^2/R^3) \end{cases} \qquad (II.9a)$$

for $F > 2 \mu / R^3$
$$\begin{cases} \cos \dfrac{\theta_{min}}{2} = u_{min} = 1 \\[2mm] E_{min}^{F} = - 2\mu F + \mu^2/R^3 \end{cases}$$

For the "critical" value $F = 2 \mu/R^3$, both expressions (II.9a) and (II.9b)
give for E_{min}^{F} the same value $-3\mu^2/R^3$, as it should be.

We therefore already see that, when we start (for F=0) from the antipa-

rallel configuration, the optimum configuration of the complex is modified as soon as F > 0, and this modification becomes considerable except
if we keep F << $2\mu/R^3$. For F > $2\mu/R^3$, the interaction of the dipoles
with the field E(dd,F) completely overwhelms the interaction E_{dd} between
the dipoles themselves, since the optimum configuration (parallel dipoles) then is the most favourable for E(dd,F) but the least favourable
for E_{dd} ! For completing the picture, it remains only to consider the
stabilization energy corresponding to the optimum configuration, namely

$\Delta E^F_{min} = E^F_{min} - 2 E^F_d = E^F_{min} + 2\mu F$ (see eq.II.5), hence the explicit
expressions :

for F < $2\mu/R^3$ $\Delta E^F_{min} = -\dfrac{R^3}{2} F^2 + 2\mu F - \dfrac{\mu^2}{R^3}$

$\hspace{10cm}$ (II.10a)

$\hspace{3.5cm} = -\mu^2/R^3(2u^2_{min} - 4u_{min} + 1)$

for F > $2\mu/R^3$ $\Delta E^F_{min} = \mu^2/R^3$ (II.10b)

since $u_{min} = F/F_c$, with $F_c = 2\mu/R^3$.

Figures 2 (a and b) give the graphs of θ_{min} and ΔE^F_{min}, respectively, as
functions of F.

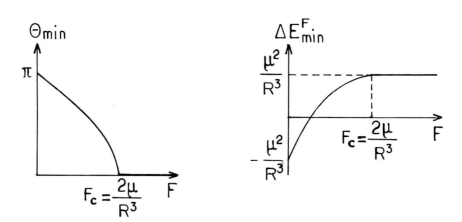

<u>FIGURE</u> 2 : Variation of θ_{min} (angle between the two dipoles in the opti-
mum configuration) and ΔE^F_{min} (corresponding stabilization energy of the
complex with respect to the isolated dipoles) as functions of the external electric field F.

These results are quite striking : ΔE_{min}^F varies monotonously from $-\mu^2/R^3$ (complex more stable than the separated dipoles) when F=0 to μ^2/R^3 (complex less stable than the separated dipoles when $F > F_c$, taking the value 0 for $u_{min} = F_0 / F_c = 1 - 1/\sqrt{2}$ (the corresponding value of θ being $\theta_0 = 2$ Arc cos $(1 - 1/\sqrt{2}) \simeq 146°$).

We can therefore conclude that, for the unperturbed configuration corresponding to antiparallel dipoles, the perburbing effect of the external electric field may become considerable, in strong contrast with the situation found in the previous case (parallel dipoles in the unperturbed configurations).

Thus, it clearly appears necessary to investigate in detail the effect of the external electric field specifically for each case, and we shall now present the first results of such a study for the complexes between the main nucleic acid bases (A,T,U,G,C) and some methyl derivatives:1-methyl-Uracil(m1U),1,3 dimethyl Uracil (m1,3U) .
        We shall essentially consider hydrogen bonded complexes, for comparison with the experimental results of Yanson et al. [8] , but some stacked configurations will also be considered for m1,3U, in order to discuss the results of Verkin et al. [16] .

III. THEORETICAL METHOD.

We shall first recall the method used for evaluating the interaction energy between two molecules in the absence of any external electric field, and afterwards we shall indicate the simple modifications which are required in the presence of the field F.

A) Interaction energy between two molecules.

Since we used here precisely the method labeled (A) in our recent work [17] , where this method and some others were described in great detail, we shall only recall the main features here, refering to our previous works [17, 29, 30] for further details.

As suggested by a perturbation-theoretical treatment (see e.g. [30]), the interaction energy between two molecules (labeled 1 and 2) will be expressed as a sum of four contributions :

$$E_{1,2} = E_{el} + E_{pol} + E_{disp} + E_{rep} \qquad (III.1)$$

E_{el} : electrostatic term ; E_{pol} : polarization term ; E_{disp} : dispersion term ; E_{rep} : repulsion term (short-range).

(1) From the electrostatic point of view, each molecule is represented as a set of charges, dipoles and quadrupoles (multi-centred multipole expansion [30, section V.A]), located at the atoms and at the middles of

some of the segments joining the atoms : retaining the middes of all these
segments would mean a larger number of centres (N(N+1)/2 for a molecule
containing N atoms) i.e. a longer computation time. As in [17] , we
keep the middles of the chemical bonds and of the segments joining two
atoms which are chemically bonded to a common third atom. These multipoles are computed from an ab initio wave function obtained through the
SCF - MO - LCAO method with a suitable minimal basis set [31, 32, 33].
It has been shown previously [29] that the use of a reduced set of centers does not result in significant changes with respect to the use of
the full set of N(N + 1)/2 centers, which was considered in earlier works
under the name Overlap Multipole Expansion [34]. Detailed formulae, noticeably those giving the interaction energy between charges, dipoles and
quadrupoles, may be found in [17].

(2) The polarization energy E_{pol} is obtained as the sum of the polarization energies $E_{pol(m)}$ of each molecule (m) :

$$E_{pol} = E_{pol(1)} + E_{pol(2)} \qquad\qquad (III . 2)$$

and the polarization energy of each molecule is evaluated as a sum of
contributions associated with "polarizable centers" (labeled by i) :

$$E_{pol(m)} = - \frac{1}{2} \sum_i {}^{(m)} \alpha_i (\vec{\xi}_i)^2 \qquad\qquad (III . 3)$$

where $\vec{\xi}_i$ denotes the electric field (here due to the other molecule only) at the center i, and α_i the polarizability of this center. As polarizable centers, we use the atoms and the middles of the chemical bonds ;
their polarizabilities $\vec{\xi}_i$ are obtained from bond polarizabilities through
a suitable sharing procedure (see [17, 29]). Each electric field $\vec{\xi}_i$ is
obtained here as the field created at the point i by all the multipoles
(labeled j, say) of the other molecule, except for the following change
(see [17, 29]) : In the denominators $(r_{ij})^2$, $(r_{ij})^3$, $(r_{ij})^4$ occurring in
the expressions of these fields, we replace the true distance r_{ij} by a
"modified distance".

$$r'_{ij} = r_{ij} + D (R_i^W + R_j^W)/2 \qquad\qquad (III . 4)$$

where R_i^W, R_j^W denote van der Waals radii associated with the centers i, j
respectively, and D is a parameter (we used D = 0.10 in the present work
This recipe conveniently reduces the absolute magnitude of the polarization energy at short distance, and, noticeably, it prevents this energy
of becoming (spuriously) infinite for r_{ij} = 0.

(3) The dispersion and short-range repulsion energies are evaluated as a
sum of atom-atom terms of Buckingham type (6-exp.), by using a formula
adapted from the one proposed by Kitaigorodskii (see [19] and references
therein). In the case of interaction terms between a Hydrogen atom and a
heavier atom, the parameters undergo some modification at short distance
(between 2.6 and 1.8 Å), and this refinement allows for a satisfactory
representation of equilibrium distances and energies of hydrogen bonded

complexes. Further details and parameter values may be found in refs. [17, 19, 20].

B) <u>Interaction energy in the presence of an external electric field.</u>

As previously seen in the case of the simplified model of two dipoles (section II), we have two modifications : first, the interaction energy $E_{1,2}^F$ of the two molecules in the complex embedded in the external field $\vec{F}$, the zero energy being the same as previously, namely the two separated molecules in the absence of any field ; and second, the energies E_m^F (m = 1,2) of the separated molecules, now also embedded in the field $\vec{F}$, with respect to the same origin of the energy. Then, we shall get the stabilization energy $\Delta E_{1,2}^F$ of the complex with respect to the isolated molecules in the presence of the field $\vec{F}$ as the difference (cf. eq. (II. 5)) :

$$\Delta E_{1,2}^F = E_{1,2}^F - (E_1^F + E_2^F) \qquad (III . 5)$$

(1) <u>evaluation of $E_{1,2}^F$</u>

$E_{1,2}^F$ is given by a formula essentially similar with eq. (III.1) (case $\vec{F}$ = 0), namely :

$$E_{1,2}^F = E_{el}^F + E_{pol}^F + E_{disp} + E_{rep} \qquad (III . 6)$$

The dispersion (E_{disp}) and repulsion (E_{rep}) terms are given by the same formulae as previously.
The electrostatic term now becomes :

$$E_{el}^F = E_{el} - (\vec{\mu}^{(1)} + \vec{\mu}^{(2)}). \vec{F} \qquad (III . 7)$$

where we added to E_{el} (calculated as indicated in subsection A above) the energy $- (\vec{\mu}^{(1)} + \vec{\mu}^{(2)}). \vec{F}$ of the total dipole moment ($\vec{\mu}^{(1)} + \vec{\mu}^{(2)}$) of the complex in the field $\vec{F}$. Since this complex may freely orient itself, we may immediately assume (as in the two dipoles model) that it has already taken the most favourable orientation, namely ($\vec{\mu}^{(1)} + \vec{\mu}^{(2)}$) parallel to $\vec{F}$, and E_{el}^F then takes the value :

$$E_{el}^F = E_{el} - |\vec{\mu}^{(1)} + \vec{\mu}^{(2)}| F \qquad (III. 8)$$

It must be emphasized that, exactly as in the case of the two dipoles model, the second term also depends on the relative position of the two molecules in the complex, through the value of $|\vec{\mu}^{(1)} + \vec{\mu}^{(2)}|$. The total dipole moment $\vec{\mu}^{(m)}$ of each molecules (m) is obtained from the set of charges and dipoles involved in the electrostatic representation of the molecule (i labels here the centers where the multipoles are located) :

$$\vec{\mu}^{(m)} = \sum_i {}^{(m)}q_i \vec{R}_i + \sum_i {}^{(m)}\vec{\mu}_i \qquad (III. 9)$$

Finally, the polarization energy is obtained, as previously (eq. III. 2), as the sum of the polarization energies of each molecule :

$$E^F_{pol} = E^F_{pol(1)} + E^F_{pol(2)} \qquad\qquad (III . 10)$$

but now the polarization energy of each molecule (m) is obtained by using as polarizing field (on each polarizable center i) the total electric filed ($\vec{\xi}_i$ + F), where $\vec{\xi}_i$ denotes, as previously, the electric filed created at the center i by the other molecule ; we therefore have :

$$E^F_{pol(m)} = - \frac{1}{2} \sum_i{}^{(m)} \alpha_i (\vec{\xi}_i + \vec{F})^2 \qquad\qquad (III . 11)$$

This completes the evaluation of the energy of the complex in the field $\vec{F}$. It remains only to consider the energies of each molecule isolated in the field $\vec{F}$, namely :

$$E^F_m = - \vec{\mu}^{(m)} \cdot \vec{F} - \frac{1}{2} \alpha^{(m)} F^2 \qquad\qquad (III . 12)$$

and considering that each molecule takes the most favourable orientation ($\vec{\mu}^{(m)}$ parallel to $\vec{F}$), we finally get :

$$E^F_m = - \mu^{(m)} F - \frac{1}{2} \alpha^{(m)} F^2 \qquad\qquad (III . 13)$$

where $\alpha^{(m)} = \sum_i{}^{(m)} \alpha_i$ denotes the total molecular polarizability of molecule (m).

C) Main steps of the complete procedure

We first compute the energy E^F_m of each isolated molecule (m) in the field F according to formula (III . 13). Then, for every prescribed configuration of some complex, we get the "intrinsic" interaction energy $E^F_{1,2}$ according to eq. (III . 6), and the genuine stabilization energy $\Delta E^F_{1,2}$ as the difference (III . 5). As concerns the configurations, the following choices were made, as in ref.[17] :

a) for hydrogen bonded configurations, various possible associations schemes were considered with initial standard geometries, and the geometry was then varied so as to get the corresponding local minimum of $E^F_{1,2}$

b) for stacked configurations, a set of initial configurations was generated by rotating one of the molecule through some given angle θ around the z-axis (possibly after having first turned the molecule upside down through a symmetry with respect to the y-axis), and then translating it by 3.4 Å along the z-axis ; afterwards, the energy $E^F_{1,2}$ was minimized with respect to the translation variables (x, y, z) only, the resulting minimum thus being a function of θ.

As concerns the values of the electric field, if we take the formula F = $U_e/(5 R_e)$, according to refs. [8, 12] , as giving, for the electrode potential U_e, the value of the field near the endpoint of the electrode with radius R_e, we get for R_e = 1000 Å and U_e = 1.0, 2.0, 3.0 kV the values F = 0.2, 0.4, 0.6 V/Å respectively ; we therefore considered in our calculations the values f = 0.01, 0.05, 0.1, 0.2, 0.31, 0.4, 0.47, 0.6 V/Å, ranging from 0 to 0.6 V/Å.

IV. RESULTS

In this work we have studied the influence of the electric field upon the complexes of nucleic acid bases : Adenine, Guanine, Uracile (and Thymine) and Cytosine, for hydrogen-bonded and stacked configurations. For the latter case, our calculations have been performed for 1-3 dimethyl-uracile (since the experimental data [16] concern this compound and 1-3 dimethyl-thymine).

Experimental data [8] concern bases methylated on nitrogen N_1 for pyrimidines (m^1 - U, m^1 - T, m^1 - C) and on nitrogen N_9 for purines (m^9 - A and m^9 - G). Yanson et al [8] assert that their measured binding energies are independent of methylation. They argued that ΔH is independent of methylation which does not changes the H bonds. Admittedly, the methylation in the ninth position of purine would not effect the H-bonds of these molecules since, on one hand these nitrogens do not participate in any hydrogen bond between purine and pyrimidines, and on the other hand these nitrogens are not close to the atoms which are involved in some hydrogen bond. But the case might be different for N_1-methylated pyrimidines. Thus, for checking purpose we have also considered the 1-methyl-uracil and its hydrogen-bonded complexes (m^1 - U) ... (m^1 - U).

A) Individual molecules in an electric field

All data concerning these individual molecules in an electric field are summarized in table I.

From results given in table I, it appears that : (a) the sequence of dipole moments of nucleic acid bases is : G > C > U > A. The methylation (for instance of uracil in order to get thymine, 1-methyl-uracil or 1-3-dimethyl-uracil) does not change drastically the value of the dipole moment. In agreement with formula (III. 13), the electrostatic component of the energies of the isolated molecules in an electric field follows the above sequence. (b) The molecular polarizability is slightly larger for the purine bases than for the pyrimidine ones, thus the polarization energy for purine bases in an electric field is stronger than for pyrimidine bases. (c) when dealing with a very polar and very polarizable base (e.g. guanine), we may notice that the polarization energy represents 7% of the electrostatic energy in an electric field of 0.2 V/Å and 21% in an electric field of 0.6 V/Å. For these two values of the electric field, when dealing with a less polar but equally polarizable molecule such as adenine, the polarization energy represents respectively 10% and 60% of the electrostatic component. In this respect, it may be noticed that the methylation of uracil at N1 position or at C5 position (in order to give thymine) leads to a non-negligible increase of the molecular polarizability : from 10.0 $\mathring{A}^3$ for U to 11.9 and 11.7 $\mathring{A}^3$ for 1-mU and T respectively.

TABLE I : Total energy (kcal/mole), E^F, and its components (electrostatic E^F_{el} and E^F_{pol}) for the isolated molecules in various electric fields. This table also gives for each molecule, the dipole moment μ (in Debyes) and the molecular polarizability P_M (in Å³).

ELECTRIC FIELD F (V/Å)	0.10			0.20			0.40			0.60		
	E^F_{el}	E^F_{pol}	E^F	E^F_{el}	E^F_{pol}	E^F	E^F_{el}	E^F_{pol}	E^F	E^F_{el}	E^F_{pol}	E^F
GUANINE : μ = 6.9 P_M = 14.3	-3.3	-0.1	-3.4	-6.6	-0.4	-7.0	-13.2	-1.8	-15.0	-19.8	-4.1	-23.9
CYTOSINE : μ = 6.1 P_M = 10.8	-2.9	-0.1	-3.0	-5.8	-0.3	-6.1	-11.7	-1.4	-13.1	-17.5	-3.1	-20.6
URACIL : μ = 3.7 P_M = 10.0	-1.8	-0.1	-1.9	-3.6	-0.3	-3.9	-7.2	-1.3	-8.5	-10.7	-2.9	-13.6
THYMINE : μ = 3.6 P_M = 11.7	-1.7	-0.1	-1.8	-3.5	-0.3	-3.8	-6.9	-1.5	-8.4	-10.4	-3.4	-13.7
ADENINE : μ = 2.2 P_M = 13.7	-1.1	-0.1	-1.2	-2.2	-0.4	-2.6	-4.3	-1.8	-6.1	-6.5	-3.9	-10.4
1me-URACIL : μ = 3.8 P_M = 11.9	-1.8	-0.1	-1.9	-3.7	-0.4	-4.1	-7.3	-1.5	-8.8	-11.0	-3.4	-14.4
1,3 dime-U : μ = 3.7 P_M = 13.7	-1.8	-0.1	-1.9	-3.6	-0.4	-4.0	-7.2	-1.8	-9.0	-10.7	-4.0	-14.7

B) Hydrogen-bonded complexes

For each pair of bases (G-C, A-U and A-T, U-U and T-T) we have studied
the classical hydrogen-bonded dimers as given in ref. [17]. But for uracile we have also studied some hydrogen bonded dimers involving the
nitrogen N1 (dimers $\begin{Bmatrix} N_3 \; H \; \dots \; O \; (C_2) \\ (C_4) \; O \; \dots \; H \; N_1 \end{Bmatrix}$ and $\begin{Bmatrix} N_3 \; H \; \dots \; O \; (C_2) \\ (C_2) \; O \; \dots \; H \; N_1 \end{Bmatrix}$ for instance).

Figs. 3-6 gives the relative molecular orientation of the two molecules
for the different hydrogen-bonded complexes G-C (Fig. 3A1 - A2), C-C
(Fig. 4A1 - A2), A - U (Fig. 5A1 - A4) and U - U (Fig. 6A1 - A5) ; hydrogen-bonded dimers A - T and T - T have strictly the same relative orientation as complexes A - U and U - U respectively.

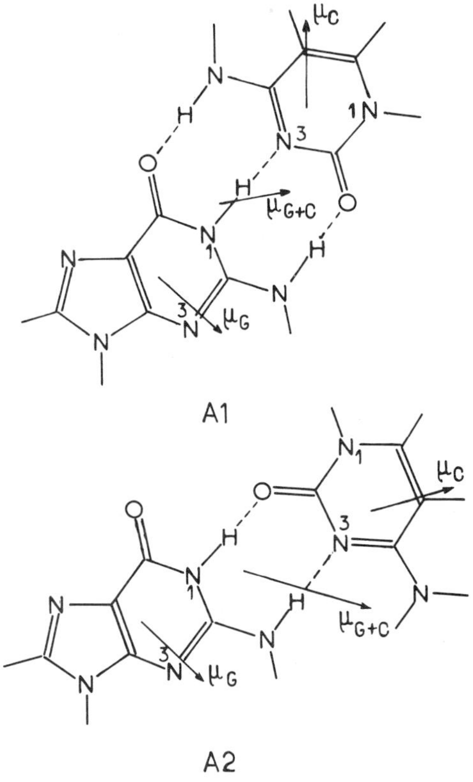

A1

A1

A2

A2

FIGURE 3. G-C, different hydrogen-   FIGURE 4. C-C, different hydrogen-
            bonded dimers.   bonded dimers.

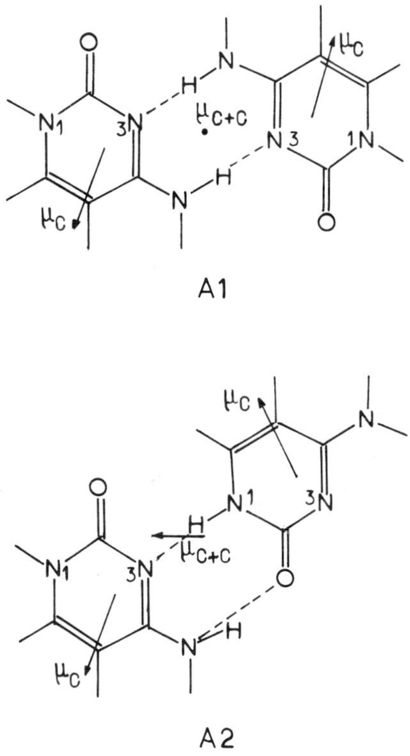

In these figures we have also indicated the dipole moment of the complex
and of its individual partners. We may notice that three kinds of complexes may be distinguished according to the value of their dipole moment

a) complexes with a zero dipole moment, such as the totally symmetrical dimers A_1 and A_2 of uracile and thymine and A_1 of cytosine (Figs. $6A_1$, A_2 and Fig. $4A_1$).

b) complexes with a dipole moment weak with respect to those of the two par tner molecules, such as Watson-Crick dimer of G-C (Fig. $3A_1$), A-U and A-T (Fig. $5A_2$, $5A_4$), dimer A_2 of C-C (Fig. $4A_2$ and dimer A_4 of U-U (Fig. $6A_4$) (and of the various methylated uracile).

c) complexes with a dipole moment of the same order of magnitude as the sum of the dipole moments of their two partners (i.e. $\vec{\mu}_1$ and $\vec{\mu}_2$ are almost parallel in the unperturbed complex) such as Hoogsteen A-U and A-T (Fig. $5A_1$), dimers $\begin{cases} N_3 - H \ldots O(C_2) \\ (C_2) - 0 \ldots H - N_1 \end{cases}$ and $\begin{cases} N_3 - H \ldots O(C_2) \\ (C_4) - 0 \ldots H N_1 \end{cases}$ of U-U and T-T (Fig. $6A_3$ and $6A_5$).

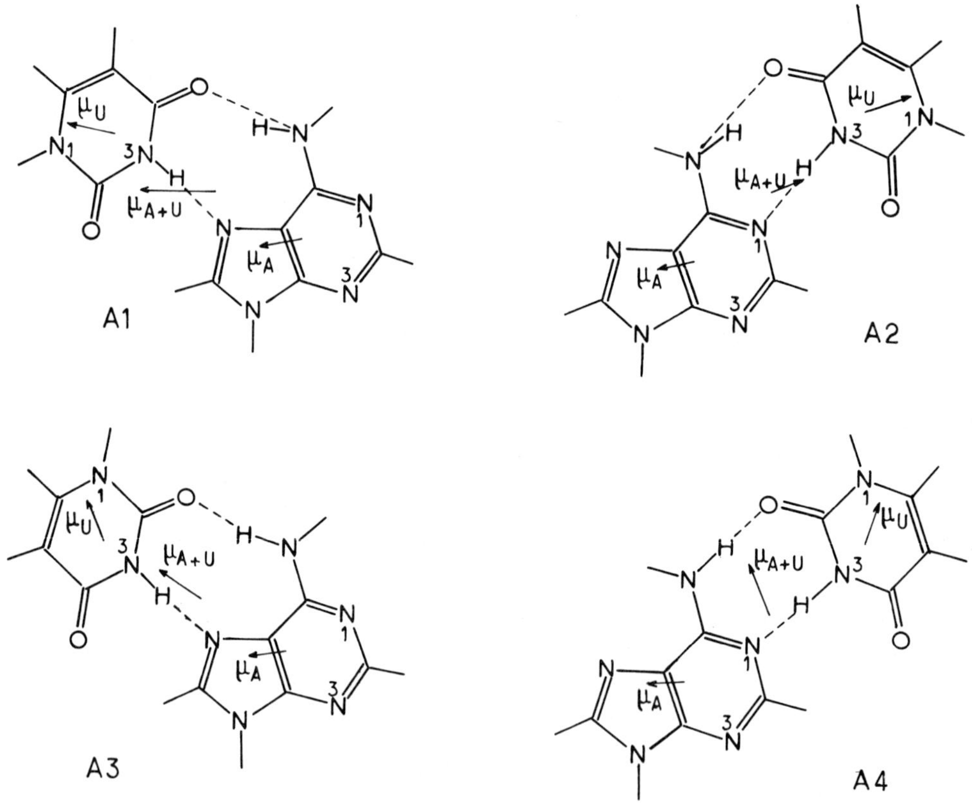

FIGURE 5. A-U, different hydrogen-bonded complexes.

From values given in table I, we may foresee some qualitatives results : when a complex has an unfavorable relative molecular orientation leading to a zero or weak dipole moment, it is destabilized by an electric field

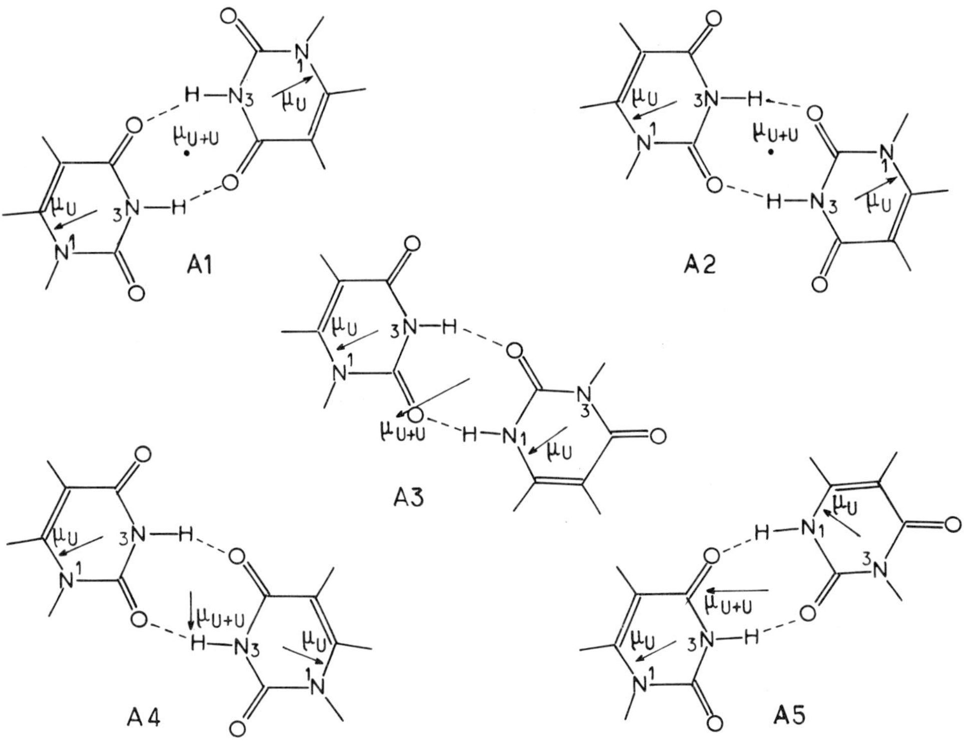

FIGURE 6. U-U, different hydrogen-bonded complexes.

(this destabilization increasing with the field strength).But for a given value of the field, this destabilization will be more pronounced for complexes formed by very polar molecules (such as guanine or cytosine) than for complexes formed by less polar ones (such as uracil or adenine). These assertions have been verified in our calculations as it will be shown hereafter.

Table II gives for each complex denoted I – II), and for each value of the electric field the sum $(E^F_I + E^F_{II})$ (corresponding to the separated molecules), and ΔE^F_{I-II}, the stabilization energy of the complex (I-II) with respect to the separated molecules in the field F (cf. eq. III-5).

Fig. 7-10 give for each kind of complex (respectively G-C, C-C, A-U and U-U) the evolution of $(E^F_I + E^F_{II})$, E^F_{I-II} (energy of the complex) and ΔE^F_{I-II} as functions of the electric field.

All these results reveal the following features :
a) a strong destabilization caused by the electric field for complexes with a zero dipole moment (dimer A1 and A2 of U-U and dimer A1 of C-C), (see fig. 10, curve 1 and Fig. 8, curve 1). We may even notice that for F > 0.3 V/Å, the symmetrical dimers A1 and A2 of U-U cannot exist, since

TABLE II. For each complex $(I - II)$: Sum of the energy of the two partner molecules $(EF_I + EF_{II})$ in an electric field F ; energy of the complex (EF_{I-II}) for each value of F; stabilization energy of the complex (ΔEF_{I-II}) All values in kcal/mole.

ELECTRIC FIELD V/Å	0.0	0.01	0.10	0.20	0.40	0.60
GUANINE-CYTOSINE						
$EF_G + EF_G$	0.0	-0.6	-6.4	-13.2	-28.1	-44.6
Watson-Crick (A1)(μ=5.9 D) EF		-24.0	-27.3	-31.0	-40.1	-50.6
ΔEF	-23.7	-23.4	-20.9	-17.8	-12.0	- 6.0
Dimer A2 (μ=11.0 D) EF		-15.0	-20.6	-28.3	-41.6	-58.9
ΔEF	-14.6	-14.4	-14.2	-15.1	-13.5	-14.3
CYTOSINE-CYTOSINE						
$EF_C + EF_C$	0.0	- 0.6	- 6.0	-12.2	-26.2	-41.4
Dimer A1 (μ=0.0 D) EF		-20.4	-20.7	-21.4	-23.5	-27.1
ΔEF	-20.3	-19.8	-14.7	- 9.2	+ 2.7	+14.3
Dimer A2 (μ=3.8 D) EF		-20.0	-22.5	-24.8	-33.2	-44.2
ΔEF	-19.9	-19.4	-16.5	-12.6	- 7.0	- 2.8
ADENINE-URACIL						
$EF_A - EF_U$	0.0	- 0.3	- 3.1	- 6.5	-14.6	-24.0
Hoogsteen A1 (μ=5.8 D) EF		-14.2	-16.9	-20.2	-28.0	-37.2
ΔEF	-13.5	-13.8	-13.8	-13.7	-13.4	-13.2
Watson-Crick A2 (μ=1.5 D) EF		-13.3	-13.9	-15.0	-18.5	-23.4
ΔEF	-13.2	-12.9	-10.8	- 8.5	- 3.9	+ 0.6
Rev. Hoogsteen A3 (μ=4.7 D) EF		-13.8	-16.1	-18.8	-25.6	-33.4
ΔEF	-13.1	-13.5	-13.0	-12.3	-11.0	- 9.4
Rev. Wat.-Crick A4 (μ=3.0 D)EF		-12.6	-14.1	-16.4	-21.3	-27.8
ΔEF	-13.1	-12.2	-11.0	- 9.9	- 6.7	- 3.8

ELECTRIC FIELD V/Å	0.0	0.01	0.10	0.20	0.40	0.60
ADENINE-THYMINE						
$EF_A + EF_T$	0.0	- 0.3	- 3.0	- 6.5	-14.5	-24.1
Hoogsteen A1 (μ=5.8 D) EF		-13.9	-16.3	-19.5	-27.1	-36.3
ΔEF	-13.6	-13.7	-13.3	-13.0	-12.6	-12.2
Watson-Crick A2 (μ=1.5 D) EF		-12.9	-13.5	-14.5	-17.9	-22.8
ΔEF	-12.9	-12.6	-10.5	- 8.0	- 3.4	+ 1.3
Rev. Hoogsteen A3 (μ=3.0 D) EF	-13.3	-13.7	-15.7	-18.4	-25.3	-33.3
ΔEF		-13.3	-12.7	-11.9	-10.8	- 9.2
Rev. Wat.-Crick A4 (μ=2.8D) EF	-12.4	-12.4	-13.6	-15.7	-20.0	-26.8
ΔEF		-12.1	-10.6	- 9.2	- 5.5	- 2.7
URACIL -URACIL						
$EF_U + EF_U$	0.0	- 0.4	- 3.8	- 7.8	-17.0	-27.2
Dimer A1 (μ=0.0 D) EF		- 9.2	- 9.4	- 9.9	-11.9	-15.1
ΔEF	- 9.2	- 8.8	- 5.6	- 2.1	+ 5.1	+12.1
Dimer A2 (μ=0.0 D) EF		- 9.3	- 9.5	-10.0	-12.0	-15.1
ΔEF	- 9.3	- 8.9	- 5.7	- 2.2	+ 5.0	+12.1
Dimer A3 (μ=6.5 D) EF		-12.1	-15.1	-18.8	-27.0	-36.7
ΔEF	-11.8	-11.7	-11.3	-11.0	-10.0	- 9.5
Dimer A4 (μ=3.3 D) EF		- 9.4	-10.9	-13.0	-18.1	-24.5
ΔEF	- 9.2	- 9.0	- 7.1	- 5.2	- 1.1	+ 2.7
Dimer A5 (μ=7.4 D) EF		-11.7	-15.0	-19.1	-28.3	-38.5
ΔEF	-11.3	-11.3	-11.2	-11.3	-11.3	-11.3

ELECTRIC FIELD V/Å		0.0	0.01	0.10	0.20	0.40	0.60
THYMINE-THYMINE							
$E^F_T + E^F_T$			- 0.4	- 3.6	- 7.6	-16.8	-27.4
Dimer A1 (μ=0.0 D)	E^F		- 9.5	- 9.7	-10.3	-12.5	-16.3
	ΔE^F	- 9.5	- 9.1	- 6.1	- 2.7	+ 4.3	+11.1
Dimer A2 (μ=0.0 D)	E^F		- 9.6	- 9.8	-10.4	-12.7	-16.5
	ΔE^F	- 9.6	- 9.2	- 6.2	- 2.8	+ 4.1	+10.9
Dimer A3 (μ=6.2 D)	E^F		-11.4	-14.3	-17.9	-26.3	-36.0
	ΔE^F	-11.1	-11.0	-10.7	-10.3	- 9.5	- 8.6
Dimer A4 (μ=2.5 D)	E^F		- 9.6	-10.8	-12.6	-17.2	-23.5
	ΔE^F	- 9.6	- 9.2	- 7.2	- 5.0	- 0.4	+ 3.9
Dimer A5 (μ=7.1 D)	E^F		-11.2	-14.3	-18.3	-27.4	-37.8
	ΔE^F	-10.9	-10.8	-10.7	-10.7	-10.6	-10.4
1-METHYL-URACIL dimers							
$E^F_{mU} + E^F_{mU}$		0.0	- 0.4	- 3.8	- 8.0	-17.6	-28.8
Dimer A1 (μ=0.0 D)	E^F		-10.1	-10.3	-10.8	-13.3	-17.1
	ΔE^F	-10.1	- 9.7	- 6.5	- 2.8	+ 4.3	+11.7
Dimer A2 (μ=0.0 D)	E^F		- 9.3	- 9.6	-10.1	-12.4	-16.1
	ΔE^F	- 9.3	- 8.9	- 5.8	- 2.1	+ 5.2	+12.7
Dimer A3 (μ=5.3 D)	E^F		- 8.0	-10.5	-13.5	-20.9	-29.9
	ΔE^F	- 7.8	- 7.6	- 6.7	- 5.5	- 3.3	- 1.1
Dimer A4 (μ=3.2 D)	E^F		-10.0	-11.3	-13.3	-18.9	-25.8
	ΔE^F	- 9.6	- 9.6	- 7.5	- 5.3	- 1.3	+ 3.0
Dimer A5 (μ=7.1 D)	E^F		- 7.4	-10.6	-14.7	-23.8	-34.3
	ΔE^F	- 7.2	- 7.0	- 6.8	- 6.7	- 6.2	- 5.5

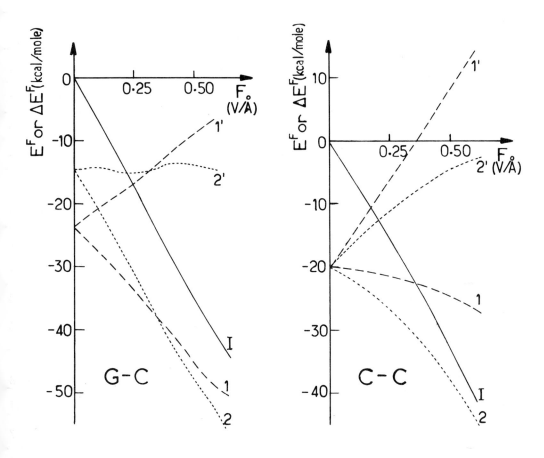

FIGURE 7.   FIGURE 8.
Hydrogen-bonded complexes (G-C, C-C).
The two figures give the evolution as a function of the external electric
field F, of $E_I^F + E_{II}^F$ (curve I), E_{I-II}^F (curves i (i = 1, n, for each A$_i$
dimer)) and ΔE_{I-II}^F (curves i' (i' = 1, n for each A$_i$ dimer)).

their stabilization energy ΔE_{U-U}^F is positive ; the same phenomenon occurs with the symmetrical dimer A1 of C-C for F > 0.35 V/Å. Furthermore,
in agreement with our previous assertions, the energy destabilization
due to the electric field is more important for dimer A1 of C-C than for
dimers A1 and A2 of U-U : 34.3 kcal/mole, 21.4 kcal/mole and 21.4 kcal/
mole, respectively, for the value 0.6 V/Å of the electric field.

An analysis of these results has shown that this destabilization proceeds by a weakening of the electrostatic component $\Delta E_{I-II}^{el(F)}$ of the total
energy of the complex : For dimer A1 of cytosine the destabilization due
to the electrostatic energy (for F = 0.6 V/Å) has been calculated as

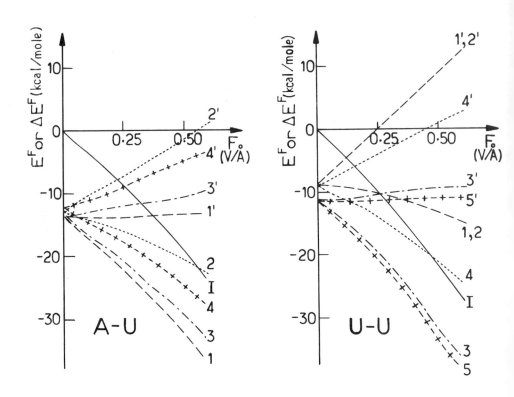

FIGURE 9.   FIGURE 10.
Hydrogen-bonded dimers A-U and U-U
Variation of energies $(E^F_I + E^F_{II})$, E^F_{I-II} and ΔE^F_{I-II} as functions of F
(see caption of fig. 7-8 for further details).

34.0 kcal/mole. We have also obtained (for the same value of the electric
field) the value 21.6 kcal/mole for the destabilization due to the electrostatic energy in the case of dimers A1 and A2 of U-U. Note that all
these results are in qualitative agreement with the results found for
the simple model of two dipoles (see section II-3 above).

Our calculations have also shown a very weak increase (about 0.4 kcal/
mole) of the polarization energy of the complex with regards to the separated molecules. This energy gain is unsignificant in comparison with the
destabilisation due to the electrostatic energy.

As concerns the geometry of these hydrogen-bonded complexes, it has appeared from our results in some cases (for dimer A1 of C-C, but not for
dimers A1 and A2 of U-U) a slight deformation : the second molecule slig-
tly rotates in order that the complex may get a non zero dipole moment
(which remains weak, however) : The two N ... H(N) lengths in dimer A1

of cytosine are equal to 1.82 Å in vacuo), and become equal to 1.75 Å
and 1.97 Å in an electric field of 0.6 V/Å.

b) A significant destabilization of the energy of the complexes having a
total dipole moment weak with respect to the dipole moment of the two
partner molecules : For instance 4' dimer A2 of cytosine (μ_{C-C} = 3.8 D
versus μ_C = 6.1 D) (see Fig. 8, curve 2'), dimer A4 of uracile (μ_{U-U} =
3.4 D versus μ_U = 3.6 D) (Fig. 8, curve 4'), Watson-Crick dimer G-C
(μ_{G-C} = 5.9 D versus μ_G = 6.9 D and μ_C = 6.1 D) (Fig. 7, curve 1'),
Watson-Crick (normal and reverse) dimers of A-U (Fig. 9, curves 2' and
4') and A-T. But in these cases the destabilization of the complex never
reaches such a high level as previously : from our results, for instance,
the energy destabilization has been evaluated as 17.0 kcal/mole for dimer A2 of cytosine and as 11.9 kcal/mole for dimer A4 of uracile.

We may notice an interesting phenomenon : in vacuo Watson-Crick and reverse Watson-Crick pairs of A-U (and A-T) have nearly the same energy :
-13.2 kcal/mole and -13.1 kcal/mole (-12.9 kcal/mole and -12.4 kcal/mole),
the Watson-Crick pair being slightly more stable than the reverse one.
But the Watson-Crick pair (μ = 1.5D) is more destabilized than the reverse one (μ = 3.0 D) by the electric field, so that the reverse pair becomes
the most stable one for F > 0.10 V/Å.

In the same way as previously the energy destabilization by the electric
field is essentially due to the electrostatic term : it amounts to 21.4
kcal/mole for dimer A2 of C-C and to 11.8 kcal/mole for dimer A4 of U-U,
for instance, but in these two cases we have noticed a gain in polarization energy of the complex (ΔE_{C-C}^{pol} = - 7.2 kcal/mole for F = 0.6 V/Å versus - 2.7 kcal/mole in vacuo and ΔF_{U-U}^{pol}(F)= - 1.5 kcal/mole for F = 0.6
V/Å versus - 1.0 kcal/mole in vacuo), but once again, this increase of
the polarization term does not compensate for the destabilization due to
the electrostatic term.

c) Practically no destabilization for complexes with a strong dipole moment such as Hoogsteen dimer of A-U (and A-T) (Fig. 9, curve 1'), dimer
A3 and A5 of U-U (and T-T) (Fig. 10, curves 3' and 5') and dimer A2 of
G-C, (Fig. 7, curve 2'). The electrostatic component of the energy decreases very slightly from F = 0.0 V/Å to F = 0.6 V/Å. Once again this
result agrees with the behaviour of the simple model in the corresponding case (see section II A above).

C) Stacked complexes

Recently Verkin et al. [16] have studied (using the same experimental
technique), the dimers of 1-3 dimethyl-uracile ($m^{1,3}$-U) and 1-3 dimethyl
thymine ($m^{1,3}$ - T) As a result, they have given an estimation of the
association enthalpy ΔH for these two stacked dimers (-3.56 kcal/mole
and -3.90 kcal/mole respectively, see the end of section I above).

In the same way as for hydrogen bonded dimers, the electric field is not

without influence upon the stacked complexes : It destabilizes very stron
gly complexes having a zero dipole moment (antiparallel complexes), it
destabilizes significantly complexes having a weak dipole moment, and it
is without influence for complexes where the dipole moments of the partners are parallel (parallel complexes).

We have studied the stacked dimer of $m^{1,3}$ - U in vacuo and in an electric
field (taking the same values as previously from 0.01 to 0.6 V/Å). We
considered two initial positions for our stacked complexes, obtained in
the following way (cf. section III.C above) : (a) translating one molecule along the z-axis (2) turning upside down this molecule (e.g. through
a symmetry operation with respect to the y-axis) and then translating it
along the z-axis. Then, in both cases, we performed rotations by angles
θ with respect to the z-axis (θ = 0°, 30°, 60°, 90°, 120°, 180°, 240°,
300°), and we minimized the energy of the complex with respect to the
variables x, y, z (coordinates of the center of the upper molecule).

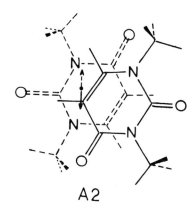

A1   A2

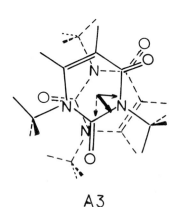

FIGURE 11. The three most stable com-
          formations of stacked com-
          plexes of 1,3-dimethyl-
          Uracil in vacuo.

A3

The results thus obtained (Table III and Figure 12 give them for the
case (1), namely molecule not turned upside down ; the results for case

(2) are qualitatively very similar) show that :

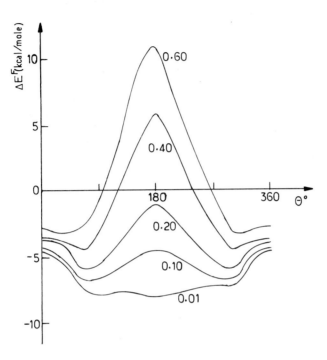

FIGURE 12. Variation of the stabilization energy ΔE^F of the dimethyl-
Uracile (m1,3U) stacked dimer as a function of the rotation
angle θ of the upper molecule around the z-axis for several
values of the electric field F : 0.01, 0.10, 0.20, 0.31, 0.60
V/Å (the molecular planes are perpendicular to the z-axis).

a) three stable complexes have been obtained in vacuo (see Fig. 11 A1-A3)
with the corresponding energies : - 9.1 kcal/mole, - 8.4 kcal/ mole,
- 8.2 kcal/mole respectively.

In the most stable stacked complex (Fig. 11A) we may notice that the
N-CH3 groups of the second molecule interact with the C- O groups of the
first one : essentially the same geometry was obtained for the stacked
U-U complex, but with the N-H instead of N-CH3 (see ref. [17] Fig. 3A1).
Furthermore, in agreement with our results in [17], our calculations
indicate a stabilization of stacked complexes when methyl group(s) are
added to the partner molecules (the minimum energy of U-U stacked complex
has been calculated as   - 6.3 kcal/mole).

The second minimum for the m1,3 - U dimer corresponds to an antiparallel
conformation (see Fig. 11-A2).

TABLE III. dmU – dmU stacked complexes : stabilization energy (ΔEF) obtained as a function of two parameters : on one hand, the value of the electric field F (from 0.1 V/Å to 0.6 V/Å) and on the other hand the rotation angle θ (around the z axis) of the upper molecule. Tables IIIa and IIIb refer respectively to two different starting conformations : (1) the two molecules in a parallel conformation,(2) the second molecule turned upside down. Underlined values correspond to local minima.

III (a)

$\theta°$	F(V/Å) 0.00	0.01	0.10	0.200	0.400	0.60
0	− 4.6	− 4.5	− 4.4	− 4.0	− 3.8	− 2.7
30	− 5.0	− 5.5	− 5.1	− 4.0	− 3.9	− 3.3
60	− 7.3	− 7.2	− 6.6	− 5.9	− 4.6	− 3.1
90	− 8.0	− 7.9	− 6.7	− 5.4	− 3.0	− 3.1
120	− 7.8	− 7.6	− 5.9	− 4.2	− 0.2	− 0.5
150	− 7.8	− 7.5	− 4.2	− 2.4	+ 3.9	+ 3.0
180	− 8.3	− 8.0	− 4.7	− 1.2	+ 5.9	+ 9.1
240	− 7.9	− 7.7	− 6.1	− 4.2	− 0.5	+12.8
300	− 7.3	− 7.2	− 6.6	− 5.9	− 4.5	+ 3.2
330	− 5.6	− 5.5	− 4.9	− 4.8	− 3.9	− 3.1

III (b)

$\theta°$	F(V/Å) 0.00	0.01	0.10	0.200	0.400	0.60
0	− 6.0	− 6.0	− 5.7	− 5.3	− 4.7	− 4.1
30	− 5.2	− 5.1	− 4.8	− 4.7	− 4.0	− 3.4
90	− 7.3	− 7.2	− 5.8	− 4.8	− 2.3	+ 0.1
120	− 8.9	− 8.7	− 7.1	− 5.3	− 1.7	+ 2.0
180	− 7.5	− 7.2	− 4.3	− 1.1	+ 5.5	+12.0
240	− 7.3	− 7.1	− 4.6	− 2.9	+ 1.1	+ 6.2
300	− 7.5	− 7.5	− 6.7	− 6.0	− 4.5	− 2.9
330	− 7.0	− 6.9	− 6.3	− 6.2	− 5.6	− 5.0

b) The three minima persist for F < 0.01 V/Å.

c) when the electric field increases from 0.01 V/Å to 0.6 V/Å, the two
initially most stable stacked conformations are strongly destabilized,
by 10.7 kcal/mole for the first one and by 20.8 kcal/mole for the second
one (see table III and Fig. 11) ; the antiparallel stacked dimer even
becomes a maximum of energy. Furthermore, as shown by Fig. 11, the position of the third energy minimum obtained in vacuo moves from.θ = 90°
for F = 0 to θ = 60° for F = 0.1 V/Å and to θ = 30° for F $\simeq$ 0.6 V/Å. As
indicated above, the same phenomenon has been obtained for stacked complexes obtained by first turning upside down one of the molecules.

Anyway, even when retaining the most favourable configuration the stabilization energy of the stacked complex in the field (ΔE^F) is markedly
weaker than the energy of the complex for zero field. Instead of - 9.1
kcal/mole, we get - 6.0 kcal/mole at F = 0.2 V/Å, - 4.7 kcal/mole at
F = 0.4 V/Å and - 4.1 kcal/mole at F = 0.6 V/Å. Thus, we obviously see
there a first possibility of explaining the rather weak values found by
Verkin et al. [16] (see the begining of the present section C).

V. DISCUSSION

The H-bonding energies that we have calculated using ab-initio multipoles
(method A in ref. [17])are in quite good agreement with experimental data
of Yanson et al. [8] , provided that the perturbing effect of the ionizing
electric field is properly taken into account.

1) A-U and A-T complexes (see figure 5) :

From our Table II, we may notice that the Hoogsteen A-U (and A-T) dimer
is not destabilized by the electric field, thus we think that this is
the one actually observed by Yanson et al. in their experiment [8] .

From ref. [8] dimer A-U is the most stable one (14.5 kcal/mole versus
13.0 kcal/mole), but from our calculations in vacuo (results A in ref.
[17]) dimer A-T is more stable than dimer A-U (13.6 kcal/mole versus 13.3
kcal/mole). The present work shows that the electric field reverses the
result found in vacuo for values of F from 0.1 V/Å to 0.6 V/Å, and the
energy difference between A-U and A-T dimers increases with the value of
the electric field : $\Delta E_{A-U} - \Delta E_{A-T}$ has been calculated as 0.5 kcal/mole
and 1.1 kcal/mole for F = 0.1 V/Å and F = 0.6 V/Å respectively. Thus our
calculations suggest that the 1.5 kcal/mole energy difference (in favor
of A-U) found experimentally may be an artifact due to the experimental
conditions (presence of a strong enough electric field).

2) G-C complexes (see figure 3) :

There are two H-bonded possible dimers for the pair G-C : The Watson-
Crick pair existing in the nucleic acids and the dimer A2 (see fig. 3A2).

The Watson-Crick pair is the most stable one : - 23.8 kcal/mole versus - 14.6 kcal/mole. These two calculated values bracket the experimental one : - 21.0 kcal/mole [8]. Watson-Crick dimer is destabilized by an electric field, but dimer A2 is not (see Table II and Fig. 7) ; Thus, for F > 0.3 V/Å dimer A2 becomes the most stable one, but its calculated energy is smaller than the 21.0 kcal/mole found experimentally ; we may therefore ask the question : what happens exactly in the experimental situation with a complex whose energy strongly varies with the value of the electric field ?

In their paper [8], Yanson et al. discussed whether the time τ' for establishing the dimerization equilibrium was actually small enough with respect to the ionization time τ, and concluded that this was actually the case. By the way, they argued that ionization took place very near to the electrode (a few Å apart), essentially because the electric field being maximum there, so is the ionization probability, too. But we think that the influence of the electric field on the complex adds a new dimension to the problem. Indeed, for complexes destabilized by this field, there is a competition between the ionization process on one hand, and the association process on the other hand : actually, the probability of the first process decreases when the distance to the electrode increases, while the reverse holds for the second process (since the complex is destabilized with respect to the separated molecules when the electric field increases). It is therefore not trivial to predict the final outcome. A detailed investigation is presently being pursued. Let us give here only a general outline of the full statistical treatment.

The association constant obtained directly from experiment is :

$$K_a^+ = \frac{N_{I-II(+)}}{N_{I(+)} \, N_{II(+)}} \tag{V . 1}$$

$$\text{with } N_{I-II(+)} = \iiint n_{I-II}[F(\vec{r})] \quad P_{I-II}^{ion} \, [\, F(\vec{r}) \,] \, d\vec{r} \tag{V . 2}$$

$$N_{I(+)} = \iiint n_I \quad [F(\vec{r})] \quad P_I^{ion} \, [\, F(\vec{r}) \,] \, d\vec{r} \tag{V . 3}$$

$$N_{II(+)} = \iiint n_{II} \quad [F(\vec{r})] \quad P_{II}^{ion} \, [\, F(\vec{r}) \,] \, d\vec{r} \tag{V . 4}$$

where $N_{I-II(+)}$, $N_{I(+)}$ and $N_{II(+)}$ denote the number density of ions formed per unit time (for the complex (I-II) and the separated molecules (I) and (II) respectively ; n_{I-II}, n_I and n_{II} denote the number densities of the corresponding neutral species (these densities depend on the electric field, since they involve Boltzmann factors associated with the energies E^F_{I-II}, E^F_I and E^F_{II} which are themselves functions of F) ; and finally P_{I-II}^{ion}, P_I^{ion}, P_{II}^{ion} denote the corresponding ionization probabilities (functions of F, too).

When a complex is not destabilized by the electric field (such as Hoogsteen A-U and A-T dimers), the stabilization energy ΔE^F_{I-II} does not depen on F and in that case the essential contributions to N^+_{I-II}, N^+_I, N^+_{II} come from the region closest to the electrode (F largest) and K^+_a may be factorized as :

$$K^+_a = \frac{n_{I-II} \ (r \ min)}{n_I \ (r \ min) \ n_{II} \ (r \ min)} \qquad \frac{P^{ion}_{I-II} \ (r \ min)}{P^{ion}_I \ (r \ min) \ P^{ion}_{II} \ (r \ min)} \qquad (V \ . \ 5)$$

or $\quad K^+_a = K_a \dfrac{P^{ion}_{I-II} \ (r \ min)}{P^{ion}_I \ (r \ min) \quad P^{ion}_{II} \ (r \ min)} \qquad\qquad\qquad (V \ . \ 6)$

where K_a denotes the association constant for the neutral species. Hence : d (log K^+_a) / d (1/T) = d (log K_a / d (1/T)) = $-\Delta H$. But the situation is less simple in the general case when a complex is more or less destabili- zed by the electric field since n_{I-II} (F) ↗ while P^{ion} (F) ↘ as F ↘ (i.e. when the distance r to the electrode increases). In that case eq. (V .1) cannot be factorized and we have to calculate explicitly N^+_{I-II}, N^+_I, N^+_{II} by integration over the distance $\vec{r}$.

3) C-C complexes (see figure 4) :

When considering Cytosine-Cytosine dimer, Yanson et al [8], because of the N1 methylation, think to have obtained A1 dimer (Fig. 4A1), and they have noticed for this pair a dependence of their computed ΔH on the elec- trode voltage as recalled in the introduction.

From our calculations (see table II et Fig. 8, curve 1') we obtained the values reported by Yanson et al. [8], namely − 17, − 16 and − 12 kcal/ mole for the electric field values F = 0.05, 0.08 and 0.2 V/Å, respecti- vely.

In actual fact, the formula F = $U_e/5 \ R_e$ would give higher values of F near the electrode for the values of U_e used in the experiment (U_e = 1.3 kV, 2 kV and 3 kV would give F = 0.26 V/Å, 0.4 V/Å and 0.6 V/Å, respec- tively). But, in the same way as for the G-C pair, we may argue that, due to the occurence of the two competing factors (ionization probabili- ty versus Boltzmann factor), the highest value of the electric field does not provide the total amount of complex ions formed, and it could there- fore be understood that the Van't Hoff plot for K^+_a ultimately gives a value of ΔH which corresponds to some "effective" (intermediate) value of the field F.

4) U-U complexes (see figure 6) :

The experimental results of Yanson et al. [8] actually concern 1-methyl- Uracil instead of Uracil itself. As concerns dimers A1 and A2, this difference should not have important consequences, since the hydrogen bonds then involve the N3 nitrogens only. Now, our results indicate that these two dimers are significantly destabilized when the electric field increases, and this behaviour would suggest that these configurations were not the one(s) observed by Yanson et al. [8], since these authors claim that their evaluated ΔH does not vary with the electrode voltage U_e.

Dimer A4 is also destabilized by the electric field, but to a lesser extent than dimers A1 and A2. Only dimers A3 or A5 are not destabilized by the electric field, and, remarkably enough, they correspond to the highest interaction energies among the five dimers (- 11.8 kcal/mole for A3 and - 11.3 kcal/mole for A5).

But these last two dimers involve N_3 H ... O (C) and N_1 H ... O (C) hydrogen bonds, and the second type of bond will be seriously affected when dealing with the N1-methyl-uracil . We may expect that this compound still gives rises to a weak N_1 CH_3 ... O (C) bond in addition to the standard N_3 H ... O (C) bond ; actually, preliminary calculations of the energy in vacuo of these two dimers gave the values - 7.2 kcal/mole and - 7.4 kcal/mole. Thus, even if they are not destabilized by the electric field, these value are from the beginning lower than the experimental one (- 9.0 kcal/mole).

The problem therefore seems to require further theoretical, and possibly experimental, investigation. Noticeably, it would be very interesting, in the light of the previous discussion, to have experimental information concerning the Uracil dimer itself (instead of the m^1-U ... m^1U dimer).

VI. CONCLUSION

We would like now to emphasize some more general aspects of this investigation.

With the presently used theoretical method, a quite satisfactory agreement is obtained between the theoretical results and the experimental data. In the present work, this agreement is of course better than in our previous work [17], since we now take into account the perturbing effect of the electric field involved in the experimental procedure. This improved agreement confirms our previous conclusions [17], according to which a crucial role is played by the accuracy of the representation of the charge distributions of the interacting molecules. The use of charges, dipoles and quadrupoles (obtained here from ab initio molecular wave functions) located essentially at the atoms and the middles of the chemical bonds (as in [17],[29]) apparently provides a satisfactory solution. The possibility of further simplification (e.g. by reducing the number of extra-atomic centers) is not excluded and would deserve further investigation.

Several years ago, it was suggested by some authors [35, 36] that the polarization energy could play an important role in stacking association. In a previous work [18], we already showed that the arguments presented in [35] from the consideration of crystal structures were not conclusive, because the calculation of the global energy minimum of a crystal lattice may give a structure in agreement with the experimental one by using an interaction energy dominated by the electrostatic and dispersion terms : the requirement for a global minimum explains the deviations with respect to a minimum calculated for a two-molecule complex (moreover it must be emphasized that, even for the two-molecule complex, the optimum geometry cannot be safely obtained by using the simplest dipole-dipole approxima- tion, because the use of a single "center of electrostatic force" is not legitimate for large molecules in the region of the van der Waals minimum, where the distance between the molecules is smaller than the size of the molecules themselves [37 - 40]). The results of our present more elabora- te method entirely support our previous conclusions : our polarization term always plays a minor role, either in hydrogen bonded or stacked con- figurations, and all results (either the energies or the configurations of the complexes) are essentially due to the interplay between the elec- trostatic, dispersion and short-range repulsion terms. In order to be fair, it must be mentioned that our present method does not explicitly includes a short-range induction term (commonly refered to as charge transfer), and the explicit consideration of such a term (which is cur- rently investigated) could increase the relative weight of an overall induction term (sum of the usual long-range "polarization" term and of the short-range "charge transfer" term), especially in hydrogen bonded complexes where the equilibrium distance between the closest atoms is rather short. But even there, this overall induction term would certainly remain smaller than the other ones, and this statement would apply even more strongly to the stacked configurations (which were the ones consi- dered in [35, 36]), since the equilibrium distances are larger there.

Adequate care must be exercized when considering a given molecule and its methyl substituted derivatives : in some cases (N_9 methylated purines for example), the perturbation due to the methyl group seems indeed small as concerns hydrogen bonded associations which involve the other side of the molecule, but in other cases (hydrogen bonded U ... U dimers), the N_1 methylation has drastic consequences since it strongly defavours two configurations which were found the most stable ones for the U ... U association ! As another example, when comparing the A ... U and A ... T complexes, the latter is found more stable in the absence of the electric field (which is consistent with the increase of dispersion energy asso- ciated with the additional methyl group of thymine), but, remarkably enough, the situation becomes reversed when the electric field increases (because A ... U is less destabilized than A ... T), in nice agreement with the experimental result. This example therefore illustrates another aspect of the rather subtle and non-trivial changes which may be brought about by methylation.

As a final remark, we would like to emphasize the fact that the experi- mental procedure may modify in a non negligible way the system under investigation and the very quantities to be measured. Although this re- mark may look fairly evident, we think that the present work shows that the analysis of such a perturbing effect may actually lead to non trivial results. From the point of view of the experimental procedure, the obvious conclusion, in the present case, is that the measurements should be systematically performed for a sequence of decreasing values of the electric field F, and the limit values (of the association constant and therefore of ΔH) should be taken for F→0. Note that this problem occurs for the specific topic considered, namely the study of intermolecular complexes (whose binding energy is rather weak), and would not necessarily occur in other areas of application of field mass spectrometry dealing with isolated molecules only (e.g. determination of molecular structures, quantitative analyses [41]).

ACKNOWLEDGEMENTS.

The authors express their thanks to Dr. C. Giessner-Prettre for having drawn their attention to some problems concerning the stacking association of 1,3-dimethyl-uracil .

REFERENCES

[1] Leonas, V.B. : Sov. Phys. - Uspekhi 15, 266 (1973).
[2] Kołos, W., Ranghino, G., Clementi, E. and Novaro, O. : Intern. J. Quantum Chem. 17, 429 (1980).
[3] Ewing, G.E. : Angew. Chem. Internat. Edit. 11, 486 (1972).
[4] Ewing, G.E. : Acc. Chem. Res. 8, 185 (1975).
[5] Blaney, B.L. and Ewing, G.E. : Ann. Rev. Phys. Chem. 27, 553 (1976).
[6] Leckenby, R.E. and Robbins, E.J. : (a) Nature 207, 1253 (1965) ; (b) Proc. Roy. Soc. (London) A 291, 389 (1966).
[7] Payzant, J.D., Cunningham, A.J. and Kebarle, P. : Can. J. Chem. 51, 3242 (1973).
[8] Yanson, I.K., Teplitsky, A.B. and Sukhodub, L.F. : Biopolymers 18, 1149 (1979).
[9] Hill, T.L. : J. Chem. Phys. 23, 617 (1955).
[10] Hill, T.L. : J. Chem. Phys. 23, 623 (1955).
[11] Stogryn, D.E. and Hirschfelder, J.O. : J. Chem. Phys. 31, 1531 (1959), and erratum, ibid. 33, 942 (1960).
[12] Beckey, H.D. : "Field Ionization Mass Spectrometry", Pergamon Press, Oxford (1971).
[13] Beckey, H.D. : "Principles of Field Ionization and Field Desorption Mass Spectrometry", Pergamon Press, Oxford (1977).
[14] Müller, E.W. and Bahadur, K. : Phys. Rev. 102, 624 (1956).
[15] Landau, L.D. and Lifschitz, E.M. : "Quantum Mechanics", Pergamon Press, Oxford (1958). Chap. X, problem at the end of section 73.
[16] Verkin, B.I., Sukhodub, L.F. and Yanson, I.K. Doklady Biophysics (Proc. Acad. Sci. USSR, Biophysics section) 245, 76 (1979).
[17] Langlet, J., Claverie, P., Caron, F. and Boeuve, J.C. : "Interactions between nucleic acid bases in hydrogen bonded and stacked configurations : the role of the molecular charge distribution", contribution to the "International Conference on Theoretical Bio-

chemistry and Biophysics", held in Goa, India, December 3-9 (1980).
Proceedings to appear in Intern. J. Quantum Chem.

[18] Caillet, J. and Claverie, P. : Biopolymers 13, 601 (1974).
[19] Caillet, J. and Claverie, P. : Acta Cryst. A 31, 448 (1975).
[20] Caillet, J., Claverie, P. and Pullman, B. : Acta Cryst. B 32, 2740 (1976).
[21] Caillet, J., Claverie, P. and Pullman, B. : Acta Cryst. A 33, 885 (1977).
[22] Caillet, J., Claverie, P. and Pullman, B. : Theoret. Chim. Acta 47, 17 (1978)
[23] Caillet, J., Claverie, P. and Pullman, B. : Acta Cryst. B 34, 3266 (1978).
[24] Caillet, J. and Claverie, P. : Acta Cryst. B 36, 2642 (1980).
[25] Claverie, P., Daudey, J.P., Langlet, J., Pullman, B., Piazzola, D. and Huron, M.J. : J. Phys. Chem. 82, 405 (1978).
[26] Langlet, J., Claverie, P., Pullman, B., Piazzola, D. and Daudey, J.P. : Theoret. Chim. Acta 46, 105 (1977).
[27] Langlet, J., Claverie, P., Pullman, B. and Piazzola, D. : Int. J. Quantum Chem. Quantum Biol. Symp. 6, 409 (1979).
[28] Langlet, J., Giessner-Prettre, C., Pullman, B., Claverie, P. and Piazzola, D. : Int. J. Quantum Chem. 18, 421 (1980).
[29] Gresh, N., Claverie, P. and Pullman, A. : Int. J. Quantum Chem., Quantum Chem. Symp. 13, 243 (1979).
[30] Claverie, P. : Chap. 2, p. 69 in "Intermolecular Interactions : from Diatomics to Biopolymers", B. Pullman ed., Wiley, New York (1978).
[31] Clementi, E., Andre, J.M., Andre, M. Cl., Klint, D. and Hahn, D. : Acta Phys. Acad. Sci. Hung. 27, 493 (1969).
[32] Dunning, T.H. Jr. : J. Chem. Phys. 53, 2823 (1970).
[33] Berthod, H. and Pullman, A. : (a) Israel J. Chem. 19, 299 (1980) (b) J. Comput. Chem. 2, 87 (1981).
[34] Goldblum, A., Perahia, D. and Pullman, A. : Int. J. Quantum Chem. 15, 121 (1979).
[35] Bugg, C.E., Thomas, J.M., Sundaralingam, M. and Rao, S.T. : Biopolymers 10, 175 (1971).
[36] Lawacsek, R. and Wagner, K.G. : Biopolymers 13, 2003 (1974).
[37] Bradley, D.F., Lifson, S. and Honig, B. : p. 77 in "Electronic Aspects of Biochemistry", ed. by B. Pullman, Academic Press, New York (1964).
[38] Pullman, B., Claverie, P. and Caillet, J. : Proc. Natl. Acad. Sci. U.S.A. 55, 904 (1966).
[39] Pollak, M. and Rein, R. : J. Chem. Phys. 47, 2045 (1967).
[40] Rein, R., Claverie, P. and Pollak, M. : Int. J. Quantum Chem. 2, 129 (1968).
[41] Beckey, H.D. : Angew. Chem. Intern. Edit. 8, 623 (1969).

QUANTITATIVE STRUCTURE ACTIVITY RELATIONSHIPS OF ANTHRACYCLINE ANTITUMOR ACTIVITY AND CARDIAC TOXICITY BASED UPON INTERCALATION CALCULATIONS

A.J. Hopfinger[1] and Y. Nakata[2]
Department of Macromolecular Science
Case Institute of Technology
Case Western Reserve University
Cleveland, Ohio 44106, USA

N. Max
Lawrence Livermore Laboratory
Livermore, California 94550, USA

Calculated physicochemical properties of the successfully predicted mode of intercalation of anthracyclines with d(CpG) dimers have been correlated with corresponding thermodynamic and biological features. It is found that the calculated intercalation energies have a high correlation to the observed apparent binding constants of eight antracyclines with DNA. This suggests the calculated intercalation energies are meaningful on a relative basis. Moreover, highly significant quantitative structure activity relationships, QSARs, relating antitumor potency and cardiotoxicity of a set of anthracyclines to physicochemical properties have been determined. The intercalation energy is again found to be a key ingredient in these correlations. Prescriptions for maximizing the difference in antitumor and cardiotoxic activities are presented.

INTRODUCTION

The anthracyclines exhibit a striking activity against a wide variety of tumors. Doxorubicin (adriamycin) and daunomycin, see fig. 1, are two particularily effective anticancer anthracyclines. However, a major limitation to anthracycline therapy is congestive, irreversible heart failure which can result from long-term usage.

Many anthracyclines are observed to bind to DNA, and the anticancer activity is considered to involve this binding process (3-4). Spectroscopic and thermodynamic studies suggest that the DNA binding involves intercalation of the anthraquinone ring between adjacent base pairs of DNA (5-7). Detailed molecular models have been proposed for the intercalation of daunomycin and doxorubicin (8-11). These models are all similar to the intercalation complex of d(CpG)$_2$-doxorubicin shown in stereo in fig. 2. This complex is characterized by the long axis of the anthraquinone ring being <u>parallel</u> to the long axes of the base-

B. Pullman (ed.), Intermolecular Forces, 431–444.
Copyright © 1981 by D. Reidel Publishing Company.

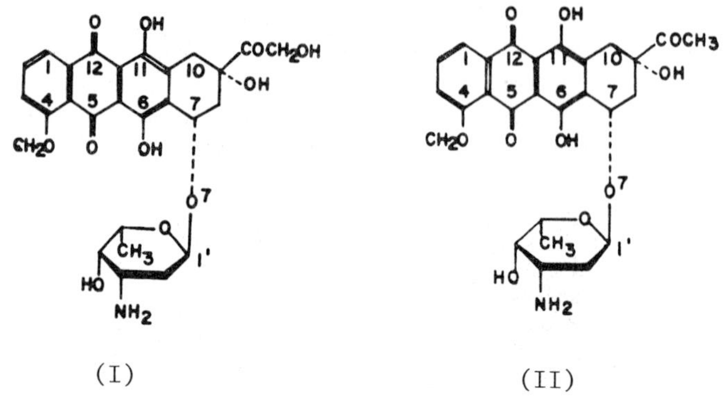

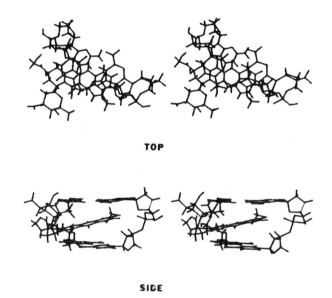

Figure 1. Chemical structures of (I) doxorubicin, (II) daunomycin.

pairs, and inserted from the <u>major</u> groove.

The structure shown in fig. 2 is an energy minimum according to molecular mechanics calculations carried out in our laboratory (12). However, the complex shown in fig. 3 is found to be 3-8 kcal/mole/complex more stable, depending upon base-pair sequence, than that shown

TOP

SIDE

Figure 2. Preferred mode of the intercalation of doxorubicin with d(CpG)$_2$ having mixed sugar ring puckering for the major groove. This is a stereo representation.

in fig. 2 (12).   The intercalation complex in fig. 3 is characterized
by insertion of the anthraquinone ring from the <u>minor</u> groove such that
the long axis of the ring is nearly <u>perpendicular</u> to the long axes of
the base pairs.   This predicted structure is supported by NMR studies
of solution complexes of nucleic acid oligomeric dimers with daunomycin (13-14).   Rich and coworkers have also reported the structure of
a d(CpGpCpGpCpG) dimer-daunomycin crystal complex (15,16).   A schematic

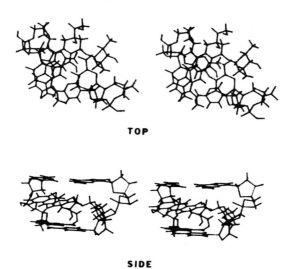

TOP

SIDE

Figure 3.   The mode of intercalation of doxorubicin into a
d(CpG) dimer for the minor groove.   This is the global
minimum energy intermolecular complex.   The nucleic acid
backbones have mixed sugar ring puckering.

representation of the crystal intercalation structure is shown in fig.
4 (16).   Obviously, the structures in fig. 3 (doxorubicin) and fig. 4
(daunomycin) are very similar thus supporting our prediction in detail.
Figure 5 is a stereo drawing of our predicted daunomycin-d(CpG)$_2$ complex.   It is again very nearly identical to figs. 3 and 4.   Thus we
conclude the mode of intercalation shown in figs. 3 - 5 is that which
occurs at least <u>in vitro</u>, and also that our calculations constitute
a correct prediction of a complicated molecular geometry.   Further, we
point out that synthetic programs to make anthracyclines based upon the
intercalation geometry shown in fig. 2 had, or have, little chance of
success.

In this regard, we have attempted to see if, in fact, our calculated properties of anthracycline-nucleic acid structures can be correlated with observed thermodynamic DNA-anthracycline binding features,
and, <u>in vivo</u> biological activities.   In this latter investigation we
have strived to construct quantitative structure activity relationships,
QSARs.

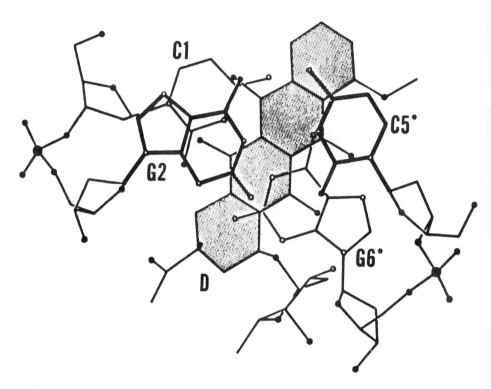

Figure 4.   Schematic illustration of the crystal complex of
daunomycin-d(CpGpCpGpCpG)$_2$ as determined by Wang et al.   Taken
from a preprint of ref. 16.

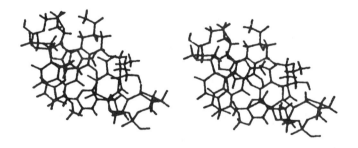

Figure 5.   Global energy minimum computed for a daunomycin-
d(CpG) dimer intercalation complex.

METHODS AND RESULTS

1. Intercalation Calculations

In all cases a d(CpG) dimer found in the ethidium ion-d(CpG)$_2$ crystal complex (17) was used to represent the intercalation site of DNA. The dinucleoside geometry was held fixed in each intermolecular energy minimization. Each anthracycline was first positioned in the doxorubicin-d(CpG) dimer minimum energy geometry. The substituents on each anthracycline, different from those of doxorubicin, were positioned in space through internal torsional bond rotations, to eliminate intra- and inter- molecular steric overlaps. In some cases the anthraquinone ring had to be partially un-intercalated to realize a sterically accept-able complex. The resulting complex, having no steric violations, was used as the starting point in an energy minimization. The degrees of freedom in this minimization are the six intermolecular dimensions and the torsional rotations of the anthracycline. It is to be noted that this procedure does <u>not</u> guarantee finding the global energy minimum for the complex. However, it is probably one of the few reasonable ways of studying the intercalation behavior of a large number of anthracyclines.

2. Comparison of Calculated Energetics to Observed Thermodynamic Properties

The apparent binding constant, K_{app}, and change in the helix-coil melt transition temperature, ΔT_m, for some DNA-anthracycline solutions have been determined. A composite listing of K_{app} and ΔT_m for eight anthracyclines taken from a review by Neidle (18) are reported in Table 1. The K_{app} and ΔT_m for two simple, and well-characterized intercalators, proflavin and ethidium bromide, are also included for a frame of refer-ence. It must be noted that the K_{app} and ΔT_m in Table 1 are composite average values. These thermodynamic properties are sensitive to both DNA source and relative concentration of the antracycline. In most cases calf-thymus DNA has been used.

Table 1 also contains the calculated intercalation energy of each anthracycline with a d(CpG) dimer. The intercalation energy has been "normalized" to a per atom of the anthracycline base. This normalization provides the relative intrinsic binding potency of an anthracycline. That is, a larger anthracycline will, by our calculations, have a large intercalation energy simply due to its size, even though it may not be an effective intercalator. The normalization process removes this anomaly, and the corresponding energies, (I.E.)s, are relative measures of complexing potency.

This is a somewhat speculative approach to quantitatively defining binding potency. However, most of the compounds in Table 1, and those in Table 2 used to construct the QSARs, contain a relatively large and constant number of atoms ($\sim$70). Thus small deviations in the number of

Table 1

Change in the DNA helix-coil transition temperature, ΔT_m, apparent binding
constant, K_{app}, and intercalation energy per intercalator atom for 10
intercalators.

No.	Compound	$\Delta T_m{}^a$	$K_{app}{}^b$	$(-I.E.\times 10)^c$
1	Proflavin	13.5	1.6×10^6	11.96
2	Ethidium Bromide	14-20	6.5×10^6	12.82
3	4-Demethoxydaunomycin	21.0	3.1×10^6	14.28
4	4'-Epi-adriamycin	12.4	2.5×10^6	14.31
5	4'-Deoxyadriamycin	8.9	2.0×10^6	13.64
6	Daunomycin	13.4	3.8×10^6	15.23
7	Adriamycin	14.8	3.0×10^6	14.43
8	N-Acetyl-daunomycin	1.0	0.002×10^6	12.62
9	3'-Epi-daunomycin	-	0.9×10^6	13.14
10	3,4'-Epi-daunomycin	-	0.9×10^6	13.30

a) For a constant intercalator concentration, $^\circ$C

b) $mole^{-1}$

c) kcal/mole/intercalator atom, for a d(CpG) dimer.

atoms (± 5) have relatively minor effects on the relative values of the
I.E. It is only when the number of atoms in an intercalating compound
varies largely from that of most of the anthracyclines that this
"normalization" procedure becomes significant.

The I.E. values of the anthracyclines in Table 1 exhibit a quantitative linear relationship with the corresponding K_{app}. This is shown
in fig. 6. The ΔT_m have a qualitative correlation with both K_{app} and
I.E., but the relationships are not nearly as striking as that demonstrated by I.E. and K_{app} in fig. 6. The I.E. and K_{app} of proflavin
and ethidium bromide do not fit into anthracycline relationship. This
may be a consequence of the normalization procedure, since proflavin
and ethidium each have significantly less atoms than an anthracycline.
However, proflavin and ethidium also exhibit, in crystal complexes with
d(CpG) dimers, different intercalation geometries from that of daunomycin (17, 19). Differences in the geometric mode of intercalation migh
also be responsible for the observed variations in K_{app} and I.E. In any

Table 2

Anthracycline data base, structure-activity descriptors, and observed and predicted antitumor and cardiac activities

No.	R	X	log P	(I.E. x10) kcal/mole/atom	$E(N^+)-E(N)$ kcal/mole	V_4 $\overset{\circ}{A}{}^3$	Antitumor Activity (log 1/C) obs.	Antitumor Activity Pred.	Cardiac Toxicity Obs.	Cardiac Toxicity Pred.
1	$-CH(CH_3)-NH_2$	NH_2	1.84	-9.91	93.3	39.1	4.34	4.27	4.27	4.23
2	$COCH_3$	$N(CH_2-\phi)_2$	5.99	-10.96	82.9	39.1	4.37	4.78	3.77	4.03
3	$COCH_2OH$	$N(CH_3)_3 \oplus$	-.95	-12.44	-	-	4.54	4.73	-	-
4	$COCH_2OCO(CH_2)_2CH_3$	$NHCOCF_3$	1.83	-11.37	51.8	39.1	4.55	4.88	5.02	5.02
5	$COCH_3$	OH	1.76	-10.98	-	-	4.64	4.71	-	-
6	H	NH_2	1.92	-11.59	93.3	39.1	5.05	4.99	4.25	4.30
7	$COCH_2OH$	$NCO(CH_2)_{10}CH$	6.57	-11.30	97.8	39.1	5.15	4.84	3.60	3.73
8	$COCH_3$	$N(CH_3)_2$	2.74	-12.38	64.0	39.1	5.17	5.41	4.74	4.76
9	$COCH_2OH$	$NCH_2\phi$	3.48	-11.80	-	-	5.18	5.22	-	-
10	$-C(CH_3)=NOH$	NH_2	2.80	-12.73	93.3	39.1	5.20	5.57	3.98	4.25
11	$-C(CH_3)=NNHCOCH_2OH$	NH_2	1.82	-13.05	93.3	39.1	5.58	5.60	4.28	4.37
12	$-C(CH_2OH)=NNHCO-$ (phenyl-Cl)	NH_2	3.97	-12.74	93.3	39.1	5.66	5.63	4.33	4.13
13	$-C(CH_2OH)=NNHCO-(CH_2)_6CH_3$	NH_2	4.26	-13.00	93.3	39.1	5.68	5.74	4.24	4.11
14	(5-Iminos)-$COCH_3$	NH_2	1.74	-13.62	93.3	39.1	5.75	5.82	4.30	4.40
15	$COCH_2OH$	$N(CH_3)_2$	2.18	-12.62	64.0	39.1	5.78	5.46	4.91	4.82
16	$C(CH_3)=NNHCO-$ (phenyl-Cl)	NH_2	4.22	-12.75	93.3	39.1	5.78	5.63	4.25	4.11
17	$C(CH_3)=NNHCO-$ (phenyl)	NH_2	3.51	-13.09	93.3	39.1	5.80	5.76	4.46	4.20
18	$COCH_2OCO-$ (pyridyl)	NH_2	1.27	-14.51	93.3	39.1	5.90	6.13	4.44	4.49

Table 2

Anthracycline data base, structure-activity descriptors, and observed and predicted antitumor and cardiac activities

No.	R	X	log P	(I.E.x10) kcal/mole/atom	$E(N^+)-E(N)$ kcal/mole	V_4 Å^3	log 1/C Antitumor Activity obs.	Pred.	Cardiac Toxicity Obs.	Pred.
19	(2,3-diCH$_3$;4-Des-OCH$_3$)-COCH$_2$OH [NO$_2$ ring]	NH$_2$	2.28	-14.51	-	-	5.94	6.27	-	-
20	C(CH$_3$)=NNHCO 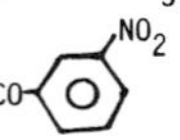	NH$_2$	3.69	-12.69	93.3	39.1	5.96	5.60	4.41	4.16
21	COCH$_3$	NH$_2$	1.83	-15.23	93.3	39.1	5.97	6.51	4.27	4.46
22	(4-des-OCH$_3$)-COCH$_3$	NH$_2$	1.85	-14.71	93.3	5.3	6.11	6.30	5.28	5.37
23	COCH$_2$OCOCH$_2$OCO(CH$_2$)$_{16}$CH$_3$	NH$_2$	1.27	-15.92	93.3	39.1	6.25	6.72	4.44	4.54
24*	(4des-CH$_3$)-CH(CH$_3$)OH	NHCOH*	1.44	-14.05	-	-	6.32	5.96	-	-
25	COCH$_2$OH	NH$_2$	1.27	-14.43	93.3	39.1	6.33	6.09	4.69	4.48
26	COCH$_2$OCO(CH$_2$)$_6$CH$_3$	NH$_2$	1.27	-14.09	93.3	39.1	6.34	5.94	4.44	4.47
27	(4-des-OCH$_3$)-COCH$_2$OH	NH$_2$	1.29	-14.23	93.3	5.3	6.46	6.01	5.45	5.41
28	(4'-epi)-COCH$_2$OH	NH$_2$	1.27	-14.31	93.3	39.1	6.50	6.04	4.44	4.48
29	(4-des-CH$_3$)-COCH$_3$	NH$_2$	1.73	-15.09	93.3	10.8	6.78	6.44	5.31	5.25

* NHCOH → NH$_2$

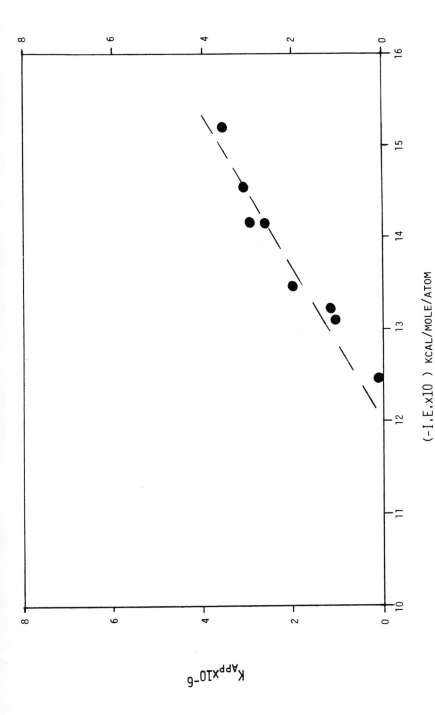

Figure 6. The "normalized" intercalation energy, I.E., versus the observed apparent binding constant, Kapp, of the eight antracyclines in Table 1.

event, it would appear from fig. 6 that I.E. is a meaningful measure
of binding strength, at least for congeneric anthracyclines.

3.   Construction of Anthracycline QSARs

Recently, Fine et al. (20) reported QSAR analyses of twenty-nine
anthracyclines of the form

anthracyclines

(III),

which are reported in Table 2. The antitumor activity of these compounds against B-16 melanoma have been measured in terms of log(1/C).
C is the molar concentration (moles/kg) producing a t/c of 125 against
B-16 melanoma introduced intraperitoneally into mice. (t) is the
increased life span of the test animal compared to that of the control
(c).

The cardiotoxicity of twenty-four of these compounds has also been
investigated. The activity is expressed in terms of a log(1/C) format
where C in this case is the concentration in (moles/kg.) of rat giving
the minimum cumulative cardiotoxicity (MCCD) (21). Fink et al. (20)
constructed a QSAR for the antitumor activity of 23 of the anthracycline
in Table 2.

$$\log(1/C) = -.41[\log P] + .48[I_0] + .81[I_1] + 6.57 \qquad (1)$$

$$N = 23 \qquad R = .874 \qquad S = .288 \qquad AE = \pm 12.0\%$$

N is the number of compounds considered, R is the correlation coefficien
S is the standard divation, and AE is the average error of prediction.
P is the water/octanol partition coefficient (log P values are reported
in Table 2), I_0 and I_1 are indicator variables having values of 0 or 1.
I_0 is set equal to 1 only if the 4-OCH$_3$ of (III) is converted to OH or
H. I_1 is set equal to 1 for the five hydrazones having "large lipohpili
moieties". Six compounds in the data base (Table 2) could not be handle

by eqn. 1 and were deleted from the analysis.   A QSAR was also developed for cardiotoxicity by Fink et al. (20).

$$\log(1/C) = -.30[\log P] + 1.01[I_o] + .69[I_1] + .74[I_2] \qquad (2)$$
$$+ 4.82$$

$$N = 21 \qquad R = .934 \qquad S = .181 \qquad AE = \pm 9.9\%$$

The descriptors in eqn. (2) are the same as in eqn. (1) and I_2 is an additional indicator variable which is assigned a value of 1 only if X is an alkylated amino group.   Three compounds could not be treated by eqn. (2) and were neglected. The main conclusion of Fink et al. (20) is that demethylation and demethoxylation at the four position of the anthracyclines appears to have a greater effect on increasing cardiotoxicity than on antitumor activity.

We have carried out a QSAR analysis of the anthracyclines in Table 2 on the basis that DNA intercalation may be involved in the biological actions of such compounds.   Succintly, we substituted I.E., calculated in the manner described earlier, for the indicator variables I_0, I_1, and I_2.   The resulting correlation equation for antitumor activity is;

$$\log(1/C) = .246[\log P] -.030[\log P]^2 -.423[I.E.x10] -.27 \qquad (3)$$

$$N = 29 \qquad R = .901 \qquad S = .289 \qquad AE = \pm 11.9\%$$

$$\text{Log } P_{opt} = 4.17$$

Log P_{opt} corresponds to that value of log P which maximizes its contribution to increasing log(1/C).   Equation (3) involves only two descriptors, log P and I.E., and can be used to explain the activity of all twenty-nine anthracyclines in Table 2 with about the same accuracy as eqn. (1).   Thus the S and AE of eqns. (1) and (3) are, respectively, about the same, but the value of R in eqn. (3) is significantly larger than that of eqn. (1).

We also attempted to use log P and I.E. to construct a QSAR for cardiotoxicity. This attempt was a complete failure. The highest correlation coefficient of a regression equation relating cardiotoxicity to log P and I.E. is only .634! This prompted us to make use of the observation by Fink et al. (18) that cardiotoxicity increased with demethylation and demethyoxylation of the 4-position. We thus included the steric volume of a 4-substitutent, V_4, as a correlation descriptor. The ability to ionize the amino group of the sugar has been postulated to be involved in the cardiotoxicity of the anthraclines (18). Consequently, we used the CNDO/2 method (22) to compute the ionization potentials of the amino nitrogens on the 24 anthracyclines for which cardiotoxicities are reported in Table 2.   The absolute values of the ionization potentials may not be significant, but we believe relative differences are meaningful. These quantities, $[E(N^+)-E(N)]$, were also used as physicochemical measures in constructing a cardiotoxic QSAR. A very significant regression equation was constructed;

$$\log(1/C) = -.103[\log P] -.042[I.E.x10] -.017[E(N^+)-E(N)] \qquad (4)$$
$$-.028[V_4] + 6.72$$

$$N = 24 \qquad R = .946 \qquad S = .145 \qquad AE = \pm 7.8\%$$

Equation (4) is capable of describing the cardiotoxic activity of all 24 compounds in Table 2 with greater accuracy than eqn. (2). Predicted activities using eqns. (3) and (4) and values of the physico-chemical descriptors, log P, I.E., V_4, and [E(N$^+$)-E(N)] are listed in Table 2.

DISCUSSION

The theoretical intercalation energies computed for a set of anthra cyclines with a d(CpG) dimer have been found to strongly correlate with corresponding observed DNA binding constants. This suggests that the intercalation calculations not only correctly predicted the geometric mode of anthracycline–DNA intercalation, but also that the associated complexing energies are meaningful on a relative basis. The fact that change in the DNA helix-coil transition temperature, due to anthracyclin intercalation, does not strongly correspond to either K_{app}, or I.E., suggests that the melt transition involves additional molecular processes besides intercalation.

A striking success of our total investigation of anthracycline-nucleic acid interactions are the QSARs expressed in eqns. (3) and (4). These highly significant correlation equations largely depend upon physicochemical properties, most notably I.E., which are ultimately based upon a postulated mode of intercalation, determined from calculations (12), that has been subsequently verified (13-16).

A comparison of eqns. (3) and (4) suggests criteria for increasing the difference between antitumor and cardiotoxic activities in the antracyclines. The coefficient in the I.E. term in eqn. (3) is ten-times larger than the corresponding coefficient in eqn. (4). Hence, we predict that the antitumor activity can be 10-fold increased over cardiotoxicity by increasing the intercalation binding energy. Secondly the antitumor activity is optimized with respect to log P at a value of log P = 4.17. Larger and smaller values of log P decrease antitumor activity. Cardiotoxicity, however, decreases as log P increases. Thus a log P of about 4 would maximize antitumor potency and, correspondingly reduce cardiotoxicity by almost one-half log(1/C) unit. This finding is contrary to eqns. (1) and (2) of Fink et al. (20) who find that increasing log P decreases both antitumor and cardiotoxic activities three to four times as strongly as that predicted in eqn. (4).

Equations (3) and (4) suggest also that cardiotoxicity can be diminished by increasing the size of the 4-substituent without an effec on antitumor potency. V_4 in eqn. (4) is equivalent to I_0 in eqns. (1) and (2). It is tempting to simply make the 4-substituent a large hydrophobic group which would increase antitumor activity and decrease cardiotoxicity. However, an inspection of figs. 2-5 clearly indicates that the wrong size and/or shape of the 4-substituent can alter the intercalation geometry and, therefore, I.E. Thus care must be taken in the selection of the 4-substituents in that V_4 and I.E. are inter-related descriptors. Fortunately, we are in the position of being able to calculate an I.E. value for any choice of a 4-substituent. Thus we have a self-consistent method of maximizing antitumor potency while minimizing cardiotoxicity.

The anthracycline QSARs represented by eqns. (3) and (4) can be judged superior to those of Fink et al. (20) on the basis of both statistical fit of the data, and that no compound in Table 2 had to treated as an outlyer. Further, the procedure used to construct eqns. (3) and (4) is totally self-contained, and is based upon actual physico-chemical properties, not statistical constructs like indicator variables.

ACKNOWLEDGEMENTS

This work was supported by a contract from the National Cancer Institute (contract NO1-CP-65927) and funds from Adria Laboratories of Columbus, Ohio.

REFERENCES

1. Address all correspondence to this author.

2. Permanent Address: Department of Chemistry, Faculty of General Studies, Gunma University, Aramaki-cho, Maebashishi 371, Japan.

3. Skovsgaard, T.: 1977, Biochem. Pharmacol., 26, p. 215.

4. Meriwether, W.D. and Bachur, N.R.: 1972, Cancer Res., 32, p. 1137.

5. DuVernay, Jr., V.H., Pachter, J.A. and Crooke, S.T.: 1979, Mol. Pharm., 16, p. 623.

6. Byrn, S.R. and Doluch, G.D.: 1979, J. Pharm. Sci., 67, p. 688.

7. Zunino, F., Gambetta, R., DiMarco, A. and Zaccara, A.: 1972, Biochim. Biophys. Acta, 277, p. 489.

8. Henry, D.W.: 1976, Cancer Chemotherapy (Sartorelli, A.C., ed.) Am. Chem. Soc., Washington, pp. 15-57.

9. Neidle, S.: 1978, Topics in Antibiotic Chemistry (Sammes, P.G. ed.) Ellis Horwood, Chichester, 3, pp. 249-278.

10. Pigram, W.J., Fuller, W. and Hamilton, L.D.: 1972, Nature New Biol. 235, pp. 17-19.

11. Neidle, S.: 1977, Cancer Treat. Rep. 61(5), pp. 923-929.

12. Nakata, Y. and Hopfinger, A.J.: 1980, Biochem. Biophys. Res. Commun., 95, p. 583.

13. Patel, D.J.: 1979, Biopolymers 18, pp. 553-569.

14. Kollman, P.A., private communication.

15. Wang, H.-J., Quigley, G.J. and Rich, A: 1980, Symp. on Struct. of Large Drugs and Biomolec., Div. Phys. Chem., paper No. 12, Am. Chem. Soc. Meeting, Houston Tex., March 24-26.

16. Ibid. submitted for publication to Biochemistry.

17. Jain, S.C., Tsai, C.-C. and Sobell, H.M.: 1977, J. Mol. Biol. 114, pp. 317-331.

18. Neidle, S.: 1979, in Progress in Medicinal Chemistry, Academic Press, New York p. 152.

19. Neidle, S., Archari, A., Taylor, G.L., Berman, H.M., Carrell, H.L., Glusker, J.P. and Stallings, W.C.: 1977, Nature (london), 269, p. 304.

20. Fink, S.I., Leo, A., Yamakawa, M., Hansch, C. and Quinn, F.R.: 1980, Il Farmaco, 35, p. 965.

21. Zbinden, Bachmann, E. and Holderegger, C.: 1978, Antibotics Chemother., 23, p. 255.

22. Pople, J.A. and Beveridge, D.L.: 1970, Approximate Molecular Orbital Theory, McGraw-Hill, New York.

A MODEL FOR DRUG-RECEPTOR INTERACTIONS: THE OPIATE RECEPTOR. A PRELIMINARY REPORT.

H. J. R. Weintraub
Department of Medicinal Chemistry and Pharmacognosy
Purdue University
West Lafayette, IN   47907

ABSTRACT

Theoretical conformational analysis studies using classical, empirical potential energy functions have been carried out on a proposed model opiate binding site, cerebroside sulfate. The methods used in these calculations are discussed. Current efforts involve the interaction of opiates with the low energy conformers of cerebroside sulfate. Preliminary results indicate that cerebroside sulfate is able to adopt a conformation which is complementary to the opiate pharmacophoric pattern.

INTRODUCTION

Recent attention has been directed toward the elucidation of the molecular structures of opiate receptors. Several hypotheses have been discussed in the literature. Loh and coworkers [1-14] have proposed that cerebroside sulfate (CS) is involved in the opiate receptor function and have presented substantial supportive experimental evidence. Other workers have argued that CS is neither necessary nor sufficient for opiate receptor function [15-18]. Abood and coworkers [19,20] have proposed that phosphatidylserine binds stereospecifically to opiates. Dennis [21] has speculated that the opiate binding site is a stable complex of acidic glycolipid (such as CS) coupled with a basic polypeptide on the order of 20 amino acids in length.

It is apparent that there is considerable controversy in the development of a model for opiate-receptor interactions. In the investigations which we are undertaking, we hope to provide a theoretical model for these interactions. The model will also provide insight into drug-receptor interactions in general. The work reported herein represents our preliminary findings in tackling this large conformational problem. Intermolecular calculations representing the opiate-receptor interaction will be reported in future communications.

B. Pullman (ed.), Intermolecular Forces, 445–464.

The cerebroside sulfate (sulfatide) model receptor proposed by Loh and coworkers has been chosen for these studies, in part, because of the large volume of available experimental evidence. We are not, however, assuming that CS is the opiate receptor; rather, our studies are aimed at determining the mode of binding between opiates and CS.

BACKGROUND

Cerebroside sulfate (sulfatide) has been shown to bind stereospecifically to opiates, consistant with in vivo and in vitro data [1-14]. A schematic representation of the structure of cerebroside sulfate (CS) is shown below.

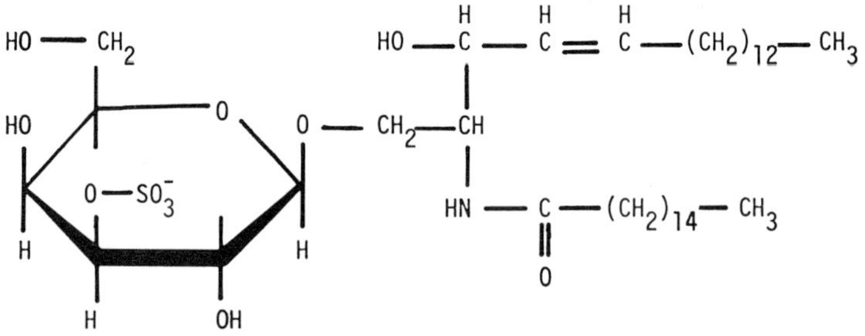

Structure I

Recently Lowney and coworkers [22] have reported the partial purification of an opiate receptor from mouse brain. This substance is a proteolipid which demonstrates stereospecific binding to levorphanol. Loh and corworkers [1-14] have established by chemical and chromatographic analyses and by narcotic binding properties that this proteolipid is "virtually identical" to cerebroside sulfate (CS). Loh has also shown that CS exhibited the highest affinity for narcotic agonists among the acidic lipids CS, phosphatidylinositol, phosphatidylserine, phosphatidic acid and triphosphoinositide. Abood and coworkers have observed stereospecific opiate binding to phosphatidylserine [19,20]. However, the binding affinity of the compounds investigated did not correlate well with their in vivo potency. In contrast, the binding affinities of various opiates to CS correlated well with their analgetic potencies [1,4,8].

Loh and coworkers have also noted that CS fulfills the structural requisites of the analgesic receptor as postulated by Beckett and Casey [23] and by Portoghese [24]. Examination of molecular models of CS indicates structural complementarity to morphine as well as similarities to the Beckett and Casey [23] receptor model [6]. It has been proposed that the major mode of opiate receptor interaction is through the formation of an electrostatic bond between the protonated nitrogen of opiates and an ionic site of the receptor model [23]. Portoghese, based on structure-activity studies, has indicated that the nitrogen atom plays an important role in the association of analgesics with their receptors. These receptors should either have some degree of flexibility or there should be more than one receptor [24]. Loh has proposed the following mode of binding [6,10]:

a) The phenyl group interacts with a complementary ceramide moiety through van der Waals bonding (through the π bonds of the ring and those of the amide);

b) The phenolic OH binds with the ceramide OH through hydrogen bonding;

c) The piperidine ring (if present) occupies the space between the ceramide and the galactose group by hydrophobic bonding, allowing,

d) The electrostatic interaction between the protonated nitrogen and the anionic sulfate group of CS.

Since CS is very similar to the purified proteolipid narcotic receptor isolated from mouse brain, we have chosen this compound as a model for an opiate receptor. By using CS as a model opiate receptor we are not assuming that CS is the opiate receptor. However, since its binding to opiates has been demonstrated, CS is a useful compound for the present modeling study. At present, empirical potential energy calculations are uniquely capable of handling this problem. The potential functions used in these calculations lend themselves equally well to intermolecular as well as intramolecular calculations, and the computational time required to examine a complex of this size is tractable only if these types of calculations are employed. Berkowitz and Loew [25] have calculated the interaction between a model anionic receptor site ($CH_3SO_3^-$) and several substituted piperidine ring systems using the CNDO method. However, a complete conformational study of CS with several opiates would not be possible using semi-empirical or more rigorous methods.

Loh and coworkers have investigated the binding affinities of a series of N-alkylnorketobemidones for CS and compared them with their pharmacological potencies [3,4]. The binding affinities of these compounds were directly proportional to their pharmacological potencies. In future communications we will report on our ongoing study of the conformational energy surfaces and binding strengths of this series of compounds interacting with CS, with respect to the change in N-alkyl chain length in order to explain the experimentally determined binding affinities [3,4].

A number of experimental approaches have been used to study drug-receptor interactions. NMR and other spectroscopic methods have been used to study enzyme-substrate interactions [26-29] and drug-DNA inter-actions [30]. We plan to use similar approaches to study the binding of opiates to CS. The knowledge gained from these studies should be useful in its own right and will provide important experimental verification for the theoretical studies.

METHODS

The CAMSEQ software system [31-33] was employed in this study. CAMSEQ utilizes a set of classical, empirical potential energy functions which have been fitted to reproduce experimental results over a variety of molecular classes [31,34-41]. The potential energy functions have been described in detail elsewhere [31,34,35,38,41] and will only briefly be reviewed at this time.

The conformational energy as a function of the torsional (dihedral) angles of the molecule is approximated by

$$E(\chi_i) = E_{steric} + E_{electrostatic} + E_{torsional} + E_{hydrogen\ bonding} + E_{solution} \tag{1}$$

where the χ_i are the relevent torsion angles of the molecule, and the individual potential energy terms are described below.

The steric nonbonded energy is approximated by the pairwise Lennard-Jones 6-12 interaction term:

$$E_{steric} = -A/r_{ij}^6 + B/r_{ij}^{12} \tag{2}$$

where A and B are the attractive and repulsive coefficients and r_{ij} is the interatomic distance for the interacting atom pair. Tables of the A and B parameters are found in references [31] and [41].

The electrostatic or coulombic energy term can be represented by a variety of functions in CAMSEQ. The Coulomb's law function was chosen for this study. This function is represented by

$$E_{electrostatic} = \frac{332 \times q_i q_j}{\varepsilon r_{ij}} \tag{3}$$

where the q_i are the partial atomic charges on the atom pair, 332 is a units conversion constant, and ε is the dielectric constant [41,42], chosen as 3.5 in this study. Partial atomic charges were estimated from CNDO/2 calculations performed on selected fragments of CS.

Torsional correction terms were based on experimentally determined values from the literature and other sources [41-50]. The actual functions used are listed in Table I.

In an effort to determine environmental effects on CS, calculations were performed on the isolated molecule (vacuum approximation) and in simulated aqueous, and 1-octanol solutions. A hydration shell model described originally by Gibson and Scheraga [47] and modified by Hopfinger [48-49] and later by Weintraub and Nichols [38] was used in these simulations.

Table I.   Torsional "Correction" Functions for sulfatide calculations.

Group	Torsional Correction Function
CX_3—CX_3 (with τ)	$2.7 \cdot SIN^2(3/2 \cdot \tau)$
CX_3—OX	$0.86 \cdot SIN^2(3/2 \cdot \tau)$
NX_2—$C{\overset{\displaystyle =O}{X}}$	$10.82 \cdot [1-COS(2 \cdot \tau)]+0.92 \cdot [1-COS(\tau)]$
CX_2—CX_3	$0.97 \cdot [1+COS(6 \cdot \tau)]$
NX_2—CX_3	$0.2 \cdot SIN^2(3/2 \cdot \tau)$
CX_3—$C{\overset{\displaystyle =O}{X}}$	$0.58 \cdot [1+COS(6 \cdot \tau)]$

The geometries used in this study were taken from several sources. Since crystallographic coordinates were not available for CS, the molecule was assembled from several fragments using CAMSEQ/M [51], a microprocessor-based highly interactive, graphics version of CAMSEQ. CS may be conveniently broken down into three fragments:
a) galactose sulfate
b) peptide unit
c) hydrocarbon fatty acid chains
Galactose sulfate was assembled from the crystal structures of β-D-Galactose [52] and an organic methyl sulfate [53]. These structures were obtained through the Cambridge Crystal Data Base, a component of the NIH-EPA-Chemical Information System [54]. The hydrocarbon chains and peptide unit were built from standard bond lengths and bond angles [55]. The various fragments were assembled using standard geometries. In these preliminary studies, the geometry of the galactose ring was held in the crystal conformation; similarly, only torsional rotations were made on the remainder of the structure; bond length and bond angle distortions were frozen at their "ideal" values [55]. Current studies involve geometry optimization at various stages of the conformational study using molecular mechanics and semi-empirical molecular orbital techniques as embodied in the MMI/MMII/MMPI [56] and MNDO [57] computer programs.

The partial atomic charges were determined with a fragment approach using CNDO/2 [58]. Figure 1 illustrates the fragments used to determine the charge distribution. There was considerable overlap in the fragments to help assure a reasonable final charge distribution as indicated in Figure 2.

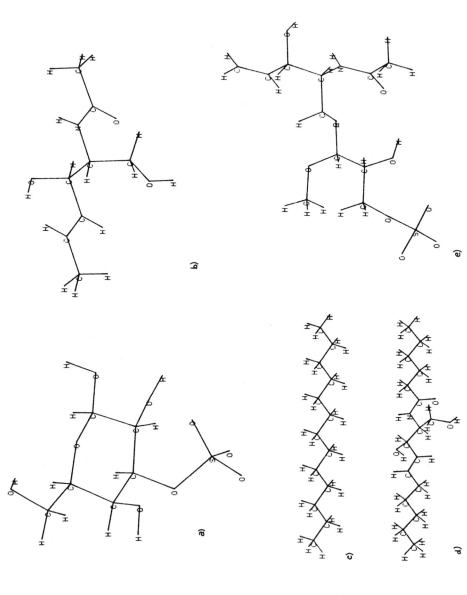

Figure 1.   Fragments used in the determination of the electronic charge distribution.

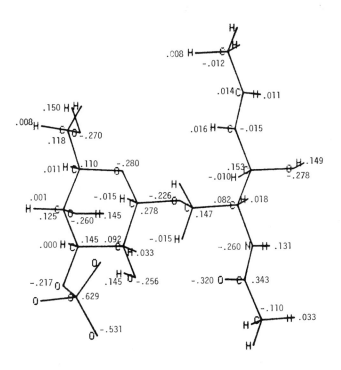

Figure 2. Charge distribution used in the calculations on the "total" cerebroside sulfate molecule. The hydrocarbon chains have been shortened for clarity.

DISCUSSION

As illustrated in Structure I above, cerebroside sulfate has a very large number of degrees of rotational freedom. The galactose sulfate fragment of the molecule has seven rotatable bonds; the fatty acid chains have a total of 28 rotatable bonds between them; the remainder of the molecule has 6 significant bonds, neglecting the amide linkage, which has partial double bond character, and the double bond in the vinyl group. Therefore, one can count 41 rotatable bonds which must be considered. It is necessary to reduce this problem to a more manageable size.

Turning to the fragment approach again, CS was broken down into five major fragments for study, as shown in Figures 1 and 3. These

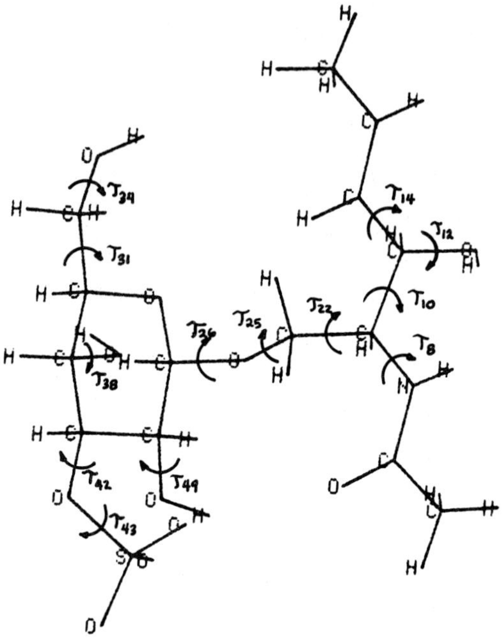

Figure 3. Additional fragment used to study the conformational energy surface of CS.

fragments were examined individually; the total molecule was then studied using the knowledge of allowed and disallowed regions of the potential energy surfaces of the respective fragments to reduce the total available conformational space to a tractable level.

The utility of this technique lies in the observation that as a molecule becomes more highly substituted, its allowed conformational space normally is reduced. The addition of substituents to a parent structure rarely opens up regions of the conformational potential energy surface which were disallowed for the parent compound. (This might occur if strong hydrogen bonding or some other strong interactions were present, however.) A comparison of the potential energy surfaces for a small fragment and a larger fragment (which includes the smaller one) will be presented in a later section.

Because of the large number of degrees of rotational freedom, there is no guarantee that all regions of low energy were detected in this study. However, through the use of the fragmentation approach, the number of highly interdependent rotatable bonds in each fragment was greatly reduced. The discussion which follows describes the determination of the conformational potential energy surfaces for each of the five major fragments illustrated in Figures 1 and 3 (above).

The simplest fragment (b) consists of the ceramide moiety, no fatty acid chains and no sugar ring. There are 8 degrees of rotational freedom: two methyl group rotations, two hydroxyl group rotations, and four other "major" rotatable bonds. Over 24,000 conformational states were explored in the determination of the conformational energy surface of this compound. The reference conformation for this fragment is indicated below:
   a) methyl groups staggered with respect to the adjacent groups
   b) hydroxyl groups set "trans" to the H on the parent carbon
   c) "backbone" torsion angles set to 180° (all trans, extended
      conformation)
Initially, τ_8, τ_{10}, and τ_{14} (refer to Figure 3) were considered to be the "major" rotatable bonds, with τ_{22} being of secondary importance, followed by the two hydroxyl rotations (τ_{12}, τ_{25}) and lastly, the methyl rotations. Upon inspection of a CPK model of this structure, it became apparent that rotation of the CH_2-OH group (τ_{22}) was critical in determining the conformational flexibility of τ_{10} and τ_8, and had little influence over the rotation of τ_{14}. Therefore the major bonds under consideration became τ_8, τ_{10}, and τ_{22}. These torsion angles were sequentially "scanned" over a 20° grid. The low energy regions of the resultant conformational energy surface were then explored while varying τ_{14} and the hydroxyl group rotations. The final conformational energy surface for this fragment was later used in the exploration of the conformational hyperspace of fragments (d), (f), and the total CS molecule. Figure 4 shows the "allowed" regions of the conformational energy surfaces associated with this fragment. We have arbitrarily defined an allowed region to be within approximately 5-6 kcal/mole of the global minimum energy conformation. The global minimum energy conformation for this fragment is shown in Figure 5. The terminal methyl groups (which represent the two hydrocarbon chains) are oriented such that the chains could adopt a nearly parallel orientation as traditionally envisioned for such a structure. We will discuss this further below. There was very little observed effect of solvent on the conformation of this fragment.

The next fragment (a) is the galactose sulfate ring structure. The reference conformation for this compound was taken to be the crystal structure of β-D-Galactose [52]. The initial orientation of the sulfate group placed the sulfur atom along the vector defined by the galactose hydroxyl hydrogen. The sulfate oxygens were oriented in a staggered conformation with respect to the ring-O bond.

This fragment has 7 rotatable bonds. Four of these are hydroxyl groups, two involve the sulfate group, and the final involves the

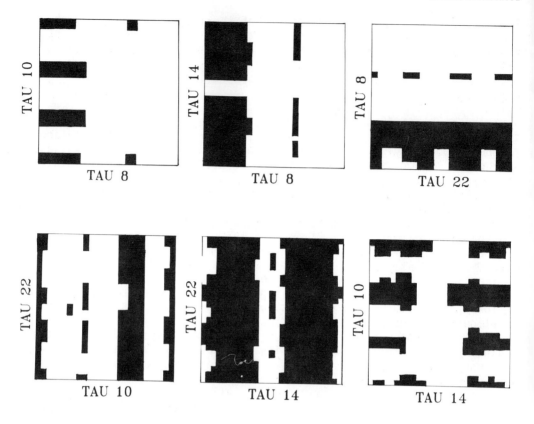

Figure 4.   Allowed conformational space for the sulfatide no hydrocarbon chains, no sugar ring fragment.   Shaded regions are within 5-6 kcal/ mole of the global minimum energy conformation.

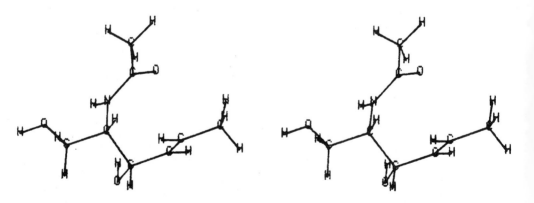

Figure 5.   Stereoscopic molecular drawing of the "no chains, no sugar" fragment of CS.   Note the orientation of the two terminal methyl groups which represent the two hydrocarbon chains.

CH_2OH substituent rotation. A CPK model of galactose sulfate was examined to determine a strategy for the exploration of the conformational energy surfaces. The scheme which was used involved rotation of all bonds, initially in groups of two or three adjacent rotations, and moving around the ring in a sequential fashion. The presence of the ring oxygen made this approach feasible, since it provided a gap across which steric interactions were not a major factor. As might be expected, free rotation was calculated for all of the hydroxyl groups. The CH_2OH group rotation had local minima at the expected $120°$ intervals. The rotation of the hydroxyl group between the CH_2OH and sulfate groups was restricted when it approached the adjacent methylene hydroxyl group. The sulfate group, as expected, preferentially adopted a conformation essentially axial to the ring, due to the steric bulk of the three sulfate oxygens. As in fragment (b), there was no appreciable solvent effect on the conformational preferences of this molecule. The lowest energy conformations for the rotations of the hydroxyl groups did not correspond to the crystal structure. They were, however, within 1-1.5 kcal/mole of the crystal structure conformer.

Fragment (c) represents the two hydrocarbon chains. Three models were used for this fragment: a C6 chain; a C13 chain; and a C13 chain with the vinyl group attached. Many authors have studied the conformational energy of a polymethylene chain [59]. A combined total of over 17,000 conformational states were calculated to assure that all reasonable care was taken to locate all local minima. Our results were consistant with experimental data [60-62].

The techniques used were selected to reduce the effective number of degrees of rotational freedom. The vinyl-containing C13 chain fragment has 14 rotatable bonds. This problem can be significantly simplified if the "equivalence condition" [41] is applied to the 10 methylene rotations in the $-(CH_2)_{11}-$ portion of the chain.

The "equivalence condition", EQC [41] has been used for homopolymer homopolypeptide, and syndiotactic polymer $-(ABABAB)-$ [63] conformational analysis. It is based on a statistical analysis of conformational states in a long polymer chain. The principle is illustrated for polymethylene below.

$$CH_3 \diagup CH_2 \diagdown CH_2 \diagup CH_2 \diagdown CH_2 \diagup CH_2 \diagdown CH_2 \diagup CH_2 \diagdown CH_3$$

As illustrated above, an "infinite" polymer of polymethylene consists of 2 methyl group rotations and an infinite number of methylene group rotations. This is the ideal case for application of the EQC. In this case, statistically, there is no way to distinguish the nth methylene rotation from the $n+1^{st}$, since the molecular environment of both of these residues is identical (if residue n is far removed from the ends of the polymer). By induction, we can extend this observation to all of the internal methylene rotations except those near the terminal methyl

groups, where "end effects" become important. Therefore, we can
approximate the conformational energy surface for a tridecamer of
polyethethylene, $H_3C-(CH_2)_{11}-CH_3$ by considering 2 distinct rotations:
i) the two methyl rotations and ii) the 10 methylene group rotations
as shown below.

A more accurate representation might include an additional rotation
that of the penultimate methylene (rotations "P" above). This would
better account for end effects. Several calculations were made representing various EQC approximations, including accounting for the pen-
penultimate rotations in the chains. Calculations using the EQC and a
repeat group of (CH_2-CH_2) were performed; this approximation involves
two sets of equivalenced rotations, one between repeat groups and one
within the repeat group. In addition, random, pseudo-minimization
calculations were made on the fragments. The results indicated that
there were several local minima; most of these were symmetry related.
Two distinct minima are illustrated in Figure 6. A comparison of the
stick drawings would lead one to believe that there are two distinct
three-dimensional structures of this fragment; however, the spacefilling drawings show that both conformers are linear, although one has
a shorter end-to-end length and a larger cross sectional area; they are
quite similar in gross appearance.

The three fragments discussed above represent the entire CS
molecule when assembled. Fragment (d) is an assembly of a portion of
each of the hydrocarbon chains (c) and the "no chain" fragment (b).
This unit was chosen because it is smaller than the full molecule and
adds only a few additional rotational degrees of freedom to (b). The
influence of the hydrocarbon chains on the conformational flexibility
of (b) was explored with this fragment. Preliminary studies indicate
that the hydrocarbon chains can assume the conformation shown in
Figure 7, in which the global minimum energy conformer is shown.

The following general considerations apply to the calculations.
A common numbering system has been used for the fragments and the total
molecule. All torsional rotation angles are measured in a clockwise
sense as indicated by arrows in Figure 3. The starting points are
defined as zero degrees rotation and are indicated in Figure 3.

The assembly of the fragments was done using the interactive
graphics capabilities of CAMSEQ/M [51]. Fragments (a), (b), and (c)
were connected and oriented into an "observable" all trans conformation

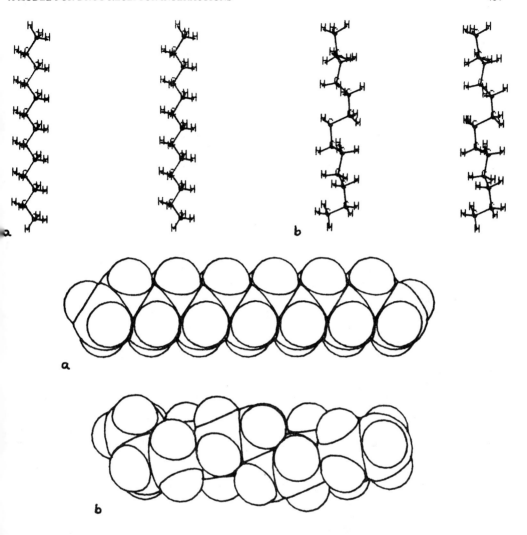

Figure 6. A comparison of the all trans extended conformer of polymethylene (a) and a low energy conformer with alternating torsion
angles of 200° and 300° (b). Below the stereoscopic pairs are spacefilling representations of the conformers [64].

(i.e. most torsion angles were set to 180°). Initially, the galactose
sulfate hydroxyl and sulfate group rotations were set to their minima
as determined in the galactose sulfate, fragment (a), conformational
studies. Likewise, the bonds in the "no chains, no sugar" fragment
(b) were set to their respective minima. An initial sequential scan
at 30° resolution was performed on torsion angles τ_{22}, τ_{25}, and τ_{26}.
Following this initial scan, the "allowed" conformational regions of
all torsional rotations in fragment (a) were explored systematically
and simultaneously with τ_{22}, τ_{25}, and τ_{26}.

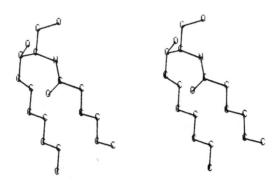

Figure 7.   Global minimum energy conformer of fragment (d).

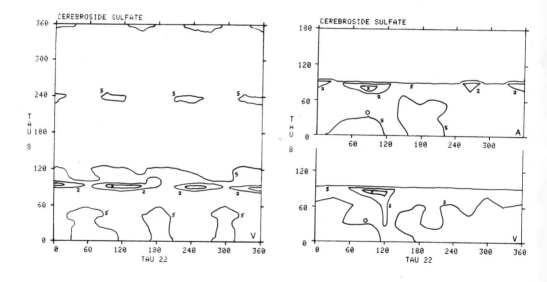

Figure 8.   Conformational isoenergy contour maps for bonds τ_{22} versus τ_8. On the left is the map for fragment (b) with no hydrocarbon chains and no sugar ring. On the right is the same plot for the full molecule. Since there were no local minima for $\tau_8 > 90°$, the map has been split to illustrate the effect of an aqueous environment on CS (top right). The contour intervals are plotted at 1, 2, and 5 kcal/mole above the global minimum, denoted by "x". The small circles on these maps and those of Figure 9 represent the conformer shown in Figure 10.

Figure 8 compares the conformational potential energy surfaces for τ_{22} versus τ_8 for the "no chain, no sugar" fragment (b) on the left, and the full molecule on the right. The full molecule map did not contain any low energy "allowed" regions for τ_8 greater than 90°. The fragment (b) map, on the other hand, had several small local minima at $\tau_8 = 240°$. Due to the complexity of the problem, we do not rule out the possible existence of these local minima. Calculations are underway to examine this region in detail. As discussed previously, the conformationally accessible region for the full molecule (right) is more restricted than that of the fragment (left).

We have analyzed the "disallowed" regions of these energy maps and related them to the structural features of the molecules. The high energy region at $\tau_8 > 90°$-120° results from the steric interaction of the amide moiety with the CH_2-0-Galactose Sulfate methylene hydrogens. The abrupt rise in energy at $\tau_8 < 0°$ is due, in part, to an interaction with the vinyl group and/or the CH_2-OH group adjacent to the vinyl. One observes that there is relatively free rotation about τ_{22}; the minimum energy conformer occurs at approximately $\tau_{22} = 120°$, $\tau_8 = 90°$. There is little difference between the vacuum (free space) energy map presented here, and the octanol energy surface (not shown). A comparison of the full molecule vacuum and aqueous environment maps indicates an increase in the conformationally accessible regions in the latter.

The conformational energy surfaces for τ_{22}, τ_{25}, and τ_{26} are shown in Figure 9. There are subtle differences between the respective free space (vacuum) and aqueous solution maps. The conformationally allowed regions are similar; the actual location of local minima is solvent dependent.

Rotation about τ_{25} changes the orientation of the sugar ring with respect to the rest of the molecule. The high energy regions between $\tau_{25} = 120°$ and $\tau_{25} = 240°$ reflect the steric interaction between the sugar ring and the ceramide moiety. The conformationally dissallowed regions for τ_{26} result from a) the interaction of the hydroxyl group (between the sulfate and the ether linkage of the sugar ring) with the methylene hydrogens of the ether linkage ($\tau_{26} > 330°$, $\tau_{26} < 300°$) and b) the interaction of the ring oxygen with the same methylene hydrogens ($\tau_{26} < 120°$).

SUMMARY AND CONCLUSIONS

This preliminary report describes our work on a multifaceted problem involving large scale intramolecular conformational energy calculations, intermolecular conformational calculations, and docking maneuvers between a drug and a model receptor.

We have chosen to use a fragmentation approach in the determination of the conformational properties of cerebroside sulfate (CS) due to the

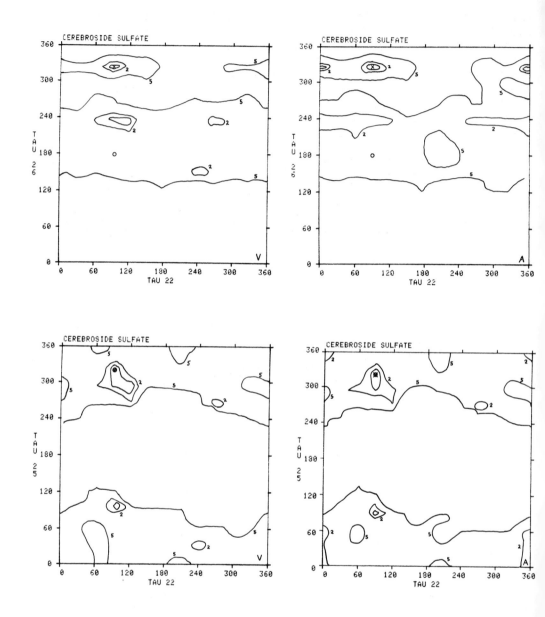

Figure 9. Conformational isoenergy contour maps for τ_{22}, τ_{25}, and τ_{26}.
"V" or "A" in the lower right-hand corner denotes vacuum (free space) and
aqueous environments, respectively. Contour intervals are plotted at
1, 2, and 5 kcal/mole above the global minimum, denoted by "x".

size of the molecule.   In this way we were able to explore a 40 dimensional conformational hyperspace with reasonable assurance of accuracy.
It cannot be guaranteed that we have not overlooked important local
minima in our analysis of the problem.   However, through the use of
extensive sequential scanning, random pseudo-minimization scanning, and
true minimization techniques, we feel that our results, though still
in their preliminary stages, are representative of the conformational
preferences of CS.   Over 76,000 individual conformational states were
analysed in these "preliminary" studies.

We have generated computer displays and built molecular models of
most of the major local minima conformers.   A qualitative model of
morphine "bound" to a low energy conformer of CS has been generated
based on the scheme proposed by Loh and coworkers [1-14] (discussed in
the Background section, above).   This complex is illustrated in Figure
10.   Our preliminary results indicate that this "binding conformation"
of CS is energetically feasible, lying within approximately 3-4 kcal/
mole of the global minimum energy conformer.

Figure 10.   Spacefilling representation of a possible binding conformation for CS.   Morphine (shaded) is shown bound to CS.   This is a
qualitative picture based on Loh's model [6,10].

We plan to further explore the conformational hyperspace of CS to
more completely describe the geometrical properties of the molecule.
Classical empirical, intermolecular calculations will be performed on
morphine and other analgetics to understand more fully the mode of
binding and relative binding afinities.   We plan to employ both semi-
empirical and molecular mechanics-type geometry optimization schemes.
In order to characterize the electronic nature of the interaction,
approximate molecular electrostatic potential surfaces will be calculated.

Future communications will report our progress in these areas. The long term goal of these studies is to develop reliable methods which can be used to study and characterize drug-receptor interactions.

ACKNOWLEDGEMENTS

The author wishes to thank Professor David E. Nichols for his stimulating discussions on opiate receptors. Many of the calculations described in this report were performed by Tiee-Leou Shieh and Ann T. McKenzie. Chris Borgert assisted with figure preparations. Thanks also to Ms. Camilla Cox for typing the manuscript. Partial support for this work, was provided by the National Institutes of General Medical Sciences, NIH, (GM25142) and McNeil Laboratories.

REFERENCES

1. H.H. Loh, T.M. Cho, Y.-C. Wu, E.L. Way, Life Sci, 14, pp. 2231 (197
2. T.M. Cho, H.H. Loh, E.L. Way, Proc. West Pharmacol. Soc., 18, pp. 176 (1975).
3. J.S. Cho, T.M. Cho, H.H. Loh, E.L. Way, Proc. West Pharmacol. Soc., 18, pp. 298 (1975).
4. H.H. Loh, T.M. Cho, Y.-C. Wu, R.A. Harris, E.L. Way, Life Sci., 16, pp. 1811 (1975).
5. H.H. Loh, T.M. Cho, Y.-C. Wu, Fed Proc., 34, pp. 815 (1975).
6. H.H. Loh, T.M. Cho, in "Tissue Responses to Addictive Drugs", D.H. Ford, D.H. Clovet, eds, Spectrum Publications, New York, pp. 355 (1976).
7. T.M. Cho, J.S. Cho, H.H. Loh, Life Sci., 18, pp. 231 (1976).
8. T.M. Cho, J.S. Cho, H.H. Loh, Life Sci., 19, pp. 117 (1976).
9. P.Y. Law, R.A. Harris, H.H. Loh, E.L. Way, J. Pharm. Exptl. Ther., 207(2), pp. 458 (1978).
10. H.H. Loh, P.Y. Law, T. Ostwald, T.M. Cho, E.L. Way, Fed. Proc., 37(2), pp. 147 (1978).
11. D.A. Johnson, R. Cooke, H.H. Loh, Mol. Pharmacol., 16, pp. 154 (197
12. T.M. Cho, J.S. Cho, H.H. Loh, Mol. Pharmacol., 16, pp. 393 (1979).
13. P.Y. Law, G. Fischer, H.H. Loh, A. Herz, Biochem. Pharmacol., 28, pp. 2557 (1979).
14. F.B. Craves, B. Zalc, L. Leybin, N. Bavmann, H.H. Loh, Science, 207, pp. 75 (1980).
15. G. Dawson, S.M. Kernes, R.J. Miller, B. Wainer, J. Biol. Chem., 253, pp. 7999 (1978).
16. E.J. Simon, J.M. Hiller, Fed. Proc., 37, pp. 141 (1978).
17. S.H. Snyder, S. Matthysse, Neurosci. Res. Progr. Bull., 13, pp. 1 (1975).
18. R.S. Zukin, R.M. Kream, Proc. Nat. Acad. Sci. USA, 76, pp. 1593 (1979).
19. L.G. Abood, W. Hoss, Eur. J. Pharmacol., 32 pp. 66 (1975).

20. L.G. Abood, F. Takeda, Eur. J. Pharmacol., 39, pp. 71 (1976).
21. S.G. Dennis, Progr. Neuro-Psychopharmacol., 4, pp. 111 (1980).
22. L.I. Lowney, K. Schulz, P.J. Lowery, A. Goldstein, Science, 183, pp. 749 (1974).
23. A.H. Beckett, A.F. Casy, J. Pharm. Pharmacol., 6, pp. 986 (1954).
24. P.S. Portoghese, J. Med. Chem., 8, pp. 609 (1965).
25. D.S. Berkowitz, G.H. Loew, in "Opiates Endogenous Opioid Peptides", H.W. Kosterlitz, (ed), North Holland Publishing Co., Amsterdam, pp. 387 (1976).
26. P.J. Andree, Biochem., 17, pp. 772 (1978).
27. D.A. Kooistra and J.H. Richards, Biochem., 17, pp. 2960 (1978).
28. M.H. O'Leary and J.R. Payne, J. Biol. Chem., (1978), 251, pp. 2248 (1976).
29. A.S. Mildvan, Acc. Chem. Res., 10, pp. 246 (1977).
30. R.L. Kastrup, M.A. Young and T.R. Krugh, Biochem., 17, pp. 4855 (1978).
31. H.J.R. Weintraub, Ph.D. dissertation, Case Western Researve University, Cleveland, Ohio (1975).
32. H.J.R. Weintraub, A.J. Hopfinger, Int. J. Quantum Chem., QBS2, pp. 203 (1975).
33. R. Potenzone, Jr., E. Cavicchi, H.J.R. Weintraub, A.J. Hopfinger, Computers and Chem, 1, pp. 187 (1977).
34. H.J.R. Weintraub and A.J. Hopfinger, J. Theor. Biol., 41, pp. 53 (1973).
35. H.J.R. Weintraub and A.J. Hopfinger, in "Jerusalem Symposium on Quantum Chemistry and Biochemistry", Vol. 7, E.D. Bergmann and B. Pullman, Eds. (Reidel, Boston, MA, 1974), pp. 131.
36. H.J.R. Weintraub et al., Int. J. Quant. Chem., Quant. Biol. Symp., 3, pp. 99 (1976).
37. H.J.R. Weintraub, Int. J. Quant. Chem., Quant. Biol. Symp., 4, pp. 111 (1977).
38. H.J.R. Weintraub and D.E. Nichols, Int. J. Quant. Chem., Quant. Biol. Symp., 5, pp. 321 (1978).
39. R. Potenzone, Jr., M.S. Thesis, Case Western Reserve University, Cleveland, OH, 1975.
40. R. Potenzone, Jr. and A.J. Hopfinger, Carbohydrate Res., 40, pp. 444 (1975).
41. A.J. Hopfinger, Conformational Properties of Macromolecules (Academic, New York, 1973).
42. T. Ooi, R.A. Scott, G. Vanderkooi, and H.A. Scheraga, J. Chem. Phys., 46, pp. 4410 (1967).
43. H.A. Scheraga, Adv. Phys. Org. Chem., 6, pp. 103 (1968).
44. D.R. Hershbach, "Bibliography for Hindered Internal Rotation and Microwave Spectroscopy", Lawrence Radiation Lab., Univ. of California, Berkeley, California (1962).
45. J.D. Roberts, "Nuclear Magnetic Resonance", McGraw Hill, New York (1959).
46. F.A. Bovey, "High Resolution NMR of Macromolecules", Academic Press, New York (1972).

47. K.D. Gibson and H.A. Scheraga, Proc. Nat. Acad. Sci. USA, 58, pp. 420 (1967).
48. A.J. Hopfinger, Macromolecules 4, pp. 731 (1971).
49. A.J. Hopfinger and R.D. Battershell, J. Med. Chem. 19, pp. 569 (1976).
50. M. Perricaudet, A. Pullman, Int. J. Peptide Protein Res., 5, pp. 99 (1973).
51. H.J.R. Weintraub, in "Computer-Assisted Drug Design", E. Olson, R.E. Christoffersen, eds, ACS Symposium Series, 112, Chapter 17, pp. 353 (1979).
52. F. Longchambon, J. Ohannesian, D. Avenel, A. Neuman, Acta Cryst B, 31 (1975).
53. E. Shefter, S. Singh, T. Brennan, P. Sackman, Crystal Struc. Comm., 3, pp. 209 (1974).
54. S.R. Heller, G.W.A. Milne, R.J. Feldman, Science, 195(4275), pp. 253 (1977).
55. L.E. Sutton, "Tables of Interatomic Distances and Configurations in Molecules and Ions," Chemical Society Spec. Publ. 11, 1958; Suppl. 18 (1965).
56. MMI/MMPI/MMII: N.L. Allinger, et al, as amended by H. Hönig, QCPE, 13, pp. 404 (1981).
57. MNDO: W. Thiel, QCPE, 11, pp. 353 (1978).
58. CNINDO/74: P.A. Dobosh, N.S. Ostlund, QCPE, 11, pp. 281 (1975).
59. P.J. Flory, "Statistical Mechanics of Chain Molecules", Interscience, New York (1969).
60. T. Shimanouchi, Disc. Faraday Soc., 49, pp. 60 (1970).
61. S. Mizushima, T. Shimanouchi, J. Am. Chem. Soc., 86, pp. 3521 (1964)
62. C.W. Bunn, Trans. Faraday Soc., 35, pp. 482 (1939).
63. G. Natta, P. Corradini, P. Ganis, Makromol. Chem., 39, pp. 238 (1960).
64. G. Smith, P. Gund, J. Chem. Inf. Comput. Sci., 4, pp. 207 (1978).

H-Bond-State and Solubility in Aqueous Systems
- A Working Hypothesis -

H. Kleeberg
Department of Physical Chemistry, University of Marburg, FRG

The infrared spectra of H_2O in the region of the H_2O-combination band $(\nu_3 + \nu_2)$ are investigated in the saturated organic and aqueous phase of binary mixtures H_2O-organic liquids (B). In agreement with the available heat of vapourization data the H_2O-spectra correlate with the H-bond energy of H_2O. The solubility of H_2O in organic liquids as well as that of the organic molecules in water increases with decreasing molar volume of B, its H-bond acceptor strength and the amount of H-bonds of H_2O, which is actually formed; the latter seems to depend on the stereochemical environment of the H-bond acceptor group.

Introduction

The processes in living organisms are characterized by highly specific - and effective - molecular interactions. During these processes covalent bonds are made and broken; intermolecular interactions are the guide with respect to which and how reactions have to occur; this shows that intermolecular forces are of extreme importance for life (see for example: hormone-acceptor-, enzyme-substrate interactions).

During the last years H_2O-interactions with biomolecules have been increasingly studied (Haly and Snaith, 1971; Ikada et al., 1980; Privalov and Mrevlishvili, 1967; Luck,1976, 1978 a; Luck and Kleeberg, 1978; Kleeberg and Luck 1979, 1981 a and b); it became clear that H_2O can not only be regarded as an inert solvent.

Special attention has been paid to the change in the H_2O-structure by non-polar groups (Ben-Naim, 1974; Franks and Ives, 1966; Bohon and Claussen, 1951; Krishnan and Friedman, 1969; Frank and Franks, 1968; Franks, 1975; Hermann, 1972; Hofman und Birnstock, 1980; Hildebrand and Scott, 1950 ; Frank and Evans, 1945).

B. Pullman (ed.), Intermolecular Forces, 465–487.
Copyright © 1981 by D. Reidel Publishing Company.

It is concluded to be mainly an entropy effect - in this connection the well defined but however misleading term ice-berg
gained in importance - and dominates much of the biochemical
literature (see for example: Némethy, 1967).

X-ray investigations indicated that H_2O forms H-bonds
with the polar groups of biopolymers at low H_2O-contents
(usually below 0.5 g H_2O/g polymer) (Finney, 1980; Ramachandran and Chandrasekharan, 1968; Luck, 1976).

Using infrared-(IR)-spectroscopic techniques we could
show, that these H-bonds of H_2O are present - to a similar
amount and strength - after dilution as well (Kleeberg and
Luck, 1979, 1980a, 1981a; Luck and Kleeberg, 1978) (polymer
concentration approximately 10 % by weight - this corresponds
roughly to the H_2O/organic matter ratio in living systems).
For an understanding of biopolymer-interactions we have to
take into account the change in H-bonding. If two biopolymer
- or parts of polymers (P_1 and P_2) - each interact with H_2O
after interacting with one another - i.e. the reaction
P_1---P_2 + H_2O $\rightleftharpoons$ P_1---H_2O + P_1---H_2O takes place - the interaction energy will depend on the total balance of interaction
energies of the species involved. This process may be regarded
similar to a solvation process. Furthermore the equilibria
between membranes or associated proteins and the surrounding
solution may be regarded as phase equilibria. Therefore investigations on solubility phenomena of low molecular weight
organic molecules may help to understand biopolymer interactions - apart from the large interest in the physico-chemical
behaviour of these solutions themselves.

Frequently solubility data are correlated empirically,
since the direct evaluation of interaction energies is difficult (Körösy, 1937; Butler, 1937; Copley, Zellhoefer and
Marvel, 1940; Marvel Copley and Ginsberg, 1940; Hildebrand
and Scott, 1950; Hermann, 1972; Reichhardt, 1973).

In order to demonstrate the change in the H-bond structure of H_2O in binary mixtures (H_2O-low molecular weight
polar and non-polar organic molecules) in comparison to water
IR-spectra were recorded in the range of 5555 - 4760 cm^{-1}
(1800 - 2100 nm), which corresponds to the wave length region
where the H_2O combination band $(\nu_3 + \nu_2)$ is observed. This
region was used on behalf of the following reasons:
1. quantitative measurements are possible for an extremely
   large range of concentrations for solutions as well as
   for systems containing polymers; this is due to the
   favourable extinction coefficient i.e. optical path
   length (Luck and Schiöberg, 1979; Luck, Schiöberg and
Siemann, 1980; Luck and Kleeberg, 1981);
2. the area of this band of H_2O $(\int \varepsilon \, d\nu)$ is nearly constant
   ($\pm$ 5 %), independent of H-bond strength of the organic

acceptor; therefore free OH-groups, weak and strong
H-bonds contribute to the spectrum with similar intensit

3. each H_2O-molecule contributes nearly exclusively to only
   one vibrational transition in this region (Luck and
   Schiöberg, 1979);

4. there is hardly any absorption of other molecules asides
   from H_2O in this wave length region (even the absorption
   band of alcohols is separated enough to interpret the
   spectrum of H_2O without severe difficulties);

5. a large variety of polymers have been investigated in
   this region, so that the influence of low molecular weight
   weight compounds and macromolecules on H_2O may be compared easily (Kleeberg and Luck, 1981a and b; Luck and
   Kleeberg, 1978, 1981).

Detailed results on the evaluation of intermolecular
interactions by IR-spectroscopic methods are given by Luck
(this volume).

METHODS

Near infrared spectra have been recorded with a Cary 17i
spectrometer. Absorbances were digitalized by a Fluke Digital
Multimeter and conserved on tapes by a Facit tape punch 4070;
these tapes were processed into the final plots. In the case
of aqueous solutions of CH_2Cl_2, 3-pentanol, 3-methyl-buta-
nol-(1) (see figure 3, 14) 10, 2 and 3 spectra were sampled
respectively. The optical path length was determined of the
empty cuevettes by the interference fringe method, and H_2O
concentrations by mixing weighted amounts (± 0.02 %) of H_2O
and organic liquids and subsequent determination of the density of the mixture (± 0.02 %) in 5 - 25 cm^3 pycnometers.

The thermal stability for spectroscopic and density measurements was obtained with a Haake FK thermostat within
± 0.1°C.

RESULTS AND DISCUSSION

1. THE ENERGY BALANCE IN DILUTE SOLUTIONS

The heat of solvation may be defined as the heat
associated with the transfer of one mole of solute from its
vapour at 1 atm to a solution at infinite dilution, i.e.
$\Delta H_{sol} = \Delta H_m - \Delta H_{vap}$ (Franks and Watson, 1969), where ΔH_m is
the heat of mixing or solution and ΔH_{vap} is the heat of
vapourization of a mole of solute from its pure liquid phase.

If we neglect structural changes of the solute molecules
and assume as an approximation that the heat of cavity for-

mation is zero (Eley, 1939; see also: Lange and Watzel, 1938
Hermann, 1972), the heat of mixing or solution will be due
to the heats of interaction of solute-solute, solute-solvent
and solvent-solvent. Furthermore we assume in this approximation that these interactions are composed of a term due to
H-bonding (subscript: H-B) and another one due to van der
Waals forces (orientation, induction and dispersion effects;
Briegleb, 1937; Hildebrand, 1950; subscript: vdW). If B is
the solute and C the solvent, the heat of mixing or solution
will be composed by the following terms:

$$-\Delta H_m = \Delta H_{H-B}^{B/B} + \Delta H_{vdW}^{B/B} + \Delta H_{H-B}^{C/C} + \Delta H_{vdW}^{C/C} - \Delta H_{H-B}^{B/C} - \Delta H_{vdW}^{B/C} \qquad (1$$

The addition of certain amounts of two liquids (water
(or H_2O; abbreviation: W) and a low molecular weight (m.w.)
organic liquid (B)), which are not totally miscible at a giv
temperature, results in phase separation. Thus H_2O (W) is
the solute in the organic and the solvent in the aqueous pha
which are in equilibrium.

1.1 Van der Waals and H-bond energy of pure H_2O

The major contribution to the lattice energy of ice
(56.1 kJoule/mole; Whalley, 1957) result from H-bonds and
van der Waals forces (Rowlinson, 1951; Briegleb, 1937). Base
on IR-spectroscopic determinations of the amount of free OH-
groups, Luck (1967, 1980b; Luck and Ditter, 1969,1970) could
describe the inner heat of vapourization - asides from other
properties of water - with ΔH_{H-B}^{W} = -33.5 and ΔH_{vdW}^{W} = -15.2
kJoule/mole H_2O quantitatively between the melting and criti
cal temperature. A Comparison of the heats of sublimation
of $H(CH_2)_mOH$ and $H(CH_2)_{m/2}O(CH_2)_{m/2}H$ after extrapolation to
m=0 indicates that the van der Waals and H-bond energy of ic
at the melting point amounts to approximately -18 and -31 kJou
per mole respectively. For this approximation the assumption
has been made that ΔH_{subl} of the extrapolated data of ethers
correspond to the van der Waals energy of H_2O and the differ
ence of ΔH_{subl} of the corresponding alcohol and ether value
to the contribution of an H-bond. The calculated heat of
sublimation of H_2O ($\Delta H_{subl,calc} = \Delta H_{ROR}^{m=0} + 2(\Delta H_{ROR}^{m=0} - \Delta H_{ROR}^{m=0})$) =
54.2 kJoule/mole) is in agreement with the experimental one
(48.5 kJoule/mole).

It is not surprising that other properties of H_2O (like
the melting, boiling and critical temperature or the molar
volume) can be decomposed into contributions of $ROR_{R=H}$ and
$ROH_{R=H}$ in the same manner with even better agreement between
calculated and experimental results.
One crucial point with respect to water, however, remains: matrix isolation studies (Luck and Schrems, 1980) indicate that not only linear H-bonds and not H-bonded OH (fre
OH-groups) are present, but H-bonds with unfavourable angles

as well (Behrens, Luck and Mann, 1978; Behrens and Luck,
1980); furthermore on behalf of H-bond cooperativity
effects (Gordy, 1939; Frank and Wen, 1957; Schrems, 1981)
the H-bond energy of a linear dimer($\overset{H}{\diagdown}$O-H---O$\overset{-H}{\diagup}_H$) for example
may differ considerably from a linear H-bond like in ice.

1.2 Interaction Energy of Organic Molecules

For organic substances studied in this work, we will
take $\Delta H_{vap}^{B} \simeq -\Delta H_{vdW}^{B/B}$. Only in alcohols (15.5 kJoule/mole OH)
and primary or secondary amins H-bonds have to be taken into
consideration. This is supported by IR-spectroscopic investigations (Luck and Ditter, 1968; Schiöberg, personal
communication).

1.3 Van der Waals Energy between H_2O and Organic Liquids

Solute: H_2O The lack in reliable data for experimental
heats of vapourization of H_2O from nonpolar solvents makes
a discussion on the contribution of ΔH_{vdW} in these solutions
difficult. The heat of vapourization of H_2O from benzene
(17.1 kJoule/mole H_2O at 25°C (Frank and Evans, 1945) corres-
ponding to $\Delta H_{vdW}^{W/B} + \Delta H_{vdW}^{B/B}$) is nearly identical with $\Delta H_{vdW}^{W/W}$
(see above), so that we take:

$$\Delta H_{vdW}^{B/B} + \Delta H_{vdW}^{W/B} = \Delta H_{vdW}^{W/W} \qquad \text{as an approximation.}$$

Solvent: H_2O For aqueous solutions an estimation of van
der Waals energies between solute-solvent may be gained from
the heats of vapourization of the guest molecules from clathrates (Stackelberg, 1949; Glew 1962).

In figure 1 the spectra of the system H_2O:tetrahydro-
furane = 17 at 20 (solution) and 0°C (clathrate) may be compared with those of water and ice. It is apparent, that the
spectrum of the solution differs more from the spectrum of
pure liquid water than that of the clathrate from the spectrum of ice. The H_2O-spectra of other clathrates (acetone,
SO_2) are similar to the spectrum of ice as well (Luck and
Ditter, 1969; Kleeberg and Luck, 1981b). This justifies one
to say that the H-bond state of H_2O in clathrates and ice is
very similar. The small differences may be explained by the
small deviations of the H-bond angle from linearity (Luck and
Ditter, 1969, 1970; Luck 1976). Hence the heat of vapourization of the guest molecules from the clathrates depends mainly on the van der Waals interaction of the guest molecules
with the H_2O-host lattice.

In table 1 the heats of vapourization of some clathrate
forming molecules from the clathrate, the aqueous solution
(both at 0°C) and the pure liquid at the temperature indicated in the second column may be compared. While the heats

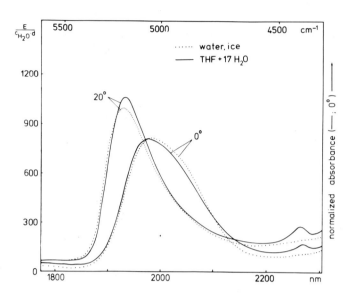

Figure 1. IR-spectra of water, ice and the system H_2O:tetrahydrofurane (THF) = 17:1 (molar ratio) at 20 (solution) and $0°C$ (clathrate). The absorbance at 4470 cm^{-1} (2240 nm) corresponds to CH-vibrations of THF.

Table 1. Heats of vapourization of clathrate forming molecules from the pure liquid (at the temperature indicated in the second column), from the solid clathrate and from aqueous solution at $0°C$; in the last column the difference between the heats of vapourization from the liquid and the clathrate is given.

	ΔH_{vap}^B [kJoule/mole] from the:			
	T (°C)	pure liquid	clathrate	aqueous solution	difference of ΔH_{vap}^B: liquid-clathrate
Br_2	0	34.4^a	34.7^b	36.6^b	- 0.3
CH_3J	0	31.5^a	30.5^b	35.6^b	- 1.0
SO_2	-10.1	25.1^a	32.2^b	31.0^b	- 7.1
CH_3Br	0	23.8^a	33.9^b	30.9^b	-10.0
C_2H_5Cl	12.2	24.7^a	36.4^b	35.1^b	-11.7

a) Landolt-Börnstein, 1962
b) Glew, 1962

of vapourization from the clathrate and from aqueous solution
are similar (the experimental error as indicated by Glew, 1962
may amount to approximately ± 2 kJoule/mole), the heat of
vapourization of the pure liquid is smaller by up to 12 kJoule
per mole. This indicates that the van der Waals energy between
B and H_2O is larger than between B/B (see last column). This
may be mainly due to the larger coordination number of B in
the clathrates as compared to the pure liquid. Per mole H_2O
the contribution of van der Waals energy amounts to less
than 1.2 kJoule.

On behalf of the various clathrate structures found (Stackelberg, 1949; Glew, 1962; Glew, Mak and Rath, 1972; Davidson, 1973), we conclude that the flexibility of H_2O in forming 4-, 5- and 6-membered H-bonded rings in order to incorporate the guest molecule is large.

2. SOLUBILITY OF NONPOLAR MOLECULES IN WATER

For different clathrates with the same host lattice
structure the melting temperature may be regarded as a measur
for the stability of the clathrate. In figure 2a the melting
temperature of several clathrates is plotted against the molar volume of the guest (pure liquid; T approximately T-mel-
ting of the clathrate). This figure shows that the stabilization of clathrates type I increases up to approximately 60
cm^3/mole guest; at higher molar volumes the host lattice
structure is changed. Melting temperatures above $0°C$ - the
melting point of ice - seem to be due to the additional van
der Waals energy per mole H_2O (see above).

In figure 2b the dependence of the reciprocal solubility
of different substances (expressed in: log (mole H_2O/mole solute) on the volume of the pure liquid solute at the same
temperature is given. The solubility has a maximum in the
region of 60 cm^3/mole solute and decreases to lower as well
as higher molecular volumes.

Figure 3a shows the spectra of CH_2Cl_2, water and a
nearly saturated solution of CH_2Cl_2. Since the concentration
of CH_2Cl_2 is 1.78 x 10^{-4} mole/cm^3 the solution spectrum can
not be distinguished from that of pure water. The difference
of these two spectra (figure 3b, upper part) is zero in the
region of the absorption band of H_2O; it shows a sharp band
in the region of CH_2Cl_2 absorption (compare figure 3b, upper
and lower part), which is shifted with respect to ν_{max} (as
indicated by the arrow in figure 3b) of liquid CH_2Cl_2.

The presence of the CH_2Cl_2-band (a combination of the
streching and bending mode of the CH_2-group) in the solution

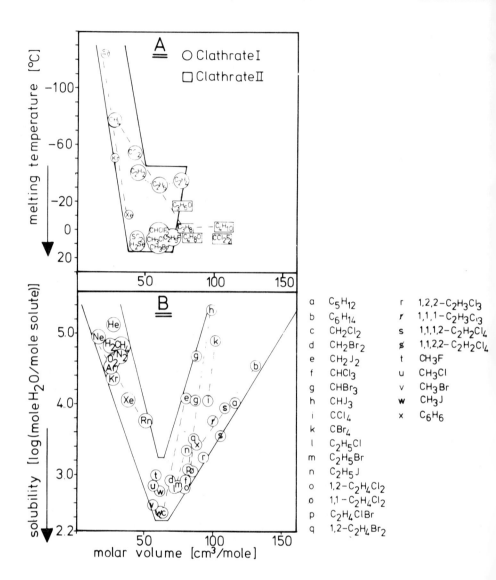

Figure 2. The dependency of the melting temperature (Davids,
1973) of different clathrates (a) and the solubility of non-
polar molecules in water at 20°C (b) on the molar volume of
the guest molecules and solute (Landolt-Börnstein, 1962) is
shown. For deteils see text.

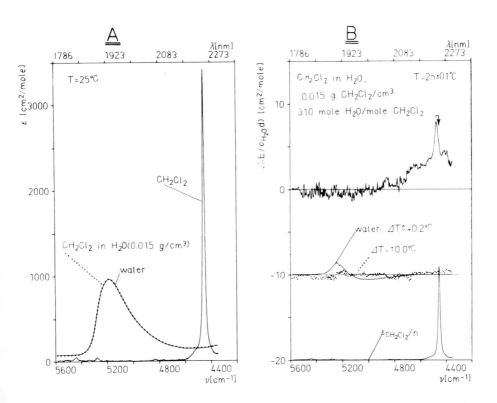

Figure 3. Spectra of water, a nearly saturated aqueous solution of CH_2Cl_2 and pure CH_2Cl_2 (a).
b shows in the upper part: the difference. between solution
and pure water spectra of 3a; lower part: the spectrum of
CH_2Cl_2 plotted corresponding to the concentration in the
aqueous solution; middle part: the difference between spectra
of water under the same conditions and the difference spectrum
of water corresponding to a temperature difference of $+0.2°C$;
the middle and lower part is shifted by -10 units.

indicates that very small changes of the H_2O-spectrum should
be detectable with the method used. The dotted line in figure
3b (middle part) shows the difference between water spectra
under the same conditions and hence is due to electrical
noise; the full line corresponds to two water spectra at a
temperature difference of $+0.2°C$. Since approximately 15 %
of the H-bonds of water are broken between 0 and $100°C$ (Luck,
1967, 1980b, Luck and Ditter, 1969) this difference corresponds to a change of 3×10^{-4} H-bonds per mole H_2O. If a
change of H-bonds - especially OH---OH$_2$ (linear) $\rightleftharpoons$ OH (free) -
of H_2O occurs on mixing with CH_2Cl_2, it must be smaller than

10^{-4} bonds/mole H_2O or, at the given concentration, below 0.03 H-bonds/mole CH_2Cl_2. Thus we can neglect - if present at all - a change in the H-bonding of H_2O in aqueous solutions of molecules like CH_2Cl_2 and consequently of nonpolar residues as well.

From these results we may conclude the following with respect to the solubility of nonpolar molecules: the increase at low molecular volumes will be due to an increase in the van der Waals energy (see figure 2). The decrease in solubility of the larger molecules (above 60 cm^3/mole) may be explained by the small amount of energy necessary to expand the cavities present in the water structure (Eley, 1939). The optimal volume of nonpolar molecules or residues for van der Waals interaction with H_2O and hence solubility is in the region of 60 cm^3/mole. It seems worth noticing, that the ratio of H_2O/solute (approximately 250) of the nonpolar substances which dissolve most easily in water corresponds to one solute molecule per H_2O cluster in pure water according to Luck and Ditter (1972).

3. H-BONDS BETWEEN H_2O AND ORGANIC ACCEPTORS

3.1 Low H_2O Concentrations

Figure 4a shows the H_2O spectra of small amounts of H_2O in different organic solvents and figure 4b the correlation between the heats of vapourization of H_2O with ν_{max} of the H_2O combination band. Figure 4b indicates that the Badger-Bauer-Rule (Badger and Bauer, 1937; Drago, O'Bryan and Vogel 1970; Glew and Rath, 1971) - ν_{max} is proportional to the H-bond energy - is correct, if the assumptions made for these solutions are reasonable; i.e.:

$$- \Delta H_{vap}^{W} = \Delta H_{vdW}^{W/W} + \Delta H_{H-B}^{B/W} \quad \text{with } \Delta H_{vdW} = -15.2 \text{ kJoule/mole } H_2$$

In the case of alcohols $\Delta H_{H-B}^{B/B}$ is assumed to be equal to $\Delta H_{H-B}^{W/W}$.

Thus we can roughly estimate the H-bond energy (kJoule per mole H-bond) of $B---H_2O$-complexes (corresponding to ν_{max} between 5250 to 5050 cm^{-1}) by:

$$\Delta H_{H-B}^{B/W} = (\nu_{max} - 5400) \times 0.058 \tag{2}$$

where 0.058 is half of the slope of the upper part of the solid line in figure 4b.

Figure 5 shows that ν_{max} of the H_2O combination band of $B---HOH---B$-complexes is only slightly influenced (see Gordy, 1940, 1941) if the molecular environment of the acceptor group is changed within biochemically reasonable limits. However, for example halogenation may change the acceptor strength considerably (Schrems, 1981).

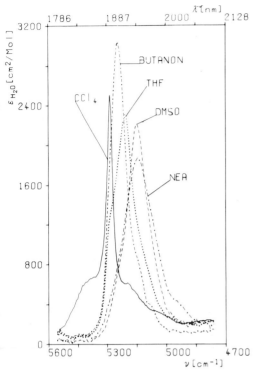

Figure 4a. Spectra of H_2O in different organic solvents; mole fraction of H_2O: 0.05, in CCl_4:0.0006; $T = 25°C$ (NEA = N-ethylacetamide).

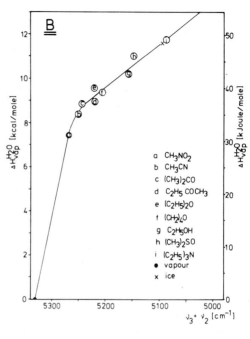

Figure 4b. Correlation between ν_{max} of the H_2O combination band and the heat of vapourization from the corresponding solutions (Landolt-Börnstein,1976) at zero H_2O concentration. For triethylamine the heat of vapourization was extrapolated from values at finite concentrations.

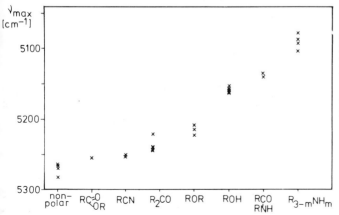

Figure5. ν_{max} of the H_2O band in different H-bond acceptor systems with a common acceptor group. (R = hydrocarbon residue).

3.2 Concentration Dependence of the H_2O Spectra

In order to understand the solubility of organic molecules in water it seems useful to follow the change of the H_2O spectra with H_2O concentration in systems which are miscible with water. Figure 6 shows the difference spectra (ε_{H_2O}(solution) $- \varepsilon_{H_2O}$(water)) of H_2O in binary solutions containing acetonitrile in an H_2O concentration range from 0.2 to 53 mole/cm^3. The spectra at low concentrations have been discussed already above.

Further addition of H_2O to binary solutions dilute in H_2O changes the shape of the spectrum and analogously that of the difference. The change indicates that these spectra are due to H_2O interactions which are different from those at low H_2O concentrations and those in pure water.

One explanation for the change in the position of ν_{max} and ν_{min} (see figure 6) may be H-bond cooperativity (Frank and Evans, 1945): at low H_2O concentrations in polar organic solvents mainly B---HOH---B-complexes are formed. In aqueous solution B---HOH---OH_2-complexes will be present preferrentially. The H_2O molecule forming one H-bond with B will be the acceptor of H-bonds of neighbouring H_2O; these H-bonds increase the acidity of the hydrogens and thus the H-bond strength of the B---HOH-bond (Gordy, 1939; Schrems, 1981)

In the cases investigated (see below) the difference spectra showed minima, maxima and $\varepsilon = 0$ at the same wave length for each organic substance at mole ratios $H_2O/B \geq 5$ until very high H_2O concentrations. This is strong evidence for the presence of one other H_2O interaction asides from those present in pure liquid water (see Luck, 1980b).

From the ratio of the absolute area under these difference spectra to that of the original spectra, which are all constant to $\pm$ 5 %, - i.e. $\int |\Delta\varepsilon| d\nu / \int \varepsilon d\nu$ - the amount of changed H-bonds of H_2O in the solution in comparison to pure water may be roughly estimated. Above 5 mole H_2O/B the H-bonds of approximately 1 mole H_2O/B are changed.

In figure 7 the difference spectra of H_2O in five different binary mixtures are recorded. The vertical line at 5185 cm^{-1} corresponds to ν_{max} of pure water at 25°C. Absorptions appearing at larger wave length correspond to weak H-bonds ($\nu_{max} \sim$ 5275 cm^{-1}; free OH-groups; compare figure 4,5). Absorptions appearing at shorter wave length are due to stronger H-bonds ($\nu_{max} \sim$ 5070 cm^{-1}; H-bonds like in ice).

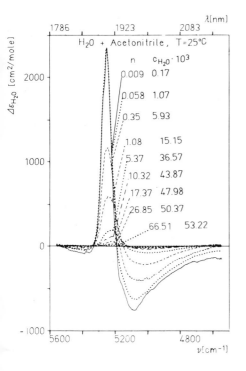

Figure 6: Difference spectra of H$_2$O-aceto-nitrile mixtures at different concentrations (n = mole H$_2$O per mole CH$_3$CN; c = mole H$_2$O/cm^3 solution; T = 25 $^\circ$C).

Figure 7: Difference spectra of aqueous solutions of aceto-nitrile, acetone, 2-butanone, tetra-hydrofurane and tri-ethylamine at 25 $^\circ$C (in the case of tri-ethylamine T = 15 $^\circ$C, above 18 $^\circ$C phase separation occurs; the numbers indicate: mole H$_2$O/mole B).

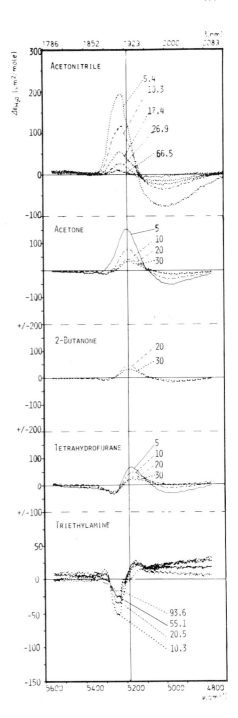

The difference spectra of H_2O-acetonitrile mixtures (figure 6,7) are positive in the wave length region of weak H-bonds and negative in the region of ice-like H-bonds. With respect to the H-bond balance of these solutions in comparison to pure water, we can say that weak H-bonds (B---H_2O) are formed and comparatively strong H_2O---H_2O H-bonds are broken. This H-bond balance between the solution and water should give rise to a positive heat of mixing or a heat of vapourization of acetonitrile from the solution smaller than from pure acetonitrile. This is actually found experimentally at nearly all concentrations. At large H_2O concentrations, however, the heat of mixing is slightly negative. We explain this by the contribution of van der Waals forces to the heat of mixing, since a favourable coordination of CH_3CN by H_2O can be achieved in this concentration range in addition to H-bonds. At mole fractions of H_2O of 0.5, 0.90 and 0.98 the heat of mixing of water and acetonitrile is +1.7, +1.34 and -1.2 kJoule/mole acetonitrile respectively (20°C, Landolt-Börnstein, 1976).

For the other solutions (figure 7) the reasoning is analogous: with stronger H-bonds formed (ν_{max} shifted to shorter wave length) the heat of mixing is more negative. For acetone and tetrahydrofurane at infinite dilution -9.9 and -14.9 kJoule/mole (Landolt-Börnstein,1976; Kister and Waldman, 1958) were found at the same temperature as indicated in figure 7.

In the case of triethylamine (figure 7, lowest part) the H-bond balance is shifted to the formation of strong H-bonds on account of weak ones. This is reflected by the heat of mixing: -16.3 kJoule/mole triethylamine at 10°C and a mole fraction of H_2O of 0.9 (Landolt-Börnstein, 1976).

4. MISCIBILITIES

4.1 Solubility of H_2O in organic solvents

H_2O will form the more H-bonds the higher the volume concentration of acceptor groups is - corresponding to the probability of H-bond formation - and consequently dissolve better if the acceptor strength is the same.

The influence of the acceptor concentration is shown in the spectra of H_2O in different alcohols at the same mole ratio (n)(figure 8). The increase in the area below the shoulder (corresponding to 1900 nm or 5270 cm^{-1}; free OH-groups of H_2O) of the H_2O absorption band parallels the decrease in the volume concentration of acceptor groups, and, as far as the higher alcohols are concerned the solubility of H_2O. At much lower H_2O concentrations the amount of free

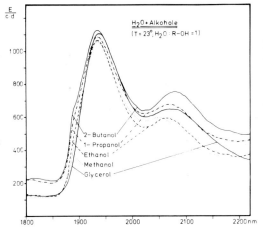

Figure 8. The amount of free OH-groups (approximately 5270 cm^{-1}) increases with the concentration of the H-bond acceptor groups.

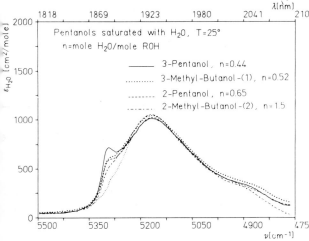

Figure 9. The amount of free OH-groups of H_2O in pentanols saturated with H_2O, depends strongly on the isomeric form of the H-bond acceptor, although its molar volume and H-bond acceptor strength is not changed. The solubility of H_2O decreases with increasing amounts of free OH-groups of H_2O (see figure 11).

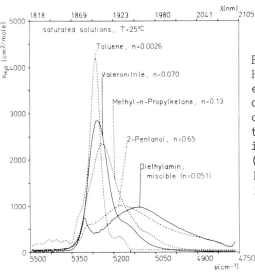

Figure 10. The solubility of H_2O depends on the H-bond energy (low wave number shift of the H_2O band) at practically constant acceptor concentration (n = H_2O/acceptor ratio in the saturated solution (Landolt-Börnstein, 1962). Molar volume of B approximately 105 cm^3/mole.

OH-groups of H_2O is similar to that in figure 8.

Why then is the solubility of H_2O in these liquids not independent from the structure of a polar molecule, if the molar volume and the H-bond acceptor strength is the same, i.e. in different isomers?

From comparison of molecular models of B---HOH---B-complexes of isomers (for example butanols, pentanols or C_5-ketones), which vary in molecular volume by less than 25%, some of these complexes look more stable than others. This means that the H-bond acceptor group is more or less shielded by the hydrocarbon chain. In other words the probability of a free OH-group of H_2O in solution to come into contact with such an acceptor group depends on the stereochemical surrounding of this group. The obvious stability of these models shows the same order like the amount of free OH-groups of H_2O as indicated by the absorbance at 5270 cm^{-1}. As shown for some pentanols in figure 9 the amount of free OH-groups of H_2O increases with decreasing solubility.

More important than these stereochemical effects for the solubility of H_2O in polar organic solvents is the energy of H-bonds formed. H_2O spectra of different saturated solutions are shown in figure 10. If we remember that ν_{max} is found at shorter wave numbers for stronger H-bonds, we see immediately that the solubility of H_2O (expressed as mole H_2O/mole B) increases with increasing H-bond strength. This correlation is shown to give a straight line in figure 11; the dependence of the solubility of H_2O on the amount of free OH-groups of H_2O in pentanols is included in figure 11 as an insert.

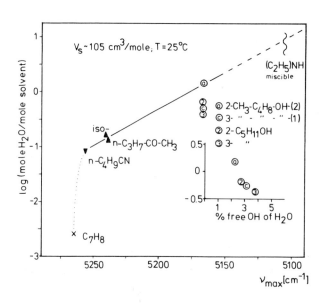

Figure 11. Correlation of the logarithm of the solubility of H_2O in different polar organic solvents with H-bond strength and amount of free OH-groups of H_2O (insert).

Thus we conclude that the solubility of H_2O increases with the average H-bond energy. This is not surprising, but has hardly been noted (see: Gordy and Stanford, 1940, 1941; Copeley, Zellhoefer and Marvel, 1940).

4.2 Solubility of polar organic molecules in water

In figure 12 the composition of the organic and aqueous phase of different organic solvents is plotted against the molar volume of the organic substance. The dependence of the solubility of molecules with the same H-bond acceptor group in water on the molar volume is parallel to that of nonpolar molecules (see figure 12 and 2). The introduction of an H-bond acceptor group into the molecule increases the solubility according to the acceptor strength (compare figure 12 and 5). The further the H_2O combination band is shifted to smaller wave numbers (stronger H-bonds) the better is the solubility of molecules with the corresponding acceptor group amine⟩ alcohols ~ ethers ~ ketones > nitriles. This means with respect to the miscibility of an organic substance with H_2O that the hydrocarbon residue may be the larger the stronger the H-bonds of B---H_2O are.

Like in the organic phases in these solutions too we have to explain the deviations in solubility of pentanols for example from one another. They have practically equal molar volumes and H-bond acceptor strength. Figure 13 shows that the amount of free OH-groups of H_2O in these saturated aqueous phases increases with the amount of H_2O which is necessary to solve one mole of pentanol. Per mole alcohol, however, the relative amounts of bound and free OH-groups of H_2O is similar.This means for the stereochemically less favourable pentanols (for example 3-methyl-butanol-(1)) more H_2O in the solution is necessary to reach the same H-bond state of H_2O in the surrounding of ROH.

4.3 Conclusions

In figure 14 the composition of the aqueous and organic phase of H_2O alcohol mixtures are plotted against each other. On behalf of the similar acceptor strength and molar volume (in each group) the correlation indicates that the sterochemistry of the H-bond acceptor group seems to be one reason for solubility differences in these groups of isomers.

Apart from the composition of the organic and aqueous phase for some substances which are totally miscible at 25 $^{\circ}C$ the critical solution temperatures are given in the row heade with 'miscible': ▽ stands for acetonitrile, ₁ower crit. solut temperature 1.4 $^{\circ}C$; ⑤ for tertiary-butanol, for which no miscibility gap has been found; ◇ for dimethylamine, upper

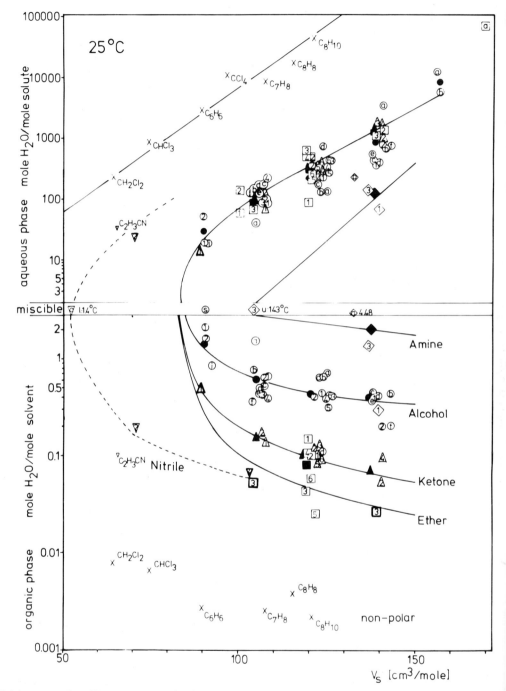

Figure 12. The composition of the saturated aqueous and organic phase of different organic substances are plotted against the molar volume of the organic compounds. The composition is expressed by: mole H$_2$O/mole organic substance. Symbols are explained in figure 14. Filled symbols correspond to average values of the respective groups of molecules.

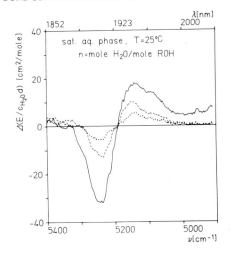

Figure 13. Difference spectra of the saturated aqueous phase with different pentanols at 25 $^{\circ}$C (n = mole H_2O per mole pentanol (Landolt-Börnstein, 1962).

——— 2-Methyl-Butanol-(2), n=39.6

····· 3-Pentanol, n=90.1

········ 3-Methyl-Butanol-(1), n=178

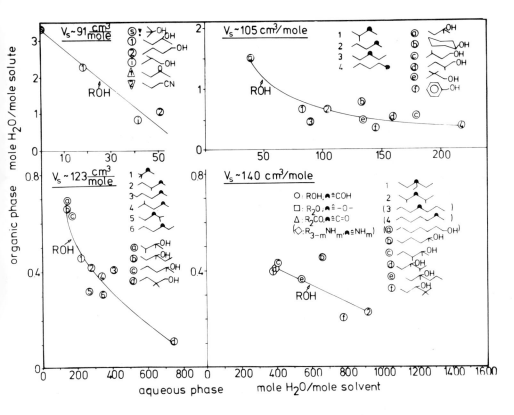

Figure 14. The composition of the aqueous phase is plotted against that of organic phase , which it is in equilibrium with, for H_2O-ROH-systems. The compositions are expressed as moles H_2O/mole ROH; T = 25 $^{\circ}$C; Landolt-Börnstein, 1962. Symbols are used in figure 12 as well.

crit. solut. temp. 143°C and ⓐ for 2,4,6,-tri-methylpyridine
4.48 mole H$_2$O/mole at saturation at 25°C and a lower crit.
solut. temp. of 5.7°C. Thus it seems that the critical solution temperatures may have similar reasons like we have
shown for the solubilities. In figure 15 the critical solution temperatures are correlated to the molar volume of the
pure organic liquid. The following tendencies seem to be
present:
1. the lower the molar volume to the lower temperatures is
   the concentration range, where both liquids are not totally miscible, shifted;
2. the weaker the acceptor strength of the polar group
   (for example acetonitrile, see chapter 3.2 or ethylenoxi
    the smaller the nonpolar residue needs to be to give
   rise to a miscibility gap; with strong acceptor groups
   (amines for example)lower and upper critical solution
   temperatures come closer together with lower molar volumes.

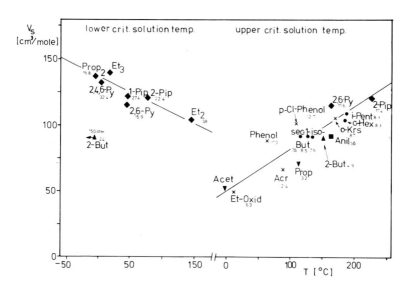

Figure 15. The lower and upper critical solution temperatures
of different organic molecules are plotted against the respec
tive molar volume of the organic liquid at 25 °C (Landolt-
Börnstein, 1962). The symbols used have the same meaning as
given in figure 14. Abbreviations:Et, Prop, But, Pent, Hex,
Acet stand for ethyl-, propyl-, butyl, pentyl-, hexyl-, and
aceto-residues and Py, Pip, Anil, Krs and Acr for pyridine,
piperidine, aniline, kresol and acroleine respectively. The
small numbers give the mole H$_2$O/B ratio at the critical solution point.

Until now only molecules with one H-bond acceptor atom have been included in the discussion. As far as solubility data are available molecules with two acceptor atoms fit well into the rules described, if their molar volume is approximately taken as: $V = V_{liquid}/2$.

Investigations of the solubility of solid compounds (electrolytes, sugars) in binary solvents (H_2O-polar organic liquids) which are regarded as totally miscible like methanol, acetone, tetrahydrofurane for example show miscibility gaps of these ternary solutions too. They seem to depend on the balance discussed above and on the change of the H_2O structure by these solid compounds (Luck, 1980a; Kleeberg and Luck, 1981b).

Summarizing we can say that the solubilities depend on a very sensitive balance of H-bonds (between H_2O molecules and polar groups and H_2O) and van der Waals forces (between organic molecules as well as with H_2O). Thus it seems reasonable that the complex systems in biology may be influenced considerably by comparatively small changes in the surrounding H_2O structure by added solutes or temperature. In these systems electrostatic interactions and differences in conformational energies have to be taken into account in addition.

This work is part of the thesis of H. Kleeberg, Marburg, 1981; it was put through in the laboratory of Prof. Luck, whose encouragement and interest is greatly acknowledged. I have to thank Otto Schrems for reading the manuscript.

REFERENCES

Badger R.M. and Bauer S.H.: 1937, J.Chem.Phys.5,pp.839-851.
Behrens A. and Luck W.A.P.:1980,J.Mol.Struct.60,pp.337-342.
Behrens A., Luck W.A.P. and Mann B.:1978,Ber.Buns.Ges.82.pp. 47-48.
Ben-Naim A.:1974, Water and Aqueous Solutions, Plenum Press, New York.
Bohon R.L. and Claussen:1951,J.Amer.Chem.Soc.,73,pp.1571- 1578.
Briegleb G.:1937,Zwischenmolekulare Kräfte und Molekülstruktur, Enke Verlag, Stuttgart.
Butler J.A.V.;1937,Trans.Farad.Soc.,pp.229-236.
Copley M.J., Zellhoefer G.F. and Marvel C.S.;1940,J.Amer.Chem. Soc.,62,pp.227-228.

Davidson D.W.:1973, in: Water a Comprehensive TReatise
            (F.Franks, ed.),Vol.II,pp.115-234,Plenum Press,
            New York.
Drago R.S., O'Bryan N. and Vogel G.C.:1970,J.Amer.Chem.Soc.,
            92,pp.3942-3929.
Eley D.D.,1939,Trans.Farad.Soc.,35,pp.1281-1293.
Finney J.:1980,Vth Internat.Symp.on SoluteSolute Solvent Inte
            actions, Florence,June; International Conference
            on Water and Ions in Biological Systems, Buchares
            June.
Frank H.S. and Evans M.W.:1945,J.Chem.Phys.,13,pp.507-532.
Frank H.S. and Franks F.:1968,J.Chem.Phys.,48,pp.4746-4757.
Frank H.S. and Wen W.-Y.:1957,Disc.Farad.Soc.,24,pp.133-140.
Franks F. and Ives D.J.G.:1966,Quart.Rev., 20,pp.1-44.
Franks F. and Quickenden M.A.J.,Reid D.S. and Watson B.:1970,
            Trans.Farad.Soc.,66pp.582-589.
Franks F. and Reid D.S.;1973,in:Water, VolII,pp.323-380.
Franks F. and Watson B.: 1969, Transa.Farad.Soc.,65,pp.2339-
            2349.
Glew D.N.:1962,J.Chem.Phys.,66,605-609.
Glew D.N.,Mak H.D. and Rath N.S.:1968,in:Hydrogen bonded
            Solvent Systems (Covington A.K. and Jones P.,eds.
            pp.195,Taylor and FrancisLtd.,London.
Glew D.N. and Rath N.S.:1971,Can.J.Chem.,49,pp.837.
Gordy W.:1939,J.Chem.Phys.,7,pp.93-102.
Gordy W.: 1941,J.Chem.Phys.,9,pp.215-223.
Gordy W. and Stanford S.C.:1940,J.Chem.Phys.,8,pp.170-177.
Gordy W. and Stanford S.C.:1941,J.Chem.Phys.,9,pp.204-214.
Gough S.R. and Davidson D.W.,1971,Can.J.Chem.,49,pp.2691.
Haly A.R. and Snaith J.W.:1971,Biopol.,10,pp.1681-1699.
Hermann R.B.,1972,J.Phys.Chem.,75,pp.363-368;76,pp.2754-2759.
Hildebrand J.H. and Scott R.L.:1950,The Solubility of Nonelec
            trolytes, Reinhold Publ.Co.,New York.
Hofmann H.-J. und BirnstockF.,1980,Z,phys.Chem.,261,pp.1212-
            1216.
Ikada Y.,Suzuki M. and JawataH.,1980,in:Water in Polymers
            (Rowland ,ed.),pp.43,ACS Symp. Series, Amer.Chem
            Soc.,Washington.
Kleeberg H., and Luck W.A.P.;1979,in:Glycoconjugates (R.Schau
            et.al.,eds.)pp.98-99,G.Thieme Verlag,Stuttgart.
Kleeberg H. and Luck W.A.P.:1980,VthInt.Symp.SoluteSoluteSol-
            vent Interactions,Florence;Inorg.Chim.A.,40,p.103.
Kleeberg H. and Luck W.A.P.:1980, Tagung d.dt.Kolloid-Ges.,
            29.9-1.10.
Kleeberg H. and Luck W.A.P.:1981a,Proceedings of the 3rdInter
            Coll.onPhysical and ChemicalInformation TRansfer
            (J.Vassileva-Popova, ed.),Varna 1980 (in press).
Kleeberg H. and Luck W.A.P.:1981b,studia biophysica (accepted
            Intern.Conf. on Water and Ions in Biological
            Systems,Bucharest,June,1980.

Kister A.T. and Waldman D.L.:1958,J.Phys.Chem.,62,pp.245-255.
Körösy F.;1937,Trans.Farad.Soc.,33,pp.416-425.
Krishnan C.V. and Friedman H.L.:1969,J.Phys.Chem.,73,pp.
       1572-1580.
Landolt-Börnstein:1962,Zahlenwerte und Funktionen, Springer
       Verlag,Berlin,Vol.II.2.
Landolt-Börnstein,NS,1976, Zahlenwerte und Funktionen, Vol.
       IV.2,Springer Verlag,Berlin.
Lange E. und Watzel R.:1938,Z.physikal.Chem.,182,pp.1-17.
Luck W.A.P.:1967,Disc.Farad.Soc.,43,pp.115-127.
Luck W.A.P.;1976, Topics.in Curr. Chem.,Vol.64,pp.113-180.
Luck W.A.P.:1978a,Progr.Colloid & Polymer Sci.,65,pp.6-28.
Luck W.A.P.:1980a,in: Water in Polymers, see Ikada et.al.
Luck W.A.P.:1980b,Angew.Chem.Int.Ed.,19,pp.28-41.
Luck W.A.P. and Ditter W.:1968,Ber.Buns.Ges.,72,pp.365-374.
Luck W.A.P. and Ditter W.:1969a,Z.F.Naturforschung,24,pp.482.
Luck W.A.P. and Ditter W.:1969b,Ber.Buns.Ges.,73,526.
Luck W.A.P. and Ditter W.,1970,J.Phys.Chem.,74,3687.
Luck W.A.P. and Ditter W.,1972,Advan.Mol.Relaxation Processes,
       3,pp.321-339.
Luck W.A.P. and Kleeberg H.;1978,in: Photosynthetic Oxygen
       Evolution (H.Metzner,ed.),p.1-29,Academic Press,
       London.
Luck W.A.P.(and Kleeberg H.),1981,ISOPOW.-II,(in press).
Luck W.A.P. and Schrems O.,1980,J.Mol.Struct.,60,pp.333-336.
Luck W.A.P. and Schiöberg D.,1979,Advan.Mol.Relaxation Proce-
       sses,14,pp.277-296.
Luck W.A.P.,Schiöberg D.,Siemann U.,1980,J.C.S.Faraday II,
       76,pp.136-147.
Marvel C.S., Copley M.J. and Ginsberg E.,1940,J.Amer.Chem.Soc.
       62,pp.3109-3112.
Némethy G., 1967, Angew.Chem.,79pp.260-271.
Privalov P.L. and Mrevlishvili G.M.,1967,Biophysics,12,
       pp.19-28.
Ramachandran G.N. and Chandrasekharan R.,1968,Biopol.,6,pp.
       1649-1658.
Rowlinson J.S.,1951,Trans.Faraday Soc.,47,pp.120-129.
Reichhardt.C.,1973,Lösungsmittel-Effekte in der org.Chem.,
       Verlag Chemie,Weinheim
Schiöberg D. and Luck W.A.P.,1979,J.C.S.Faraday TRans I,75,
       pp.762-773.
Schrems O.,1981,thesis,University of Marburg (in preparation).
Stackelberg M.v.,1949,Naturwissenschaften,36,pp.327-333,359-362.
Whalley E.,(1957,Trans.Faraday Soc.,53, pp.1578-1585.

EXPERIMENTAL STUDIES OF VARIATIONS OF THE STATE OF WATER IN LIVING CELLS

Florentina MOSORA
Institute of Physics, University of Liège,
SART-TILMAN-B-4000 LIEGE, Belgium.

ABSTRACT.

The high resolution NMR spectra and the nuclear magnetic relaxation times of water protons in frog sciatic nerves untreated and treated with 5 mM DNP, as well as the high resolution NMR spectra and the nuclear magnetic relaxation times of water protons in nerves from which this inhibitor was washed away, showed marked differences as a consequence of the change occuring in water state. Considering these results in the light of our present knowledge concerning the physical state of water in living cells, a tentative interpretation is proposed.

INTRODUCTION.

The importance of water in biological systems is obvious, since it constitutes 70 to 90 percent of the mass in most living systems. The structure and function of this simple molecule in biology, however, is not yet completely understood in spite of its fundamental importance. The state of water may affect ionic and molecular transport, maintainance of chemical gradients and osmotic pressure, reaction equilibria and kinetics, biopolymer conformation, nervous influx, etc.

If we review our present knowledge concerning the physical state of water in the living cytoplasm, we are inclined to assume that, both in the living as well as in the dead cells, water occurs under two main forms : free water and bound water. This conclusion has begun to emerge in principal, from the studies performed by nuclear magnetic resonance (NMR) spectroscopy, which is one of the quantitative methods available to study the physical properties of water. NMR spectrometers comprise a magnet and a high frequency radio transmitter which produce fields perpendicular to each other. The hydrogen nuclei of water molecules will absorb

B. Pullman (ed.), Intermolecular Forces, 489–498.

energy when placed in a strong magnetic field at a specific
resonance frequency. In pulsed NMR, the water hydrogen protons absorb energy during a brief (microseconds) pulse of
radio-frequency energy. Following this brief pulse, the
excited system relaxes back to equilibrium with certain time constants (time spin-echoes), known as T_1, the spin-lattice (longitudinal) relaxation time, and T_2, the spin-spin
(transversal) relaxation time. Thus, NMR method is particularly well suited to the study of the motional freedom of
observable nuclei (in the present case the hydrogen nuclei)
because the width of the resonance signals depends on this
motional freedom (1,2). That is, when the mobility of the
tested nuclei increases, then the width of the NMR signal
decreases.Furthermore, the relaxation times enable us to
determine more precisely the value of line widths, resolve
new lines which can not be observed by means of high resolution NMR spectra and supply supplementary data. Moreover,
from pulsed NMR it is interesting to determine the self-
diffusion constants by plotting the relative amplitude of
the spin echoes in the presence and absence of a linear
field gradient versus τ^3 where τ is the time between 90°
and 180° pulses.

Water in living tissues produces NMR signals which
differ from those observed in the case of dilute aqueous
ionic solutions. The differences are : the broadening of
the steady-state absorbed signal (3-6), the shortening of
the spin-lattice relaxation time T_1 and T_2 (4, 7-14), and a
slight decrease of the self-diffusion coefficient of water
(8-10, 15-17). The other studies showed the changes of NMR
relaxation times in function of the pathological state of
different tissues (18,19) or under the action of various
substances (20,21).

Although there is a general agreement that these
particularities are not due to experimental artifacts (16)
there are two opposite interpretations explained on the
basis of different theories (association-induction, coacer-
vate, membrane theories, etc.) : either water in the cells
is in its major more ordered than a liquid solution (4, 22-
27) or most of the water is liquid with the exception of a
very small fraction bound to the proteins which rapidly exchanges with the liquid phase and accounts for the shortening of the relaxation times (8,9,28,29). That is, the difference of interpretation arises from the different propor-
tions of these two categories of water, independently of
the state of the biological material, dead or alive.

We thought that a study concerning the variations
of the state of water in living cells provoked by certain
agents until the irreversible modifications of the biological material can give supplementary informations in view
of elucidating this difficult biophysical problem.

We utilized the action of metabolic inhibitors to study by means of NMR method (high resolution spectra and relaxation times) the variations of the state of water protons in frog sciatic nerve. We will present here only a brief summary of these study (30-38) and a tentative interpretation.

MATERIAL AND METHODS.

The system chosen for investigation was the sciatic nerve of the frog (Rana ridibunda Pall.) and the metabolic inhibitor was 2,4 dinitrophenol (DNP).

The details of the experimental procedure were described in previous papers (35,36).

The NMR spectra of water protons in DNP treated and untreated nerves were recorded with a high resolution nuclear magnetic resonance spectrometer Varian A-60A. Normal nerves, removed from the Ringer solution in which they have been immersed after dissection, were blotted on a filter paper and then fixed in the sample tube in a longitudinal central position. After the tracing of the NMR spectrum, the nerves were soaked in the inhibitor for the required period of time, after which, they were fixed in the same manner in the sample tube, in order to record the new spectrum.

The nuclear magnetic relaxation times for protons of nerve water were determined by a pulse NMR spectrometer which employed pulses with peak to peak height of approximately 150 V with frequency of 19MHz (33). In this case, a group of nerves was placed in the sample tube, in an irregular coiled position, to fill it to height of 3-4 cm.

The temperature in the sample tube, in both cases, was 37°-38°C.

RESULTS AND DISCUSSION.

The 5mM DNP concentration, with which these investigations were performed, was chosen after following the influence of different DNP concentrations upon the functionality of frog nerve (30). For instance, the immersion of the frog nerves in 5 mM DNP during a period of 9-10 min caused the appearance of primary effects characterized by small changes in the action potential. Subsequent immersions in this solution during a period of 20 min induced secondary effects characterized by reversible or irreversible loss of excitability.

In comparison with the narrow NMR spectra(1.3-1.4 Hz) of distilled water, Ringer, DNP or other solutions employed in our experiments (Fig.1), the NMR spectra of nerve water -nerves fixed in the sample tube in a longitudinal central position- are much broader and differ from one nerve to another both in amplitude and width, as shown in Fig. 2. The broadening of NMR signals is due to the fact that a part of

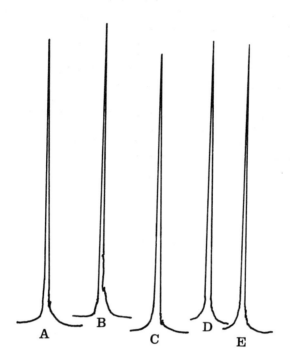

Fig. 1. High resolution NMR spectra of distilled water (A), Ringer solution (B), 5 mM DNP solution (C), 5 mM sucrose solution (D) and 5 mM alanine solution (E). The width at one-half amplitude is 1.3-1.4 Hz. Sweep width : 500 Hz.

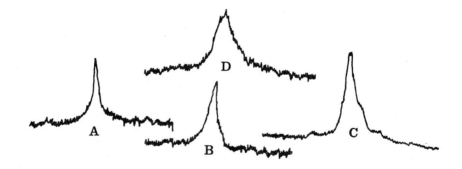

Fig. 2. High resolution NMR spectra of water in four frog sciatic nerves. The width at one-half amplitude is : 7 Hz (A), 10 Hz (B), 12 Hz (C), 24 Hz (D) Sweep width : 500 Hz.

water contained in the living cells possesses a certain degree of organization different from ordinary water. If the
organized water molecules or its hydrogen nuclei interchanged
rapidly with the free water molecules from the interior of
the nerve, in a time comparable with that of the NMR scale,
the obtained signal will be a single peak with the width
reflecting the average vicinity of the hydrogen nuclei.

Fig. 3 shows that if the nerve is submitted to DNP
action for 10 min causing only small modifications of the
action potential of the nerve, the width of NMR spectrum
of its water remains inchanged (12 Hz) with respect to the
width of initial spectrum of the same untreated nerve.
Comparing the integrals of both spectra (A,B) we see that

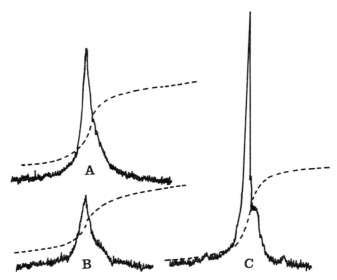

Fig. 3. High resolution NMR spectra and associated
integrals of frog sciatic nerve water in the case : of
untreated nerve (A), of the same nerve treated with 5 mM
DNP for 9-10 min (B) and of the same nerve washed and reim-
mersed in Ringer for 30 min (C). The width at one-half amplitude is 12 Hz (A,B). Sweep width : 500 Hz.

total water of the nerve decreases by 13% under the influence of DNP. The fact that the line width is unmodified, shows
that the ratio of the free to organized water remained the
same as that which existed at the initial moment. As the
total quantity of the nerve water diminishes by 13% owing to
the partial exit of water to the surronding medium, the conclusion can be drawn that, in this case, a part of the organized water has been transformed into free water.

After the removal of the inhibitor and the reimmersion of the nerve into the Ringer solution the total amount

of water increased with 13% as compared to the initial moment and the organisation of the nerve water is restored in another order. Indeed, though the nerve position to the applied field remained unchanged, the initial and final spectra, however, are similar to those that appear in the positions of the nerve varying in relation to the applied field as shown in Fig. 4.

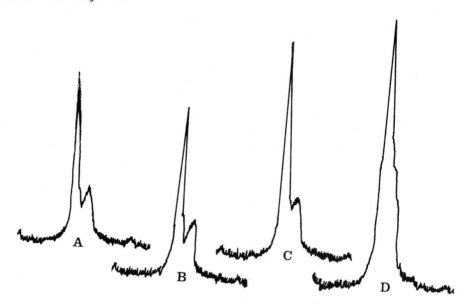

Fig. 4. High resolution NMR spectra of water protons in the same frog sciatic nerve, for different positions of the nerve axis related to the applied field. These positions vary from parallel (A) to perpendicular orientation (D). Sweep width : 500 Hz.

In order to verify if the changes produced by DNP in the sate of nerve water are not accidental, we studied the action of 5 mM sucrose and 5 mM alanine solutions on nerves immersed into them. We observed that they have no effect even when their action lasted for 4 hours.

The relaxation times T_2 and T_1 for protons of water in different groups of frog sciatic nerves presented values varying between 0.01-0.05 sec (T_2) and 0.12-0.30 sec (T_1), values considerably smaller than those obtained for the same relaxation times of the various bathing solutions : Ringer, DNP, alanine (0.11-0.13 sec for T_2 and greater than 1 sec for T_1). These shorter values are a consequence of a certain degree of organization of a part of axoplasm water.

The frog sciatic nerves treated in the same manner

as in precedent experiment with 5 mM DNP, show, for the protons of their water, a T_2 increased proportionately with passage from primary to secondary effects induced by the inhibitor. In nerves undergoing the secondary effects, the

Table 1. Spin-spin and spin-lattice relaxation times before (T_2, T_1) and after (T_2', T_1') immersion in 5 mM DNP and after removal of the inhibitor and reimmersion in Ringer (T_2'', T_1'').

| | Group of nerves | | | |
	A	B	C	D
T_2 (sec)	0.04305	0.017409	0.0212175	0.011266
T_1 (sec)	0.129	0.301	0.212	0.136
Time of immersion in DNP (min)	20	10	15	22
T_2' (sec)	0.080305	0.20183	0.050017	0.040752
T_1' (sec)	0.172	0.264	0.235	0.150
Time of reimmersion in Ringer (min)	10	20	21	52
T_2'' (sec)	0.038985	0.0302875	0.026958	0.014806
T_1'' (sec)	0.164	0.264	0.230	0.142

values of relaxation times T_2 for water protons are twice as high as those obtained for water protons in untreated nerves. After the removal of the inhibitor and after the reimmersion of the respective groups of nerves in Ringer, the relaxation time T_2, either returned to its initial values (group of nerves ACD), or presented an increase suggesting possible changes in water state going beyond the threshold of reversibility (group of nerves B). The relaxation time T_1 remained unchanged in the case of the DNP treatment (Table 1).

The effect produced by DNP is specific as the 5 mM alanine and 5 mM sucrose solutions caused no variations in the relaxation times T_1 and T_2 of water protons, even after 200 min of action.

The invariance of T_1 and the changes in the values of T_2 caused by primary and secondary effects induced by DNP on the nerve can be explained by means of a two-phase model (39). This model places, on the one hand, a fraction of intracellular water in a "solid" phase whose molecules are characterized by a random thermal rotational motion greatly restricted and, on the other hand, the remaining fraction in a "liquid" phase. The appearance of relaxation times identical for both fractions can be accounted for by the assumption that exchange of water molecules between the two phases occurs at a faster rate than the relaxation rates for either separate phase. Since the "solid" phase contributes mainly to the transverse relaxation rate and since the "liquid" phase is chiefly responsable for the longitudinal relaxation rate, the increase of T_2 values after the inhibitor action demonstrates the release of a part of water from the "solid" phase. Likewise, the return of T_2 to its initial values shows the reintegration of the free water in the "solid" phase. Certainly, in both cases, the longitudinal relaxation time T_1 was not able to change as it is intimately linked with "liquid" phase".

It is noteworthy that, although the results obtained reveal the existence only of two phases of water in the nerve, it is fairly certain that the living cell contains a third fraction which probably consists of two or more subfractions with an even higher degree of "solidity" than the "solid" phase (4) and whose T_2 values being below 2.10^{-3} sec cannot be measured because of instrumental limitations.

As seen from the above experiments, under the action of a metabolic inhibitor, a partial conversion of the water from organized into free state took place. Since it is difficult to understand how a metabolic inhibitor provokes the release of the bound water of the cytoplasm, as this water is defined by the actual theories (polarized water multilayers on cell proteins, water retained by colloids, etc.), our results seem to be better interpreted if we considered not only two main states of water : free and bound, but, three.

Indeed, in the last years, a new conception has appeared : the biostructure theory, according to which the water occurs within living tissues under three main states : free, bound and biostructured, while it occurs under two states only within the dead tissues : free and bound (40). This new category is the water integrated in the biostructured matter of the living tissues. The biostructured matter is desintegrated with death of the cell and releases not only the water it contains, but also the other chemical combinations it is consisted from.

Considering our results in the light of the biostructure theory, the conclusion can be drawn that DNP changes the ratio between the quantity of water integrated in the biostructure and the quantity of free water in the nerve. Under the action of this inhibitor the former diminishes, owing to the partial breakdown of the biostructure, in favour of free water. After removing of the inhibitor the biostructure is restored, reintegrating a part of the free water, but in an order different from the initial one. The admission and issue of free water from the nerve, for different duration of immersion are accounted for by the variation of the osmotic balance between the interior and the bathing solution. This disturbance is provoked by the release from the desintegrated biostructure of water and various substances in different quantities.

Work is presently in progress to study other biological materials and the action of different other agents as hyper and hypoosmotic solutions, in order to establish the quantitative aspects of the variation of these categories of cell water. These new results will probably be ascertained in our present tentative interpretation.

REFERENCES.

1. Kavanau, J.L.:1964, "Water and solute-water interactions", Holdenday, Inc. San Francisco, pp. 39.
2. Emsley, J.W., Feeney, J., Sutcliffe, L.H.:1965, "High resolution magnetic resonance spectroscopy", Pergamon Press Oxford, 1, pp. 30.
3. Hazlewood, C.F., Nichols, B.L., Chamberlain, N.F.:1969, Nature, 222, pp. 747.
4. Cope, F.W.: 1969, Biophys. J., 9,pp. 303.
5. Civan, M.M., Shporer, M.:1972, Biophys. J., 12, pp. 404.
6. Villey, D., Martin, G.:1974, Physiol. Chem. Physics, 6, pp. 339.
7. Abetsedarska, L.A., Miftakhutdinova, F.G., Fedotov, V.D.: 1968, Biophysics, 13, pp. 750.
8. Finch, E.D., Harmon, J.F., Muller, B.H.:1971, Arch. Biochem. Biophys., 147, pp. 299.
9. Hansen, J.R.:1971, Biochim. Biophys. Acta, 230, pp. 482.
10.Chang, D.C., Hazlewood, C.F., Nichols, B.L., Rorschach, H.E.:1972, Nature, 235, pp. 170.
11.Finch, E.D., Schneider, A.S.:1975, Biochim. Biophys. Acta, 406, pp. 146.
12.Beall, P.T., Hazlewood, C.F.:1976, Science, 192, pp. 904.
13.Eisenstadt, M., Fabry, M.E.:1978, J. of Magn. Res., 29, pp. 591.
14.Ader, R., Cohen, J.S.:1979, J. of Magn. Res., 34, pp. 349.
15.Woessner, D.E., Snowden, B.S.:1970, J. Colloid Interf. Sci., 34, pp. 290.
16.Hazlewood, C.F., Nichols, B.L., Chang, D.C., Brown, B.: 1971, Hohns Hopkins Med. J., 128, pp. 117.
17.James, T.L., Gillen, K.T.:1972,Biochim. Biophys. Acta, 286, pp. 10.
18.Ranade, S., Shah, S., Korgaonkar, K.S., Kasturi, S.R., Ghaughule, R.S., Vijayaraghavan, R.:1976, Physiol. Chem. Physics, 8, pp. 131.
19.Beall, P.T., Asch, B.B., Chang, D.C., Medina, D., Hazlewood, C.F.:1980, J.N.C.I., 64, pp. 335.
20.Updall, J.N., Alvarez, L.A., Chang, D.C., Soriano, H., Nichols, B.L., Hazlewood, C.F.:1977, Physiol.Chem. Physics, 9, pp. 13.
21.Raaphorst, C.P., Law, P., Kruuv, J.:1978, Physiol. Chem. Phys., 10, pp. 177.
22.Ling, G.N.:1962,"A physical theory of the living state: an association-induction hypothesis", Blaisdell Publ. Co. Waltham Mass.
23.Fenichel, I.R., Horowith, S.B.:1965, Acta Physiol. Scand.

60, suppl. 221, pp.1.
24.Fenichel, I.R., Horowith, S.B.:1965, Ann. N.Y. Acad.
Sci., 125, pp. 290.
25.Hechter, O.:1965, Ann. N.Y. Acad. Sci., 125, pp. 625.
26.Troshin, A.S.:1966, "Problems of cell permeability",
Translation ed. by Middas, W.F. Oxford, Pergamon Press.
27.Ling, G.N., Miller, C., Ochsenfeld, M.M.:1973, Ann. N.Y.
Acad. Sci., 204, pp. 6.
28.Walter, J.A., Hope, A.B.:1971, Prog. Biophys. Mol. Biol.,
23, pp. 1.
29.Outhred, R.K., George, E.P.:1973, Biophys. J.,13, pp. 97.
30.Mosora, F.:1971, St. Cerc. Biochim., 14, pp. 63.
31.Mosora, F.:1971, St. Cerc. Biochim., 14, pp. 409.
32.Mosora, F.:1971, Rev. Roum. Biochim., 8, pp. 97.
33.Mosora, F.:1971-1972, Biochem. exp. Biol., 10, pp. 25.
34.Mosora, F.:1972, Biochem. exp. Biol., 3, pp. 235.
35.Mosora, F.:1972, Rev. Roum. Biochim., 9, pp. 51.
36.Mosora, F.:1972, Rev. Roum. Biochim., 9, pp. 147.
37.Mosora, F.:1973, Natura, 4, pp. 85.
38.Mosora, F.:1975, Bull. Soc. Roy. Sci. Liège, 9-10, pp.
619.
39.Zimmerman, J.R., Brittin, W.E.:1957, J. Phys. Chem., 61,
1328.
40.Macovschi, E.:1976, Izd. Akademii Rumynskoi S.R., pp.
98 and 99.

STRUCTURAL VARIATIONS IN A HOMOLOGOUS SERIES OF FLUORINATED TETRA-CYANO-p-QUINODIMETHANES

Thomas J. Kistenmacher, F. Mitchell Wiygul and Thomas J. Emge
Department of Chemistry, The Johns Hopkins University,
Baltimore, Maryland 21218, U.S.A.

ABSTRACT

The various intermolecular interactions observed in the crystalline structures of the tetracyano-p-quinodimethanes TCNQ, TCNQF, 2,5-TCNQF2, and TCNQF4 are examined. These intermolecular interactions include: (1) the quinone ring-over-cyano group overlap in TCNQ; (2) the antiparallel cyano-cyano coupling in TCNQF and 2,5-TCNQF2; and (3) the donor/acceptor(acid-base) contacts in TCNQF4. Results are also presented on an initial attempt at the resolution of the twinned nature of the crystal structure of TCNQF. The electrostatic(Madelung) energy has been evaluated for eight dipole-ordered arrays and appears to indicate only limited model differentiation.

INTRODUCTION

The solid-state chemistry and physics of organic charge-transfer complexes have enjoyed a renaissance of major importance in the last decade. The electron acceptor tetracyano-p-quinodimethane, TCNQ, and its derivatives and analogs continue to play a central role in the chemistry of this renaissance.

Recently, our laboratory in collaboration with others has been investigating the chemistry and physics of several charge-transfer complexes utilizing fluorinated TCNQ acceptors. In this context, synthetic routes to progressively fluorinated TCNQ acceptors have been developed to the degree that the monofluoro,[1] TCNQF, one of the difluoro,[2] 2,5-TCNQF2, and the tetrafluoro,[3-4] TCNQF4, derivatives of TCNQ are known.

In addition to the charge-transfer chemistry of fluorinated TCNQ derivatives, we have been investigating other properties of these compounds. In particular, we have been studying the intermolecular interactions in the crystalline structures of the neutral acceptors and the relationship of these interactions to those observed in their charge-

B. Pullman (ed.), Intermolecular Forces, 499–512.

transfer complexes. Thus far, we have determined the crystal structures of TCNQF, 2,5-TCNQF2, and TCNQF4. In this paper, we compare the
variety of intermolecular interactions found in these compounds to
that found in the unsubstituted parent TCNQ.

MOLECULAR PROPERTIES

Molecular structures for the four compounds under study are presented in Figure 1. In these illustrations, the shaded bonds indicate

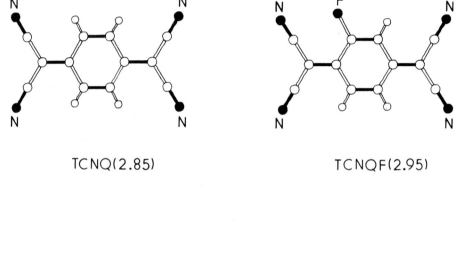

TCNQ(2.85)   TCNQF(2.95)

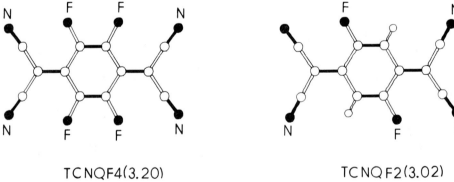

TCNQF4(3.20)   TCNQF2(3.02)

Figure 1. Molecular structure and electron affinity(in eV) for TCNQ,
           TCNQF, 2,5-TCNQF2 and TCNQF4.

bond orders significantly greater than one and emphasize the dominant
quinone resonance contribution to their electronic structure. Also

given in Figure 1 are the electron affinities for each compound.[1,2-4]
Important to the charge-transfer chemistry of this family of TCNQ
derivatives is the recognition that there is a monotonic increase in
electron affinity as the number of fluoro substituents is increased.
In addition, this systematic variation in electron affinity is
achieved without a similar degree of variation in molecular size.
This latter point is perhaps essential to the diversity observed in
the intermolecular interactions in the crystalline structures of the
neutral molecules and their charge-transfer complexes.

Secondly, we comment briefly on the molecular geometry of these
compounds as observed in the crystalline state, see Figure 2.   It is

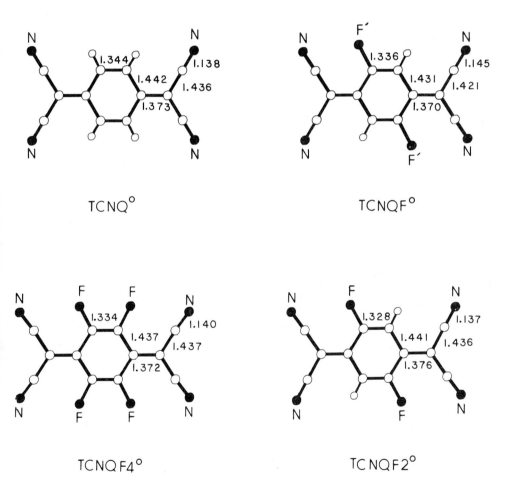

Figure 2.   Observed molecular geometries(solid state) for TCNQ, TCNQF,
            2,5-TCNQF2 and TCNQF4.

easily recognized that the intramolecular bond lengths are virtually
insensitive to the degree of fluorine atom substitution. The effects
of the number of fluoro substituents on the intramolecular bond angles
are greater, see refs. 4-7. The alterations in the bond angles in
TCNQF4 are dramatic enough to allow it to be added to the ranks of
"crowded" organic molecules.

Lastly, we have investigated the charge distribution in TCNQ,
TCNQF, 2,5-TCNQF2 and TCNQF4 by the INDO semi-empirical molecular
orbital approximation.[8] Results of these computations, employing the
observed crystalline molecular geometries, are given in Figure 3. In

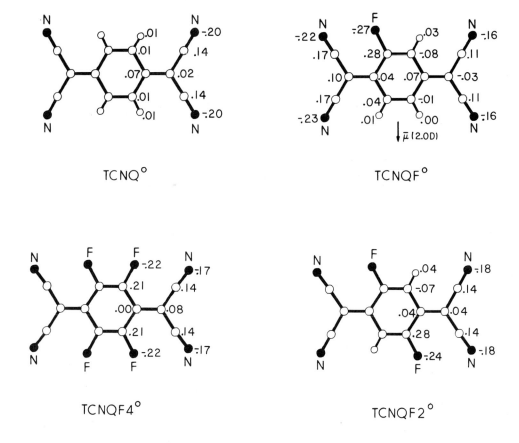

Figure 3. Net atomic charges for TCNQ, TCNQF, 2,5-TCNQF2 and TCNQF4
           as given by the INDO approximation.

all of these compounds, the N atoms of the malonitrile moiety and the
fluoro substituents bear negative charges on the order of 0.15-0.27e.
The abundance of bond moments, particularly in the fluoro derivatives,

is noteworthy and possibly determinative of the many specific interactions in the crystalline structures of these compounds. Of the presently known fluorinated TCNQ acceptors, TCNQF is the sole example of an asymmetric molecule with a permanent molecular dipole moment. From the INDO net atomic charges, with a nominal contribution from atomic rehybridization,[6] we have derived a molecular dipole moment for TCNQF of 2.0 D. An interesting aspect of the calculated dipole moment is its orientation relative to the molecular framework, Figure 3. The major contribution to the magnitude is the C-F bond moment, estimated to be about 1.9 D. However, there is sufficient polarization of the malonitrile moiety near the fluoro substituent so that $\bar{\mu}$(calc) is canted by about 34° to the C-F bond and lies within 5° of being parallel to the short in-plane molecular axis. We will return to the apparent influence of the permanent dipole moment of TCNQF on its crystal structure in a later section.

CRYSTALLINE STRUCTURES: BASIC ELEMENTS

As noted earlier, the crystalline structures of TCNQ,[5] TCNQF,[6] 2,5-TCNQF2,[7] and TCNQF4[4] have all been determined. In Table I, we present a summary of the primary crystallographic data for each of these systems. Below we briefly describe the relevant intermolecular interactions found in each of these crystalline structures; particular attention will be paid to the role of the bond (or permanent molecular) moments in the specific intermolecular interactions.

a) TCNQ. The crystal structure of the parent compound TCNQ was elucidated several years ago by Long, Sparks and Trueblood.[5] The herringbone array has the symmetry of space group C2/c and the four centrosymmetric molecules are arranged as indicated in the (010) and (001) projections of Figure 4. Two types of intermolecular interactions present in the structure of TCNQ are of interest here: (1) a cyano N atom-to-hydrogen bearing C atom; and (2) the parallel stacking of molecules in a quinone ring-over-cyano group pattern [emphasized in the (001) projection of Figure 4]. The former of these two interactions is observed at 3.18A and lies just short of the expected sum of the van der Waals radii(3.25A).[9] While this interaction can be classified as donor/acceptor in character, based on the charge distribution given in Figure 3, neither the N···C contact distance nor the intermolecular dihedral angle at 48° (maximal interaction is observed for dihedral angles approaching 90°)[4] is suggestive of strong coupling. The crystal packing is dominated by the stacking of parallel molecules along the crystallographic [110] direction, Figure 4. The mean separation between planes is short at 3.45A, and the extended one-dimensional array is characterized by the interaction of the substantial C≡N bond moment, Figure 5, and the polarizable quinone ring system.

b) TCNQF. As noted above, TCNQF is the sole compound considered here whose molecular structure is asymmetric, leading to a permanent

Table I.   Primary crystallographic data.

	TCNQ	TCNQF	2,5-TCNQF2	TCNQF4
a(A)	8.906(6)	7.596(3)	10.208(4)	14.678(7)
b(A)	7.060(4)	8.204(4)	6.026(2)	9.337(5)
c(A)	16.395(5)	8.428(2)	8.836(3)	8.174(2)
α(deg)	90.00	90.00	90.00	90.00
β(deg)	98.54(4)	90.90(3)	106.64(3)	90.00
γ(deg)	90.00	90.00	90.00	90.00
V(A^3)	1019.4	525.2	520.8	1120.0
Z	4	2	2	4
V/Z(A^3)	254.9	262.6	260.4	280.0
Space group	C2/c	"P2$_1$/n"	C2/m	Pbca
Site symmetry	$\overline{1}$	"$\overline{1}$"	2/m	$\overline{1}$

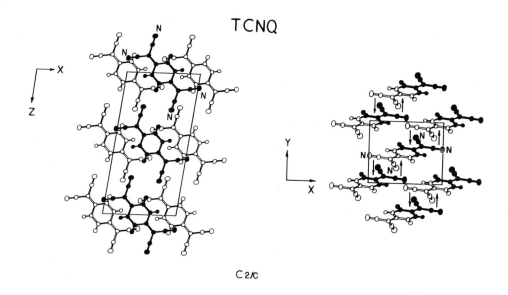

TCNQ

C2/c

Figure 4.   The (010) and (001) projections of the structure of TCNQ.

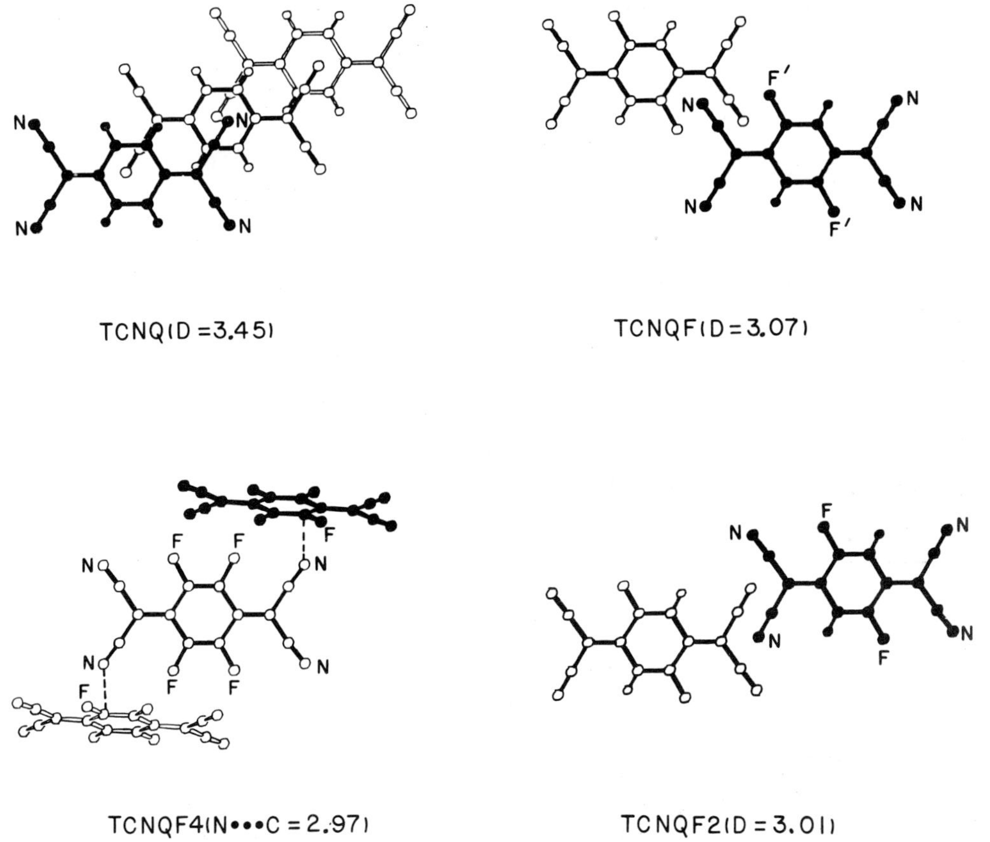

TCNQ(D =3.45)   TCNQF(D = 3.07)

TCNQF4(N•••C = 2.97)   TCNQF2(D = 3.01)

Figure 5.   Dominant intermolecular interactions in the crystal struc-
             tures of TCNQ, TCNQF, 2,5-TCNQF2 and TCNQF4.

molecular dipole moment. Analysis[6] of the crystal structure of TCNQF
suggests that dipolar fields do play a critical role in the determination of the primary structural motif, although the analysis is not
without its complications.

The crystal chemistry of TCNQF is entangled by the presence of
centrosymmetric twinning.[6] In the space group (P2$_1$/n) of the twinned
crystals, the presence of two formula units per cell demands $\bar{1}(C_i)$
molecular symmetry, inconsistent with the non-centrosymmetrical molecular structure of TCNQF. The centrosymmetrically-averaged crystal
structure of TCNQF is presented in the (001) and (100) projections of
Figure 6. The structure can be qualitatively separated into layers
parallel to the (001) plane. Each layer consists of parallel, but not
coplanar, TCNQF molecules with interplanar spacing of 3.07A. Interlayer extension along c and intralayer coupling along a are achieved

TCNQF

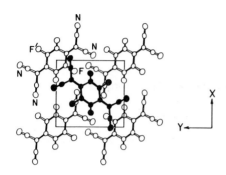

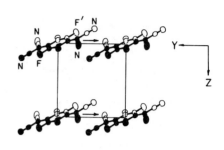

P2₁/N

Figure 6.   The (001) and (100) projections of the twinned structure of
           TCNQF.

primarily through dispersive and dipolar interactions.

It is, however, the rather intimate nature of the intermolecular
coupling along the b axis [emphasized in the (100) projection of
Figure 6 and viewed normal to the molecular plane in Figure 5] that
has captured our attention.[6] About crystallographic inversion centers,
one finds tightly-coupled cyano groups of symmetry-related TCNQF
molecules in antiparallel orientations. The contacts in this intimate
coupling are as follows: (1) C···C = 3.35A (van der Waals sum = 3.50A);
and (2) N···C = 3.31A (van der Waals sum = 3.25A). We have postulated[6]
that the one-dimensional array of antiparallel cyano bond moments may
provide a means through which interaction of permanent molecular
moments may be achieved. We will return to the subject of the untwinned structure (and the role of the permanent molecular dipole
moment) of TCNQF later.

c) 2,5-TCNQF2. The crystal structure[7] of 2,5-TCNQF2 is both similar
and contrastible to that found for TCNQF. Crystals of 2,5-TCNQF2 are
untwinned, display internal symmetry consistent with space group C2/m,
and show strong coupling of cyano bond moments.

The presence of two formula units of 2,5-TCNQF2 per cell in space
group C2/m demands 2/m(C₂ₕ) molecular symmetry - the full symmetry of
the 2,5-TCNQF2 molecule. The crystal structure of 2,5-TCNQF2, displayed in the (010) and (100) projections of Figure 7, consists of
interacting molecular layers parallel to (010), with a separation
between molecular planes of 3.01A(|b|/2). Two specific intermolecular
interactions in the structure of 2,5-TCNQF2 are of primary interest.

2,5-TCNQF2

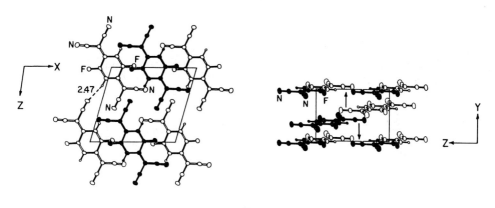

C 2/M

Figure 7. The (010) and (100) projections of the structure of 2,5-TCNQF2.

First, within the layers parallel to (010), there is a close CH(qui-none)···N(cyano) contact (2.47A, van der Waals sum = 2.70A) which is nominally donor/acceptor (hydrogen bond-like) in character. In this regard, we note (see Figure 3) that the quinone ring hydrogen atoms of 2,5-TCNQF2 are moderately electropositive, with a formal charge of +0.04e.

Secondly, interlayer extension is dominated by antiparallel coupling of cyano group moments (see Figures 5 and 7). Contact dis-tances in this coupling are C···C = 3.35A and C···N = 3.31A, identical to those shown by the antiparallel cyano group coupling mode in the crystal structure of TCNQF, Figures 5 and 6. As indicated in Figure 5, however, the local moment coupling in the structure of TCNQF involves an internal interleaving of molecules while that in the structure of 2,5-TCNQF2 is completely external. Each of these modes offers the same degree of local moment coupling as indicated by the equivalence of the contact distances. The mode present in the structure of TCNQF allows, however, a closer distance between molecular centers (8.20A vs 9.35A for 2,5-TCNQF2), and the potential for coupling of permanent molecular moments.

d) TCNQF4. The last of the crystalline structures described here is that for the perfluoro substituted compound, TCNQF4. The crystal structure of TCNQF4 has the symmetry of space group Pbca, and the four centrosymmetric molecules are arranged as indicated in the (010) and (100) projections of Figure 8.

The structure of TCNQF4 displays two interesting intermolecular

TCNQF4

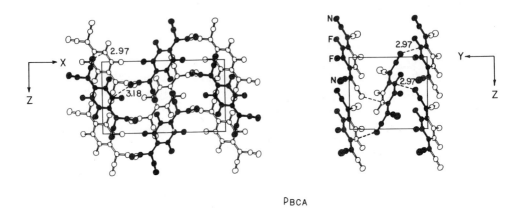

PBCA

Figure 8.   The (010) and (100) projections of the structure of TCNQF4.

contacts of the same type, but of different strength. Each of these
interactions is between a cyano N atom and a fluorinated C atom of the
quinone ring, Figure 8. The first of these has a contact distance at
3.18A, identical to that noted earlier in the structure of TCNQ. The
dihedral angle between planes for molecules involved in this interaction is 44.1°, a value close to that (48.0°) between interacting
molecules in the structure of TCNQ. As discussed earlier, neither the
contact distance nor the interplanar angle is suggestive of a strong
interaction.

A second, stronger, interaction of this type (N···C-F) is found
at 2.97A - less than the sum of the van der Waals radii by 0.28A,
Figure 8. In addition, the interplanar dihedral angle between molecules involved in this apparently stronger interaction is 73.6°, (see
Figure 5), a value near to that found for similarly strong interactions
in parabamic acid (71°),[10] chloranil (68°),[11] and barbituric acid
(67°).[12] Given the nature of the charge distribution for the TCNQF4
molecule, see Figure 3, we characterize both of these N···C-F interactions as donor/acceptor(acid-base) in character. As such, the TCNQF4
molecule can be described as being amphoteric: that is, it acts as
both a base(donor) through its electron-rich cyano groups and an acid
(acceptor) through its electron-deficient fluorinated ring C atoms.
Bürgi and coworkers[13] have made similar deductions based on analyses
of an extensive number of structures showing interactions of this type.

Finally, we emphasize that the stronger of the two N···C-F intermolecular interactions allows, qualitatively, a reduction in the
dimensionality of the structure of TCNQF4. It can be seen from Figure
8 that the strong N···C-F contacts lie within planes parallel to (100).

The weaker contacts serve to couple these planes of strongly interacting molecules. Thus, the structure of TCNQF4 can be considered as composed of strongly interacting two-dimensional layers weakly extended to three dimensions.

CRYSTAL COHESION: A START

In the previous section we have examined in some detail the diversity of the specific intermolecular interactions found in the crystal structures of TCNQ, TCNQF, 2,5-TCNQF2 and TCNQF4. These intermolecular interactions range over the ring-to-cyano group overlap in TCNQ, the antiparallel cyano-cyano coupling in TCNQF and 2,5-TCNQF2, to the donor/acceptor(acid-base) interactions in TCNQF4; see Figure 5 for a summation. We have noted the tendency of bond moments (or the combination of bond moments and permanent molecular moment in TCNQF) to dominate the variety of modes by which self-association is achieved. Only in the case of TCNQF4 does the negative inductive effect of the fluoro substituents promote donor/acceptor(acid-base) type interactions to a key role.

This series of structures, possessing a diversity of intermolecular forces, is a nearly ideal set for a systematic investigation of various contributions to the crystal stability. In this regard, we have made a small start, focusing on the contribution of the electrostatic(Madelung) energy to the crystal cohesion in TCNQF. Given the derived molecular charge distribution and attendant molecular dipole moment presented in Figure 4, we have computed, employing the lattice summation technique of Bertaut,[14] the electrostatic contribution to the crystal stability for eight dipolar models for the untwinned structure of TCNQF (illustrations of these eight models are presented in Figure 9).

These eight models arise from the consideration of parallel(P) or antiparallel(A) ordering of moments along the $\underline{a}$ axis(P_x or A_x) and the $\underline{b}$ axis(P_y or A_y) and the two possible ways(Z_+ or Z_-) for extending the dipolar array along the $\underline{c}$ axis. Each of these dipole-ordered arrays can be assigned a space group [$P2_1$, Pn or $C\bar{1}(P\bar{1})$], and all generate the macroscopic twin motif(space group $P2_1/n$, Figure 6) upon addition of an inversion center at the centroid of the quininoid ring of the TCNQF molecule. The results of the Madelung calculations for these eight models are presented in Table II. We note that the electrostatic energies are small as expected for an uncharged organic structure, and there is little differentiation between models based solely on the electrostatic energy. Surely the addition of other terms, such as polarization energies, must be invoked to rationalize the stability of the crystal structure of TCNQF. However, these additional terms may not allow a complete differentiation between untwinned motifs. This is a question which we are presently pursuing.

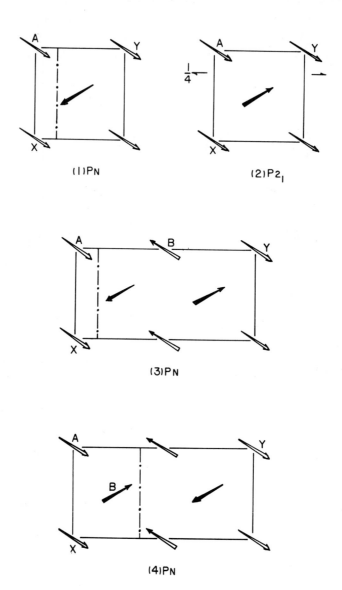

Figure 9.   Illustrations for eight dipolar motifs for the untwinned
            structure of TCNQF. In each case the space group of the
            dipolar array is indicated. In most models (3-8), two
            independent molecules (labelled A and B) are needed to
            generate the cell contents. Primitive lattices can be
            chosen for models 7 and 8 and the cell outline for the
            primitive cell is enclosed in dashed lines.

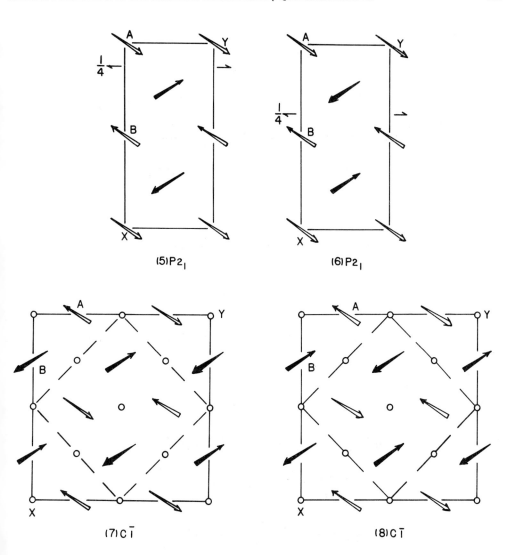

Figure 9, concluded.

ACKNOWLEDGMENT

We gratefully acknowledge support of this study by the National Science Foundation under grant DMR 78-23957. We also express our appreciation to Professors Dwaine O. Cowan (Johns Hopkins University) and John P. Ferraris (University of Texas at Dallas) for supplying the materials upon which these crystallographic studies were based.

Table II.   Madelung energies for eight dipolar models for the
            untwinned structure of TCNQF.

Model	E_M(kJ/mole)
(1) $Pn(P_x P_y Z_+)(a,b,c)$	-18.72
(2) $P2_1(P_x P_y Z_-)(a,b,c)$	-15.05
(3) $Pn(P_x A_y Z_+)(a,2b,c)$	-18.55
(4) $Pn(P_x A_y Z_-)(a,2b,c)$	-18.55
(5) $P2_1(A_x P_y Z_+)(2a,b,c)$	-15.22
(6) $P\bar{2}_1(A_x P_y Z_-)(2a,b,c)$	-15.22
(7) $C\bar{1}(A_x A_y Z_+)(2a,2b,c)$	-18.25
(8) $C\bar{1}(A_x A_y Z_-)(2a,2b,c)$	-18.25

REFERENCES

(1) Ferraris, J. P. and Saito, G.: 1978, JCS, Chem. Commun., 992.
(2) Saito, G. and Ferraris, J. P.: 1979, JCS, Chem. Commun., 1027.
(3) Whelan, R. C. and Martin, E. L.: 1975, J. Org. Chem. 40, 3101.
(4) Emge, T. J., Maxfield, M., Cowan, D. O., and Kistenmacher, T. J.:
    1981, Mol. Cryst. Liq. Cryst., in press.
(5) Long, R. E., Sparks, R. A., and Trueblood, K. N.: 1965, Acta
    Crystallogr. 18, 932.
(6) Wiygul, F. M., Emge, T. J., Ferraris, J. P., and Kistenmacher,
    T. J.: 1981, Mol. Cryst. Liq. Cryst., in press.
(7) Wiygul, F. M., Ferraris, J. P., Emge, T. J., and Kistenmacher,
    T. J.: 1981, to be published.
(8) a) Pople, J. A. and Segal, G. A.: 1966, J. Chem. Phys. 44, 3289;
    b) Chung-Phillips, A.: 1975, QCPE 12, 274.
(9) Van der Waals radii employed here are as follows: H, 1.2A; F,
    1.35A; N, 1.50A; C, 1.75A.   Values are from Pauling, L.: 1960,
    "The Nature of the Chemical Bond", Cornell University Press, 3rd.
    Ed.
(10) Craven, B. M. and McMullan, R. K.: 1979, Acta Crystallogr. B35,
    934.
(11) Bolton, W.: 1963, Acta Crystallogr. 16, 166.
(12) a) Chu, S. S. C., Jeffrey, G. A., and Sakurai, T.: 1962, Acta
    Crystallogr. 15, 661; b) van Weperen, K. J. and Visser, G. J.:
    1972, Acta Crystallogr. B28, 338.
(13) a) Bürgi, H. B.: 1975, Angew. Internat. Edit. 14, 460; b) Bürgi,
    H. B., Dunitz, J. D., and Shefter, E.: 1974, Acta Crystallogr.
    B30, 1517.
(14) a) Bertaut, E. F.: 1952, J. Phys. Radium 13, 499; b) We have em-
    ployed a highly-modified version of the program of Blake, A. B.:
    1972, QCPE 12, 222.

INTERLAYER PROPERTIES OF EXPANDED SILICATF STRUCTURES – NEW CALCULATIONAL APPROACHES CONCERNING INTERCALATION.

H DONALD B JENKINS
(Department of Chemistry and Molecular
Sciences, University of Warwick, Coventry
CV4 7AL, West Midlands, UK.

Calculations made by Jenkins and Hartman are reported for the electrostatic energy of potassium vermiculite containing a single layer of water molecules intercalated into an expanded phlogopite. The intercalation energy comprises the energy involved in expanding the silicate and arranging the water molecules in the gap. It emerges that the intercalation is mainly determined by the expansion energy. Both the expansion energy and the arrangement energy are functions of the silicate layer charge. For large layer charges no inter-calation occurs due to the large expansion energy required and for low layer charges the repulsive interaction of the water and silicate layers prevents intercalation. However for intermediate layer charges the large $K-H_2O$ interaction assists the formation of the intercalate.

1   INTRODUCTION

Despite the existence of covalent contributions to the bonding in silicate structures, studies which regard the forces within the structure as consisting of interactions between positively and negatively charged ions have generated valuable insights into the relative stabilities and energetics cf these systems. The success of purely electrostatic models to predict, for example, the atomic positions of oxygen in garnets (1) or M(2) sites in olivines (2), the relative stabilities of Mg-, Fe- and Ni- olivines and spinels (3) and the hydroxyl orientations in muscovite (4) or the cation site preferences in amphiboles (5) have lent credence to this type of approach.

The advent of the modern digital computers has extended the sizes of the lattices for which electrostatic calculations of this nature can be satisfactorily made.

B. Pullman (ed.), Intermolecular Forces, 513–530.
Copyright © 1981 by D. Reidel Publishing Company.

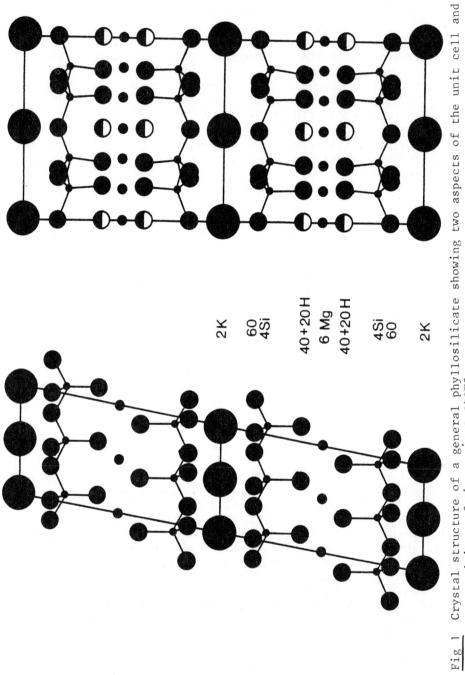

2K

6O
4Si

40+20H

6 Mg

40+20H

4Si
6O

2K

Fig 1   Crystal structure of a general phyllosilicate showing two aspects of the unit cell and the positions of the various atoms

Interest in intercalation compounds has increased dramatically in recent years and one of the reasons has been both the number and variety of reactions found capable of taking place within the interlayer and the number of varying types of molecule that can be taken up in such materials as guest molecules. The reactions that can occur in the inter-layer are often highly specific (eg: the biologically important formation of polypeptides). Certain organic molecules can be converted into commercially important products, often in high yields, within intercalates (eg conversion of trans 4-4' diaminostilbene into aniline (6)). More recently still, facile syntheses of esters (7) and novel intermolecular elimination of ammonia from amines (8) within the interlamellar regions of cation exchanged montmorillonites have been reported and the interest continues.

Quite recently (9, 10, 11, 12) we have embarked on a programme of study of silicates with the following objectives in mind:
(i) consideration of computational aspects of the theory involving the treatment of large structures at the electro-static level; (ii) to examine the energetics of expansion of silicate structures (which can be regarded as the step preceeding intercalation); (iii) to set up a model of vermiculites containing both single and double layers of intercalated water as specific case studies and (iv) refine-ment of the electrostatic models to incorporate repulsion and van der Waals terms, thus giving a route to the direct provision of thermodynamic information. The present paper offers a progress report (April 1981) on these studies.

2 COMPUTATIONAL ASPECTS

If we consider a general phyllosilicate $KX_2X'T_4O_{10}(OH)_2$ where X' corresponds to a trioctahedrally occupied site (extra site occupied by atom in a trioctahedral mica but vacant in the dioctahedral case), X corresponds to a dioctahedrally occupied site and T refers to the tetrahedral site (T_4 is either Si_4 or Si_3Al etc) the basic structural framework of the monoclinic cell is shown in Fig 1.

A conventional computation of the electrostatic energy of this mineral would input the cell parameters and atomic co-ordinates into a suitable program (eg: MADPROG (13) or LATEN (14)) with assigned charges (usually formal oxidation state charges) $q_K=+1$, $q_X=q_{Mg}=+2$, $q_{X'}=q_{Mg}=+2$, $q_T=q_{Si}=+4$, $q_O=q_O=-2$, $q_H=+1$ (where q_O and q_O are the charges on the oxygen of the silicate layer and of the hydroxyl oxygen respectively). Using the crystal structure data for this mineral (15) an electrostatic energy of some 74915 kJ mol^{-1}

is computed. Quite recently however, Jenkins and Hartman (9) recognising that the electrostatic energy for $KX_2X'T_4O_{10}(OH)_2$ can be written in a form:

$$U_{elec} = \sum_{i=0}^{2} \sum_{j=0}^{(2-i)} \sum_{k=0}^{(2-j)} \sum_{\ell=0}^{(2-k)} \sum_{m=0}^{(2-\ell)} \sum_{n=0}^{(2-m)} \sum_{p=0}^{(2-n)} A_{ijklmnp} q_K^i q_X^j q_{X'}^k q_T^\ell q_O^m q_{O'}^n q_H^p$$

(1)

where the charge dependent portion of the electrostatic energy can be separated out and the residual coefficients, $A_{ijklmnpq}$, can be separately computed, undertook a study of phllosilicates of this type.

Considering the 1M-Al mica, $KAl_2(Si_3Al)O_{10}(OH)_2$, structure as reported by Siderenko et al (16) (having C2 symmetry) and modifying this by the introduction of a mirror plane and addition of two atoms X' in positions (0, 1/2, 1/2) and (1/2, 0, 1/2) (the trioctahedral site positions) we obtain (9) the values for the coefficients, $A_{ijklmnp}$, given in Table 1.

Table 1 $A_{ijklmnp}/(kJ \ mol^{-1} e^{-1})$ coefficients for hypothetical mica, $KX_2X'T_4O_{10}(OH)_2$

ijklmnp	inter-action	$A_{ijklmnp}/$ (kJ $mol^{-1}e^{-1}$)	ijklmnp	inter-action	$A_{ijklmnp}/$ (kJ $mol^{-1}e^{-1}$)
2000000	KK	254.4	0010010	X'O'	-421.0
1100000	KX	619.2	0010001	X'H	-772.3
1010000	KX'	309.0	0002000	TT	1814.1
1001000	KT	276.2	0001100	TO	-980.0
1000100	KO	795.8	0001010	TO'	833.9
1000010	KO'	524.2	0001001	TH	716.7
1000001	KH	486.5	0000200	OO	6290.2
0200000	XX	300.8	0000110	OO'	1228.5
0110000	XX	-416.0	0000101	OH	1091.0
0101000	XT	884.4	0000020	O'O'	486.3
0100100	XO	319.8	0000011	O'H	-1695.5
0100010	XO'	-1008.2	0000002	HH	592.7
0100001	XH	-515.8			
0020000	X'X'	254.2			
0011000	X'T	455.9			
0010100	X'O	423.4			

This approach generates a degree of flexibility for Madelung calculations not previously recognised and offers a powerful extension of the general method enabling, inter alia, the possibility of study of site substitution effects in isostructural silicate lattices.

If one considers the adaptation of this approach to
the types of problem cited in the introduction as illustrating
the success of the electrostatic approach, we can see, for
example, that in order to predict atomic positions within a
structure then the approach just proposed has some attraction.
Suppose we have to predict the most likely position of a
cation M^{z+} within a given structural framework containing n
different atoms whose atomic positions are precisely known.
Selecting m possible positional co-ordinates for a given M^{z+}
cation (labelled M_1, M_2...M_m, to distinguish them) we can
then carry out a single calculation determining (m+n)!
Jenkins-Hartman coefficients, A, for the (m+n)! interactions.
If we then calculate the energy of the n cases taking
$q_{M_i}=z_i$ (i=1,2,...n) while taking $q_{M_j}=0$ (for all j≠i) we can
determine the electrostatic energy as a function of M^{z+} ion
position and hence assign the position corresponding to the
most stable situation. This type of approach is being adopted for the study of interlayer cation positions in the model
potassium vermiculite calculation discussed below. The advantage of the approach lies in the fact that the basic fixed
framework interactions are computed once only.

The separation of structurally dependent coefficients,
$A_{ijklmnp}$ from the product of the magnitudes of the interacting charges enables us to examine the electrostatic energy
change on substitution of one atom for another without corresponding relaxation of the positions of the surrounding atoms.
Comparison of the energies of the various phyllosilicates,
calculated using precise crystal structures in each case, with
the energies of the same phyllosilicates calculated using the
scheme of Table 2 have enabled Jenkins and Hartman to evaluate
energies of relaxation of the structures, energies brought
about by changes in angle between hydroxyl groups and the
mica layer, energies of transition of trioctahedral to
dioctahedral phyllosilicates, site potentials and surface
energies. The approach has been employed too in the sections
that follow.

3   EXPANSION ENERGIES OF DI- AND TRIOCTAHEDRAL PHYLLOSILICATES

In the phyllosilicates, $K_xX_2X'(Si_{4-x}Al_x)O_{10}(OH)_2$, the
influence of the interlayer charge, $q_K=x$ on the expandability
of micas has been the subject of several studies. Appelo(17,
18) has considered the expansion of the di- and trioctahedral
micas for the case where the interlayer potassium ions remain,
during the expansion, in a plane equidistant from the two
neighbouring mica layers. (Fig 2)

Giese (19) has considered the cases where the potassium
ions remain all on one side of the separating layers and where

TABLE 1. GENERIC SCHEME SHOWING HOW A GENERAL LATTICE ENERGY CALCULATION ON THE MINERAL $KX_2X'T_4O_{10}(OH)_2$ CAN BE USED, WHEN PARAMETRIZED IN THE FORM OF EQUATION (2), TO GENERATE RESULTS FOR A VARIETY OF RELATED SILICATES

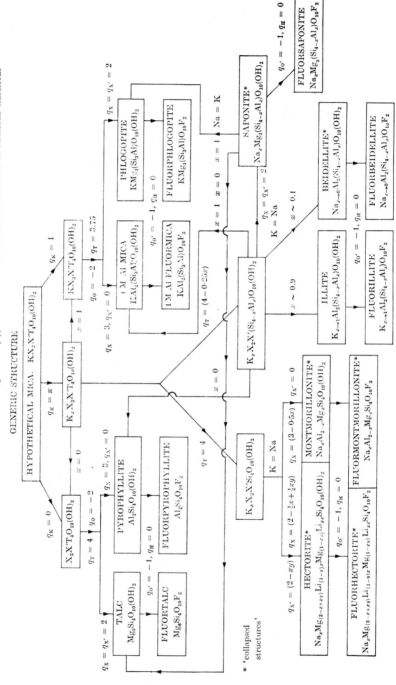

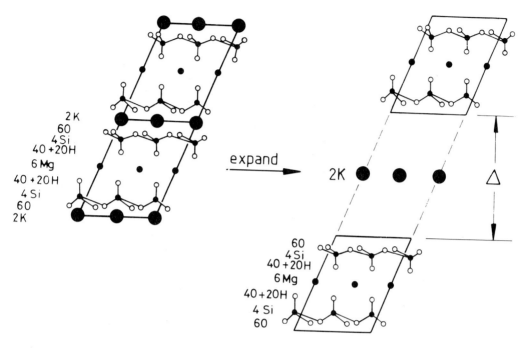

Fig 2 Expansion of a phyllosilicate while retaining the potassium ions in the interlayer midway between the separating silicate units. A=interlayer separation on expansion, Δ=0 corresponds to normal phyllosilicate.

the ions remained half on one side and half on the other side of the layers in an ordered arrangement (Fig 3). The inherent importance of this topic in relation to intercalation properties of micas has led us to examine this question as a first step for an extended study on intercalation of micas.

We have considered the cases where the layer charge is in the tetrahedral sites, $K_xX_2X'(Si_{4-x}Al_x)O_{10}(OH)_2$, or in the octahedral sites, $K_xAl_{2-x}Mg_xSi_4O_{10}(OH)_2$, both formulas being derived from the 'generic' $K_xX_2X'T_4O_{10}(OH)_2$, where $T_4 = (Si_{4-x}Al_x)$ and where X is the octahedral cation present in both di- and trioctahedral micas while X' is the octahedral cation present only in trioctahedral micas. Calculations have been made for the unexpanded micas and the micas expanded by 1.0Å, 2.5Å and 4.5Å. The mode of expansion adopted was such that the silicate layers move apart in the direction of c*, while the K ions remain exactly midway between the layers. (Fig 2). For an expansion by ΔÅ we define the expansion energy as $\Delta U_{elec} = U_{elec}(\text{unexpanded}) - U_{elec}(\text{expanded})$.

$$(2)$$

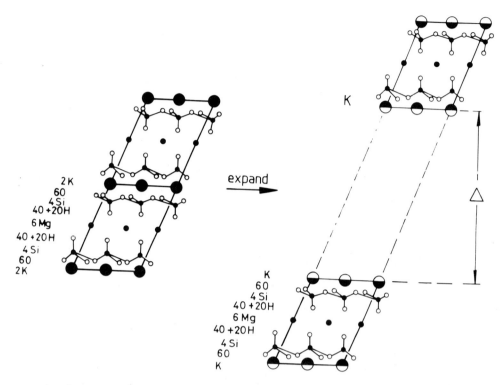

2K
60
4 Si
4O +2OH
6 Mg
4O +2OH
4 Si
60
2 K

expand

K

K
60
4 Si
4O +2OH
6 Mg
4O +2OH
4 Si
60
K

Fig 3   Expansion of a phyllosilicate with division of the
potassium ions in the interlayer such that half move with
the upper silicate unit and half with the lower. Δ= interlayer separation on expansion, $\Delta=0$ corresponds to normal
phyllosilicates.

        From our calculations (10) we find:
(1)   It takes more energy to expand a mica when the layer
charge is in the octahedral sites than when it is in the
tetrahedral sites, the difference being about 27 kJ mol^{-1} at
most.
(2)   The expansion energy of the fluormicas is larger than
that of the OH-micas by a few kJ mol^{-1}.
(3)   When the substitution is tetrahedral, the dioctahedral
micas have a slightly larger expansion energy than the
trioctahedral micas, the difference being about 1-2 kJ mol^{-1}.

Fig 4 shows ΔU_{elec}^{Δ} as a function of x for the three chosen
values of Δ in the cases of tetrahedral and octahedral substitution.

        We now turn to the mode of expansion. Giese (19,20) and
Jenkins and Hartman (9) took a different mode of expansion in
which the layers were moved apart in the direction of the

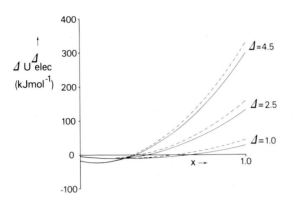

<u>Fig 4</u>   Expansion energy ΔU_{elec}^{Δ} (kJ mol^{-1}) as a function of layer charge x. Full lines: tetrahedral substitution, broken lines: octahedral substitution.

c-axis (oblique expansion) and in which the K^{+} ions were divided between the layers thereby leaving their positions with respect to the nearest layer unchanged, (Fig 3).

Fig 5 shows the effect of the different modes of expansion. It is seen that the latter mode gives the more stable configuration in the interlayer. At low interlayer separation the expansion energies are almost identical. At 2.5Å separation the gap is about 60 kJ mol^{-1}, suggesting that intercalation of water might stabilize the structure in which the potassium ions are in the middle, through the hydration energy. Work has been carried out on such a model of a 12.5Å vermiculite.

4   CALCULATIONS ON A MODEL POTASSIUM VERMICULITE

Telluria, Slade and Radoslovich (21) have reported an x-ray diffraction study of a barium vermiculite having a triclinic unit cell all with parameters (a=5.33Å, b=9.26Å, c=12.47Å, α=100.75° and β=93.5°) and containing a single layer of water molecules in the interlayer, and having the following features: (i) in each unit cell four water molecules are arranged in a hexagonal pattern (similar to the arrangement of C atoms in a graphite layer), (ii) above and below these water molecules the interlayer Ba^{2+} ions are located,

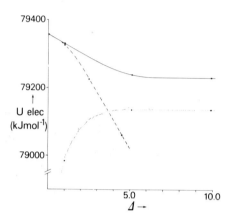

<u>Fig 5</u> U_{elec}(kJ mol^{-1}) as a function of the separation dist-
ance Δ of the layers. Full line: K ions divided between the
layers. Broken line: K ions midway between the layers.
Dotted line: K ions divided between the layers and having a
charge $+\frac{1}{2}$.

(iii) the layers of Ba^{2+} ions lie upon the silicate layer
so that the Ba^{2+} ions fit, approximately, into the ditrigonal
holes caused by the arrangement of oxygen atoms.

Adopting these features of the intercalated water
geometry we can model a potassium vermiculite, $K_{2x}Mg_6(Si_{4-x}$
$Al_x)_2O_{20}(OH)_4.(H_2O)_4$ by taking phlogopite based on McCauley,
Newnham and Gibbs structure (15) and expanding this structure
in a direction perpendicular to the (001) plane such that a=
5.308 Å, b=9.183Å, c=12.608Å, β=98.08° and incorporating the
H_2O in the interlayer (Fig 6).

For the case where x=1 the calculation is straightforward and is based on the approach discussed in section 2
of this paper. We can consider variable positions of the K^{+}
interlayer ions above and below the H_2O molecules by perform-
ing the calculation including all the positions $K_A, K_B, K_C...$
etc and then employing stepwise elimination of the coefficients in the final calculation.

For vermiculites in which x < 1 specific consideration
must be given to the arrangement of the interlayer cations

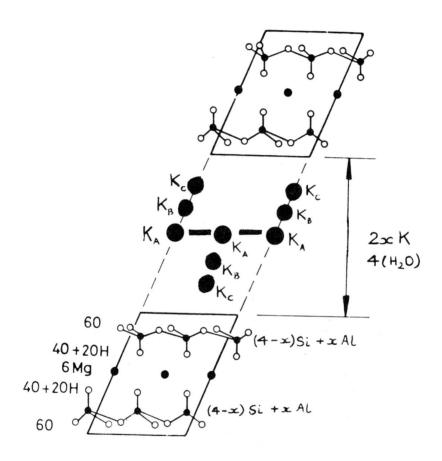

Fig 6. Potassium vermiculite showing the interlayer arrangement taken for the K+ ions and for the water molecules (illustrated by solid horizontal lines) midway between the layers.

and the water molecules and these interactions are then calculated separately. When x is low we have isolated K^+ ions
surrounded by large numbers of water molecules (Fig 7).

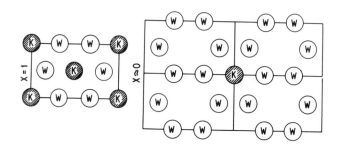

<u>Fig 7</u>   For x=1, the vermiculite $K_2Mg_6(Si_3Al)_2O_{20}(OH)_4 \cdot (H_2O)_4$
has the hexagonal arrangement of $H_2O(W)$ molecules around each
K^+ ion and a full complement of K^+ ions in the interlayer.
For intermediate values of x, each situation involves a specific interlayer arrangement. When x→0, we have few K^+ ions
surrounded by large numbers of $H_2O(W)$ molecules. The figure
shows the two extreme arrangements.

    We have in each case to give careful consideration to the
arrangement of the water dipoles. Figures 8,9,10 and 11 show
this explicit consideration of the interlayer geometry for the
cases where x=1, thus corresponding to $K_2Mg_6(Si_3Al)_2O_{20}(OH)_4 \cdot$
$(H_2O)_4$, where x = 15/16 corresponding to $K_{1.88}Mg_6(Si_{3.06}Al_{0.94})_2$
$O_{20}(OH)_4 \cdot (H_2O)_4$, where x = $\frac{3}{4}$ corresponding to $K_{1.50}Mg_6$
$(Si_{3.25}Al_{0.75})_2O_{20}(OH)_4 \cdot (H_2O)_4$, and where x=1/3 corresponding to
the case of $K_{0.67}Mg_6(Si_{3.67}Al_{0.33})_2O_{20}(OH)_4 \cdot (H_2O)_4$,. The
arrows indicating the water molecules point from the oxygen
atom towards the hydrogen atoms.

    As the interlayer charges, $q_K=x$, are decreased from unity
the vermiculite is correspondingly modified (i) in the interlayer by geometrical changes caused by creation of vacant
cation sites and the corresponding adjustments (Figures 9,10
and 11) caused to the orientations of the water molecule dipoles and (ii) in the silicate framework by corresponding
occupational changes in the tetrahedral layer sites ($q_T=$
4-0.25x) required to maintain electroneutrality. The latter
modifications are readily incorporated using our 'generic'
approach, the former changes affect the $K-H_2O$, H_2O-H_2O and KK
interaction energies within the interlayer. For the purposes
of calculations when x<1 the interlayer cations are assumed
to be located in the central position within the interlayer
(position K_A in Figure 6) throughout.

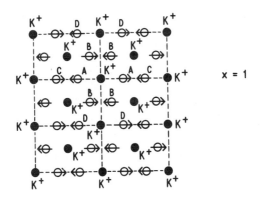

x = 1

Fig 8 The diagram illustrates explicit consideration of the interlayer arrangement of water dipoles and K^+ ions for x=1 and hence for $K_2Mg_6(Si_3Al)_2O_{20}(HO)_4 \cdot (H_2O)_4$. Arrows representing water molecules point from oxygen atoms in the direction of the hydrogen atoms.

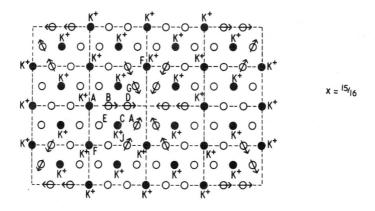

x = $^{15}/_{16}$

Fig 9 The diagram illustrates explicit consideration of the interlayer arrangement of water dipoles and K^+ ions for x = 15/16 and hence for $K_{1.88}Mg_6(Si_{3.06}Al_{0.54})_2O_{20}(OH)_4 \cdot (H_2O)_4$.

Arrows representing water molecules point from oxygen atoms
in the direction of the hydrogen atoms.

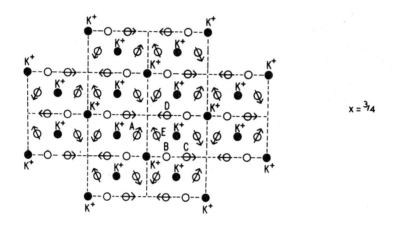

Fig 10 The diagram illustrates explicit consideration and the
interlayer arrangement of water dipoles and K^+ ions for $x=\frac{3}{4}$,
$K_{1.50}Mg_6(Si_{3.25}Al_{0.75})_2O_{20}(OH)_4 \cdot (H_2O)_4$. Arrows representing
water molecules point from oxygen atoms in the direction of
the hydrogen atoms.

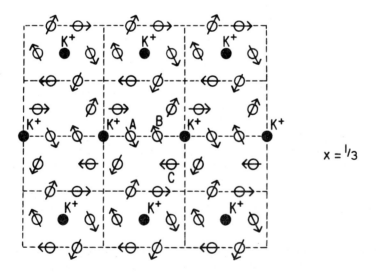

Fig 11 The diagram illustrates explicit consideration and
the interlayer arrangement of water dipoles and K^+ ions for

$x = 1/3$, $K_{0.67}Mg_6(Si_{3.67}Al_{0.33})_2O_{20}(OH)_4 \cdot (H_2O)_4$. Arrows representing water molecules point from oxygen atoms in the direction of the hydrogen atoms.

5   INTERCALATION ENERGIES

The intercalation energy of the process: $\Delta U_{inter}(g)$

$2K_xMg_3(Si_{4-x}Al_x)O_{10}(OH)_4(c) + 4H_2O(g) \xrightarrow{\hspace{2cm}}$
PHLOGOPITE
$K_{2x}Mg_6(Si_{4-x}Al_x)_2O_{20}(OH)_4(H_2O)_4(c)$
VERMICULITE INTERCALATE

is found to be:

$$\Delta U_{inter}(g) = U_W + \Delta U_{exp}^{\Delta} \tag{3}$$

where ΔU_{exp}^{Δ} is the expansion energy of phlogopite as given by equation (2) and U_W is given by:

$$U_W = -E(H_2O-H_2O)-E(K-H_2O)-E(H_2O\text{-silicate layer}) \tag{4}$$

where $E(H_2O-H_2O)$, $E(K-H_2O)$ and $E(H_2O\text{-silicate})$ are the H_2O-H_2O and $K-H_2O$ interlayer interactions and $E(H_2O\text{-silicate})$ is the interaction of the water molecules with the silicate layer.

For the process: $\Delta U_{inter}(\ell)$
$2K_xMg_3(Si_{4-x}Al_x)O_{10}(OH)_4(c) + H_2O(\ell) \xrightarrow{\hspace{2cm}}$
PHLOGOPITE
$K_{2x}Mg_6(Si_{4-x}Al_x)_2O_{20}(OH)_4(H_2O)_4(c)$
VERMICULITE INTERCALATE

where liquid water is taken up by the expanded phlogopite lattice:

$$\Delta U_{inter}(\ell) = \Delta U_{inter}(g) + 4\,\Delta H_{vap}(H_2O)(\ell) \tag{5}$$

where $\Delta H_{vap}(H_2O)(\ell)$ is the enthalpy of vapourisation of water, which has a value of 44 kJ mol^{-1}.

Figure 12 shows the variation of $\Delta U_{inter}(g)$ with x for various choices of the charge $q_{H''}$ on the hydrogen atoms of the intercalated water molecules from which we see that for $q_{H''} \geq 1/3$ no uptake of water is predicted as possible and for $q_{H''}=\frac{1}{2}$ (a value commonly chosen to model water in crystal structure work on hydrates (22)) no intercalation is possible when $0.1 \leq x$ or $x \geq 0.8$.

Figure 13 shows the behaviour of the individual contributions, U_W and ΔU_{exp}^{Δ} of equation (3) for the case where $q_{H''}=1/3$. The behaviour of ΔU_{exp}^{Δ} is of course similar to that illustrated in Figure 5 while that for U_W is a complex and

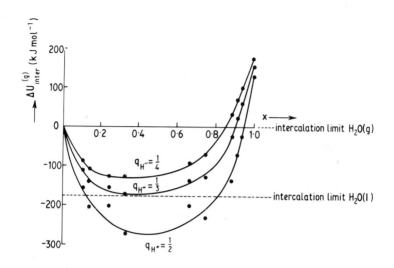

<u>Fig 12</u> Variation of $\Delta U_{inter}(g)$ and $\Delta U_{inter}(\ell)$ as functions of x for hydrogen atom charges $q_{H''}$ of $\frac{1}{4}$, 1/3 and $\frac{1}{2}$.

irregular variation, due of course to the various arrangements adopted for the water dipoles. $\Delta U_{inter}(g)$ however emerges as a relatively smooth function of x.

The following conclusions emerge:
(1) At large values of x, the large expansion energy required, ΔU_{exp}^{Δ} , to separate the interlayer of the phlogopite is greater than the stability provided by U_W and $\Delta U_{inter}(g)$ (and hence $\Delta U_{inter}(\ell)$) rises above the intercalation limit and no uptake of water takes place.
(2) At low values of x, the expansion energy is unimportant and very small and $\Delta U_{inter}(g) \approx U_W$. The predominant term in U_W is the $K-H_2O$ interaction while the interaction of the water molecules with the silicate layer is vanishingly small.
(3) The interaction of water molecules and the silicate layer becomes a relatively important factor only when x>0.5.
(4) For intermediate values of x intercalation occurs due to the large $K-H_2O$ interaction energy.

Much work on these systems remains and we are currently examining many further questions: the effect of inclusion of repulsion energy and dispersion energy on these calculations,

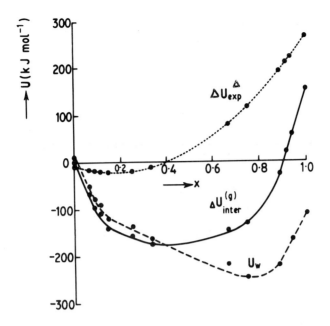

Fig 13 Variation of ΔU_{inter} (g) and the component terms U_W and ΔU_{exp}^{Δ} as a function of interlayer charge, $q_K = x$ for $q_{H''} = 1/3$.

the closer estimation of $q_{H''}$ (the hydrogen atom charge on the water molecules) necessary to interpret the results more fully.

The fact that layer charge is important to determine whether intercalation takes place has already been recognised experimentally (23). On the basis of the present work it seems clear that this is the case and that factors regarding the choice between tetrahedral or octahedral substitution are very much of secondary importance.

6 ACKNOWLEDGEMENTS

The Author wishes to acknowledge the kind invitation of the Edmond de Rothschild Foundation to present this paper at the Israel Academy of Sciences in Jerusalem in April 1981. The provision of a travel grant by the Department of Chemistry and Molecular Sciences of the University of Warwick is also greatly appreciated.

The work described was carried out by the author and his colleague Professor Piet Hartman of the University of Utrecht

whose contribution to this work is here acknowledged. The
provision of a six month visiting fellowship by the ZWO
(Netherlands Organisation for Pure Science) to carry out much
of this work at the University of Leiden, The Netherlands and
the assistance of the Science Research Council is also
acknowledged.

REFERENCES

1   Born, L, and Zemann, J:1964, Beitr Mineral Petrog, 10,
    pp 2-23
2   Born, L: 1964, N Jb Mineral Monat 1964, pp 81-92
3   Gaffney, E S and Ahrens, T J: 1970, Phys Earth Planet
    Interiors, 3, pp205-213
4   Giese, Jnr, R F :1971, Science 172, pp 263-264
5   Whittaker, E J W: 1971, Amer Mineral, 56, pp 980-995
6   Tenakoon, D T B, Thomas, J M, Tricker, M J and Graham, S H,:
    1974, J Chem Soc Chem Comm, pp124-125
7   Ballantine, J A, Purnell, H, Rayanakorn, M, Thomas, J M and
    Williams, K J: 1981, J Chem Soc Chem Comm, p8
8   Ballantine, J A, Purnell, H, Rayanakorn, M, Thomas J M and
    Williams, K J: 1981, J Chem Soc Chem Comm, p9
9   Jenkins, H D B and Hartman, P: 1979, Phil Trans Royal Soc
    (London), A293, pp169-208
10 Jenkins, H D B, and Hartman, P: 1980, Phys Chem Minerals,
    6, pp313-325
11 Jenkins, H D B: 1980 Proc Daresbury Study Weekend, S.R.C,
    Daresbury Laboratory, Warrington, DL/SCI/RIS, pp45-55.
12 Jenkins, H D B and Hartman, P: 1981, submitted for public-
    ation.
13 Jenkins, H D B and Pratt, K F: 1978, Comp Phys Communica-
    tions, 13, pp 341-348
14 Jenkins, H D B and Pratt, K F: 1980, Comp Phys Communica-
    tions
15 McCauley, J W, Newnham, R E and Gibbs, G V: 1973, Amer
    Mineralogist, 58, pp249-254
16 Siderewko, O V, Zvyagin, B B and Soboleva, S V: 1978, Sov
    Phys Cryst, 20, pp332-335
17 Appelo, C A J: 1978, Amer Mineralogist, 63, pp 782-792
18 Appelo, C A J: 1978, 'Aspects of mica-related clay minerals
    in hydrogeochemistry', Thesis, Vrije Universiteit, Amsterdam.
19 Giese, Jnr, R F: 1974, Nature (Phys Sci), 248, pp580-581
20 Giese, Jnr, R F: 1975, Z Kristallogr, 141, pp 138-144
21 Telleria, M I, Slade, P G and Radoslovich, E W: 1977, Clays
    and Clay Minerals, 25, pp 119-125
22 Baur, W H: 1961, Acta Cryst, 14, pp 209-213
23 Schulz, L G: 1969, Clays and Clay Minerals, 17, pp115-149

INTERMOLECULAR FORCES AND LATTICE DYNAMICS OF MOLECULAR CRYSTALS.

Salvatore Califano
Istituto di Chimica Fisica – Laboratorio di Spettroscopia
Molecolare,Via Gino Capponi 9, Florence Italy

Abstract.Intermolecular potentials including the atom-atom and multi-pole multipole interactions,are used for the calculation of dynamical properties of molecular crystals.Specific examples of harmonic,anharmonic and multi-phonon calculations are discussed.

Thanks to the progress made in recent years in lattice dynamics theory, it is possible today to make full use of the large body of information carried by the infrared,Raman and neutron scattering spectra of molecular crystals in order to develop intermolecular potentials able to reproduce a variety of physical properties. It is the purpose of this exposition to illustrate some of the recent results obtained in this field by discussing harmonic and anharmonic lattice frequency calculations as well as multiphonon processes.

Consider a piece of crystal,large enough to fulfill the cyclic boundary conditions,containing L unit cell,Z molecules per unit cell and N atoms per molecule. For each molecule we define 3N molecular coordinates $R_\ell^{m\mu}$ (m = 1,2,...,L;μ = 1,2,...,Z;ℓ = 1,2,...,3N) such that three of them describe translations of the molecular center of mass,three describe rotations of the molecule about the principal axes of inertia and 3N - 6 describe internal vibrations.The crystal potential,expanded in powers of these coordinates,is given by

$$V = \sum_{m\mu}\sum_{\ell} F_\ell(m\mu)R_\ell^{m\mu} + \frac{1}{2}\sum_{mn}\sum_{\mu\nu}\sum_{\ell\ell'} F_{\ell\ell'}\binom{m\mu}{n\nu}R_\ell^{m\mu}R_{\ell'}^{n\nu} +$$

$$+ \frac{1}{6}\sum_{mnp}\sum_{\mu\nu\pi}\sum_{\ell\ell'\ell''} F_{\ell\ell'\ell''}\binom{m\mu}{n\nu}{p\pi}R_\ell^{m\mu}R_{\ell'}^{n\nu}R_{\ell''}^{p\pi} + \ldots.$$

where
$F_\ell(m\mu) = (\partial V/\partial R_\ell^{m\mu})_0 = 0$ (equilibrium conditions of the crystal)

B. Pullman (ed.), Intermolecular Forces, 531–545.
Copyright © 1981 by D. Reidel Publishing Company.

$$F_{\ell\ell'}\binom{m\mu}{n\nu} = (\partial^2 V / \partial R_\ell^{m\mu} \partial R_{\ell'}^{n\nu})_0$$

$$F_{\ell\ell'\ell''}\binom{m\mu}{n\nu}{p\pi} = (\partial^3 V / \partial R_\ell^{m\mu} \partial R_{\ell'}^{n\nu} \partial R_{\ell''}^{p\pi})_0$$

In the harmonic approximation we neglect cubic and higher terms of the potential and, by using the translational symmetry, we define crystal normal coordinates [1]

$$Q_j(\mathbf{k}) = L^{-\frac{1}{2}} \sum_{\ell\mu} E^*(\ell\mu | j\mathbf{k}) \sum_m R_\ell^{m\mu} e^{-i\mathbf{k}\cdot\mathbf{r}_m}$$

where $\mathbf{k}$ is the wavevector, $\mathbf{r}_m$ is the position vector of unit cell m with respect to an origin unit cell and $E(\ell\mu | j\mathbf{k})$ are the coefficients of the system of linear equations

$$\sum_{\ell'} \sum_\nu D_{\ell\ell'}\binom{\mu}{\nu}|\mathbf{k}) E(\ell'\nu | j\mathbf{k}) = \omega_j^2(\mathbf{k}) E(\ell\mu | j\mathbf{k})$$

In this equation the quantities $D_{\ell\ell'}\binom{\mu}{\nu}|\mathbf{k})$ are the elements of the dynamical matrix, given by [1]

$$D_{\ell\ell'}\binom{\mu}{\nu}|\mathbf{k}) = \sum_n F_{\ell\ell'}\binom{o\mu}{n\nu} e^{i\mathbf{k}\cdot\mathbf{r}_n}$$

In order to solve this system of equations we need the force constants $F_{\ell\ell'}\binom{m\mu}{n\nu}$, which are the second derivatives of the crystal potential with respect to the molecular coordinates.

A crystal potential which correctly accounts for the occurrence in the crystal of well defined entities, the molecules themselves, bound together by weak intermolecular forces, is conveniently written in the form

$$V = W + \Phi$$

where W represents the internal potential of all molecules in the crystal and Φ the interaction potential among them.

The intramolecular potential W is a sum of the type

$$W = \sum_{m\mu} W(m\mu)$$

where $W(m\mu)$ represents the internal potential of molecule μ in the unit cell m. The intramolecular potential is normally expanded in power series of internal coordinates $S_t^{m\mu}$ and in the harmonic approximation is given by

$$W(m\mu) = \frac{1}{2} \sum_{tt'} F_{tt'}^{m\mu} S_t^{m\mu} S_{t'}^{m\mu}$$

These quadratic force constants are then determined by means of well-known refinement procedures, through a best fit of the internal phonon frequencies.

If we consider only pairwise interactions between molecules, the intermolecular potential Φ can be written in the form

$$\Phi = \frac{1}{2} \sum_{m\mu} \sum_{n\nu}' V\begin{pmatrix} m\mu \\ n\nu \end{pmatrix}$$

where $V\begin{pmatrix} m\mu \\ n\nu \end{pmatrix}$ represents the interaction potential between molecule $m\mu$ and molecule $n\nu$ and the prime indicates that the sum omits the term with $m\mu = n\nu$.

The pairwise intermolecular interactions are the dominant part of the intermolecular potential. For normal polyatomic molecules three-body forces are negligibly small and can be confidently neglected. Only in the case of molecules with very high polarizabilities these interactions can be of some importance and can be taken into account.

The connection between the lattice dynamics and the theory of intermolecular forces is then based on the possibility of using analytical forms of intermolecular potentials for the calculation of phonon frequencies and in general of all dynamical properties of the molecular crystals. Agreement with the spectroscopic observables is then used as a check of the validity of the potential chosen. Refinement procedures are used, as for the internal potential, to improve the agreement between calculated and observed quantities up to the best possible fit. For this purpose the phonon frequencies constitute not only a large collection of data, but also a set of very specific parameters, selectively sensitive to different terms of the interaction potential, a fact that is not yet fully apreciated in the literature. In order to avoid convergence of the refined potentials toward physically unreasonable minima, the crystal structure and the crystal energy are normally used as constraints in the refinement process.

The theory of intermolecular forces distinguishes a number of different physical mechanisms which contribute to the intermolecular interaction between two molecules. Perturbation theory indicates actually that the interaction potential between two molecules A and B, can be represented as the sum of at least four different contributions

$$V^{AB} = V_{el}^{AB} + V_{ind}^{AB} + V_{disp}^{AB} + V_{rep}^{AB}$$

where the four terms indicate the electrostatic, the induction, the dispersion interactions, respectively.

The electrostatic interaction potential between the charge distributions on the two molecules can be described correctly by means of classical electrostatic theory. The simplest approach is to localize effective charges on the atoms and to use Coulomb's law. In this way the electrostatic potential is partitioned into a sum of atom-atom contributions

$$V_{el} = \sum_i \sum_j q_i q_j / r_{ij} \qquad \begin{array}{l} \text{i belonging to molecule A} \\ \text{j belonging to molecule B} \end{array}$$

where the q's are the effective atomic charges and r_{ij} the atom-atom
distance. Alternatively the charges can be placed in the bonds or can be
replaced by point dipoles.

A different approach is the well-known expansion of the potential
in terms of multipole-multipole interactions, using multipoles located
at the centers of mass of the molecules. A convenient form of the multipole expansion is due to Jansen   and utilizes a very compact tensor
formalism. Accordingly the electrostatic interaction potential is given
by [2]

$$V^{AB}_{el} = \sum_m \sum_n (-1)^m \left[(2m - 1)!!(2n - 1)!! \right]^{-1} V^{AB}_{mn}$$

where

$$(2m - 1)!! = (2m - 1)(2m - 3)(2m - 5) \ldots (1)$$

and V^{AB}_{mn} represents the interaction potential between a multipole M^A_m of
order m on molecule A and a multipole M^B_n of order n on molecule B

$$V^{AB}_{mn} = M^A_m \cdot T^{AB}_{m+n} \cdot M^B_n$$

the tensor T^{AB}_{m+n} of order m+n being defined as [2]

$$T^{AB}_{m+n} = \nabla^{m+n} (1/R_{AB})$$

where R_{AB} is the intermolecular distance. The dot means tensor product.
For a pair of neutral molecules the electrostatic potential is then

$$V^{AB}_{el} = - M^A_1 \cdot T^{AB}_2 \cdot M^B_1 + \frac{1}{3}(M^A_2 \cdot T^{AB}_3 \cdot M^B_1 - M^A_1 \cdot T^{AB}_3 \cdot M^B_2) + \frac{1}{9}(M^A_2 \cdot T^{AB}_4 \cdot M^B_2) +$$

$$- \frac{1}{15}(M^A_1 \cdot T^{AB}_4 \cdot M^B_3 + M^A_3 \cdot T^{AB}_4 \cdot M^B_1) + \ldots.$$

where all quantities are defined in the crystal-fixed system. The space-fixed multipoles are easily expressed in terms of molecular multipoles defined in a molecule-fixed frame. The electrostatic potential assumes then the form

$$
\begin{aligned}
V^{AB}_{el} = &- \sum_{\alpha\beta}\sum_{\rho\tau} \Gamma^{A}_{\alpha\rho} \Gamma^{B}_{\beta\tau} T^{AB}_{\alpha\beta} \mu_{\rho} \mu_{\tau} \; + \; \frac{1}{3} \sum_{\alpha\beta\gamma}\sum_{\rho\tau\sigma} (\; \Gamma^{A}_{\alpha\rho} \Gamma^{A}_{\beta\tau} \Gamma^{B}_{\gamma\sigma} \; + \\
&- \Gamma^{B}_{\alpha\rho} \Gamma^{B}_{\beta\tau} \Gamma^{A}_{\gamma\sigma}) T^{AB}_{\alpha\beta\gamma} \mu_{\sigma} \Theta_{\rho\tau} \; + \; \frac{1}{9} \sum_{\alpha\beta\gamma\delta}\sum_{\rho\tau\sigma\upsilon} \Gamma^{A}_{\alpha\rho} \Gamma^{A}_{\beta\tau} \Gamma^{B}_{\gamma\sigma} \Gamma^{B}_{\delta\upsilon} T^{AB}_{\alpha\beta\gamma\delta} \Theta_{\rho\tau} \Theta_{\sigma\upsilon} \; + \\
&- \frac{1}{15} \sum_{\alpha\beta\gamma\delta}\sum_{\rho\tau\sigma\upsilon} (\; \Gamma^{A}_{\alpha\rho} \Gamma^{B}_{\beta\tau} \Gamma^{B}_{\gamma\sigma} \Gamma^{B}_{\delta\upsilon} \; - \; \Gamma^{B}_{\alpha\rho} \Gamma^{A}_{\beta\tau} \Gamma^{A}_{\gamma\sigma} \Gamma^{A}_{\delta\upsilon}) T^{AB}_{\alpha\beta\gamma\delta} \mu_{\rho} \Omega_{\tau\sigma\upsilon} \; + \; \ldots
\end{aligned}
$$

where μ_{ρ} ,$\Theta_{\rho\tau}$,$\Omega_{\rho\tau\sigma}$ are components of the molecular dipole,quadrupole and octopole respectively,$\Gamma_{\alpha\rho}$ represents the direction cosine between the molecule-fixed axis ρ and the space-fixed axis α and

$$
T^{AB}_{\alpha\beta\ldots} = \partial/\partial R_{\alpha} . \partial/\partial R_{\beta} \; \ldots \; 1/R_{AB}
$$

R_{α} being a Cartesian component of R_{AB} in the crystal-fixed frame.

This form of the electrostatic potential is very convenient for lattice dynamics calculations since the different elements in each product depend on only one type of molecular coordinate. In particular the T tensor elements depend only on the intermolecular distance and thus vary only with the translational coordinates, the direction cosines depend only on the molecular orientation and thus vary only with the rotational coordinates and the molecular multipoles depend only on the molecular structure and vary only with the internal coordinates.Furthermore the molecular multipoles are often accessible by experiments and the number of independent multipole components is normally very small for molecules with some symmetry.

The induction potential can be also treated classically in terms of molecular polarizabilities. Using the same formalism as before, the induction potential can be written in the form [3]

$$
\begin{aligned}
V^{AB}_{ind} = \sum_{C}{}' \Big[&- M^{A}_1 \cdot T^{AB}_2 \cdot \alpha^{B} \cdot T^{BC}_2 \cdot M^{C}_1 \; + \; \frac{1}{3} (\; M^{A}_2 \cdot T^{AB}_3 \cdot \alpha^{B} \cdot T^{BC}_2 \cdot M^{C}_1 \; + \\
&- M^{A}_1 \cdot T^{AB}_2 \cdot \alpha^{B} \cdot T^{BC}_3 \cdot M^{C}_2) \; + \; \frac{1}{9} (\; M^{A}_2 \cdot T^{AB}_3 \cdot \alpha^{B} \cdot T^{BC}_3 \cdot M^{C}_2) \; + \\
&- \frac{1}{15} (\; M^{A}_1 \cdot T^{AB}_2 \cdot \alpha^{B} \cdot T^{BC}_4 \cdot M^{C}_3 \; + \; M^{A}_3 \cdot T^{AB}_4 \cdot \alpha^{B} \cdot T^{BC}_2 \cdot M^{C}_1) \; + \; \ldots \Big]
\end{aligned}
$$

where the prime indicates that in the sum $C \neq B$. For simplicity only the interaction of induced dipoles is considered in this form of the induction potential. The symbol α represents here the polarizability

expressed in the space-fixed system. When C = A the potential represents
the normal two-body polarization interaction. When C ≠ A, the potential
describes a three-body interaction in which the multipoles on molecule
A interact with the dipole induced on B by all multipoles in the crystal.

In general these contributions are very small and the effect on
the lattice frequencies is within few wavenumbers.

The last two terms of the intermolecular potential are the most
difficult ones to represent in analytical form. The theory furnishes
only a general indication on the functional form of the repulsive and
dispersion interactions and suggests specific dependences on the distance
between interacting elements. In recent years accurate quantum-mechanical calculations of these interactions have been made for some simple
molecular systems with ab-initio methods. In order to apply these potentials to the calculation of phonon frequencies, it is common practice
to fit functional forms to the calculated interaction energies for a
variety of intermolecular distances and of relative molecular orientations. These ab-initio calculations are extremely promising but are
far for the moment from being easily generalized to complex molecular
systems. In particular the calculated interaction energies are extremely
sensitive to the dimension of the orbital basis chosen and to the cut-
off of configuration interactions. For this reason semiempirical potentials are normally used in lattice dynamics calculations. The term semi-
empirical means here only that they respect the functional form suggested
by the theory but that the parameters are obtained from a best fit to
crystal properties.

For polyatomic molecules it is difficult to develop functional
forms of general validity, without using the drastic approximation that
the total repulsive and dispersion potential between two molecules can
be partitioned into additive contributions involving the interactions
between all atoms on a molecule and all atoms on the other. The repulsive
and dispersion potentials are then written in the form

$$V^{AB}_{rep} = \sum_{ij} V\left(^{Ai}_{Bj}\right)_{rep} \qquad \text{i belonging to A}$$
$$V^{AB}_{disp} = \sum_{ij} V\left(^{Ai}_{Bj}\right)_{disp} \qquad \text{j belonging to B}$$

i.e. as the sum over all atom pairs of atom-atom interactions.

The additivity of the atom-atom contributions is easily questioned
on theoretical grounds. There is no doubt, however, that the atom-atom
approach furnishes one of the simplest and yet more powerful types of

interaction potential. Atom-atom potentials are transferable within classes of similar molecular crystals and have been used with success for the calculation of a large number of crystal properties, including crystal structures and energies, harmonic and anharmonic phonon frequencies, phonon lifetimes and phonon-phonon interaction and decay processes.

A general form of the atom-atom potential that includes all the r dependences suggested by second order perturbation theory is of the type

$$V(^{Ai}_{Bj}) = V(^{Ai}_{Bj})_{rep} + V(^{Ai}_{Bj})_{disp} =$$

$$= A_{ij} \exp(- B_{ij} . r_{ij}) - f(r_{ij}) [c_6^{ij} r_{ij}^{-6} + c_8^{ij} r_{ij}^{-8} + c_{10}^{ij} r_{ij}^{-10}]$$

where $f(r_{ij})$ is a damping function that is included in order to avoid that the potential goes to $- \infty$ when $r \to 0$. The coefficients C_6, C_8 and C_{10} are empirical parameters. A convenient form of the damping function[4] is

$$f(r_{ij}) = 1 \qquad \text{for } r_{ij} \geqslant r_{ij}^0$$

$$f(r_{ij}) = \exp [- \alpha(r_{ij}^0 . r_{ij}^{-1} - 1)^m] \qquad \text{for } r_{ij} < r_{ij}^0$$

where α and m are integers and r_{ij}^0 is the value of the atom-atom distance at the potential minimum.

In actual calculations it is difficult to determine the coefficients C_6, C_8 and C_{10} for all pairs of atom-atom contacts in a polyatomic molecule. In particular, for normal Van der Waals distances, the r^{-6} term is the leading one in the attractive dispersion potential and it obscures the effect of the less important terms in r^{-8} and r^{-10}. For this reason the simplified form

$$V(^{Ai}_{Bj})_{disp} = - c_6^{ij} r_{ij}^{-6}$$

is often utilized to represent the dispersion part of the interaction potential between two atoms belonging to different molecules.

In principle the coefficients of the atom-atom potential are isotropic only for interactions between single atoms, as for instance in rare gas crystals. For atoms involved in chemical bonds the coefficients may be strongly anisotropic and vary with the relative orientation of the two bonds. The anisotropy may be introduced in the potential by mul-

tiplying the coefficients by an expansion in terms of Legendre polynomials. Quantum-mechanical calculations show that in order to ensure the
correct angular dependence, the expansion must be extended at least to
the polynomial of order 4. The angular dependence can be taken into account in a very efficient and automatic way by extending to the atom-
atom potential the tensor formalism utilized for the electrostatic interaction. The coefficients A_{ij} and C^{ij} are then written in the form [5]

$$A_{ij} = \sum_m \sum_n s_m^i \cdot t'_{m+n} \cdot s_n^j$$

$$C^{ij} = \sum_m^\infty \sum_n^\infty p_m^i \cdot t''_{m+n} \cdot p_n^j$$

where

$$t'_{m+n} = [(\partial/\partial r_{ij})^{m+n}(e^{-B_{ij}r_{ij}})](-B_{ij})^{m+n}$$

$$t''_{m+n} = [(\partial/\partial r_{ij})^{m+n}(r_{ij}^{-6})](r_{ij}^{m+n+6})$$

r_{ij} being the vectorial distance between the atoms and s_m, p_m formal
tensors on the atoms.

In the last few years we have made extensive applications of the
intermolecular potentials described above to the calculation of several
dynamical properties of molecular crystals. We describe here some of the
results obtained.

A crystal that we have studied in detail is that of ammonia[6], which
crystallizes in the cubic system, space group $P2_13$ (T^4), with four molec
ules per unit cell, located on sites of C_3 symmetry. The atom-atom part
of the interaction potential used includes four contributions. Three of
them describe the normal H...H, H...N and N...N interactions while the
fourth one is used specifically for the weak N...H hydrogen bonds betwee
each molecule and the six nextneighbors. The electrostatic potential was
represented by a multipole expansion including terms up to the dipole-
octopole and quadrupole-quadrupole. Owing to the C_{3v} molecular symmetry,
there is only one dipole component, one independent quadrupole component
and two independent octopole components. By orienting the z axis of the
molecular frame along the three-fold symmetry axis of the molecule and
by using the trace relations

$$\Theta_{xx} = \Theta_{yy} = -\frac{1}{2}\Theta_{zz} \quad ; \quad \Omega_{zxx} = \Omega_{zyy} = -\frac{1}{2}\Omega_{zzz} \quad ; \quad \Omega_{xyy} = -\Omega_{xxx}$$

the electrostatic potential becomes

$$V_{el}^{AB} = - \sum_{\alpha\beta} \mu_z^2 \Lambda_{\alpha z}^A \Lambda_{\beta z}^B T_{\alpha\beta}^{AB} - \frac{1}{3} \sum_{\alpha\beta\gamma} \sum_\sigma \mu_z \Theta_{\sigma\sigma} (\Lambda_{\alpha z}^A \Lambda_{\beta\sigma}^B \Lambda_{\gamma\sigma}^B - \Lambda_{\alpha z}^B \Lambda_{\beta\sigma}^A \Lambda_{\gamma\sigma}^A) T_{\alpha\beta\gamma}^{AB} +$$

$$+ \frac{1}{9} \sum_{\alpha\beta\gamma\delta} \sum_{\tau\sigma} \Theta_{\sigma\sigma} \Theta_{\tau\tau} \Lambda_{\alpha\sigma}^A \Lambda_{\beta\sigma}^A \Lambda_{\gamma\tau}^B \Lambda_{\delta\tau}^B T_{\alpha\beta\gamma\delta}^{AB} +$$

$$- \sum_{\alpha\beta\gamma\delta} \{ \frac{1}{6} \mu_z \Omega_{zzz} \Lambda_{\alpha z}^A \Lambda_{\beta z}^B (\Lambda_{\gamma z}^A \Lambda_{\delta z}^A + \Lambda_{\gamma z}^B \Lambda_{\delta z}^B) +$$

$$+ \frac{1}{15} \mu_z \Omega_{xxx} | \Lambda_{\alpha z}^A \Lambda_{\beta x}^B (\Lambda_{\gamma x}^B \Lambda_{\delta x}^B - 3 \Lambda_{\gamma y}^B \Lambda_{\delta y}^B) +$$

$$+ \Lambda_{\alpha z}^B \Lambda_{\beta x}^A (\Lambda_{\gamma x}^A \Lambda_{\delta x}^A - 3 \Lambda_{\gamma y}^A \Lambda_{\delta y}^A) | \} T_{\alpha\beta\gamma\delta}^{AB}$$

The induction potential used includes the effect of both two- and three-body interactions as discussed previously. For reason of space we do not show here the explicit form of the potential.

The results of a set of different calculations of the lattice frequencies, made in the harmonic approximation, are collected in table 1. Column 1 shows the contribution of the atom-atom potential to the lattice frequencies. Columns 2 to 4 show the additional contributions of the various terms of the electrostatic potential. Column 5 shows finally the contribution of the induction potential.

Table 1
Lattice frequencies of crystalline NH_3

Symm.	1	2	3	4	5	exp.
A	95	84	117	143	137	---
A	196	194	203	306	304	310
E	122	122	119	122	113	107
E	192	195	278	299	314	298
F	132	135	125	130	132	140
F	186	183	187	181	176	183
F	128	129	130	273	273	260
F	203	206	327	360	353	358
F	265	316	483	524	534	533

1) atom-atom
2) 1 + dipole-dipole
3) 2 + dipole-quadrupole and quadrupole-quadrupole
4) 3 + dipole-octopole
5) 4 + induction potential

The specific influence of the potential contributions on the lattice frequencies is clearly seen.

Another interesting example is that of naphthalene[7]. This is one of the few organic crystals for which a large number of experimental data is available. The structure is known at various temperatures and the infrared and Raman Spectra have been investigated in a broad range of temperatures. Furthermore the complete set of phonon dispersion curves has been recently measured at 6° K. Naphthalene crystallizes in the monoclinic system, space group $P2_1/a$ (C_{2h}^5), with two molecules per unit cell on C_i sites. For this reason the molecule has no permanent dipole and the leading term in the electrostatic potential is the quadrupole-quadrupole interaction. The influence of this term on the lattice frequencies is shown in table 2.

Table 2
Lattice frequencies of Naphthalene

	1	2	exp.		1	2	exp.
A_u	101	97	99	B_u	62	70	66
	45	46	53		101	124	125
A_g	101	106	109	B_g	70	75	71
	73	73	74		42	44	46
	48	49	51				

1) atom-atom
2) atom-atom + quadrupole-quadrupole.

We notice that the atom-atom potential produces three frequencies with the same value whereas the quadrupole-quadrupole term differenciates them in the correct order. The importance of the electrostatic term is better seen if the complete set of phonon dispersion curves is displayed.

Once the harmonic solution has been found, the techniques of many-body theory can be used to calculate the anharmonic shifts and the phonon bandwidths. Two approaches are particularly convenient: The Hamiltonian renormalization and the Green's functions method.

The Hamiltonian renormalization procedure renormalizes[8] to higher orders the phonon creation and annihilation operators used in the quantum field description of the phonon states. The renormalization is done by using the commutation relations between these operators and the Hamiltonian operator. The Green's function method[9] uses in this case temperature Green's functions defined in terms of correlation functions of time dependent phonon operators. Both methods furnish the same expression for the anharmonic phonon frequency

$$\omega_j(\mathbf{k}) = \omega_j^0(\mathbf{k}) + \omega_j^2(\mathbf{k})$$

where $\omega_j^0(\mathbf{k})$ is the harmonic frequency of a phonon belonging to the jth branch with wavevector $\mathbf{k}$ and $\omega_j^2(\mathbf{k})$ is the correction arising from second order perturbation effects. The anharmonic correction is given by

$$\omega_j^2(\mathbf{k}) = \Delta_j(\mathbf{k}) - i\Gamma_j(\mathbf{k})$$

and thus the anharmonic description corresponds to bands with Lorentian shapes of frequency

$$\omega_j(\mathbf{k}) = \omega_j^0(\mathbf{k}) + \Delta_j(\mathbf{k})$$

and bandwidth $2\Gamma_j(\mathbf{k})$.

The anharmonic shift $\Delta_j(\mathbf{k})$ is given by

$$
\begin{aligned}
\hbar\Delta_\kappa = & 12\sum_\kappa B(\kappa,-\kappa,\kappa_1,-\kappa_1)(2\bar{n}_\kappa + 1) + \\
& - 18\hbar^{-1}\sum_{\kappa_1\kappa_2} \{ B(\kappa,\kappa_1,\kappa_2)B(-\kappa,-\kappa_1,-\kappa_2)\left[\frac{(\bar{n}_{\kappa_1} + \bar{n}_{\kappa_2} + 1)}{(\omega_\kappa + \omega_{\kappa_1} + \omega_{\kappa_2})_p} + \right. \\
& \left. + \frac{(\bar{n}_{\kappa_2} - \bar{n}_{\kappa_1})}{(\omega_\kappa + \omega_{\kappa_1} - \omega_{\kappa_2})_p} + \frac{(\bar{n}_{\kappa_1} - \bar{n}_{\kappa_2})}{(\omega_\kappa - \omega_{\kappa_1} + \omega_{\kappa_2})_p} - \frac{(\bar{n}_{\kappa_1} + \bar{n}_{\kappa_2} + 1)}{(\omega_\kappa - \omega_{\kappa_1} - \omega_{\kappa_2})_p} \right] + \\
& + 2B(\kappa,-\kappa,\kappa_1)B(\kappa_2,-\kappa_2,-\kappa_1)\frac{(2\bar{n}_\kappa + 1)}{(\omega_{\kappa_1})} \}
\end{aligned}
$$

and the half-bandwidth by

$$
\begin{aligned}
\hbar\Gamma_\kappa = & 18\hbar^{-1}\pi \sum_{\kappa_1}\sum_{\kappa_2} B(\kappa,\kappa_1,\kappa_2)B(-\kappa,-\kappa_1,-\kappa_2)\{(\bar{n}_{\kappa_1} + \bar{n}_{\kappa_2} + 1)\times \\
& \times [\delta(\omega_\kappa - \omega_{\kappa_1} - \omega_{\kappa_2}) - \delta(\omega_\kappa + \omega_{\kappa_1} + \omega_{\kappa_2})] + \\
& + (\bar{n}_{\kappa_1} - \bar{n}_{\kappa_2})[\delta(\omega_\kappa + \omega_{\kappa_1} - \omega_{\kappa_2}) - \delta(\omega_\kappa - \omega_{\kappa_1} + \omega_{\kappa_2})] \}
\end{aligned}
$$

where κ is a comprehensive symbol for jk, p means principal part of the sum and $\bar{n}_\kappa$ is the thermal average of the phonon occupation number

$$\bar{n}_\kappa = < n_\kappa > = 1/(e^{\beta\hbar\omega_\kappa} - 1) \qquad (\beta = 1/KT)$$

The coefficients B in these expressions are

$$B(\kappa,\kappa_1,\kappa_2) = \frac{1}{6} \left(\frac{h^3}{8\omega_\kappa \omega_{\kappa_1} \omega_{\kappa_2}}\right)^{\frac{1}{2}} L^{-\frac{3}{2}} \sum_{mnp} \sum_{\mu\nu\pi} \sum_{\ell\ell_1\ell_2} F_{\ell\ell_1\ell_2} \binom{m\mu}{n\nu} \times$$

$$\times E(\ell\mu|\kappa)E(\ell_1\nu|\kappa_1)E(\ell_2\pi|\kappa_2)e^{i(\mathbf{k}\cdot\mathbf{r}_m + \mathbf{k}_1\cdot\mathbf{r}_n + \mathbf{k}_2\cdot\mathbf{r}_p)}$$

$$B(\kappa,\kappa_1,\kappa_2,\kappa_3) = \frac{1}{24} \left(\frac{h^4}{16\ \omega_\kappa \omega_{\kappa_1} \omega_{\kappa_2} \omega_{\kappa_3}}\right)^{\frac{1}{2}} L^{-2} \sum_{mnpy} \sum_{\mu\nu\pi\upsilon} \sum_{\ell\ell_1\ell_2\ell_3} F_{\ell\ell_1\ell_2\ell_3} \binom{m\mu}{y\upsilon} \times$$

$$\times E(\ell\mu|\kappa)E(\ell_1\nu|\kappa_1)E(\ell_2\pi|\kappa_2)E(\ell_3\upsilon|\kappa_3)e^{i(\mathbf{k}\cdot\mathbf{r}_m + \mathbf{k}_1\cdot\mathbf{r}_n + \mathbf{k}_2\cdot\mathbf{r}_p + \mathbf{k}_3\cdot\mathbf{r}_y)}$$

and involve therefore cubic and quartic terms of the potential.

In actual calculations the principal part of a sum and the Dirac delta functions are represented by the symbolic identities

$$\delta(x) = \lim_{\varepsilon\to 0} \frac{1}{\pi} \frac{\varepsilon}{x^2 + \varepsilon^2}$$

$$\frac{1}{(x)_p} = \lim_{\varepsilon\to 0} \frac{x}{x^2 + \varepsilon^2}$$

where ε is a small but finite quantity. The value of ε is chosen depending on the resolution desired.

The anharmonic shifts are very sensitive to the intermolecular potential utilized in the calculation of the harmonic frequencies and of the B coefficients. In table 3 we show three different anharmonic calculations for the α-phase of solid N_2.

Table 3

Anharmonic lattice calculations for α-N_2

Symm.		A			B			C	
	Exp.	har.	anh.	diff.	har.	anh.	diff.	har.	anh.	diff.
E_g	32.3	33.5	36.7	+3.2	40.8	39.5	−1.3	33.6	40.2	+6.6
T_g	36.3	37.7	40.8	+3.1	50.7	48.5	−2.2	37.8	42.5	+4.7
T_g	59.7	45.7	47.6	+1.9	74.3	70.3	−4.0	45.8	51.8	+6.0
T_u	48.4	46.0	48.8	+2.8	52.0	48.4	−3.6	45.5	50.9	+5.9
T_u	69.4	67.0	70.1	+3.1	77.5	72.0	−5.5	67.1	80.0	+12.9
A_u	47.8	43.2	44.3	+1.1	52.4	48.8	−2.6	42.6	47.8	+5.2
E_u	54.0	50.9	52.0	+1.1	57.6	53.5	−4.1	50.4	54.3	+3.9

A- Self-consistent phonon method. Atom-atom potential. Ref.10
B- Self-consistent phonon method. Ab initio potential. Ref.11
C- Green's function method. Atom-atom potential. Ref.12

The first two calculations were made using a simplified approach, the self-consistent phonon method, which makes use of "effective" force constants obtained by adding to the quadratic terms the averaged effect of the quartic terms of the potential. The third calculation was instead made with the Green's function method. It is interesting to observe that, within the same method, different potentials produce anharmonic shifts that are all positive in one and all negative in the other case. Furthermore the self-consistent phonon shifts are much smaller than those obtained with the Green's function method. This proves the importance of the cubic terms of the potential, neglected in the approximate self-consistent calculations.

Owing to the fact that anharmonic shifts ore not observable quantities, it is more convenient to calculate the phonon bandwidths, that can be directly measured in the spectra. Bandwidths calculated for crystalline ammonia,[13] using the mixed atom-atom plus multipole-multipole potential discussed before, are shown in table 4

Table 4
Calculated and observed bandwidths of NH_3

| | | ω | | 2Γ | |
		calc.	obs.	calc.	obs.
F	LO	530		17.4		
	TO	523	533	21.9		14
F	LO	435	431	15.3	18	27
	TO	360	{.366 / 360	4.3	17	21
A		305	313	1.7		9
E		299	298	1.9	2.4	8
F	LO	274 }	260	1.5 }	2	5
	TO	273		1.3		
F LO + TO		181	184	17.6		12
A		143		4.4		
F LO + TO		131	138	4.1	2.3	6
E		122	107	2.7	1.1	4.5

The agreement is in general very satisfactory. Broad bands have large calculated widths and sharp bands have small calculated widths. The only exception is the band at 360 cm^{-1} which has an apparent width much larger of the calculated value. This band has, however an unexplain-

ed doublet structure probably due to multiphonon absorptions and thus the
observed width cannot be considered as significant.

We discuss now the case of two-phonon bands in molecular crystals
which furnish also relevant information on the intermolecular potential.

Two-phonon bands appear normally as broad and often very structured absorptions on both sides of fundamental crystal bands and are due
to the contemporary excitation of two quanta in the crystal. Selection
rules for these processes require that

$$\mathbf{k}_1 + \mathbf{k}_2 = \mathbf{q}_{photon} \approx 0$$

$$\omega_{j_1}(\mathbf{k}_1) + \omega_{j_2}(\mathbf{k}_2) = \nu_{photon}$$

where $\mathbf{q}$ is the momentum of the incident photon and ν the relative frequency. The momentum selection rules show clearly that the whole Brillo-
uin zone is involved in the process since there is no restriction on the
individual $\mathbf{k}$ vectors but only on their sum.

Two possible absorption mechanisms can be envisaged. These are
shown schematically in fig.1.

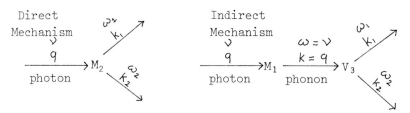

Fig.1 - Schematic representation of two-phonon processes.

The direct mechanism corresponds to the production of the two phonons
by direct absorption of a photon and involves terms higher than the linear one in the expansion of the electric moment (electrical anharmonicity). The indirect mechanism involves instead the production of a transverse phonon first, by direct absorption of a photon through the linear
term of the electric moment, followed by the decay of the phonon into
two phonons by anharmonic coupling (mechanical anharmonicity).

In our laboratory we have done extensive calculations of two-phonon processes. Both the Hamiltonian perturbation theory and the Green's
function formalism were used to analyze the two-phonon bands observed
in a variety of molecular crystals including CO_2, N_2O, SF_6, SiF_4, OCS, HCl
and HBr. For reasons of space it is impossible to illustrate here all
the results obtained. We shall therefore limit ourselves to illustrate

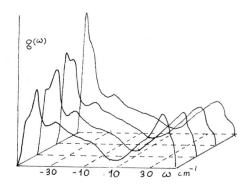

Fig.2 - Renormalized density of states for the ω_2 + ω_3 combination of crystalline SF_6.

the case of the ω_2 + ω_3 combination of SF_6. Fig.2 shows the calculated[14] density of states,computed from the imaginary part of the Green's function, for different values of the intramolecular anharmonicity constant. A good agreement with the experimental band shape is obtained for a value of the anharmonicity constant close to that found in the gas phase.

References.

1) Califano S.,Schettino V. and Neto N. Lattice Dynamics of Molecular Crystals, Springer Verlag 1981 in press.
2) Neto N.,Righini R.,Califano S.and Walmsley S.H. 1978 Chem.Phys.29,167
3) Schettino V. and Giua R.to be published.
4) Klein M.L. and Righini R.1978 J.Chem.Phys.68,5553
5) Burgos E.,Righini R. and Califano S.,to be published
6) Righini R.,Neto N.,Califano S. and Walmsley S.H.1978 Chem.Phys.33,345
7) Righini R.,Califano S. and Walmsley S.H.1980 Chem.Phys.50,113
8) Wallace D.Thermodynamics of Crystals 1972 Wiley (London)
9) Abricosov A.A.,Gorkov L.P.and Dzyaloshinsky I.E. Methods of Quantum field theory in Statistical Physics,1975 Dover (New York)
10) Raich J.C.,Gillis N.S. and Anderson A.B.1974 J.Chem.Phys.61,1399
11) Luty T.,Van der Avoird A. and Berns R.M.1980 J.Chem.Phys.73,5305
12) Kobashi K. 1978 Mol.Phys.36,225
13) Della Valle R.G.,Fracassi P.F. Righini R.,Califano S. and Walmsley S.H.1979 Chem.Phys.44,189
14) Salvi P.R. and Schettino V.1979 Chem.Phys.40,413

THE MOTION OF PARTICLES AHEAD OF A SOLIDIFICATION FRONT

Dieter Langbein
Battelle-Institut e.V., Frankfurt/Main

1 INTRODUCTION

Fig. 1 shows particles of magnesium oxide and zink, which are piled up at the solidification front of salol and thymol, respectively. The particle diameter is about 5μm. The solidification rate was 10μm/s. Fig. 1 is taken from a paper of Uhlmann, Chalmers and Jackson (1964).

Fig. 2 has been obtained under microgravity within the German space rocket program TEXUS. It shows particles of tungsten, which have been equally distributed in a matrix of silver before melting and resolidification. In the absence of gravity they have moved to the surface of the sample due to interface and surface forces or by melting and solidification forces. Fig. 2 is taken from the Final TEXUS Report of Walter and Ziegler (1978).

What then is the force driving a particle ahead of a solidification front? This is obviously intermolecular repulsion.

In general, intermolecular forces are attractive. However, if the force between a solid and its melt is stronger than that between the solid and a suspended particle, an effective repulsion of the particle results.

The question is whether this repulsion can be strong enough to drive the particle ahead of the solidification front, which in Fig. 3 is assumed to move upwards with rate v. In order for the particle to move also upwards, melt must continuously flow into the narrow interspace between particle and solidification front. This flow causes viscous forces, i.e., the intermolecular repulsion must be strong enough to overcome the viscous force of the melt. This force is generally proportional to the moving rate v. The particle

B. Pullman (ed.), Intermolecular Forces, 547–562.
Copyright © 1981 by D. Reidel Publishing Company.

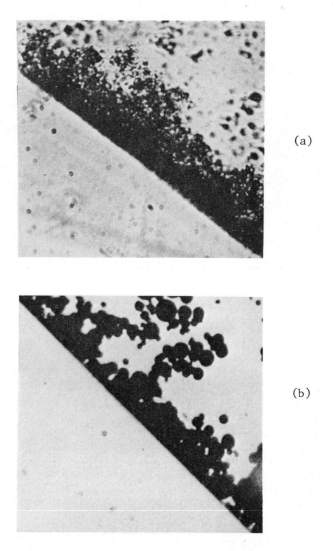

(a)

(b)

Fig. 1:   Pile-ups of particles at a solid-liquid interface:
          (a) salol+MgO, (b) thymol+Zn (taken at ✕360, shown
          here at ✕234; Uhlmann, Chalmers and Jackson 1964)

is more easily captured by the solidification front, if the
latter moves fast than if it moves slowly.

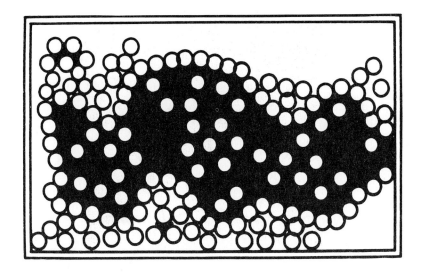

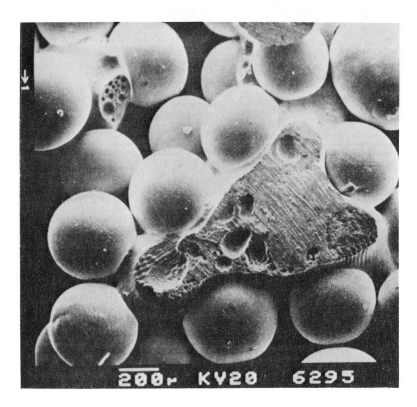

Fig. 2: Grouping of silver and W-particles
(Walter and Ziegler 1978)

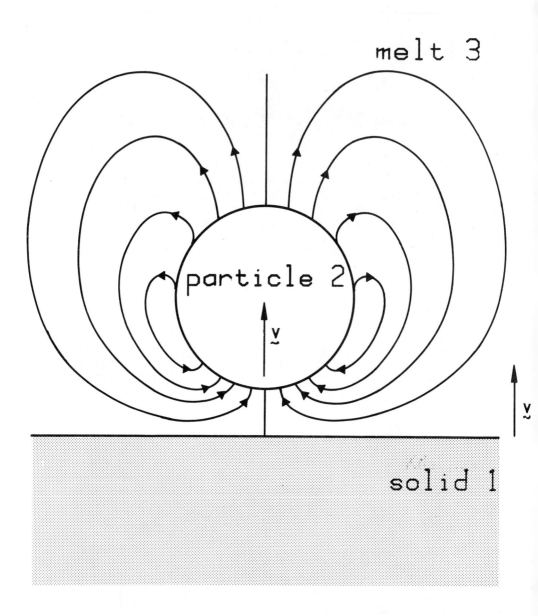

Fig. 3:   The motion of a particle ahead of a
          solidification front

2 THE VISCOUS FORCE

If the particle would move freely, without a solidification front being close by, the viscous force of the melt would be given by Stokes law

$$F = 6 \pi \eta r v , \tag{1}$$

where η is the dynamic viscosity of the melt and r the radius of the particle under investigation.

The presence of the solidification front strongly increases the viscous force. The range of intermolecular forces is at best $0.1 \mu m$, i.e., the spacing between particle and solidification front is even narrower than that distance. Since the characteristic lengths entering the flow profile of the melt are the radius r of the particle and its distance d from the solidification front, the increase in viscous force as compared to Eq. (1) is generally given by a function f (r/d). One obtains

$$F_{vsc} = 6 \pi \eta r v f(r/d) \tag{2}$$

Within the spacing the melt may be assumed to show a quadratic flow profile. The flow velocity is zero at the solidification front and at the surface of the particle and takes its maximum in between. This quadratic flow profile renders

$$f(r/d) \propto (r/d)^2 \tag{3}$$

for distances d small compared to the radius r of the particle. The viscous force required for the melt to flow into the narrow interspace between particle and solidification front increases quadratically with the reciprocal distance 1/d.

3 THE INTERMOLECULAR FORCE

The repulsive intermolecular force between particle and solid, on the other hand, increases proportional to the reciprocal distance and proportional to the fourth power of the reciprocal distance at distances d small and large compared with the particle radius r, respectively. By integrating the van der Waals energy between all atoms of a spherical particle with radius r and a half-space at distance d one obtains

$$E_{vdW} = A(\ln((2r+d)/d) - 2r/(r+d))/3 \tag{4}$$

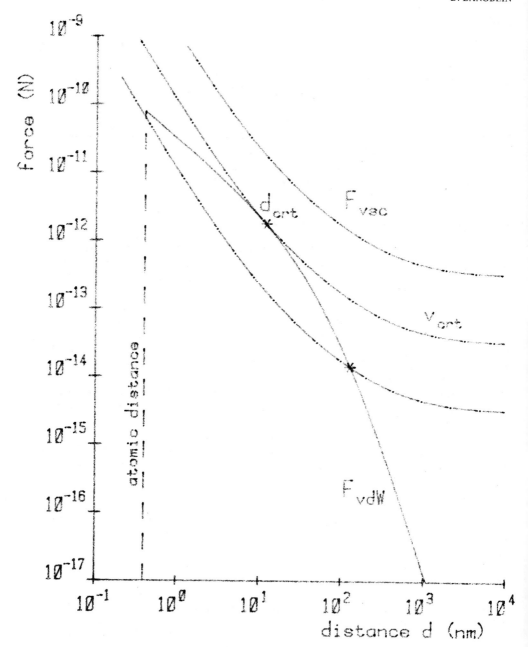

Fig. 4: Van der Waals repulsion and viscous force versus distance

where

$$A = \pi^2 \, n_1 n_2 c \tag{5}$$

is given by the numbers n_1 and n_2 of atoms per volume of the solid and the particle and c is the London-van der Waals constant.

Differentiation of Eq. (4) with respect to the distance d yields the van der Waals force

$$F_{vdW} = (2A/3) \, r^3/d(r+d)^2 \, (r+2d) \tag{6}$$

4 THE CRITICAL SOLIDIFICATION RATE

The solid line in Fig. 4 shows the van der Waals force according to Eq. (6) versus distance d. The dash-dot lines in Fig. 4 represent the viscous force of the melt according to Eq. (2). The viscous force is proportional to the moving rate v of the solidification front, i.e. the three dash-dot lines in Fig. 4 correspond to three different solidification rates. The lower line represents a low solidification rate. There exists a region of distances, at which the viscous force is smaller than the van der Waals force, i.e. the particle will be driven ahead of the solidification front and assume the distance given by the right hand crossing point between the lower dash-dot line and the solid line.

The upper dash-dot line in Fig. 4 corresponds to a high solidification rate v. It shows no crossing point with the van der Waals force, i.e. at this solidification rate the particle under investigation will be captured by the solidification front.

The solidification rate corresponding to the middle dash-dot line is choosen such that this line just touches the solid line. For this critical solidification rate v_{crt} there exists just one distance d_{crt}, where the van der Waals repulsion is able to balance the viscous force.

Thus, from Fig. 4 one obtains the dependence of the particle distance d on the solidification rate and the critical solidification rate v_{crt} as well. The particle distance increases proportional to the square root of the difference between the critical and the actual solidification rate

$$d - d_{crt} \propto \sqrt{(v_{crt} - v)} \tag{7}$$

The stable particle distance, in addition, depends on its radius r. Inspection of Eqs. (2) and (6) for the viscous force and the van der Waals force reveals that both forces

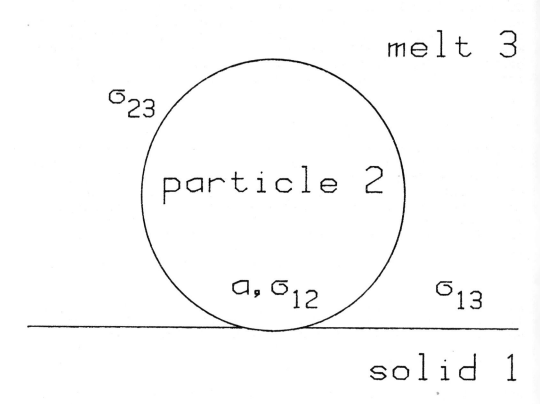

Fig.5 : Interface energies

increase proportional to the volume $(4\pi/3)r^3$ of the particle. This means rescaling of the force axis in Fig. 4. However, the van der Waals force according to Eq. (6) has several terms r in the denominator, whereas the viscous force according to Eqs. (2) and (3) has not. The van der Waals force thus falls behind the viscous force, i.e. the stable particle distance and the critical solidification rate v_{crt} decrease with increasing particle radius r.

5 MACROSCOPIC INVESTIGATIONS

Looking for a macroscopic condition for a particle being driven by a melting or solidification front, it is adequate to consider interface energies or tensions. The interface energy between two phases 1 and 2 is a measure for the difference in van der Waals attraction between the respective atoms. An atom of phase 1, which is located at the interface with phase 2, lacks part of the energy of condensation, which it would experience inside phase 1, i.e. the interface energy is the missing energy of condensation of atoms 1 and 2.

Fig. 5 shows a particle touching the solidification front. If the area of contact equals a, the change in interface energy as compared to the separated state is given by

$$E_{int} = a\ (\sigma_{12} - \sigma_{13} - \sigma_{23}) \tag{8}$$

where σ_{12} is the interface energy between the solid 1 and the particle 2, σ_{13} is that between the solid 1 and the melt 3, and σ_{23} is that between the particle 2 and the melt 3. If E_{int} according to Eq. (8) is positive, i.e. if the interface energy between solid and particle is higher than those between solid and melt and between particle and melt, the particle is repelled. If E_{int} is negative, the particle is attracted.

However, in the case of contact it is already too late for repulsion, the melt has no chance to penetrate between particle and solid. In spite of E_{int} being positive, there is no kinematic realization of this energy gain. - Other objections against these macroscopic considerations result from the fact, that the interface energies σ_{ij} include electrostatic and chemical contributions.

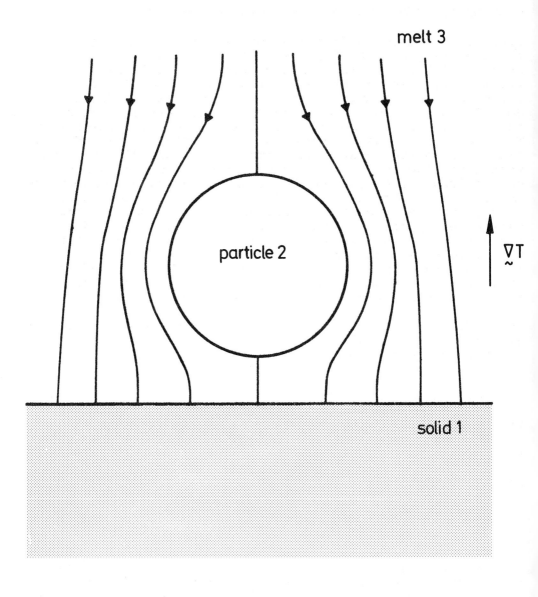

Fig. 6: The flow of the melt in a coordinate system moved with particle 2

6 DEFORMATION OF THE SOLIDIFICATION FRONT

Since directed solidification necessarily requires a temperature gradient $\underline{\nabla}T$, the temperature in the melt increases with increasing distance from the solidification front. Fig. 6 shows the same situation as Fig. 3, but in a coordinate system moving with the solidification front. Solidification front and particle are at rest and the melt is flowing towards the solidification front with velocity v. The flow around the suspended particle entails that the melt flowing into the interspace stems from a more distant region than that flowing to the uncovered regions of the solidification front. The melt moving to the interspace, consequently, is warmer than that arriving outside the interspace. This means that the solidification front, in being primarily determined by the isotherms in the melt, falls behind in the vicinity of the suspended particle. A situation as shown in Fig. 7 results. The suspended particle causes an increase in temperature in the interspace and a deformation of the solidification front, i.e. Eqs. (2) and (3) have to be reinvestigated and Fig. 4 has to be modified accordingly.

An additional deformation of the isotherms and of the solidification front results if the thermal conductivity of the particle differs from that of the melt.

7 LIQUID PARTICLES

If the particle under investigation is liquid rather than solid, the above investigations must be further modified. Firstly, the boundary conditions for the flow of the melt at the surface of the particle change. There is convection also within the particle. The Stokes equation (2) for the viscous force now reads

$$F = 6 \pi \eta_3 r v (\eta_2 + 3 \eta_3/2)/(\eta_2 + \eta_3) \qquad (9)$$

i.e. the flow inside and outside the particle is governed by the effective viscosity $\eta_3(\eta_2 + 3\eta_3/2)/(\eta_2 + \eta_3)$, where η_2 and η_3 are the dynamic viscosities of particle and melt, respectively.

Secondly, there arises interface convection. Interface convection is a consequence of the fact that interface regions differing in temperature or concentration differ in interface tension. This causes a driving force directed from regions with low interface tension to regions with high interface tension. Since interface tension generally decreases with increasing temperature, interface convection

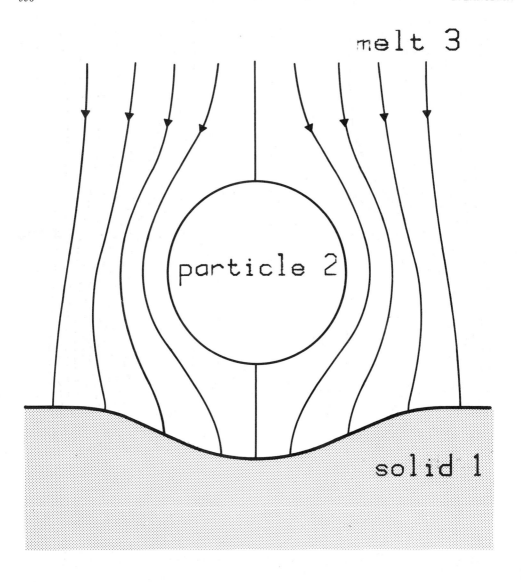

<u>Fig. 7:</u> The deformation of the solidification front
          behind a solid particle

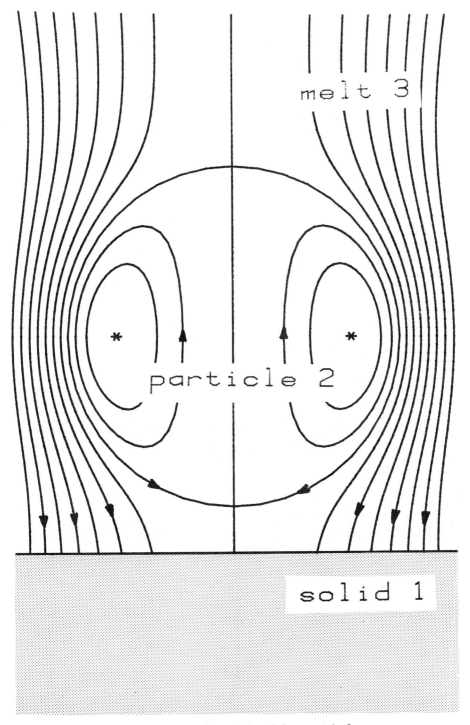

Fig. '8:   The motion of a liquid particle
           due to interface convection

at the particle-melt interface is directed towards the
solidification front. It supports the flow of the melt into
the interspace and the movement of the particle off the
solidification front. The resulting flow profile is shown
in Fig. 8.

Interface convection is governed by the balance of forces
tangential to the interface, i.e.

$$\nabla \sigma = \frac{d\sigma}{dT} \nabla T + \frac{d\sigma}{dC} \nabla C = \eta_2 \frac{\partial}{\partial z} u_2 - \eta_3 \frac{\partial}{\partial z} u_3 \qquad (10)$$

The driving force, the change of interface tension with
temperature and concentration, is balanced by the viscous
forces in the particle and in the melt. z is the coordinate
normal to the interface. u_2 and u_3 are the tangential flow
velocities in the particle 2 and in the melt 3. Interface
convection usually causes a faster movement of the suspended
particle than does van der Waals repulsion. The critical
velocity for liquid particles being driven ahead of a solidification front is much higher than that for solid particles.

8 CONCLUSIONS

Fig. 9 demonstrates the combined effect of intermolecular
forces, the separation of an immiscible melt due to interface and solidificationforces. An AlIn-sample containing
60 atom-% aluminum and 40 atom-% indium has been melted and
resolidified under microgravity during the SPAR II mission.
Aluminum and indium are immiscible; the critical point of
the phase diagram is at 830 °C, see Fig. 10.

The dark regions in Fig. 9 show indium, the light ones aluminum. Whereas under terrestrial conditions indium sinks to
the bottom of the cartrigde due to its higher density, it
now is deposited outwards. This result can be attributed to
the lower interface energy of indium to the cartridge,
which consists out of aluminum oxide. Aluminum, on the other
hand, has the higher melting point, i.e. it solidifies
first. The solidification front of aluminum run inwards and
has pushed the remaining indium to the center.

The time available for cooling in the SPAR II mission was
about 150 s. This time obviously was sufficient to push the
indium ahead of the solidification front, i.e. the speed of
the indium precipitates was about 0.1 mm/s. This proves
that in addition to the repulsion interface convection has
been active.

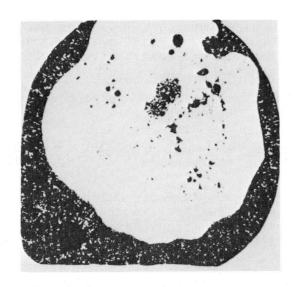

<u>Fig. 9:</u>   Cross-section of an AlIn-sample melted and
resolidified under microgravity (Ahlborn and
Löhberg 1976)

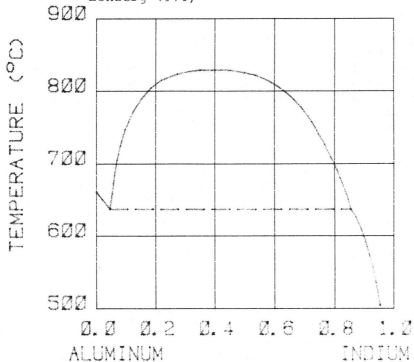

<u>Fig. 10:</u>   Phase diagram Aluminium-Indium

The present investigations demonstrate the opportunity,
which experiments under microgravity offer for studying
intermolecular forces. The absence of free convection enables investigations on weaker contributions to convection
and repulsion. Among these the shrinking of the melt during
solidification, the intermolecular attraction between
regions of different temperature and concentration, and
interdiffusion should be mentioned.

9 LITERATURE

Ahlborn, H., Löhberg, K.:
Ergebnisse von Raketenversuchen zur Entmischung flüssiger
Aluminium-Indium-Legierungen
Statusseminar Spacelab-Nutzung, Werkstofforschung und
Verfahrenstechnik 12.1, 1976

Langbein, D.:
Theoretische Untersuchungen zur Entmischung nicht misch-
barer Legierungen
Schlußbericht für den BMFT, 01 QV 558-AK-SN/N-SLN 7902-9
Battelle-Institut e.V., Frankfurt/Main, Juni 1980

Langbein, D.:
Mikrokonvektion an Erstarrungsfronten
Workshop: Erstarrungsfrontdynamik, Aachen, März 1981

Lee, S.H., Chadwick, R.S., Leal, L.G.:
Motion of a sphere in the presence of a plane interface
Part 1. An approximate solution by generalization of the
method of Lorentz,
1979, J. Fluid. Mech. 93, pp. 705-726

Pötschke, J.:
Wechselwirkung der Erstarrungsfront mit Fremdteilchen
Workshop: Erstarrungsfrontdynamik, Aachen, März 1981

Uhlmann, D.R., Chalmers, B.:
Interaction between particles and a solid interface
1964, J. Appl. Phys. 35, pp. 2986-2993

Walter, H.U., Ziegler, G.:
Stability of multicomponent mixtures
Shuttle/Spacelab Utilization
Final Report Project TEXUS II 1978, pp. 27-47

Grundlagenuntersuchungen zur Herstellung künstlicher
Verbundwerkstoffe
1979, Z. Flugwiss. Weltraumforsch. 3, pp. 311-320

SUBJECT INDEX